The National Economic Environment

The National Economic Environment

Vernon G. Lippitt, Ph.D.
Graduate School of Management
University of Rochester

McGraw-Hill Book Company

New York/St. Louis/San Francisco/Auckland/Düsseldorf/Johannesburg
Kuala Lumpur/London/Mexico/Montreal/New Delhi/Panama/Paris
São Paulo/Singapore/Sydney/Tokyo/Toronto

The National Economic Environment

1 2 3 4 5 6 7 8 9 0   S G D O   7 9 8 7 6 5

This book was set in Times Roman.
The editors were J. S. Dietrich and Richard S. Laufer;
the designer was J. Paul Kirouac, A Good Thing, Inc.;
the production supervisor was Leroy A. Young.
The drawings were done by J & R Services, Inc.
The printer was Segerdahl Corporation;
the binder, R. R. Donnelley & Sons Company.

Library of Congress Cataloging in Publication Data

Lippitt, Vernon G.
    The national economic environment.

    1. Macroeconomics. I. Title.
HB171.5.L732   330   74-18051.
ISBN 0-07-037972-6

To Ruth,
who also chooses
to suffer
for the joys
of creating

Contents

PART TWO   AGGREGATE SUPPLY AND DEMAND FOR PRODUCTS AND LABOR

PART FOUR POLICY: TARGETS AND INSTRUMENTS

PART FIVE   FORECASTING THE ECONOMIC ENVIRONMENT

Preface

In twelve years of teaching macroeconomics, I have come to feel that some of my ways of viewing and analyzing the economic system would enable students to understand macroeconomic phenomena more quickly and adequately than most available texts do. The notes which I have prepared have gone through two editions and the minds of more than 10 classes of students. Now it seems worthwhile to present them to a wider audience.

This text is intended for use in a one-semester, intermediate-level macroeconomics course in an economics curriculum or in an MBA program. It assumes knowledge from a "principles" course in micro- and macroeconomics and utilizes algebra, graphs of functions, and concepts of marginal ratios, maximization of profits, and market equilibrium.

My concern throughout has been to emphasize real-life economic decision making of persons, acting as consumers, as business executives, and as government policy makers. Macroeconomic sectors are identified as groups of decision makers with similar goals and motivations who respond similarly to market variables. The economic behavior of macroeconomic sectors is developed from microeconomic analysis of the decision-making units grouped in those sectors.

This viewpoint led me to split the business part of the economy into two sectors:

1 Business Capital Accounts. This sector involves the balance sheets of business units and the decision making related to capital-funds flows and investment expenditures.
2 Production Units. This sector involves decisions on production levels and current-account purchases of intermediate goods and factor services; its flows are summarized by operating statements of firms.

Analysis of the behavior of the Production Units Sector (Chapter 6) leads to an aggregate supply curve which I think is an improvement over that in other texts. It relates expected revenue curves (based on sales forecasts) to levels of real output chosen for profit maximization. It is similar to Keynes's aggregate demand curve, but the use of real output instead of employment permits one to split a given level of money aggregate demand into its price and real-flow components directly.

The text uses current-dollar incomes, expenditures, wage rates, and money stock from the start, because I believe these are the measures of flows and stocks which are usually the relevant items for sector decision making.   Money flows are converted to real flows and prices via the aggregate supply curve, and the production function when labor markets are analyzed.   The prices so determined are used to deflate GNP expenditures and the money stock when real values are required in analysis or forecasting.

National income accounts are presented early in the text (Chapter 4) and are illustrated by circular-flow diagrams which I believe are unique and the clearest to be found in macroeconomic texts.   Realistic, simple equation models based on national-account data are introduced early and are used to bring out the sector interactions and system properties of the economy.   Dimensions and units of economic variables are emphasized—for flows versus stocks, for real versus money values, for rates of change, and for ratio variables.   The distinction between " desired " and " actual " values of flows and stocks is emphasized early in the analysis to bring out concepts of disequilibrium and dynamic changes in decisions.   Feedback and time lags, with their dynamic effects, are brought into the basic analysis, instead of being added later in a section on "economic dynamics."

The *uses* of macroeconomic analysis are emphasized—both in forecasting and planning by private decision makers and in the solution of economic policy problems by government officials.   Chapters 16 through 20 relate macroeconomic analysis to policy in the areas of full employment, inflation, balance of payments, and economic growth.   Chapters 21 through 24 illustrate uses in forecasting by means of a step-by-step development of an economic forecast for 1974, including GNP components, sector incomes, employment, and prices.

Problems are included in the text, many of them involving actual data for the United States economy.   An instructor's manual is available.   It provides:

1 Solutions to problems
2 Descriptions of a computerized policy project, which the author has found useful in providing a valuable summary of course material and an introduction to the practical use of macroeconomic theory for policy purposes
3 Description of a forecasting project in which a class can develop and reconcile its own forecast of GNP, income flows, employment, and prices for a year ahead.

The publisher has cautioned me repeatedly to avoid any unnecessary symbols and equations and to keep the manuscript as short as possible.   I have tried to meet these requests, though recognizing that separate symbols for different concepts may shorten and clarify the presentation.   An appendix at the end of the book lists the symbols used, with their units, and guides the reader to the location in the text where the symbol is first explained.

As to length, the inclusion of chapters on economic forecasting unavoidably expanded the length of this text compared with others.   Many instructors will choose not to assign those chapters in their entirety. Those who do use them may find it necessary to omit other material. Reading assignments may be abbreviated by omission of background material in Chapter 2, skipping some of the Chapter appendixes, perhaps omitting Chapter 14, and skimming or leaving out Chapters 19 and 20. Such selective shortening should leave time in a one-semester course either for a class forecasting project or for the policy exercise to drive home the theoretical concepts in a meaningful and memorable manner.

My thanks for helpful comments and suggestions go, of course, to the students who have been required to read this manuscript and also to two professionals who reviewed it in considerable detail—Professor Ogden O. Allsbrook, Jr., at the University of Georgia, and Professor Lawrence Ritter, at New York University. I trust that publication of the text will bring me more suggestions from users, so that I shall be able to improve my creation should it survive in a second edition.

Vernon G. Lippitt

The
National
Economic
Environment

Part One

Introduction to Macroeconomics and the National Economic Accounts

Chapter 1

Nature and Usefulness of Macroeconomics

WHAT IS MACROECONOMICS ABOUT?

Recent issues of business and government publications have contained items such as the following:

"Gross national product increased $14.7 billion in the first quarter of 1974 to a seasonally adjusted annual rate of $1,352.2 billion, reflecting an inflation rate of 11.5 percent and a decline of 6.3 percent in real GNP."

"Corporate profits before taxes rose $12.5 billion in the first quarter.   Including the inventory valuation adjustment, corporate profits fell $3 billion."

"Industrial production is being held back for two reasons: soft demand in about half industry and shortages plaguing the other half."

"Retail sales in July were at an annual rate of $46.3 billion—up $1.8 billion."

"There has been much discussion of the possibility of cutting the fiscal '75 Federal budget in order to apply some 'stiff anti-inflationary medicine' to the economy."

"The seasonally adjusted money stock increased at an annual rate of 5.3 percent in May."

"The B.L.S. broad wholesale price index leaped 3.7% in July, bringing it to 20.4% above last year."

"Total loans and investments at all commercial banks increased at an annual rate of 10.2 percent in May.   Interest rates generally increased."

"The dollar's value and the price of gold both declined in European markets."

These news items referred to macroeconomic phenomena.   They involved aggregate measures of production, income flows, sales, employment and unemployment, prices, wage rates, or interest rates for large numbers of persons or businesses in the nation, or they dealt with actions by government and monetary authorities which affect the actions of consumers and producers throughout the country.

Other items in the news were concerned with particular companies or product markets, that is, with microeconomic phenomena.   *Microeconomic analysis* deals with the economic state and behavior of individual decision-making units—consumers, firms, government units, and suppliers of services of labor, capital, or natural resources—or with the behavior of groups of such units closely linked in specific markets, involving commodities or services which are close substitutes. *Macroeconomic analysis* deals with the economic state and behavior of nationwide groups of consumers and production units,with total government taxing and spending, and with financial markets and the banking system as a whole. It aims at an understanding of the determination of the overall levels of demand and supply, income flows, wealth, money supply, employment, productivity, wage rates, prices, and interest rates.

The distinction between micro- and macroeconomic analysis is quite analogous to differences of level of aggregation in physical, biological, psychological, and social systems. In physics we may analyze the structure and behavior of individual atoms and molecules, or we may study flow of fluids or properties of structures or machines as a whole. In biology there is cellular physiology and there is analysis of tissues, organs, and whole organisms. Psychologists may focus on individual motivation and behavior, or they may study group dynamics, organizational structure and behavior, or mob psychology. In the field of social science the level of aggregation ranges from the individual or family unit to the community, state, or world order embracing all mankind. In all these areas of phenomena the high-order groupings of individual units display characteristics and behavior that are qualitatively different from those of the component parts.

The national economy is a complex system of economic decision-making units. The interaction or feedback among units leads to system properties and behavior which differ from the microeconomic responses of individual units, for example, amplification of an initial rise in demand, business cycles, changes in aggregate money supply, inflation, or mass unemployment.

In macroeconomics we attempt to develop organized knowledge about the structure, the behavior rules, and the time delays which correctly describe large aggregates of economic decision-making units. We hope to arrive at an understanding of this system, that is, at a comprehension of the relations among the variables which describe the inputs and outputs of the parts of this system and of the system as a whole. Whether there are regularities in the relations among variables at the highly aggregate level used in macroeconomics is not obvious. We are analyzing a system with complex interactions, and human behavior and learning are involved. Much of this book deals with the attempt to make sense out of the macroeconomic system, to see if there are "laws" describing the relations among its variables. If there are, it may be possible to understand the macroeconomic system to some extent. Given understanding, we may be able to predict and perhaps even control the behavior of the system, at least in some degree.

The following questions about the macroeconomic system and economic policy may throw further light on the scope of the subject and suggest some key areas which you may expect to understand better after studying this book.

What are the principal economic goals of our nation?

How is aggregate economic activity measured?

How is capacity output of the economy defined? What determines it? What are possible causes of business cycles?

What causes unemployment? What can be done about it?

What are some of the main feedback channels in an economic system?

If saving and investment are always equal, how can the economy get out of balance?

What determines the distribution of incomes in the economy?

Evaluate the claim: "If wage rates rise as fast as labor productivity, labor is getting all the gains from rising productivity."

What are the principal determinants of consumption expenditures?

What are the principal determinants of investment expenditures?

How can a change in investment or government spending cause total spending to change by double the amount of the original change?

What causes inflation? What are its principal effects?

What is "money," and how does the supply of money affect economic activity?

How might higher interest rates help reduce inflationary pressures in an economy? Will lower interest rates stimulate the economy?

What is the logic of a statement that cutting tax rates can increase government revenues? Does the 1964 tax cut seem to have done so?

Explain clearly what is meant by "full-employment government surplus." How might it be harmful?

What determines the sustainable growth rate of an economy?

What is meant by automatic and discretionary fiscal policy?

What are the principal elements in monetary policy?

What policy variables can a central government use to curb inflation? To reverse a recession? To accelerate economic growth rate?

What are the principal causes of the current United States imbalance of payments? Why worry about it?

WHY WE SHOULD UNDERSTAND OUR ECONOMIC ENVIRONMENT

Perhaps some of the preceding questions aroused your intellectual curiosity. If so, you have one motivation for studying macroeconomics—the desire to understand more about the world you live in, the desire to organize perceptions and information into meaningful patterns in the field of economic phenomena. The satisfaction of this urge to understand is an end in itself.

Besides the "liberal arts" justification for studying macroeconomic phenomena, there are applied or instrumental goals. In general, *knowledge of one's environment may be used for adaptation or for control.*

Individual persons usually must adapt to the national economic environment in their activities as consumers, suppliers of labor and property services, savers, and investors, because their individual resources and buying power are miniscule in aggregate markets. But they can adapt more successfully to an uncontrollable environment if they can understand and predict business-cycle movements, inflation, and government policy actions. Such knowledge can improve their decision making regarding allocation of income to consumption and saving; major purchases and sales of durable goods, real estate, and securities; timing of borrowing; and perhaps job changes.

Individual business managers can improve their decisions on the basis of macroeconomic analysis—again primarily via more successful adaptation. Managers aim to maximize realization of company goals within the constraints of internal resources and external environment. The individual business executive has control only over the employees, facilities, and funds within the company—and not always complete control even there. He or she may have some influence over customers through advertising and sales promotion, but relatively little over competitors, financial institutions, government, and the state of the national economy. Where managers control, they issue orders; where they do not control, they must forecast and adapt. So knowledge of macroeconomics helps business owners and managers forecast those external economic forces which affect their operations. These forces may include, for example, customer incomes and industry competition which affect industry sales of their products and hence their own company's sales; prices of input materials and wage rates which affect their firms' costs; or availability and costs of capital funds which affect their decisions on investment and external financing. If forecasts of the external economic environment are accurate and timely, the business manager can adapt more successfully to uncontrollable environmental forces. This is true for officials in state and local governments and in nonprofit institutions also.

Officials in the federal government or the Federal Reserve System are able to exert some control over the course of national business activity. So they certainly need to understand macroeconomics for policy purposes. Only when the causal relations among variables, that is, the economic behavior of major sectors of the economy, are understood can government authorities make intelligent decisions regarding the type and magnitude of changes they should make in variables they control in order to achieve desired outcomes in the economy at large. For example, they need to know how a change in personal income tax rates will affect consumer demand; how a rise in interest rates will affect plant and equipment expenditures, state and local bond financing, or the availability of mortgage funds for financing home construction; or how a change in government spending will affect aggregate employment of labor. Knowledge of the specified relations needs to be quantitative too, and should include estimates of the time lags involved.

Finally, it should be noted that individual persons do have some influence on the course of events in the national economy. Acting as individuals or through consumer groups, unions, business associations, and other organizations, they can affect the level of economic understanding and the climate of public opinion, which are important elements in a nation's economic development—at least in an open society.   By participating in public forums and educational activities, they can influence public opinion; by lobbying and by giving testimony before legislative committees and regulatory agencies, they can influence legislation and economic policy actions of local, state, and national government agencies.   And we should not underrate the power of creative insights.   Keynes suggested:†   "The ideas of economists and political philosophers, both when they are right and when they are wrong, are more powerful than is commonly understood.   Indeed the world is ruled by little else."   Of course, results may be long in coming.   But it seems reasonable that they will come sooner and be more beneficial if the identification of problems and the analysis of alternatives are based on a sound understanding of the economic environment and are presented logically and clearly.

PROBLEMS

1 Compare the President's Council of Economic Advisers and the board of directors of a corporation with respect to the following characteristics:

   a Principal goals, or target variables

   b Principal instruments, or policy variables that they can control to achieve their goals

   c Principal uncontrollable variables (or actions by others) which they need to forecast and to which they must adapt

   d Contributions which macroeconomic analysis and forecasting can make to their decision making

† John M. Keynes, *The General Theory of Employment, Interest and Money*, p. 383, Harcourt Brace Jovanovich, New York, 1936.

Chapter 2
Beginning the Analysis

OBJECTIVES AND METHODS

We set out to achieve an understanding of the national economy with two purposes in mind: (1) to be able to forecast the national economic environment, so that we can adapt more successfully (profitably) to uncertain prospective events in that environment which are uncontrollable by the individual; and (2) to understand the economic system well enough so that we can evaluate government policy actions chosen to direct the economy toward the nation's economic objectives.

What do we need to learn about the economic system in order to achieve these two purposes? To understand any complex system—an automobile, an animal, a business firm, or the economy—one needs to be able to answer a few basic questions about it:

1 What are the active component parts of the system?
2 How do they "behave," that is, how are inputs (stimuli) related to outputs (actions)?
3 How do the parts interact?
4 How, then, does the system as a whole perform?
5 Can we develop a quantitative model of the performance of the system— to help us understand, control, and improve it?

This chapter presents an introduction to answers to the first four questions. Chapter 3 begins the development of a model of the system. The rest of the book elaborates on the answers and applies the understanding to some problems of economic policy and forecasting.

ACTIVE COMPONENTS OF THE SYSTEM

An automobile contains thousands of individual parts but its operation can be understood by breaking it down into a few major subsystems and then analyzing their actions and interactions. Such subsystems include the fuel system, electrical system, motor, power transmission and brake system, and steering mechanisms. The nation's economic system is composed of several hundred million economic decision-making units—currently 210 million persons or 68 million household consumer units, some 5 million businesses and 90 million members of the labor force in production units, and tens of thousands of government units—federal, state, and local. To simplify the analysis of such a complex system, economists classify the decision-making units into a few main categories based on similarity of their economic behavior. These component parts of the system are called *sectors*, and they include Production Units, the Personal Sector, the Government Sector, Business Capital Accounts, the Foreign Sector, and Financial Institutions.

Of course the decision makers in all sectors are persons—with internal motivation, memory, expectations, and ability to reason. So the economic system is more complicated than an automobile; analysis, prediction, and control of the macroeconomic system are much more difficult. Imagine an automobile whose parts grow over time, whose responses to the brakes and steering wheel vary with past history, whose accelerator-carburetor settings change cyclically, and whose transmission exhibits unpredictable changes in slippage and gear shifts! This is admittedly a little exaggeration. While the economic system is a changing human and social system and while economic forecasters are usually in error, still there is sufficient regularity in the growth of the economy and sufficient stability in the economic behavior of persons and businesses that we can develop some "laws" for the national economy. After all, that's what this book is all about.

ECONOMIC BEHAVIOR OF SECTORS

There is a common thread to economic decision making in all sectors. As individuals or organizations, we have goals we desire to achieve, but we have limited time, energy, knowledge, physical and financial resources, and authority for reaching these goals. So we weigh the probable costs and benefits of alternative actions relative to our goals. Then we make decisions, that is, we choose and carry out actions we believe will achieve our objectives as fully as possible under the constraints imposed by our limited resources and by the environment within which we must act.

For each of the six sectors mentioned, let us examine a little further its economic behavior and how it fits into the nation's system for using scarce resources to meet human wants.

PRODUCTION UNITS

This sector consists largely of the production operations of businesses† as summarized in their operating statements.

For purposes of economic analysis, businesses are assumed to make choices aimed at survival and at maximization of long-term profits. A business can realize its objectives by performing a needed social function efficiently. That function is *economic production*—the conversion of scarce resources into goods and

---

† Both corporate and noncorporate businesses are included, and all industries—farming, manufacturing, construction, trade, and other services. Also included are productive services rendered by individuals through domestic nonprofit consumer organizations, households, and governments, and services provided to the Foreign Sector.

services desired by other sectors. If you can devise a method for combining a flow of inputs of materials, labor services, and services of capital funds to produce a flow of output for which customers will pay you more than the costs of your inputs plus taxes, you will have a profitable business.

The *resources* which businesses have available to achieve their goals include balance-sheet items (cash and other financial assets, plant and equipment, inventories) and many intangibles, such as the knowledge, initiative, and dedication of managers and other employees; ownership of contracts for use of natural resources and patents; the reputation of products and of the company among customers and the public; good relations with financial institutions and government officials; effective company organization, communication, and policies; and a strong distribution network. Adequacy of these resources for achieving company objectives will be determined by the quantity and quality of those internal resources and by external factors, such as national economic conditions, customer incomes and preferences, actions of competitors, market prices of output, and costs of inputs.

Given the goals of businesses to be achieved through the economic activity of production, and given the resources available to businesses, what decision making is required? There are *long-term decisions* regarding what industry to enter, what products to produce, what production techniques to use (production function), how to organize the company, and how to select plant location, channels of distribution, sources of financing, and other factors. Since we are discussing current production activities in this section, we shall assume that long-term decisions have been made and that the current resources of firms are given. What current operating decisions must be made, and how can optimal actions be chosen?

The *short-term production decisions*, of course, are those specifying the volume of output to be produced and the price to be charged for products. Microeconomic theory of the firm indicates that profits will be maximized if output is increased so long as an added unit of output increases revenues more than costs, that is, so long as marginal revenue (MR) exceeds marginal cost (MC).

We may summarize the short-run production behavior of businesses as follows: Managers assemble information on (1) forecasts of sales (demand)—schedules of price or revenue as a function of quantity sold; (2) the production function (output as a function of input flows); and (3) prices of inputs, especially purchased materials and labor costs. Given these information inputs, managers choose the production level which they believe will probably maximize profits. As the output flow is marketed, managers discover whether their forecasts were accurate and their plans optimal. If not, they will experience a shortfall of profits below maximum, along with an undesired change in inventories or an unplanned change in prices at which the output can be sold. In the next period, forecasts and production schedules will be revised in the light of new information.

This microeconomic description of a firm's production behavior is carried over intact to the Production Units Sector as a whole, skipping lightly over many difficult questions of aggregation. The aggregate flow of output is assumed to depend on producers' forecasts of aggregate demand, on the aggregate production function, and on prices of inputs, notably the wage rate.

If we accept this description of the economic decision making of the Production Units Sector, we can deduce a partial answer to our third question; specifically, how this particular sector interacts with the rest of the system. In the first place, it interacts with the demand sectors (Personal, Government, Foreign, and Business Capital Accounts Sectors) as it sells its output of goods and services. This interaction in product markets determines market prices and any unexpected changes in producers' inventories. Second, it interacts with these same sectors in factor markets as businesses purchase their inputs of services of labor, capital funds, and natural resources. The demand-supply interaction in these markets determines wage rates, interest rates, prices of raw materials, rental rates, and other cost factors. Third, the Production Units Sector interacts with the Government Sector as the units pay taxes or receive subsidies, and as they adapt their decisions to government laws and regulations or lobby to change governmental decisions. Fourth, the Production Units (current operations of businesses) interact with Business Capital Accounts (balance sheets) through the flow of profits and of capital consumption allowances from current to capital account.†

These interactions bring out a very important concept of circular flow in the national economy. *The production units receive their sales revenue from the other demand sectors, and they pay all this stream of receipts back to those sectors as production costs, profits, and taxes. The payments by the Production Units Sector provide the demand sectors with incomes that are equal to the value of the goods and services produced.*

PERSONAL SECTOR

The Personal Sector contains domestic consumer units—individuals or family units plus nonprofit organizations serving consumers, such as private clubs, hospitals, educational institutions, and welfare organizations.

We assume that consumer units make choices which will maximize their own welfare or "utility" in the long run. Many of our satisfactions come from aesthetic, spiritual, recreational, social, organizational, and political activities. Others arise from consuming a flow of goods and services which are fashioned by production units from inputs of scarce labor services, natural resources, and capital goods. The latter sphere of consumer activities is referred to as *consumption* and is a subject for economic analysis. Of course, the two types of satisfactions are not independent. Economic production and consumption decisions affect the health, education, leisure time, and facilities available to persons for noneconomic enjoyments.

What decisions or actions are required for persons to meet "economic" needs? From Robinson Crusoe on up to our most complicated economies, no one has discovered how to consume without producing—not for long, anyway.

---

† In addition, production units purchase materials and supplies from one another on current account. These intrasector transactions are canceled out in aggregating over firms to derive measures of total national production.

So consumer units have to decide how they will contribute to production in order to earn a claim on part of current output.   Given their income flow from Production Units, the consumer units decide how to allocate that income to current consumption of various goods and services and how to provide for future consumption (saving).   And they need to decide what wealth items to acquire with the funds saved.   Should new saving flows go into cash, savings accounts, insurance, bonds, stocks, real estate?   Should the current mix of wealth items in the portfolio be changed?   Should funds be borrowed for current expenditures, for additions to the portfolio of wealth items, or for investment in one's human capital via education?

Economists analyze the consumer decision on how much labor services to supply by assuming that individuals will increase the hours they work per week or year until the disadvantages of an added hour of work become greater than the benefits from the income earned during the added hour.   The disadvantages of working longer hours may include fatigue, boredom, and sacrifice of leisure time, and the increment of such "disutility" per added hour worked is assumed to increase with the number of hours.   On the other hand, the increment to utility derived from the income earned by an additional hour of work is assumed to decline as work hours increase; that is, the marginal needs met by added income become less intense as income rises.   At the balance point, the consumer is willing to supply labor services just up to the extent that marginal disutility of an added hour of work rises to equal the marginal utility derived from the added income.   From this microeconomic analysis, we derive a supply curve for labor supplied by the Personal Sector for the economy as a whole.   In the long run, labor supplies depend on such factors as (1) the size of the population; (2) the age-sex breakdown of the population; (3) social norms for schooling, women's working, retirement age, standard workweek, and similar conditions; (4) income aspirations and provisions for unemployment compensation and welfare; and (5) real wage rates (money wages deflated by a consumer price index).   For short-term changes the first four determinants are assumed given, and the labor supply is taken to vary directly with the fifth determinant, real wage rates.

In analyzing how consumers allocate their incomes to various current goods and services and to saving, economists apply marginal analysis again, with maximization of utility taken as the goal.   In current consumption decisions, expenditures are increased over the whole range of consumer goods in such a way as to keep the incremental utility per added dollar spent the same for all products purchased.   (Were this not done, a consumer would increase the utility derived from a given total expenditure by switching a dollar from one product to another.) Relative prices of goods affect this allocation because, if one good becomes cheaper, the marginal utility from $1 spent on that good rises, and more of that good will be purchased.

There must also be a balance between expected benefits from $1 spent and those expected from $1 saved out of current income.   The expected marginal utility of $1 saved, presumably for future consumer expenditures, depends on how much a person discounts future satisfactions as compared with present ones, on one's expectations on future income and prices of consumer goods, and on the rate of interest on funds saved. The interest rate determines how much $1 saved now will increase in value by the future date at which consumer goods will be purchased.

Finally, the portfolio decisions of consumer units are also assumed to be made so as to maximize long-run utility, appropriately discounted. Each person is assumed to weigh the prospective liquidity, yield, and risk of various assets and liabilities and to arrive at an optimum portfolio or balance sheet of real and financial items.

We assume that all units in the consumer sector respond similarly to market stimuli; so we aggregate their incomes, expenditures, savings and balance sheets for purposes of macroeconomic analysis. The result is that short-run Personal Sector consumption, saving, and portfolio decisions are believed to depend primarily on aggregate income and wealth of the sector, on the price level for consumer goods, on interest rates, and on expected future levels of income, prices, and interest rates. In the longer run, the size of the population, its age distribution, the number of families in various stages of their life cycle, and other demographic characteristics would need to be taken into account.

How does the Personal Sector interact with other sectors? It acts as a supplier to the Production Units Sector in markets for labor services, capital funds, and natural resources, and it is a customer for Production Units in markets for goods and services. Its members interact with the Government Sector as government employees, as taxpayers, as recipients of government social security, welfare, and subsidy payments, and as purchasers of government securities. The sector interacts with Business Capital Accounts and the Financial Sector as it invests new saving or readjusts its portfolio of financial assets, monetary and nonmonetary. And it interacts with the Foreign Sector via imports, gifts to foreigners, and investments in foreign securities.

GOVERNMENT SECTOR

Units in the Government Sector include federal, state, and local government organizations, not only the legislative, executive, and judicial branches that we usually think of, but also public school districts, municipal water, street, and sewage systems, park districts, and other public agencies. Excluded are government enterprises which cover most of their costs by service charges rather than taxes, such as the Post Office, the Tennessee Valley Authority, and municipal utilities. (These are included in the private Production Units Sector.)

As for goals or motivations of government units, perhaps the most relevant viewpoint is to regard them as collective consumption units making purchases of goods and services in order to promote the welfare of the population (present and future) served by the governments. Among the goals of government decision makers are the following:

High standard of living for the population, involving both consumption and leisure

Economic power to influence foreign countries and achieve national security

Full employment of labor—balance between aggregate demand and capacity to produce

Reasonable stability of production, employment, and incomes

Reasonable stability of prices

Growth of capacity, or of potential economic output

Growth of output per man-hour

Equitable distribution of income

Reasonable balance of international payments

Preservation of desirable physical environment

Preservation of desirable social and human environment—economic freedom, solution of urban problems, education, health, and the role of law

These various goals may overlap or be in conflict at times, and trade-offs may be required because of limited resources. So we may postulate a cost-benefits or cost-effectiveness analysis to guide decision making in government, analogous to marginal cost and revenue analysis for profit maximization in private businesses. But noneconomic considerations must be granted a large role also, for example, national security, the self-interest of pressure groups, and the voting strengths of groups benefiting from government programs. It is difficult to conceive of a " social utility " function from which government decisions are derived by a maximization procedure.

Also note that market mechanisms are sometimes not available to guide government decision makers. Government units do obtain labor services and purchase many goods and services in private markets. But sometimes the government is the sole buyer of, for example, military equipment and highway construction, or the sole supplier of such items as marine navigational aids, nuclear fuel, and postal service. And the determination of tax structures or schedules of benefits for unemployed persons, veterans, and welfare recipients is not an economic market phenomenon. Probably not everyone is paying in taxes just what he or she feels is the value of government services received!

As a consequence of the special nature of many government transactions, economists have not gotten far in developing causal relations to explain and predict the behavior of the Government Sector in obtaining and spending income. Such flows are considered to be exogenous, except that some tax revenues can be related to the private income streams on which they are levied, if tax rates are constant. The best we can usually do is to estimate government expenditures on the basis of budgets and trends.

The interrelations of government units with other sectors are manifold. The Government Sector redistributes about 10 percent of Personal Sector income by taxing some persons and making transfer payments to others. In addition, it taxes persons and businesses enough to purchase over 20 percent of the current output of the Production Units Sector, including the hiring of about 15 percent of the nation's labor force. The federal government carries on a large volume of transactions with the Foreign Sector—buying and selling, borrowing and lending, and giving economic and military assistance. And the Government Sector is an active participant in capital funds and money markets involving the Financial Institutions Sector. In addition to its interaction via flow of funds with other sectors, the Government Sector is intricately involved in the decision making of other sectors through the legal and regulatory framework for private transactions and through provision of subsidies and services to Production Units. Examples of such influences include minimum-wage legislation, farm subsidies, air and water navigation aids, tariffs, and quotas on imports.

BUSINESS CAPITAL ACCOUNTS

This sector is essentially a consolidation of the flows of funds to and from the balance sheets of all production units in the economy. Capital-account transactions cover investment spending for plant and equipment, residential construction, and changes in business inventories; they include the financing of that investment via internal financing (capital-consumption allowances, retained earnings, changes in stock of cash) and external financing (equity issues, bond or mortgage debt, bank borrowing, and so on).

Capital-account decisions by businesses are assumed to be motivated by goals of survival and long-term profit maximization, the same as for current-account operating decisions. But they involve long-term commitments of resources with profits realized over an extended future. Hence these decisions require long-term forecasts of revenues, costs, and net cash returns associated with given projects, leading to estimates of expected present values or rates of return for the projects. The microeconomic theory of demand for investment goods suggests that the important variables influencing aggregate investment spending are: expected rates of return on investment projects, cash flow of businesses, and availability and costs (interest rates) of external financing.

The Business Capital Accounts Sector interacts with all other sectors of the economy. It buys investment goods from the production units in product markets and receives internal cash flow from production operations. It tries to attract capital funds from the Personal Sector and from Financial Institutions, and it competes with the Government Sector in the capital-funds markets. It makes real and financial investments abroad and may compete with the Foreign Sector for capital funds in domestic and foreign markets.

FOREIGN SECTOR

The Foreign Sector consists of economic decision-making units which are residents of other nations. Consumers, businesses, governments, and financial units are all lumped together. Their decisions are assumed to be guided by goals and motivations similar to those of comparable institutions in the United States economy, except for countries where central government goals and plans dominate business and financial behavior.

Transactions with foreign countries include: (1) exchange of goods and services on current account, (2) transfers (private gifts and government grants), and (3) capital transactions in real property and financial claims.

FINANCIAL INSTITUTIONS

This sector includes the U.S. Treasury, the central bank (Federal Reserve System), commercial and saving banks, savings and loan associations, insurance companies, credit corporations, pension funds, and dealers in stocks and bonds. These institutions channel funds from savers to borrowers, and the Treasury and the banking system can create and destroy money. Their principal service is making liquid funds available to would-be spenders, funds which they attract from savers or create themselves (as do commercial banks) and which they exchange for non-liquid promises to pay in the future (loans, mortgages, notes, bonds, stocks).

The private businesses in this sector operate as profit-making firms. Their managers are concerned with what combinations of labor and capital inputs to use, with pricing of their services, and with growth and innovation in their production and marketing activities. But they are more heavily involved in decision making regarding their balance sheets than are managers in nonfinancial enterprises. So managers in financial institutions spend more time in analyzing the yield, liquidity, and risk of financial assets and liabilities. And they pay more attention to the capital gains component of the return on assets, and hence to price changes of financial assets which lead to these capital gains, than do managers of nonfinancial businesses. Thus decision making in the private Financial Institutions Sector is assumed to be guided by profit-maximizing principles applied to portfolio management as well as to current operations.

The public units in the Financial Institutions Sector include the United States Treasury and the Federal Reserve System, as noted earlier. Managers in the Federal Reserve System also are concerned largely with portfolio management, that is, with controlling their assets and liabilities so as to achieve the desired money supply, interest rates, or credit conditions in the economy. Their goal is not profit maximization, but contribution to the economic welfare of the population as a whole, along the lines of the goals listed for the Government Sector. The monetary authorities in the United States are largely independent of the national government, and their actions may reflect an independent judgment regarding both the importance of the nation's various economic goals and the best way of achieving those goals in a given situation.

Obviously, the Financial Institutions Sector interacts with all other sectors of the economy in markets for capital funds. The institutions are the channel for exchanges of funds and of financial claims between units who save out of their current income and those who want to obtain funds to spend more than their current income. And they make available those completely liquid claims against the monetary authorities which constitute the money supply of a modern economy.

TABLE 2-1 Areas for Economic Research and Analysis (*Growth and change are involved in each area as well as the current situation*)

Main Topics for Research and Analysis	Total Economy	Types of Decision-making Units in the Economy				
		Persons: Consumers and Labor Force	Businesses	Governments	Foreign Units	Financial Institutions
A Structure and characteristics	Ownership and control of natural resources and business Legal framework—business, finance, and labor law Freedom of markets—competition Tax structure and burden	Population and its breakdown by age, sex, place of residence, education, occupation Households and their distributions by income level, family type, age of head, etc.	Industry breakdown—economic characteristics and location Forms of business organization—corporate and non-corporate Company distributions by size Business births and failures	Federal State Local—county, town, city, school districts, etc.	Consumers Businesses Governments Total economies (Structural detail as listed in other columns)	Federal Reserve Banks U.S. Treasury Commercial banks Savings institutions Federal credit agencies Stock and bond market institutions Pension funds and insurance companies
B Psychology and motivations	Level of confidence in economic opportunities Urge to improve self and society Acceptance of change Occupational and social mobility	Consumers: Desire for goods and services and saving Consumer motivations—social pressures, sales persuasion Expectations Level of confidence Acceptance of innovations Reactions to price and credit changes Reactions to taxes Labor force: Social value of work vs. leisure Employee attitudes and motivations Competitive spirit Willingness to take risks Motivations of union leaders Acceptance of innovations	Strength of profit motive Stimuli for innovation Competitive pressures Businessmen's confidence—attitudes, expectations, plans Reactions to taxes and government regulations Employee and public relations	Structure and burden of taxes Motives underlying fiscal policy and debt management Political pressures for spending vs. tax reduction Patent system Antimonopoly laws	Nationalism Urge for economic development Political stability and alignment in world politics (Also items in other columns)	Motives underlying monetary policy Confidence of investors—attitudes, expectations, intentions Regulations and practices governing financial institutions Relative strength of desires for security, yield, liquidity

C Natural resources, energy, and materials	Sources of energy Metallic ores Nonmetallic minerals Climate Terrain—rivers, harbors, mountains, etc. Arable land Fresh water supply Wood	Land for home sites Recreation areas Trends in per capita use of energy and materials	Natural resources owned for business use—terms of tenure Trends in sources and uses of energy and materials	Government-owned land, forests, mineral deposits, parks, roads Flood control and irrigation projects Trends in government control of sources of energy and materials	(Items in preceding columns)	(In business sector)
D Human resources	Population—age and sex composition Level of education and occupational specialization Urbanization Health and vigor Morality and discipline	Consumers: Consumption standards Consumer education Labor force: Participation by age and sex groups Health Education and ingenuity Occupational specialization Industrial discipline Labor organizations	Supply and competence of managers and professional employees Supply and ingenuity, skill, and effort of other employees Distributors and suppliers Training programs Pension and welfare programs	Public education Social security Public health and welfare programs	Immigration laws (Also items in preceding columns)	(In business sector)
E Stocks of fixed capital and inventories	Total fixed capital—growth relative to total output or on a per capita or per worker basis Total inventories in relation to output Production and distribution facilities	Consumers and Labor Force: Residential housing Stocks of durable goods Private schools, hospitals, clubs, institutions, etc.	Plant and equipment—growth relative to output and on per worker basis Patents, copyrights, etc. Inventories—working stocks and finished goods	Highways and streets Waterways, dams Public buildings Military equipment and structures Publicly owned utilities Stockpiles of farm products and strategic materials	(Items in other columns)	Buildings for own use Land, buildings, and equipment leased to other businesses
F Current physical inputs and outputs	Inputs: Total man-hours supplied by labor force Total using up of fixed capital and inventories Total inputs of energy and materials Outputs: Total current output—GNP and components at constant prices Industrial production Productivity measures and trends	Inputs: Purchase of goods, services, and houses Using up of owned capital goods Education and training Outputs: Services of labor—energy, skill, knowledge, judgment Services of capital funds	Inputs: Services of labor Services of capital funds Using up of plant and equipment Energy and materials Purchases of plant and equipment Outputs: Goods and services Productivity measures and trends	Inputs: Services of employees Services of borrowed funds Using up of buildings and equipment Materials and supplies for government operations and stockpiles Outputs: Marketed goods and services Public services provided free of charge	Inputs: U.S. exports of goods and services (Also same items as in preceding columns) Outputs: U.S. imports of goods and services (Also items in preceding columns)	(In business sector)

TABLE 2-1 (continued)

Types of Decision-making Units in the Economy

Main Topics for Research and Analysis	Total Economy	Persons: Consumers and Labor Force	Businesses	Governments	Foreign Units	Financial Institutions
G Prices and other terms of exchange	Overall price indexes—inflation or deflation Overall level of interest and profit rates Overall index of wage rates or labor costs Methods of pricing Types of competition	Consumers: Prices for consumer goods Terms for purchasing on credit Labor force: Wage rates and their structure Hours, holidays, vacations, and fringe benefits Standard workweek and overtime Union contracts	Prices for inputs and outputs Terms for purchasing on credit Pricing of distribution services Terms of contracts or sales Costs per unit of output Profit ratios	Prices for inputs and outputs marketed Rates of taxation Rates and terms for transfer payments Terms on government borrowing or on loans granted or guaranteed	Prices of U.S. imports and exports Tariff rates Quantitative restrictions Currency regulations (Also items in other columns)	Federal Reserve discount rate Bank reserve ratios Interest rates and yields on all types of securities and financial paper Availability of funds—maturity and other terms Regulations and practices in security markets
H Incomes, expenditures, and saving	Incomes: Personal and disposable income at current prices National income and components at current prices Total capital consumption allowances Total taxes Expenditures: Total GNP and components at current prices Saving	Incomes: Wages and salaries Rent, interest, and dividends Transfer payments Expenditure: Consumer durables, nondurables, and services Taxes Saving	Receipts: Orders received Receipts from sales to consumers, businesses, governments, and foreign markets Rent, interest, and dividends Expenditures: Wages and salaries and fringe benefits Rent, interest, and dividends Purchases of materials and supplies Purchases of plant and equipment Taxes Saving: Depreciation allowances Retained earnings	Receipts: Taxes—personal and corporate income and property taxes, social security, indirect taxes, tariffs Sale of services, fees Expenditures: Wages and salaries Rent and interest Purchases of materials and supplies Purchases of land, buildings, equipment Transfer payments Surplus or deficit	Receipts: Payments or claims received for U.S. imports Gifts and transfers—private and government (Also items in preceding columns) Expenditures: Payments or claims given for U.S. exports Gifts and transfers (Also items in preceding columns)	(In business sector)
I Monetary and financial areas	Total money supply and flow of funds Total of debt owed in the economy Stocks and bonds held in the economy Annual flow of borrowing and repayment of debt Bank debts	Personal monetary assets Personal debt—short and long term Annual flows of borrowing and repayment of debt	Business monetary assets Business debts Equities Annual flow of borrowing and repayment of debt Business financial policies	Government monetary assets Government debts Fiscal policy Monetary policy Debt management Government loan and loan guarantee activities	Reserves of gold and foreign securities Claims on United States Foreign assets owned by U.S. citizens and businesses (Also items in other columns)	Assets and liabilities of financial institutions Monetary policy Banking policies Reserves and free reserves Annual flow of loans and receipts from repayment of debt

THE OVERALL SYSTEM

The preceding pages presented a reasonably brief introduction to the nature and economic behavior of the six sectors used in macroeconomic analysis, and to their interaction in markets for current output, for factor services, and for capital funds. Table 2-1 classifies the information one would need in order to analyze a national economy and to forecast its changes. The table is not to be studied in detail, but it should serve to give a spaceship view of the whole area of study and to impress you with the complexity of the national economic environment.

The column headings in Table 2-1 refer to the whole economy or to the sectors already discussed, except that Production Units and Business Capital Accounts are here combined into one sector called Businesses. The rows of the table classify the types of information needed for understanding or forecasting the economy. For example, suppose that you are on a United Nations team charged with making recommendations for economic policies of a less-developed country, or suppose that you are forecasting prospective markets for your company's products in a foreign country. You will need information on the legal and economic systems of the country, on the demographic attributes of its population, and on the structure of its Business Sector, Government Sector, and capital markets (row A). Also, you will need to know something about the cultural values of the nation— strength of tradition versus rewards for innovation, emphasis on education and personal advancement, labor-management relations, orientation of business leaders toward growth and risk, and attitudes toward saving and investment (row B). Then you will need to assess the economic resources of the nation— natural, human, and man-made—which may limit the production possibilities and efficiency of the economy (rows C, D, and E). The flows of factor services and of outputs from production, along with market demand and income flows, including redistribution of incomes by the government, would provide needed information on standards of living and market opportunities (rows F, G, and H). The items in row G provide the bridges between what economists term the "real" and the "monetary" aspects of economic activity. Finally, it will be important to know whether money supply, credit availability, markets for capital funds, and the banking and financial institutions are adequate to support growth of production, markets, and capital stocks (row I).

This two-way classification for areas of macroeconomic research and analysis does not bring out the *time dimension* explicitly. You would certainly want to study data on growth and change in each of the above areas, as well as the current situation. Rates of change and time lags are a vital "third dimension" in economic analysis oriented either toward forecasting or toward policy purposes.

THE NATION'S ECONOMIC POWER AND WELFARE

What measures or variables would be most meaningful in assessing the economic power or influence of a nation, or its ability to provide for the economic welfare of its citizens?

NATIONAL WEALTH

As is suggested by the title of Adam Smith's pioneer work, *Wealth of Nations,* early economists focused attention on the stock of wealth owned by a nation or its citizens.   Wealth was defined by John Stuart Mill† as " all useful and agreeable things which possess exchangeable value excluding human beings themselves."   He suggested that a society, or mankind as a whole, is made prosperous by the production and distribution of material wealth.   In Mill's opinion, labor is to be considered productive only to the extent that it leads directly or indirectly to an increase in the community's stock of material products.   As may be imagined, the production of consumer services and some government services were considered by Mill to waste productive services and impoverish the nation.   Mill stated explicitly:‡ "All labour is, in the language of political economy, unproductive, which ends in immediate enjoyment, without any increase in the accumulated stock of permanent means of enjoyment"—wealth.   From this point of view, the economic power of a nation and the welfare of its citizens depend primarily on the nation's productive

† John Stuart Mill, *Principles of Political Economy,* p. 6, Longmans, Green & Co., London, 1896.
‡ Ibid., p. 31.

TABLE 2-2   Wealth, People, and Production

	1900	1929	1948	1958	1966	Growth Rate 1900–1966 (% per year)
Total U.S. wealth						
$ Billions at actual prices	$88	$439	$928	$1,703	$2,460	5.2
$ Billions at 1947–1949 prices	$315	$778	$883	$1,244	$1,550	2.4
Population (millions)	76.8	122.4	147.9	175.6	197.7	1.4
Per capita wealth						
Dollars at actual prices	$1,145	$3,586	$6,274	$9,698	$12,443	3.7
Dollars at 1947–1949 prices	$4,099	$6,355	$5,970	$7,084	$7,840	1.0
Gross national product						
$ Billions per year at 1947–1949 prices	$62	$158	$252	$348	$512	3.2
Ratio of total wealth to GNP at 1947–1949 prices	5.1	4.9	3.5	3.6	3.0	−0.8
GNP per capita (dollars at 1947–1949 prices)	$807	$1,290	$1,705	$1,980	$2,590	1.8

Sources: Wealth estimates are from R. W. Goldsmith, *The National Wealth of the United States in the Postwar Period,* tables A-1 and A-2, National Bureau of Economic Research, New York, 1962; extended to 1966 by J. W. Kendrick, assisted by A. Japha. Gross national product estimates are from the Department of Commerce, Bureau of Economic Analysis series extrapolated from 1909 back to 1900 by John Kendrick's constant dollar GNP series in *Productivity Trends in the United States,* National Bureau of Economic Research, New York, 1961. Figures at 1958 price levels were converted to 1947–1949 base by use of the overall GNP price deflator.

potential as measured by its stock of wealth, and economic growth consists in saving and investing to increase that stock.

The United States has done well on this criterion. Table 2-2 indicates that the nation's total wealth increased from $88 billion in 1900 to $2,460 billion in 1966, when measured at actual prices in each year. If wealth items are revalued at constant prices (1947–1949 level), then the increase was from $315 billion in 1900 to $1,550 billion in 1966. In terms of compound percent per year rates of growth, wealth increased 5.2 percent per year for the current-dollar measure and 2.4 percent per year for the constant-dollar measure. The stock of wealth has grown faster than population, so that per capita wealth in constant dollars rose from $4,099 in 1900 to $7,840 in 1966, a growth rate of 1.0 percent per year compounded. However, aggregate output of the economy (as measured by constant-dollar gross national product), grew 3.2 percent per year from 1900 to 1966, faster than the 2.4 percent per year growth in constant-dollar wealth. So, the ratio of wealth to flow of output actually declined, indicating that our stock of wealth was being used more productively.

Table 2-3 shows the forms of wealth in the United States balance sheet in various years from 1900 to 1966. Note the marked decline in nonreproducible assets (land) as a percentage of the total. Structures increased as a share of the total, with residential and public structures rising and business structures declining. Producer and consumer durable goods both rose as a percentage of total

TABLE 2-3   The Forms Wealth Takes (as a percentage of total)

	1900	1929	1948	1958	1966
Structures	**39.8**	**43.2**	**48.4**	**49.0**	**49.6**
Residential	19.8	21.8	25.2	24.2	24.5
Business	16.4	14.8	12.4	13.4	12.2
Public and other	3.5	6.6	10.7	11.4	12.2
Other reproducibles	**27.5**	**28.1**	**30.9**	**31.4**	**31.6**
Producer durables	7.4	8.7	9.4	11.7	11.6
Consumer durables	7.0	9.6	9.2	10.5	11.8
Business inventories	11.3	8.7	9.3	7.6	7.4
Monetary metals	1.8	1.1	3.0	1.5	0.7
Nonreproducible assets	**35.3**	**25.8**	**19.3**	**18.3**	**16.8**
Agricultural land	18.4	8.7	6.4	5.9	5.9
Business land	3.6	5.2	4.2	4.0	3.9
Residential land	8.4	8.2	5.0	5.5	4.6
Public land and other	4.9	3.8	3.7	2.8	2.4
Net foreign assets	**−2.6**	**2.8**	**1.4**	**1.4**	**2.0**
Total wealth	**100.0**	**100.0**	**100.0**	**100.0**	**100.0**

Source: R. W. Goldsmith and R. E. Lipsey, *Studies in the National Balance Sheet of the United States,* vol. I, table 11, National Bureau of Economic Research, New York, 1963; extended to 1966 by J. W. Kendrick, assisted by A. Japha. Military goods are excluded.

wealth, but business inventories and monetary metals declined.   Net foreign assets rose, as the United States changed from a debtor to a creditor nation.

FLOW OF GOODS AND SERVICES PRODUCED

But is a nation's wealth the best measure of that nation's economic strength or standard of living?   Modern economic analysis places more emphasis on the flow of goods and services which can be produced by the available inputs to production—natural resources, labor, and capital.   The classical economists seem to have assumed that the potential (or capacity) flow of production obtainable from a nation's resources is closely related to the value of the nation's wealth.   It is not clear that this will always be true, since the amount and quality of the labor force used with a given stock of wealth affect the capacity output obtainable from use of the stock.   Also, changes in technology (or " quality " of capital goods) may lead to changes in productive capacity which are not directly proportional to changes in capital stock, unless capital goods are priced in such a way as to incorporate all changes in productivity of capital goods into their price.

As a particular example, Table 2-2 indicates that total United States wealth at 1947–1949 prices rose 2.4 percent per year compounded from 1900 to 1966.   But real GNP rose 3.2 percent per year, and the rise in potential output was presumably close to 3.2 percent per year.   Does the rise in wealth or the rise in potential output provide a better measure of the growth in economic power of the United States economy?   Modern economists would prefer the measure of capacity to produce a flow of output.   Certainly this quantity is more readily measurable than is national wealth, and it is widely used in international comparisons between nations.   It places production of consumer and government services on the same footing as production of material goods, and it emphasizes the productivity of capital goods as well as their quantity or market value.

Figure 2-1 pictures the determinants of national economic potential output and shows the relation between real stocks and flows.   At the top of the diagram are the persons in the nation.   Their economic well-being is the ultimate end of economic activity.   The age distribution of the population, along with social rules and conventions regarding normal hours of work, education, and participation in the labor force by age, sex, and racial groups, plus real wage rates, determines the potential supply of labor services to production units.   The demand for labor input by employers determines what flow of labor services will be employed and what left unemployed.

The output flow produced by the employed services of labor depends on other factors as well as on the man-hours worked per year.

1   The quality of the labor force is important—as measured by health, education, motivation and discipline, initiative, and managerial capabilities.   Technological competence and scientific research and development capabilities are vital.

2   The political, social, and economic organizations are important also—the efficiency of the market system, monetary and banking arrangements, enforcement

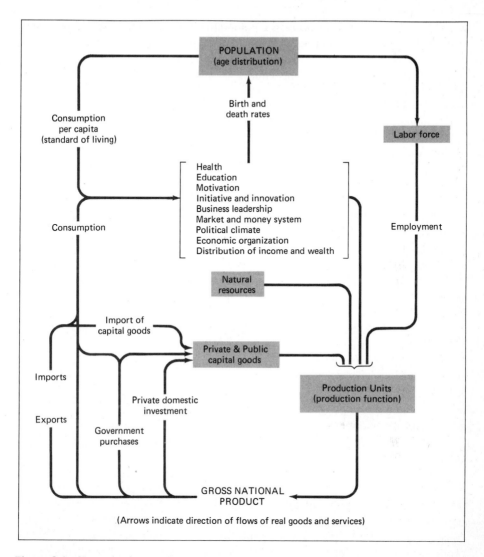

Figure 2-1   The cycle of economic growth—or poverty.

of contracts, mobility of resources, freedom of entry into various occupations and industries, communications and transportation networks, and other factors.

3   The natural resources available to producers, and their costs, are important. They include energy sources (fossil fuels, water power, nuclear energy); pure water; rich mineral ores for metals and chemicals; arable and fertile farm land; forests and wildlife; and coastlines, rivers, lakes, and their marine life.

4   Capital goods—the produced means of production—and the technological

knowledge and production processes built into them are, of course, vital   This category covers publicly owned facilities, such as highways and bridges; harbors, canals, and airports with their navigation aids; postal systems; public housing; schools and government buildings; military hardware; and computers and other office equipment.   It includes the vast array of privately owned plant and equipment; communication and public utility networks; commercial properties and office buildings; improved rural and urban land; farm machinery and buildings; and residential structures.

As shown in Fig. 2-1, all these inputs flow into Production Units and, in accordance with transformation processes summarized by a production function, are converted into the aggregate flow of millions of kinds of goods and services which are called *gross national product* (GNP).

The aggregate flow of output may be broken into several streams classified by the purchasing sector or by the use to which the goods are put—consumption or investment.   Some of the output goes abroad as exports, and imports augment the domestically produced flow of consumer and capital goods.   Aside from those flows which add to the stock of capital goods (wealth) of the economy, the rest of the commodities and services available to the economy flow into consumption by individuals and government.   Real consumption per capita is perhaps the best measure of the living standard of a population and provides a basis for comparisons between nations, or between different times for the same nation.

A high standard of living promotes the personal, political, social, and economic characteristics that make for high labor productivity, or output per man-hour worked.   High productivity, of course, makes possible a large output of goods per capita, and hence permits both a high rate of consumption and a high rate of accumulation of capital goods, each of which promotes further gains in labor productivity.   The economic cycle can be a beneficent one, leading to a high rate of economic growth.

It can also be a vicious cycle.   In underdeveloped countries, low standards of living reduce the health, education, and economic motivation of the population so that the "quality" of the labor force is low.   The low average age of the population also works against development of an experienced labor force.   The consequent low labor productivity leads to small output per capita.   A large proportion of the output must be used to support the low standard of consumption, leaving a small proportion for investment.   The consequent slow increase in the stock of capital goods in turn slows the rate of growth of labor productivity. Since low standards of living are often associated with high birth rates, per capita output and consumption tend to rise slowly also because of the high growth rate of population.   Truly, "the curse of the poor is their poverty."

Total output, viewed from the production side, depends on labor productivity times man-hours per year of labor services utilized, and the latter flow depends on average hours per year worked by employed persons times the number of persons employed.   The production or supply factors determining output per capita may be expressed in algebraic notation as follows:

Let Y_R = real output flow (GNP in $ per year at 1958 prices)
POP = midyear population of the United States
LF = number of persons in the labor force (year average)
E = number of persons employed (year average)
H = average hours worked per year per employed person
$L = H \times E$ = total flow of labor services (man-hours per year)
APL $= \dfrac{Y_R}{L}$ = average labor productivity ($ at 1958 prices per man-hour)

Then

$$Y_R = \text{APL} \times H \times E \qquad\qquad (2\text{-}1)$$

Output per capita may be expressed as

$$\frac{Y_R}{\text{POP}} = \text{APL} \times H \times \frac{E}{\text{LF}} \times \frac{\text{LF}}{\text{POP}} \qquad\qquad (2\text{-}2)$$

For the year 1970, approximate values of the variables were:

Y_R = $724 billion per year at 1958 prices (real GNP)
POP = 205 million persons
LF = 85.9 million persons
E = 81.8 million employed persons
H = 1,961 man-hours per year per employee in nonagricultural establishments

From these values we calculate:

$$\frac{Y_R}{\text{POP}} = \$3,530 \text{ per person per year at 1958 prices}$$

$$L = 1,961 \times 81.8 = 160 \text{ billion man-hours per year}$$

$$\text{APL} = \frac{724}{160} = \$4.51 \text{ per man-hour at 1958 prices}$$

$$\frac{E}{\text{LF}} = \frac{81.8}{85.9} = 0.952$$

$$\frac{\text{LF}}{\text{POP}} = \frac{85.9}{205} = 0.419$$

So, using Eq. (2-2), output per capita and the factors determining it become:

$3,530 per person per year *equals* $4.51 per man-hour *times* 1,961 man-hours per year per employed person *times* 0.952 *times* 0.419 ($ figures are at 1958 prices)

ECONOMIC GROWTH

Equation (2-2) points up four ways of increasing output per capita: (1) increase labor productivity; (2) increase annual hours worked per employed person; (3)

increase the percentage of the labor force which is actually employed, or (4) increase the percentage of the population which is in the labor force.

The last three methods have their limitations. We do want to keep employment high in relation to labor force, but not so high as to lead to undesirably large rises in wage rates and in prices. Also, we do not want everyone in the population to be in the labor force and hard at work. We want to keep young persons out of employment so they can get an education; we want mothers of young families to have the freedom to serve as homemakers if they wish; and we want elderly persons to be able to retire and enjoy life without the pressures of full-time work (in which their productivity would decline, in any event). So there are limits to the proportion of the population which we wish to have employed, consistent with a high standard of living. The wartime (1944) ratio of 47.7 percent of the population in the labor force was higher than normally desirable; about 40 percent seems normal for the type of age distribution in this country and for our norms regarding education, labor force participation of women, and retirement. The ratio was 40.5 percent in 1929, 42.1 percent in 1949, and close to 40 percent after 1960.

For average man-hours worked per year by employed persons, we want to avoid too low a figure, which would imply slack demand and either short workweeks or layoffs. On the other hand, we want to avoid excessively long hours; remember that *leisure is a desired good itself*. A tight labor market would involve overworking the labor force and dangers of rising costs and prices. Our current social mores and labor legislation indicate that 40 hours a week with 2 or 3 weeks' vacation per year is the normal desired work year for the average full-time worker.

The remaining determinant of output per capita, average productivity of labor, seems not to have a ceiling which results in undesirable consequences, at least not in a free economy. It might have such a ceiling if increased output per man-hour were achieved by forcing workers to expend greater effort per man-hour. This is not the case in our economy. Increases in average productivity of labor arise largely from innovations in productive techniques, from employment of more capital goods per worker, or from better organization of production processes. (This suggests the point that productivity measures are simply ratios of output to some one input; the ratios have no causal implications regarding the importance of the input in producing the output. For example, a rise in labor productivity cannot be taken to imply an increase in the intensity or quality of labor services provided per hour worked.)

Labor productivity thus is the most important key to the economic power and welfare of a nation. Productivity increases offer the citizens of the nation the opportunity to increase output of goods from the same amount of work or to increase their standard of living via more leisure time for education, retirement, or personal recreational and cultural pursuits. The economic processes by which productivity gains are distributed will be analyzed later. But it seems clear that productivity gains are what provide the greatest potential for economic progress by a nation and its citizens.

Table 2-4 summarizes the growth rates for the United States economy since

TABLE 2-4 Growth Rates of Private GNP, Population, Employment, and Productivity†

	1902–1929	1929–1947	1947–1970
	(% per year compounded)		
Output (private GNP at 1958 prices, or Y_R)	3.0	2.2	3.8
Population (POP)	2.25	0.9	1.55
Output per capita (Y_R/POP)	0.75	1.3	2.25
Output per man-hour worked (APL)	1.7	2.0	3.1
Private man-hours worked per year (L)	1.3	0.2	0.7
Employed persons (private) (E)	1.7	0.8	1.0
Hours worked per employed person per year (H)	−0.4	−0.6	−0.3
Private employed persons per capita (E/POP)	−0.55	−0.1	−0.5

† Approximate relations among growth rates (g) are as follows:

$g(Y_R) \simeq g(L) + g(\text{APL})$
$g(Y_R/\text{POP}) \simeq g(Y_R) - g(\text{POP})$
$g(L) \simeq g(E) + g(H)$
$g(E/\text{POP}) \simeq g(E) - g(\text{POP})$

Sources: (1) *Long Term Economic Growth*, 1860–1965, Bureau of the Census, October 1966; (2) *Economic Report of the President*, February 1971, tables C-34, C-21, C-22, C-8, C-3; and (3) "Indexes of Output Per Manhour, Hourly Compensation and Unit Labor Costs in the Private Sector of the Economy, 1947–1967," May 1968, and later editions, Bureau of Labor Statistics.

1902 in output per capita and its determinants.   Constant-dollar private GNP is the output variable used here because only in the private sectors do we have data for calculating measures of labor productivity.†   Since the middle period (1929–1947) covered the Great Depression and World War II, comparisons may well be limited to the first and last periods.   Real output grew faster between 1947 and 1970 than early in the century, and this, coupled with slower population growth, led to a much more rapid rise in output per capita, 2.25 percent per year compared with 0.75 percent per year.   As is shown in the bottom two lines in the table, the average hours worked and the ratio of employment to population declined similarly in both periods.   So the faster rise in per capita output resulted almost entirely from a more rapid growth in output per man-hour worked (APL)—3.1 percent per year between 1947 and 1970 as against 1.7 percent per year between 1902 and 1929.

Comparative growth rates for other countries in recent years are given in Table 2-5.   The most noteworthy comparison is that the growth rates of output were similar for developed and less-developed countries, but that the markedly higher population growth rates in underdeveloped countries reduced their growth in output per capita to rates well below those in the more advanced economies. In the category of developed nations, the "younger" nations have higher rates of growth in output than do the more mature economies of the United States and Europe.   This higher rate carries through into output per capita, because rates of population growth are not greatly different among the developed nations.

---

† Productivity of government employees cannot be measured accurately.   Physical measurements of their output are hard to develop, and there is no market price for their output to use as a deflator to convert money value of their services to a constant price series.

TABLE 2-5 Growth Rates of Real Gross National Product in Developed and Less-developed Countries, 1960 to 1967

	Real GNP (Y_R)	Population (POP)	Real GNP Per Capita (Y_R/POP)
	(% per year compounded)		
Developed countries	4.9	1.2	3.7
United States	4.7	1.4	3.3
Europe	4.2	1.0	3.2
Other countries†	7.7	1.4	6.3
Japan	10.4	1.1	9.3
Less-developed countries	5.0	2.5	2.5
Latin America	4.6	2.9	1.7
Near East	6.4	2.5	3.9
South Asia	4.3	2.4	1.9
East Asia	5.2	2.7	2.5
Africa	3.6	2.3	1.3

† Canada, Australia, New Zealand, South Africa, and Japan.
Source: Economic Report of the President, January 1969, p. 124.

Data are not readily available to permit international comparisons of rates of change in output per man-hour, average hours worked per year, and other production factors for GNP as a whole, especially for the less-developed countries. In the manufacturing sector of industrialized nations, however, data indicate the growth rates for output per man-hour for all employees in manufacturing, as shown in Table 2-6.   The United States and the United Kingdom, large countries with mature industries, are lowest on the list, but other mature economies in Europe showed strong rises in productivity, ranging from 5.7 percent to 7.2 percent a year. Among the more recently industrialized economies, Canada stands out on the low

TABLE 2-6 Output per Man-Hour for Employees in Manufacturing, 1960–1970

	Indexes (1967 = 100)		Percent per Year Growth Rate
	1960	1970	
United States	80.5	108.4	3.0
Canada	76.1	112.3	4.0
France	69.4	121.6	5.8
West Germany	66.2	115.3	5.7
Italy	65.1	116.8	6.0
Japan†	52.6	151.8	11.2
Sweden	62.5	125.8	7.2
United Kingdom	79.8	111.8	3.4

† Mining and manufacturing.
Source: Monthly Labor Review, August 1971.

side, and Japan is by far the highest. The "newer" economies gain from the rapid increase in scale of industry, from import of modern technology, and perhaps from shifts in the industrial mix toward products with high output per man-hour. "Older" economies must depend more on their own research and development plus high levels of investment in capital goods embodying advanced technology.

PROBLEMS

1 List the four major domestic sectors of economic decision-making units which comprise an economy.   With respect to each sector, indicate:

    a Its principal goals, or target variables

    b The principal stocks of resources it controls

    c The principal flows of economic goods and services it controls

    d The principal policies or decision-making rules it uses to maximize achievement of its objectives

    e The principal types of economic interaction between it and other sectors

2 For each of the major sectors of the economy, list the chief types of wealth which it *owns*, including negative items (liabilities or equity accounts). For each item indicate whether it is a *real* (R) or a *financial* (F) wealth item, and also whether it is an item of total national wealth (N).

3 Analysts in the Bureau of Labor Statistics have combined population projections with estimates of labor force participation rates by age and sex groups to forecast the U.S. labor force in 1980. Their figure is 101 million persons, in contrast with 86 million in 1970.

    a Given this figure for the labor force, can you make a rough projection of what the real GNP would be in 1980 if the economy were operating at "full employment," that is, with 4 percent unemployment, in that year? Real GNP in 1970 was about $727 billions per year at 1958 prices, with total employment at 81.8 millions.

    b What additional estimates do you need to make to arrive at your estimate?

    c What additional information would you like to have to improve your rough estimate?

4 If you were asked by government officials of a developing country to recommend how they should attempt to measure the "strength" of their economy, what measures would you recommend? What data would the government need to obtain to calculate these measures of economic strength?

Chapter 3

Introduction to Macroeconomic Models

MODEL FOR A SECTOR'S FLOWS OF FUNDS†

Chapter 2 presented a qualitative description of the nature and characteristics of the major sectors of decision-making units involved in the macroeconomic system, and we noted that sectors interact in several types of markets. Now we move into a more detailed and quantitative analysis of how the decision making and inter-actions of the sectors determine the behavior of the macroeconomic system which we wish to understand and forecast.

In each sector, decision makers receive inputs of information about their re-sources and incomes, and about the state of markets and the economic environment, including prices, interest rates, credit conditions, wage rates, and tax rates. Based on such input data (past as well as current), the sector decision makers forecast future states of the economy and choose their own actions to reach their goals as fully as may be possible. The decisions control the flows of funds for the sector—inflows of current income or borrowing and outflows of current expenditures and saving. The economic "behavior" of the sector may be described as *functional relations between input variables to the decision-making process and output decisions expressed by the sector's flows of funds.*

As an example, consider the Personal Sector. Given their resources of labor and property, and taking into account current and prospective wage rates, prices, and interest rates, consumer units decide on the flow of services which they wish to supply to the Production Units Sector to obtain current income. However, it is the Production Units which decide how large a flow of inputs they will purchase and which thus determine the aggregate income flows (Y_S). Given existing tax and transfer policies of the Government Sector, the disposable income of the Personal Sector will be some function of total incomes. A linear equation will prove quite adequate.

$$Y_d = e + f Y_S$$

where Y_d = disposable (after-tax) personal income

Y_S = money flow of gross national product or incomes, determined by Production Units

e, f = constants in the short run

---

† An appendix at the end of the book presents a summary of symbols used throughout the text, along with their names, units of measure, and reference pages for definitions.

The Personal Sector has more control over its flows of purchases, saving, and borrowing. Many input variables influence its decisions—personal disposable income and wealth, prices of goods and services, credit availability and interest rates, and expectations of all these variables, plus current stocks of consumer durable goods. In a simple, short-run model it is often assumed that income flows are the dominant influence on spending decisions, and we write:

$$C^* = a + bY_d$$

where C^* = desired flow of expenditures for consumer goods and services
a, b = behavioral constants (which might change with other variables in a longer-run analysis)

A SIMPLE MODEL FOR DETERMINING PURCHASES OF CURRENT OUTPUT

DESCRIPTION OF MODEL

Now let us build a simple macroeconomic model by representing the sectors in their proper relations to one another and by writing equations to determine the flows in the system. We focus attention here on product markets and on income flows from current production, assuming that equilibria in markets for labor and capital funds have determined wage and interest rates. Even in product markets we simplify the analysis by dealing only with money flows without breaking them down into price and physical flow components. Finally, we assume that the economy is in flow equilibrium. As we shall see later, this implies that purchases made by each sector are equal to the flow *desired* by that sector under the existing conditions of incomes, credit availability, wages, and prices, and of interest rates, wealth, and expectations regarding these variables.

SCHEMATIC DIAGRAM AND SYSTEM VARIABLES

Figure 3-1 presents a simple schematic diagram showing the current-account flows of the various sectors of the economy. Production units decide how much to produce, and they pay out incomes to obtain services of labor and property in factor markets. The incomes received by the four demand sectors are:

Y_d = disposable (after-tax) income of the Personal Sector
T_n = net tax receipts of the Government Sector, that is, total tax revenues less payments for interest on government debt and for income re-distribution (transfers)
GRE = gross retained earnings of Business Capital Accounts, that is, capital-consumption allowances and retained profits.
TR_f = transfers to the Foreign Sector, consisting of private gifts and government grants from the United States

The sum of sector incomes equals Y_S, the value of current production of goods and services.

Given these income flows and the information inputs to their decision-making

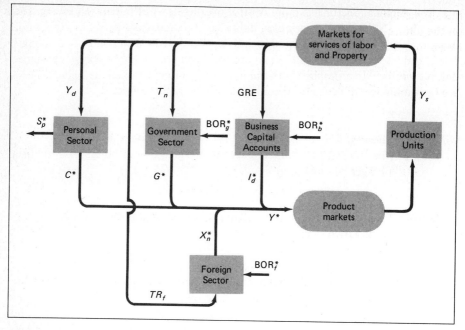

Figure 3-1 Simple diagram for flows of income and purchases of current output.

functions, the demand sectors determine how much they wish to spend for current output and how much to save or borrow. The desired† purchases are:

C^* = expenditures for consumer goods and services
G^* = government purchases of goods and services
I_d^* = gross domestic investment expenditures by U.S. production units
X_n^* = net exports to foreigners, that is, U.S. exports less U.S. imports

The sum of these desired expenditure streams is Y^*, which we shall call *aggregate demand.*

Will aggregate demand (Y^*) just equal the value of sales revenues which the producers expected when setting production schedules? Can we write a set of equations which will determine an equilibrium level of Y^* and Y_s at which both the Production Units Sector and the demand sectors will be satisfied? Let us try.

EQUATION MODEL

We wish to find an equilibrium value for aggregate demand Y^*. To do so, we need a set containing as many equations as there are endogenous variables‡ in

---

† Flows currently desired by the decision makers in a sector are designated by an asterisk.
‡ Endogenous variables are those whose values are determined by the behavior of the system, as distinguished from exogenous variables, whose values are determined by "outside" causal forces not included as variables in the system.

this simple macroeconomic system. There are several types of equations, depending on the nature of the linkages, forces, or constraints determining the relations between the variables.

1 *Definitions or accounting identities.* For example, aggregate demand equals the sum of desired purchases by all demand sectors.

$$Y^* = C^* + G^* + I_d^* + X_n^*$$

2 *Technologically determined relations.* For example, aggregate real output equals a function of technology of production (A), labor inputs (L), and capital stock used (K).

$$Y_R = A \times L^\alpha \times K^\beta \quad \text{(not used in this model)}$$

3 *Institutionally determined relations.* For example, personal disposable income equals a function of total incomes.

$$Y_d = e + fY_S$$

4 *Behavorial relations.* These relate a sector's economic activity to input variables to its decision making. For example,

$$C^* = a + bY_d$$

5 *Equilibrium conditions.* For example, aggregate demand equals value of current production determined by producers.

$$Y^* = Y_S$$

In developing our equation model we shall assume the equations are exact. In empirical work, equations of types 2, 3, and 4 may involve error terms.†

The logic for developing a set of equations which describes this economic system and permits a solution for aggregate demand is as follows:

1 Write an equation relating the variable to be determined to other variables of the system.

$$Y^* = C^* + G^* + I_d^* + X_n^*$$

2 Write additional equations to explain the endogenous variables (other than Y^*) brought in at step 1.

$$C^* = a + bY_d \quad \text{the simple consumption function}$$

Purchases by other sectors are assumed exogenously determined by variables not included in the simple model. That is,

$$G^* = G_0, \; I_d^* = I_{d,0} \quad \text{and} \quad X_n^* = X_{n,0}$$

---

† There may be errors in measuring the variables; the assumed form of functional relationship may not be entirely correct; behavior may change over time; or variables other than those in the equation may influence values of the dependent variable (for example, other variables than Y_d influence C^*).

3   Write additional equations to explain any new endogenous variables brought in at step 2.

$$Y_d = e + f Y_S$$

Continue this process until each endogenous variable has been "explained" by an equation of its own, independent of other equations in the set. For our model we explain Y_S by imposing the equilibrium condition:

$$Y_S = Y^*$$

This process should result in as many equations as there are endogenous variables, so that it will normally be possible to solve for the desired variable, Y^* in our case. The following table summarizes the set of equations, with an indication for each of its type and the variables it adds to the model.

		New Variables	
Type	Equation	Exogenous	Endogenous
Definition	$Y^* = C^* + G^* + I_d^* + X_n^*$	$G_0, I_{d,0}, X_{n,0}$	Y^*, C^*
Behavioral	$C^* = a + b Y_d$		Y_d
Institutional	$Y_d = e + f Y_S$		Y_S
Equilibrium condition	$Y_S = Y^*$		

There are four linear equations and four endogenous variables; so the set of equations should be solvable.

SOLUTION OF THE MODEL

To solve such a set of equations is relatively easy. Just use the equations following the first one to eliminate endogenous variables other than Y^* from the first equation. The result will be an equation containing Y^*, exogenous variables, and the structural constants a, b, e, f. This may be solved for Y^* in terms of the exogenous variables and structural constants.

Thus,

$$Y^* = (a + b Y_d) + G_0 + I_{d,0} + X_{n,0}$$
$$Y^* = a + b(e + f Y_s) + G_0 + I_{d,0} + X_{n,0}$$

In equilibrium, Y^* equals Y_S; so we designate the solution as Y with no modifier, to indicate either demand or supply.

$$Y(1 - bf) = a + be + G_0 + I_{d,0} + X_{n,0}$$

$$Y = \frac{1}{1 - bf} (a + be + G_0 + I_{d,0} + X_{n,0})$$

This simple economic model illustrates the general method of approach to describing the economy by a set of equations and to solving the equations for whatever endogenous variable we wish to calculate.   Even this simple economy displays a very important feedback path in the system and its effect on aggregate gross national product.   If we start the economy in an initial flow equilibrium and then increase some exogenous expenditure, say G_0, aggregate demand will rise initially by the increment in G_0.   As producers increase output to meet the added demand, incomes rise and consumer expenditures increase.   Increased production of consumer goods will lead to further rises in incomes and a new increase in C^*.   This continuous feedback between expenditures, output, and incomes continues, with decreasing increments, until a new flow equilibrium is reached.

Our solution equation permits us to calculate how much equilibrium Y increases for any given rise in government purchases (δG_0).   (The symbol δ is used to designate changes from an initial equilibrium to a final equilibrium value of a variable.)

Original equilibrium: $Y_\alpha = \dfrac{1}{1 - bf}(a + be + G_{0,\alpha} + I_{d,0} + X_{n,0})$

Final equilibrium: $Y_\beta = \dfrac{1}{1 - bf}(a + be + G_{0,\beta} + I_{d,0} + X_{n,0})$

Change: $\delta Y = Y_\beta - Y_\alpha = \dfrac{1}{1 - bf}(G_{0,\beta} - G_{0,\alpha})$

$$= \dfrac{1}{1 - bf}\delta G_0$$

For the United States economy, $b \simeq 0.9$ and $f \simeq 0.6$, so that $1/(1 - bf) \simeq 1/(1 - 0.54) = 2.2$.   So a rise in government purchases by \$1 billion per year would raise GNP by some \$2.2 billion per year in our model, that is,

$\delta Y = 2.2\delta G_0 = 2.2(1) = \$2.2$ billion per year

Note that, of the \$2.2 billion per year rise in GNP, only \$1 billion per year comes from the initiating rise in government spending.   The remainder results from the rise in consumer expenditures induced by increased disposable personal income Y_d flowing through the feedback path from Production Units to the Personal Sector.   Already we can answer one of the questions regarding the macroeconomic system which was raised in Chap. 1: "How can a change in investment or government spending cause total spending to change by double the amount of the original change?"

This amplification of an initial stimulus, called *multiplier effect* in economics, is entirely analogous to amplification in an electric or a mechanical system.   And it arises from the same phenomenon—feedback of part of the output of a system to its input channel.   Amplification is a system property; it arises from interaction of the component parts and is different from the response of any individual components.

MACROECONOMIC SYSTEM BEHAVIOR

As was indicated by our simple model, the macroeconomic analyst must describe and formulate the relations between the sectors as they interact in the overall economic system. Outputs of one sector become inputs to another sector, so that we have the possibility of feedback loops and the amplification of changes in the system. Dynamic changes in the system may lead to the development of expectations which modify the behavioral responses of the sectors. Both the feedback loops and the adjustment to changed conditions may involve time delays in communications, transportation, data gathering, decision making, and execution which affect the responses of the system as a whole. *The complex structure and interaction within the system give rise to phenomena which are different from those found in microeconomic analysis.* A few examples may be suggested.

1 In microeconomics, the consumer demand curve assumes that the consumer unit income is given and that it is independent of quantity demanded and of market price. In macroeconomics, however, changes in quantity demanded by the consumer sector and in overall prices will alter production levels, employment, and incomes—and hence will lead to shifts of the demand curves.

2 In microeconomics, a producer can adjust inventory holdings by cutting production while assuming that demand for the product will be unaffected. In macroeconomics, widespread cuts in production to adjust inventories will reduce incomes and hence shift demand curves.

3 In microeconomics, a consumer unit can elect to save more of its income without affecting the level of that income. Thereby the unit's wealth grows faster. In macroeconomics, if all consumer units in the economy elect to increase their saving out of a given income (and concomitantly reduce their demand for current output), producers may well cut production and thus reduce income flowing to the consumer sector. National wealth may not increase faster; indeed, investment may be reduced, slowing the accumulation of capital goods.

4 In microeconomics, if one bank increases its loans and other banks do not do so, the initiating bank will lose reserves as the money loaned is paid to recipients who deposit it in other banks. In macroeconomics, if all banks increase loans together, the transfer of funds among banks need not lead to any bank's losing reserves.

5 In microeconomics, a government which lowers tax rates will normally suffer a reduction in revenues. In macroeconomics, an income tax cut may lead to an expansion of economic activity which will increase the income tax base enough to yield larger revenues at the lower rate.

6 In microeconomics, a single union may raise the real wage rate of its members to the full extent of the rise in money wage rates it obtains. In macroeconomics, all-round increases in money wage rates may lead to a general price rise and reduce the gain in real wage rate.

7 Price rises for a given good lead to reduction in purchases of that good and

a shift of buying to substitutes, in microeconomics. However, in macroeconomics, price rises for all goods may fail to reduce overall purchases because the price rises lead to income rises, and these may offset any substitution effect (here, a substitution of saving for present purchases).

Consideration of such examples as these suggests several reasons why macroeconomic variables follow different laws than do microeconomic phenomena.

1 *In macroeconomic analysis, we cannot assume so many other variables to be constant as we did in microeconomics.* The *ceteris paribus* assumptions are different. We must be careful to specify our assumptions at each stage as we develop models of the economy.

2 *There are many feedback paths in macroeconomics.* Changed actions (output) of one sector alter the inputs to another sector; then the adjustment made by the second sector may alter inputs to a third sector, or to the first sector again. Such linkages open the way to amplification of changes, and to fluctuations and possible instability of the system in response to economic "shocks."

3 *The cumulative effects of similar actions at the same time by many economic units lead to reinforced macroeconomic effects*—something like the resonant response of a bridge when many persons walk across it in concert at the proper cadence.

4 *Substitution effects are less possible in the macroeconomic sphere, because aggregates often include many of the substitute items.* For example, demand shifts among products may not affect aggregate consumer expenditures very much.

In summary, we shall find much more interaction between economic units in macroeconomic analysis than in microeconomics. So we must watch more carefully our assumptions that "other things are held constant." For the macroeconomic system, as for many complex mechanical, electrical, and biological systems, *the whole is greater than, and different from, the sum of the parts.*

VARIABLES AND MEASUREMENTS

To develop quantitatively our understanding of the sectors of the economic system and of their interactions, we need to define clearly the variables which will be used to describe the inputs, outputs, and current state of each sector and to describe market conditions in the economic environment. To this we now turn our attention.

The measurement of any quantity involves specification of two elements: (1) a unit of measure, and (2) a number that indicates the number of times the unit of measure is contained within the quantity being measured. Thus a certain distance may be described as 660 feet or as $\frac{1}{8}$ mile; a given inventory may be 20,000 pounds or 10 tons of coal. Nevertheless, the qualitative nature of the item being measured remains unchanged.   It is a *length*, whether measured in feet or in miles; it is a *weight*, whether measured in pounds or tons.   This unchanging nature or quality of a measured quantity is called its *dimension*.   In the realm of mechanics, it has been found that all measurable quantities may be described by the dimensions of *mass* (M), *length* (L), and *time* (T), or combinations of them.   For example, velocity equals length traversed per unit of time; hence, dimensionally V is equivalent to L/T, which may be written LT^{-1}.

In economics also, attention to units of measure and to dimensions of quantities will help clarify concepts and relations between variables.   Table 3-1 summarizes the units and dimensions for 20 important economic variables.   As shown in the "dimensional equivalents" column, all 20 can be expressed dimensionally in terms of six basic quantities or dimensions: POP (persons), E (employed persons), M (money), Q_G (quantity of goods), K (capital goods), T (time).   The dimensions of other quantities are combinations of these fundamental dimensions, because the other quantities are derived as the product or quotient of two or more of the six basic quantities.   The dimensions of ratio variables are simply the dimensions of the numerator variable divided by the dimensions of the denominator variable. The symbol *deq* in the right-hand column stands for "is dimensionally equivalent to."   Note that "per" always refers to division; for example, a price in dollars *per* pound is a ratio of market value to weight.

Probably the most important distinction to keep in mind in classifying economic variables is the difference between *stocks* and *flows*.   *A stock is a quantity of something (money value or physical units) measured at a point in time*, like the amount of water in a reservoir at noon yesterday, or the amount of goods in inventory at year-end 1971, or the U.S. labor force for the month of July, or total national wealth or money supply at midyear.

A flow is measured as the quantity of physical goods or money value moving past a boundary during some measurement interval of time.   The boundary may be a physical one (mouth of a stream entering a reservoir or a traffic-counter cable stretched across a highway), or it may be an ownership boundary (legal dividing line between a retail store and a customer which is crossed when a good is sold). Note that the measurement of a flow always involves a unit of time; flow is a ratio of quantity divided by the time required for that quantity to move past the given boundary, for example, 2,000 gallons per minute, 400 cars per hour, or $30 billion per month.

Economic "activities" or transactions are measured as flows in macroeconomics. To a single decision-making unit, some of the individual transactions may appear quite large, like the purchase of a house or the production of a supersonic transport plane.   But when we aggregate transactions for all units in a sector during some reporting period of time, the "lumpiness" vanishes, and the activity is described and analyzed more usefully as a flow.

A few comments may be helpful regarding some of the quantities in Table 3-1. In line 5, the *quantity of labor services* performed, or human effort, depends on the number of workers and the length of time each works.   Their product equals

TABLE 3-1 Units and Dimensions for Economic Quantities

Economic Quantity	Units	Dimensions	Dimensional Equivalents (*deq*)
Stocks:			
1. Population	Persons	Persons	POP
2. Employment	Employed persons	Employed persons	E
3. Stock of cash or value of goods	Billion dollars	Money	M
4. Inventory of goods	Tons of goods	Weight of goods	Q_G
5. Labor services	Man-hours worked	Human work performed	Q_L deq ET
6. Services of capital funds	Dollar-years	Service of capital funds	Q_{SC} deq MT
7. Capital goods	Units of capital goods	Capital goods	K
Flows:			
8. Flow of payments receipts	Billion dollars per year, or billdar†	Money per unit of time	F_M deq $\dfrac{M}{T}$ deq MT^{-1}
9. Rate of change of inventory	Tons of goods per month	Weight of goods per unit of time	F_G deq $Q_G T^{-1}$
10. Flow of labor services	Man-hours worked per year	Human effort per unit of time	L deq $Q_L T^{-1}$ deq E
11. Flow of services of capital funds	Dollar-years per year	Money times time, per unit of time	F_{SC} deq $Q_{SC} T^{-1}$ deq M
12. Real investment	Units of capital goods per year	Capital goods per year	I_R deq KT^{-1}
Ratio variables:			
13. Price	Dollars per ton	Money per physical unit	P deq $MQ_G{}^{-1}$
14. Wage rates	Dollars per man-hour worked	Money per unit of human work performed	W deq $MQ_L{}^{-1}$
15. Interest rate	Dollars per dollar-year	1/time	i deq $\dfrac{M}{MT}$ deq T^{-1}
16. Average productivity of labor	Units of output per man-hour	Output per human work performed	APL deq Q_G/Q_L deq $Q_G Q_L{}^{-1}$
17. Tax rate	Dollars per year of tax payment divided by dollars per year of income	(Dimensionless)	τ deq $\dfrac{M}{T}\Big/\dfrac{M}{T}$, or τ is dimensionless
18. Marginal propensity to consume	Change in dollars per year of expenditures divided by change in dollars per year of income	(Dimensionless)	MPC deq F_M/F_M, or MPC is dimensionless
19. Inventory to sales ratio (money values)	Dollars divided by dollars per month	Time	INV/Sales deq M/F_M deg T
20. Velocity of money	Dollars per year per dollar	1/time	V deq F_M/M deq T^{-1}

† Money flow units will be abbreviated to *billdar* in this book, an acronym for *bill*ion dollars *a*nnual *r*ate.

labor services (Q_L), measured in units of man-hours.   In line 10, the *flow of labor services* into production is measured as man-hours worked per year; for example,

> *L equals ET/T equals* 79.5 million persons employed *times* 2,000 hours worked per year per employed person *equals* 159 billion man-hours worked per year for the United States in 1968.

A similar situation exists for lines 6 and 11.   *Services of capital funds* (Q_{SC}) are the product of money borrowed *times* the period of time it is used, that is, $Q_{SC} = M \times T$. So, Q_{SC} equals 60 million dollars times 2 years or 120 million dollar-years.   The *flow of services of capital funds* (F_{SC}) might then be calculated as an annual rate:

$$F_{SC} = \frac{Q_{SC}}{T}$$

which equals 120 million dollar-years *divided by* 2 years, or 60 million dollar-years per year, which is numerically equal to M

The flow of services (F_{SC}) from a stock (M) is numerically equal to the stock itself if the time units used in calculating the quantity of services and the rate of flow are the same.

Interest rate (line 15) is interesting.   It is the ratio of money payment to services of capital funds purchased, or dollars per dollar-year.   The money units cancel out and leave interest (i) with dimensions T^{-1}, for example, percent per year.

Labor productivity, or APL (line 16), is the ratio of a quantity of physical output to quantity of labor service input used in producing the output.   It might also be calculated as the ratio of the flow of output to the flow of inputs of labor services, that is,

$$\text{APL } deq \; \frac{Y_R}{T} \div \frac{Q_L}{T} \; deq \; \frac{Y_R}{Q_L}$$

Some ratio variables (lines 17 and 18) have quantities with the same dimensions in the numerator and denominator; so these ratio variables turn out to be dimensionless.   They are expressed as decimal fractions or percentages.

Ratios of stock to flow variables will have the dimensions of T (for example, line 19); so they express stocks as number of months' flow.   Ratios of flow to stock variables will have the dimension of T^{-1} (as in line 20); so they indicate turnover rates for the stock—say, number of times per year.

If the list of economic quantities were extended beyond that of Table 3-1, it would be necessary to introduce another dimension for each qualitatively different variable which needs to be distinguished separately for purposes of the problem at hand.   Thus *capital goods* are shown as being a basic quantity or dimension in line 7.   In some problems, it would be necessary to treat plant and equipment separately, or even to segregate different types of machines.   If so, new units and dimensions are introduced, say, "plant," "equipment," or "lathes."   Similarly, for some problems in estimating productivity, qualitatively different types of labor services (man-hours) would need to be distinguished; for example, "managerial man-hours" and "production-worker man-hours" might be treated separately.

An appendix at the end of the book presents a summary table of symbols used throughout this text.

AGGREGATION OF GOODS AND SERVICES

Frequently in macroeconomics we wish to calculate aggregates of dissimilar goods and services; say, to combine all plant and equipment in an aggregate of capital goods, or to combine consumer durables, nondurables, and services in a composite of consumer purchases.   If we wish to obtain an *aggregate of their money value,* the solution is simple; just add the current market values of all the goods and services to be combined, using the dollar as the common unit of measure.

However, suppose we wish to measure the complex *aggregate in physical terms,* so that we can make comparisons of aggregates of goods at different time periods with effects of price changes removed from the comparison.   How do you combine apples and oranges, or refrigerators and haircuts?   The national income accountants have attempted to meet this problem by using *constant-dollar* or *constant-price* measures.   Essentially, they *revalue all goods and services at the prices of some base period of time,* at present the year 1958.   *Then the base-period values are added together for all the items which are being lumped together,* and the composite total is expressed as so many "dollars at 1958 prices," or less precisely, as so many "1958 dollars."

In effect, this calculation simply measures the physical quantity of each good in a new unit of measure equal to the amount of each good purchasable for \$1 in the base period.   This *base-year unit* is a *unit of physical quantity.*   The physical amounts of various goods purchasable for \$1 in the base year were market equivalents to customers as they evaluated alternative goods in allocating their expenditures.   The total number of 1958 "dollars-worth" units for all goods in year t becomes the measure of the physical quantity of the whole collection of goods in that year.   Let us see how this interpretation checks with the constant-price measure described in the preceding paragraph.

Since the base-year unit for any good is the quantity purchasable for \$1 in the base year, it follows that *the market value of any amount of a good at base-year prices equals the number of base-year units in that physical amount of the good.*   Let Q be the measure of a given amount of a good in "natural" units (pounds, automobiles, barrels, or other measure), and let Q_B be the measure of that same amount of the good in base-year units.   Let P_B equal the price of the good in dollars per natural unit in the base year.   Then

$$Q_B = \text{VAL}_B = Q \times P_B$$

For example, suppose we wish to combine 9 million passenger cars and 800 million haircuts.   Assume that in the base year 1958, the average prices were \$2,500 per car (C) and \$2 per haircut (H).

$$Q_{B,\,C} + Q_{B,\,H} = 9 \times 2.5 \times 10^9 + 0.8 \times 2 \times 10^9 = (22.5 + 1.6) \times 10^9$$
$$= \$24.1 \text{ billion at 1958 prices} = 24.1 \text{ billion base-year units}$$

For various goods (j) in gross national product in year t, the market-value flow of a good ($F_{M,j}$) is usually reported instead of physical quantity flow ($F_{Q,j}$). Then the money flows are converted to base-year values by dividing by the appropriate price indexes, and the resulting flows in base-year units are summed over all goods.

$$F_{Q,B} = \sum_j F_{QB,j} = \sum_j \frac{F_{M,j}}{P_j/P_{B,j}}$$

where the price ratio $P_j/P_{B,j}$ may be identified as the dollar price in year t of the base-year unit. This formula may be applied to aggregate consumer expenditures as follows:

	(1) 1970 Value Flow ($F_{M,j}$)	(2) 1970 Price Index ($P_j/P_{B,j}$)	(3) 1970 Quantity Flow ($F_{QB,j}$)
	($ billion per year)	(1958 = 100)	($ billion per year at 1958 prices)
Durables	$ 91.3	108.9	$ 83.8
Nondurables	263.8	127.8	206.5
Services	262.6	140.2	187.2
Total	$617.7	129.4 (calc).	$477.5

From the reported figures in columns 1 and 2, the quantity flows can be calculated by components of consumer spending (column 3). Total quantity flow (477.5) divided into total value flow (617.7) yields the composite price index for consumer goods and services (129.4). Since prices rose from 1958 to 1970, expenditures at 1958 prices are less than expenditures at 1970 prices for the same physical flow of goods.

If similar calculations are made each year, it becomes possible to decompose the time series for total current-dollar expenditures into two components—one representing aggregate quantity flow, and one the composite price index. For an aggregate of many goods, this is the analog of the simple relation for a single good: $F_M = P \times F_Q$.

To be sure, the use of prices of some base period is not a perfect solution to the physical aggregation of dissimilar goods. If prices of some different base period were used in calculating constant-price market values, then the size of the base-year unit would change for each good whose price changed between the base periods, and the total constant-dollar market flow in a given year (t) would change in the same direction as the price change between base years. This follows from the relation $Q_{B,t} = Q_t \times P_B$. However, if prices (hence, the size of base-year units) varied in the same proportion for all goods between the base years, the aggregate quantity measure for all goods would show exactly the same pattern of movement when expressed in base-year units for alternative base years. We get into trouble only when the sizes of base-year units (or prices) for different goods do not change proportionately from one base year to another. With such disproportionate changes, the aggregate quantity series would show some differences in movement when evaluated at prices of different base years. The national accountants keep this error small by updating the base period to reflect a reasonably current set of relative prices for various goods. In the postwar period the base years have been successively 1939, 1947, 1954, and 1958. A shift is probably due soon, based on data from more recent censuses of manufactures and trade.

PROBLEMS

1 List important items of input information flowing to the decision-making process of the Personal Sector, and tell what flows would be controlled by the decisions reached.

2 In the simplified economy described on page 34, assume that investment spending I_d^*, instead of being exogenously determined, is related to Y_S as described by a linear equation:

$$I_d^* = g' + h'Y_S$$

where g' and h' are known coefficients.

a Write the set of equations describing this simple economy, identifying exogenous and endogenous variables as you do so.

b If there are enough equations to solve the system of equations, carry through the solution for Y_S in terms of exogenous variables and known coefficients. If there are not enough equations, add those necessary and then solve for Y_S.

c Derive the equation for personal saving S_p^* as a function of Y_S, where $S_p^* = Y_d - C^*$.

d Suppose the investment function shifted up by an increment δg to its constant term, so that $I_{d\alpha}^* = g'_\alpha + h'Y_S$ is the new investment function, with $g'_\alpha = g' + \delta g'$.
Find the rise in national product (δY_S) from the initial to the new equilibrium value, as a function of $\delta g'$ and the exogenous variables and coefficients.

e Tell where in this economy you can identify feedback effects from the production-income flows of the economy to the demand-expenditure flows.

f What coefficients express the magnitude of feedback effects? Could their values be such as to lead to huge increases in national product (δY_S) for a small initial stimulus ($\delta g'$)? Explain.

3 Functional relations between variables in macroeconomics may express definitions, behavioral relations, or technological and institutional constraints. Except for definitions, these relations will be stochastic, that is, subject to uncertainty and involving an error term, but they are often represented by exact equations.

Since straight-line relations are often used, at least over limited ranges of the variables, it is helpful to have the properties of linear functions in mind. The following chart and basic relations summarize important concepts, as applied to the consumption function.

C = personal-consumption expenditures
Y_d = disposable personal income

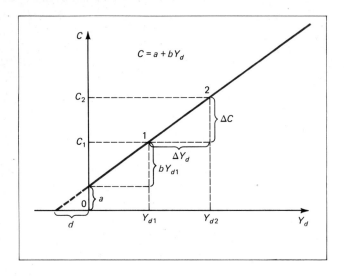

Given the values of C and Y_d at points 1 and 2 on the line, the equation of the consumption-function line may be calculated in any of three ways.

(1) Two-point method:

$$C - C_1 = b(Y_d - Y_{d1})$$

or

$$C = C_1 + b(Y_d - Y_{d1})$$

where $b = \Delta C / \Delta Y_d = (C_2 - C_1)/(Y_{d2} - Y_{d1})$

(2) Intercept-slope method:

$$C = a + bY_d$$

where $b = \dfrac{\Delta C}{\Delta Y_d}$

a = intercept on C-axis, determined graphically or from the relation
$$a = C_1 - bY_{d1}$$

(3) Two-intercept method:

$$\frac{C}{a} + \frac{Y_d}{d} = 1$$

where a and d are determined graphically.

Consider the two points on the consumption function for 1948 and 1968, years in which the saving rate was about the same.

	Consumer Expenditure (C) in billdar†	Net Disposable Income (Y_d) in billdar†
1968	$535.8	$576.2
1948	173.6	186.9

† "Billdar" is an acronym for "*billion dollars annual rate*." In a few years we shall have to start using *trilldar*!

a Calculate the equation for the line through these two points.

b The slope is called the marginal propensity to consume, MPC. Calculate it.

c The slope of a line from the origin to a point on the consumption function is called the average propensity to consume, APC. Calculate it for points 1 and 2.

d Is the intercept a positive, negative, or zero? What are the implications of that value for the relative magnitudes of APC and MPC?

e The variation of C with respect to Y_d may also be expressed in terms of the income elasticity of expenditures:

$$E_{C/Y_d} = \frac{\Delta C/C}{\Delta Y_d/Y_d} = \frac{\Delta C/\Delta Y_d}{C/Y_d} = \frac{\text{MPC}}{\text{APC}}$$

Calculate this elasticity at point 2.

f Find the equation of the personal saving function $S_p = f(Y_d)$, and the value of the marginal propensity to save, MPS.

g If the saving function were to shift upward by 5 billdar at each level of income, find the new equations for the saving- and consumption-function lines.

h What effects, if any, would this shift of the saving function have on the 1968 values of C, S_p, MPC, APC, E_{C/Y_d}, MPS, and APS? Assume Y_d stays constant.

i Instead of the shift in the saving function specified in part **g**, suppose persons raised their savings by 10 percent at every level of income. Find the new equations for the saving- and consumption-function lines.

j What effects, if any, would this proportional change in the saving function have on the 1968 values of C, S_p, MPC, APC, E_{C/Y_d}, MPS, and APS? Assume Y_d stays constant.

k In view of the preceding results, comment on the practicability of deriving "the consumption function" from historical time-series data.

4 Consider a closed economy (no Foreign Sector). In terms of money flows (billdar),

$$GNP = C + G + I$$

where C = personal-consumption expenditures
G = government purchases of goods and services
I = gross private investment

Each GNP component may be expressed as a real flow (in billdar at 1958 prices) times a price index (1958 = 100). Thus,

$$C = C_R P_C \qquad G = G_R P_G \qquad \text{and} \qquad I = I_R P_I$$

Real GNP would be defined as

$$GNP_R = C_R + G_R + I_R$$

The implicit price index for GNP is defined as

$$P_Y = \frac{GNP}{GNP_R}$$

a Record values for current- and constant-dollar GNP and its three main components for the latest year available, along with the associated price indexes. (See U.S. Commerce Department, *Survey of Current Business*, July 1974, "U.S. National Income and Product Accounts," tables 1.1, 1.2, or use data from Table 4-5 in the next chapter, letting $I = I_d + X_n$.)

b The GNP price index is a weighted average of the price indexes of its components. What are the weights? Check them by calculating the weighted average and comparing it with the reported GNP price index.

c Discuss any problems you see in the above method of calculating the GNP price index. For example, would it be possible for P_Y to change from one year to the next even if P_C, P_G, and P_I were all constant?

5 Make the following calculations for changes from 1948 to the latest year for which GNP figures are available.

a What is the percentage rise in current-dollar GNP, in constant-dollar GNP, and in the GNP price index?

b What is the relationship among the three percentage rises calculated in part **a**? Can you say what proportion of the total rise has been in real output and what proportion in price rise?

c Calculate the annual compound percentage rise in GNP, in constant-dollar GNP, and in the GNP price index.

d What is the relationship among the 3 percent per year rates of change calculated in part **c**? Can you say what proportion of the growth rate of GNP has been in real output and what proportion in price rise?

6 List the units of measure for each of the following variables:

a Money wage rate

b Consumer expenditures

c Flow of labor services

d Real stock of capital goods

e Inventory investment

f Income tax rate

g Services of capital funds

h Productivity of labor

i Real GNP

j "Velocity" of money

k Rate of change of money supply

l GNP price level

Chapter 4
National Output, Incomes, and Expenditures

INTRODUCTION

During the past forty years or so, economists have made great advances in defining the sectors of economic decision-making units and in measuring the flows of funds and of goods and services among them in our national economy. The National Bureau of Economic Research pioneered in efforts at measuring national income during the 1930s. Then, just before and during World War II, economists in the United States Department of Commerce completed the development of comprehensive national income and product accounts going back to 1929, with accounting breakdowns showing incomes, expenditures, and saving for the Consumer, Government, Business, and Foreign Sectors. National income and product statistics on a quarterly basis are now published each month in the *Survey of Current Business* (issued by the Department of Commerce), and they appear, revised and updated, in the "National Income" issue of the *Survey* in July or August of each year.

The national product and income accounts provide (1) the statistical basis for government policy actions aimed at achieving national economic goals; and (2) the framework used by business executives in making economic forecasts on which company planning is based.

Students in the fields of economics and business certainly need to develop a clear understanding of the national income and product accounts. In this book, data from these accounts will frequently be used to give an empirical content to the macroeconomic theory developed here, and to illustrate uses and inadequacies of that theory as it now stands.

GROSS NATIONAL PRODUCT AND INCOME BY SECTORS

Gross national product (GNP) is a measure of the total national output of goods and services valued at actual market prices during a specified period of time. It is the *flow of current production* averaged over the time period and measured in billion dollars per year, or *billdar*. The adjective "gross" indicates that the valuation of output at market prices includes current-period charges for depreciation and other consumption of capital goods. The adjective "national" limits the production to that attributable to services of labor and property supplied by residents of the nation, whether these services are rendered in this country or abroad. GNP is reported both at actual prices of the accounting period and at prices of a base year (1958, at present).

During the specified time period, all the value of current production is allocated by producers as costs or profits payable to their suppliers of labor and property services (the "factors of production" in economic terminology) or to various nonfactor charges, such as indirect taxes, capital consumption allowances, and business transfer payments to persons. Since these payments of costs, profits, and nonfactor charges constitute income to their recipients, we may call this total distribution gross national income (GNY). The flow of GNY to recipients in the Personal, Government, Business, and Foreign Sectors provides those demand sectors with purchasing power equal to the market value of current output.

The four demand sectors decide how much of the current output they wish to purchase, and they make expenditures for the desired goods and services in markets for current output. Thus the production units receive a flow of revenue from sales which, it is hoped, covers their costs and nonfactor charges plus the profits which they expected when they decided how much to produce.

This description suggests a circular flow in the economy. Managers in production units decide how much they will produce, thus determining real GNP. The costs and profits paid out as a result of the production activity provide incomes to persons in the demand sectors, and these sectors then make expenditures to buy the flow of goods and services produced. This circular flow was illustrated in Fig. 3-1 in the preceding chapter.

PRODUCTION UNITS

The Production Units Sector is an aggregation of the current accounts (operating statements) of economic units which organize factor services in the production of the nation's output. Business units account for over 80 percent of the production activity of our economy. They include the following:

1 Private corporate businesses.
2 Private unincorporated businesses, such as farm proprietors, legal and medical proprietors, and partnerships.
3 Government enterprises, or government units, such as the Post Office, municipal utilities, or the TVA, whose operating costs are covered to a substantial extent by sale of the goods and services they produce.
4 Persons who own their own homes. In order to make the accounting treatment for such owner-occupied homes comparable to that for similar rented dwelling units, such persons are assumed to own the dwellings and to produce housing services as the proprietors of unincorporated businesses. The homeowner is assumed to keep two sets of books, one as a consumer unit and one as a business unit. As a consumer he pays (imputed)

rent to his home-owning enterprise, and this rent is allocated to costs and proprietor's income just as would cash rent received by an enterprise owning residential property for rent.

Some production takes place outside business units. The largest amount arises from the services of employees in the general government sector, that is, employees of federal, state, and local government units aside from those working for the government enterprises described in item (3). General government employees include military personnel; civilians in the legislative, executive, and judicial branches; teachers in public schools and universities; and state, county, and municipal workers of all kinds. Such employees are considered to produce services desired by consumers and businesses, with the government acting as a kind of collective social agency for securing the output of those services. Compensation of general government employees flows into the Production Units Sector as part of the GNP stream and flows out again through the GNY stream to the Personal Sector as wage and salary income for labor services provided by these employees.

In a similar fashion, employees in the Personal Sector produce part of the GNP. The Personal Sector includes not only individuals and families but also private nonprofit organizations and institutions serving consumers, such as private hospitals and schools, religious and welfare agencies, and private clubs. When these units employ workers, they are considered to buy the desired services in the stream of personal consumption expenditures flowing to the Production Units Sector and hence through the gross national income channel to the Personal Sector as wage and salary income.

Finally, some of our GNP originates abroad. Recall that GNP includes all production attributed to services of labor and property supplied by residents of the nation, even when the services are rendered abroad. Labor compensation received by United States residents working for foreigners is counted in our GNP, although the amount is negligible. More important is the flow of interest and profit-type earnings received by American persons and businesses for services of their capital employed in other countries. These payments flow from foreign units to Production Units through the exports and GNP channels and are paid out from the Production Units Sector through the GNY channel as income to the U.S. suppliers of the labor and property services.

The relative magnitude of these sources of GNP is shown in Table 4-1 for the year 1970, where industry detail has been added for the business production sector. In the United States economy, only 5 percent of our productive activity occurs in primary industries (agriculture, forestry, fisheries, and mining); 33 percent occurs in goods production and utilities (construction, manufacturing, gas and electric utilities, sanitary services); and 47 percent occurs in service industries (transportation, communication, trade, finance and insurance, real estate, law, medicine, private education, recreation, and other personal and business services). Rounding off the figures in the lower part of Table 4-1, $84\frac{1}{2}$ percent of GNP

TABLE 4-1 GNP Originating (or Value Added) by Industry, 1970

Industry	Billdar	Percent of Total
Agriculture, forestry, fisheries	$ 30.8	3.2%
Mining	16.8	1.7
Primary industries, subtotal	$ 47.6	4.9%
Contract construction	$ 45.8	4.7 %
Manufacturing	253.2	25.9
Electric, gas, and sanitary services	22.3	2.3
Goods production and utilities, subtotal	$321.3	32.8 %
Transportation	$ 38.2	3.9 %
Communication	22.5	2.3
Wholesale and retail trade	167.3	17.1
Finance, insurance, and real estate	133.7	13.7
Other services, excluding households and institutions	82.6	8.4
Government enterprises	14.4	1.5
Services produced in business, subtotal	$458.7	46.9%
GNP originating in business (above)	$827.6	84.6%
Compensation of household and institution employees	31.7	3.2
Compensation of general government employees	114.4	11.7
Income originating in rest of the world	4.6	0.5
Total GNY (reported)	$978.6	100.0%
Plus statistical discrepancy	−4.5	
Total GNP	$974.1	

Source: U.S. Department of Commerce *Survey of Current Business*, July 1972, "U.S. National Income and Product Account," table 1.22.

originated in business units, 12 percent arose from general government employment, 3 percent originated in households and institutions, and 1/2 percent came from abroad. It is clear that the United States is a mature economy, with services and government output a major part of our GNP, and primary industry production a very small part.

VALUE ADDED

Government data indicate that manufacturers' shipments in 1970 amounted to 653 billdar; yet the GNP originating in manufacturing is given as 253 billdar in Table 4-1. Why this discrepancy?

This leads to the questions: How do the national income accountants measure the contribution to GNP made by an individual company? And how do they aggregate these contributions for an industry or for the economy as a whole?

In the first place, recall that flows of GNP and GNY cover only transactions involving production and purchase of *current output* of goods and services. For the production units of the economy, only the flows of receipts and payments related to *current operations* are involved. Transactions involving the exchange of real wealth items produced in previous periods are excluded from GNP and

GNY flows, for example, the value of land, secondhand houses, and used autos traded currently but produced in earlier periods. Also, the value of purchases and sales of financial assets is excluded. Bonds, stocks, mortgages, loans, and so on, exchanged by persons, governments, banks, and businesses, are not outputs of current productive activity. (The fees paid for the services of agents who facilitate these transactions, and interest or rent payments for use of capital funds or property, are payments for current productive services and are included in GNP or GNY. But the principal value of transactions in preexisting goods and in financial assets and liabilities is not a part of GNP or income flows.)

So, we look to the current operating statements of business units for our measurement of GNP and GNY. For an individual production unit during an accounting period, we have the identity:

Revenues from sales *equals* total cost of goods sold *plus* profits

But this is not quite what we want on two counts. (1) We want the value of goods *produced*, not of goods *sold*. The two quantities differ to the extent that inventories change. (2) We want a measure of the value of production which *originates* in this particular production unit, excluding the cost of materials, supplies, and services produced by suppliers. The cost of these purchased components (called "intermediate products") must be covered in the sales price of the particular firm's output, but this cost does not represent the firm's contribution to productive activity. With these two modifications, the left side of the above equality becomes:

Market value of current production originating in a firm *equals* revenue from sales *plus* value of change in inventories *minus* value of intermediate products purchased

The right side of the above equality becomes:

Charges against the value of current production originating in a firm *equals* total cost of goods produced *plus* profits *minus* value of intermediate products purchased

The two new equations are alternative measures of *value added* in production by the particular production unit. *Value added is the excess of the market value of a production unit's output over the value of inputs of intermediate products purchased from suppliers.* Or, value added is the excess of total production costs and profits over the value of inputs of intermediate products purchased from suppliers. It is this value added which constitutes the contribution of a production unit to GNP, and the summation of value-added flows for all production units during a given time period equals the GNP for that time period.

Figure 4-1 illustrates how a final product, which passes through several stages of production from raw material to finished good, experiences increments of value at each stage of production. Of course, the actual situation is much more complicated. Each stage of production may purchase intermediate products from

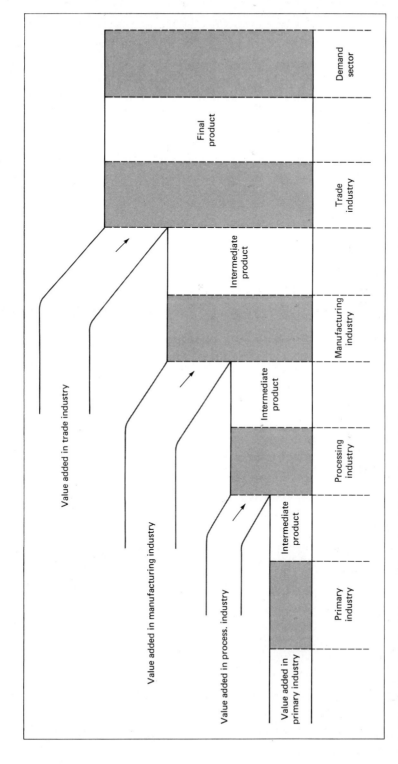

Figure 4-1 Relation between market value of final product and value added in successive stages of production

many other industries, and the production at various stages occurs at various time periods. However, the basic concept is that *the market value of a final product equals the sum of value-added increments in all the industries which contributed to its production.* A good analogy is the addition of flows by various rivulets, creeks, and streams to the flow of a major river finally emptying into the ocean.

It is clear that total sales reported by all production units involved in producing the good would far exceed the market value of the final product because of the duplicated inclusion of the value of intermediate products in sales from one stage of production to a later one. The total sales figure would be comparable to the sum of the volume flows for our river as measured when it crossed each state line in its course.

Let us break down the value-added flows into their finer cost components. We might write, for instance:

> Charges against the value of current production originating in a firm *equals* compensation of employees *plus* payments for services of property and borrowed funds *plus* profits before taxes *plus* depreciation *plus* indirect taxes *plus* other nonfactor charges

These are the total costs of production (excluding intermediate products) plus profits. The first three items on the right side of the equation are payments to the factors of production, that is, payments for the services of labor, capital funds, and natural resources in production. Together they constitute a flow of income payments which is called *national income.* The official definition is this: "*National Income is the aggregate earnings of labor and property which arise in the current production of goods and services by the Nation's economy.* Thus it measures the total factor costs of the goods and services produced by the economy. ... Earnings are recorded in the forms in which they accrue to residents of the Nation, inclusive of taxes on those earnings."†

The remaining charges are not payments for currently rendered services in production, but are charges which must be covered by revenue from sales if the profit is to be earned. They include depreciation charges, indirect taxes (excise taxes, state and local property taxes, and the like, but not taxes on incomes), and miscellaneous other small nonfactor charges, such as business transfers to persons (gifts to Community Chest and other nonprofit agencies, or consumer bad debts) and surplus of government enterprises (treated like excise taxes) less government subsidies to businesses.

At first thought, it seems odd that these nonfactor charges should be part of value added. They do not represent payments to persons or groups who are adding value from a production point of view, except perhaps for the depreciation element. Their inclusion in value added becomes clear if we take the viewpoint of purchasers of the output of the production units. The difference between the

† U.S. Department of Commerce, *National Income*, 1954 ed., p. 58.

market value of the given output and the value of intermediate products incorporated in that output does represent the increment in value arising from the activities of the production unit *as viewed by the purchaser of the output.* The market value (price) which the purchaser is willing to pay exceeds the value of intermediate products by an amount which covers both factor and nonfactor charges. This is the relevant value-added concept.

Now, let us aggregate over all production units of the economy both their value of current output and their allocations of that value to the various categories of charges against current production. The results are shown in Table 4-2, where the left side represents inflows to the bottom of the Production Units Sector in Fig. 3-1 and the right side represents payments out of the top of that sector box.

The revenues of the production units net of intermediate products may be divided into final demand and change in business inventories. Final demand is composed of purchases of current output by users who do not resell the products directly or incorporate them as an element in current production of goods and services to be sold. Thus, final demand is the sum of final goods purchases of current output by the Personal, Government, Business Capital Accounts, and Foreign Sectors reduced by the amount of those purchases that are spent on imports.

To these final sales one must add the value of the change in business inventories to arrive at market value of total national production (GNP). Inventories are

TABLE 4-2 Aggregate Market Value of, and Charges Against, Gross National Product, 1970

Revenue of Production Units (excluding sales of intermediate products)	Billdar	Payments by Production Units (excluding purchases of intermediate products)	Billdar
Final demand for national output:		*National income:*	
Personal consumption expenditures	615.8	Compensation of employees	601.9
		Proprietors' income	66.9
Domestic fixed investment (to Business Capital Account)	132.5	Rental income of persons	23.3
		Corporate profits and inventory valuation adjustment	70.8
Government purchases of goods and services	219.4	Net interest	33.0
Exports	62.9	Subtotal	795.9
Less: Imports	−59.3		
		Nonfactor charges:	
Change in business inventories (to Business Capital Account)	2.8	Capital consumption allowances	87.6
		Indirect taxes	92.9
		Business transfer payments	3.9
		Less: Subsidies less current surplus of government enterprises	−1.7
		Subtotal	182.7
		Statistical discrepancy	−4.5
Market value of national production (GNP)	974.1	Charges against national production (GNY)	974.1

Source: U.S. Department of Commerce, *Survey of Current Business,* July 1971, "U.S. National Income and Product Accounts," table A-1.

considered bought by Business Capital Accounts, since they do show up as an increase in an asset on the balance sheet. They are lumped with domestic fixed investment (plant and equipment plus residential construction) in the national income and product accounts to yield a component of GNP called gross private domestic investment.

On the right-hand side of Table 4-2 are shown the flows of costs, profits, and other charges which allocate the value of current production as income to some demand sector. The first five items constitute *national income*, payments for the services of the factors of production in generating current output. These income flows are before payment of income taxes. Corporate profits are corrected by national accountants because corporate accounting for changes in book value of inventories can introduce an element of capital gain or loss into reported profits. No figures for capital gains are permitted in the national accounts, only those for value flows arising from production activities. So the *inventory valuation adjustment* (IVA) is used to adjust reported corporate profits to the national accounts concept. In addition to national income payments, there are nonfactor charges against revenues as mentioned above for the individual production unit. When the revenues and payments sides of the production units accounts are aggregated for the entire economy, a mismatch will usually occur, which is added to the income side of the accounts as a *statistical discrepancy* (SD).

After correction for the statistical discrepancy, the two sides of the accounts in Table 4-2 balance, and we have the *accounting identity between GNP and GNY*, or between the value of current output and the allocation of that value as current income to final demand sectors. Recall also that these totals are the sum of the value-added flows for all production units in the economy for the accounting period under consideration.

MAJOR FLOWS IN NATIONAL INCOME AND PRODUCT ACCOUNTS

As was mentioned previously, payments by the Production Units Sector constitute a flow of gross national income (GNY) to the demand sectors of the economy. In Table 4-2, we may assign the payments as follows:

1 Personal Sector receives:
   a Compensation of employees
   b Proprietors' income
   c Rental income of persons
   d Net interest
   e Business transfer payments

2 Business Capital Accounts Sector receives:
   a Corporate profits adjusted for inventory valuation
   b Capital consumption allowances

3 Government Sector receives:
  a Indirect taxes
  b Surplus of government enterprises less subsidies paid to Production Units
4 Unknown—statistical discrepancy

However, these initial income payments from the Production Units Sector are redistributed by taxes, transfers (or gifts), and interest on government and consumer debt before we arrive at the net *disposable income* of each of the demand sectors. *The disposable income of each sector is its flow of current receipts which is available for purchase of current output or for saving.* The redistribution of initial income payments to arrive at sector disposable incomes is shown in full detail in the appendix to this chapter and is illustrated in the circular-flow diagram of Fig. 4-2.

This figure, "Major Flows in National Income and Product Accounts," is a basic aid to analyzing the macroeconomic system and should be thoroughly understood. Arrows indicate the direction in which money flows between sectors; real goods and services are exchanged in the opposite direction. Money and real goods and services exchange ownership instantaneously between the sectors linked by a channel; there are no funds or goods "in the pipeline." The financial institutions are not shown on this diagram, and reservoirs of bank deposits and financial assets and liabilities owned by the sectors are off the chart but connected to it through the open-ended channels for saving and borrowing.

If we start from the top of the Production Units Sector in the right-hand part of the diagram, we can trace the allocation of gross national income counterclockwise to the demand sectors. Initially, the nonfactor payments are diverted from the main channel: capital consumption allowances to the Business Capital Accounts, indirect taxes, surplus of government enterprises less subsidies to the Government Sector, business transfer payments to the Personal Sector, and statistical discrepancy to an unknown recipient who is considered to save it. After these subtractions, the remaining flow becomes national income payments for current services of factors of production. This is distributed to the three demand sectors as shown. Instead of taking corporate profits and IVA into the Business Capital Accounts and pulling corporate-profits taxes and dividends back out again, the diagram is simplified to show only the undistributed corporate profits and IVA flowing to the Business Capital Accounts. Corporate-profits taxes and dividends flow to the Government and Personal Sectors directly out of the national income flow. Contributions for social insurance also flow directly to the Government Sector.

A flow of *national income to persons* in the upper left part of Fig. 4-2 is augmented by streams of transfer and interest payments to become total *personal income* (Y_p). Then deduction of personal tax and nontax payments yields *disposable personal income* (Y_d), which flows into the top of the Personal Sector box. The two small outlays for personal interest payments on consumer debt and transfers to foreigners are siphoned off the top of that box because they are not GNP

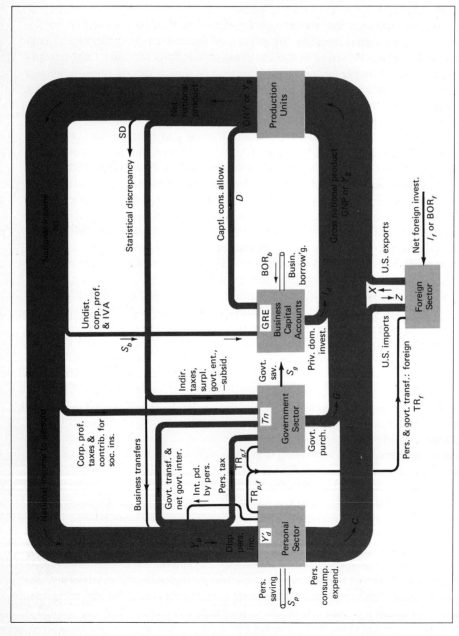

Figure 4-2 Major flows in national income and product accounts.

expenditures or saving. The resultant net flow into the Personal Sector may be called *net disposable personal income* and is designated Y_d' at the top of the box. It is used up in personal consumption expenditures (C) flowing out of the bottom of the box to buy goods and services and in personal saving (S_p) flowing out the left side to buy bank accounts, other financial assets, or existing real assets.

For the Government Sector (including federal, state, and local units), gross tax revenues include personal taxes, corporate profits taxes, contributions for social insurance, indirect taxes, and surplus of government enterprises less subsidies to business. These receipts are offset in part by government transfers and interest payments to the Personal and Foreign Sectors, leaving net receipts for the Government Sector, designated by T_n across the top of the box. Net receipts are allocated to government purchases of goods and services (G) and to government saving (S_g), presumably retirement of government debt or increase in its bank deposits.

For the Business Capital Accounts, the inflow from current operations corresponds to cash flow of a corporation, that is, undistributed corporate profits plus depreciation allowances. The undistributed profits constitute business saving (S_b) because they lead to an increase in net worth of the sector. The depreciation or capital consumption allowances (D) are just an offset to the loss in value of capital assets on the balance sheet. The two flows constitute gross retained earnings (GRE) and provide funds for gross private domestic investment (I_d) in the GNP account. Normally, investment expenditures exceed gross retained earnings; so businesses borrow on capital account, as indicated by the pipeline for business borrowing (BOR_b) flowing into the side of the sector box. This external source of funds supplements the internal flow of funds (GRE) to provide financing of the investment expenditures. It should be noted that *this borrowing does not constitute dissaving, because the funds are used to purchase capital assets; net worth of the sector is not decreased by the borrowing.* (Borrowing by the Personal and Government Sectors is dissaving because the funds are spent for goods and services consumed currently, resulting in a decline in net worth.)

The Foreign Sector receives funds on current account from United States imports (Z) and spends them to buy U.S. exports (X). The net addition to current expenditure on U.S. production is the net export surplus ($X_n = X - Z$), which may be negative. To finance a net export surplus from the United States, foreigners acquire additional funds in two ways: (1) via transfers (gifts) from the U.S. Personal and Government Sectors, and (2) by borrowing from United States residents. Transfer payments do not leave the United States donor with any claim against foreigners, but the nation's lenders do acquire claims against the foreign borrowers. The increase in such claims constitutes an increase in the wealth of the United States. Hence the increase may be termed *net foreign investment,* and that is the label used on the arrow representing borrowed funds flowing into the right side of the Foreign Sector box in Fig. 4-2. In summary, foreigners obtain funds to pay for United States exports from three sources: U.S. imports, U.S. transfers, and foreign borrowing from U.S. nationals. In symbols,

$$X = Z + TR_f + BOR_f \quad \text{or} \quad I_f = BOR_f = X_n - TR_f$$

The streams of expenditures from the four demand sectors flow counter-clockwise in the lower part of Fig. 4-2 and join to equal total GNP. Though it may not be so evident as in the tabular presentation in the chapter appendix, the closed channels distributing income from the top of the Production Units Sector over to the demand sectors and the income redistributions between those sectors guarantee that *the total of net incomes flowing to the final demand sectors will equal gross national income.* This latter equals the total GNP expenditure flows from the four demand sectors. Hence the sum of the flows through the sides of the four boxes must be zero, taking outflows as positive and inflows as negative. In symbols,

$$S_p + S_g = BOR_b + BOR_f$$

The preceding flow-chart presentation of the national economic accounts may be summarized by the following tables of sector accounts for 1970. Table 4-3 shows the net income, the GNP expenditures, and the saving or borrowing for each of the four demand sectors. In Table 4-4, the sector expenditures are totaled to gross national product and the sector incomes are totaled to gross national Income.

TABLE 4-3   Sector Accounts for the United States Economy, 1970

Item	Symbol	Billdar
Personal Sector		
Net disposable personal income	Y'_d	669.9
Personal consumption expenditures	C	615.8
Personal saving	S_p	54.1
Government Sector		
Government net receipts	T_n	206.3
Government purchases of goods and services	G	219.4
Government saving	S_g	−13.1
Business Capital Accounts		
Gross retained earnings	GRE	99.3
Gross private domestic investment	I_d	135.3
Business borrowing	BOR_b	36.0
Foreign Sector		
Net transfers to foreigners	TR_f	3.1
Net exports of goods and services	X_n	3.6
Net foreign investment by United States (= net borrowing by foreigners)	I_f	0.4

Source: U.S. Department of Commerce, *Survey of Current Business,* July 1971, "U.S. National Income and Product Accounts," table 7-1. More details are shown in tables A-2, A-3, and A-4.

TABLE 4-4 Components of Gross National Product and Gross National Income, 1970

Item	Symbol	Billdar
Gross national product		
Personal consumption expenditures	C	615.8
Government purchases of goods and services	G	219.4
Gross private domestic investment	I_d	135.3
Net exports of goods and services	X_n	3.6
Total	GNP	974.1
Gross national income		
Net disposable personal income	Y'_d	669.9
Government net receipts	T_n	206.3
Gross retained earnings	GRE	99.3
Net transfers to foreigners	TR_f	3.1
Statistical discrepancy	SD	−4.5
Total	GNY	974.1

Source: U.S. Department of Commerce, *Survey of Current Business*, July 1971, "U.S. National Income and Product Accounts," table 7-1. More details are shown in table 1-1 and in tables for various sectors.

If we wish to analyze the flow of real goods and services in the economy, we may use a table of constant-dollar values for GNP components, provided by the national accountants. This table is developed by deflating detailed components of GNP expenditures by appropriate price indexes and then summing the resulting constant-dollar flows (in units of billdar at 1958 prices) for major components and for GNP in total. Implicit price indexes for GNP and major components may be calculated by dividing current-dollar values by constant-dollar values. The data are summarized in Table 4-5. No price index is shown for net exports. Exports and imports are deflated separately and subtracted; a price index for this difference would have no meaning. The overall price index rose 35.3 percent from 1958 to

TABLE 4-5 Current- and Constant-Dollar Flows and Implicit Price Indexes for Major Components of GNP, 1970

Item	Symbol	Billdar	Price Index (1958 = 100)	Billdar at 1958 Prices
Personal consumption expenditures	C	615.8	129.4	475.9
Gross private domestic investment	I_d	135.3	132.4	102.2
Net exports of goods and services	X_n	3.6	—	2.4
Government purchases of goods and services	G	219.4	157.3	139.4
Gross national product	GNP	974.1	135.3	720.0

Source: U.S. Department of Commerce, *Survey of Current Business*, July 1971, "U.S. National Income and Product Accounts," tables 1-1, 1-2, and 8-1.

1970; so the valuation of output at 1958 prices is considerably below the value at actual prices in 1970.

No similar deflation is carried out on the gross national income components, except that disposable personal income is sometimes deflated by the implicit price index for consumer expenditures to obtain a measure of the real purchasing power of that income.

BASIC IDENTITIES

From the national economic accounts there emerge several basic identities which provide a framework for thinking and analysis in macroeconomics. They express overall balances or constraints among the major flows linking sectors in the economy; hence they provide some of the basic equations needed for describing the economic system in terms of an interrelated set of functional relations.

The first identity expresses the equality between expenditures to purchase current production (GNP) and income payments arising from that productive activity (GNY). As may be appreciated from the flow diagram in Fig. 4-2, the total value of current production may be measured by four possible approaches, two utilizing the accounts of production units and two utilizing the accounts of the final demand sectors. These approaches are:

1 Receipts from sales to demand sectors by the production units, plus change in business inventories. This approach is illustrated by the left side of the Production Units account in Table 4-2.

2 Payments by production units for charges against the value of current production. This measurement appears at the right side of the Production Units account in Table 4-2 and can be broken down by industry as shown in Table 4-1.

3 Receipts of income by the final demand sectors. These are traced from the payments made by production units in Fig. 4-2, and are summarized as income components by sector in Tables 4-3 and 4-4.

4 Expenditures by the final demand sectors to purchase current production. These GNP expenditures are also summarized by sector in Tables 4-3 and 4-4.

Approaches (1) and (4) lead to GNP measures; approaches (2) and (3) yield measures of GNY. All provide useful perspectives, and the particular viewpoint adopted should be selected to fit the problem under consideration.

The first basic identity arises from the income and expenditures of final demand sectors, approaches (3) and (4) above. (To simplify the equation, let us assume that the statistical discrepancy is zero.)

$$\text{GNP} = \text{GNY}$$

$$C + G + I_d + X_n = Y_d' + T_n + \text{GRE} + \text{TR}_f \tag{4-1}$$

Aggregate expenditures to purchase current production equal aggregate incomes flowing to the demand sectors. *Purchases of current output do not equal income for each sector separately, but in the aggregate they must do so.* This aggregate equality is identically true in every interval of time, however short. Any mismatch between *current* production (as decided by production units) and *final* purchases (as decided by the demand sectors) results in unplanned change in business inventories, which is included as a component of gross private domestic investment (I_d).

A second basic identity is useful in problems where consumer spending is taken as the stable base of the circular flow in the economy and where we wish to study the necessary relations between the remaining streams. It may be found by subtracting C from total expenditures on the left side of Eq. (4-1) and by subtracting C from Y_d' on the right side. Noting that Y_d' minus C equals S_p (personal saving), we obtain:

$$G + I_d + X_n = S_p + T_n + \text{GRE} + \text{TR}_f \tag{4-2}$$

The left-hand side will be called *injections* (INJ) and the right-hand side *withdrawals* (WDL). *Injections are all expenditures for current output except for consumer purchases.* Often these expenditures are considered more volatile than consumer spending and are determined by considerations of national policy or by exogenous considerations not closely related to current income. So they are highlighted in some analyses.

Withdrawals are diversions of components of gross national income from the consumption expenditure stream, either to sectors other than the Personal Sector or to personal saving. In total, they must always be balanced by injections of spending. The necessary equality may be appreciated visually in the circular-flow diagram of Fig. 4-2. If we start from the lower left corner with the stream for C and go around the circuit counterclockwise to the Production Units Sector and back to the starting point by way of the Personal Sector, then the injections flows widen the main stream and the withdrawal flows narrow it. In the complete circuit, their effects must just offset each other.

$$\text{INJ} = \text{WDL}$$

A third basic identity is *the equality of aggregate saving and aggregate investment.* It may be derived from Eq. (4-2) by two adjustments. First, subtract G from both sides and identify T_n minus G as S_g (government saving) on the right side. Second, subtract TR_f from both sides and identify X_n minus TR_f as I_f (net foreign investment) on the left side. The result is:

$$I_d + I_f = S_p + S_g + \text{GRE} \tag{4-3}\dagger$$

---

† This relation implies the equality between saving and borrowing.

$\quad (I_d - \text{GRE}) + I_f = S_p + S_g$

$\quad\ \ \text{BOR}_b + \text{BOR}_f = S_p + S_g$

This identity applies to flows of funds through capital markets.

This is an equality between gross investment and gross saving of the nation. It could be written:

$$I = S$$

If we wished net flows, subtraction of capital consumption allowances (D) from both sides would yield:

$$I_{d,n} + I_f = S_p + S_g + S_b \qquad\qquad (4\text{-}4)$$

or

$$I_n = S_n$$

Saving is the difference between current income of a sector or nation and its current consumption of goods and services.† Saving flows divert funds from current account to capital account for the acquisition of financial or real assets.

Investment is the difference between current production of a nation and its current consumption of goods and services. Investment expenditures channel funds from capital account to current account for acquisition of currently produced investment goods or of net exports. Business Capital Accounts is the only domestic capital-account sector in the national income and product accounts. Its receipts (GRE) flow from current operations to the balance sheet of the Business Sector, and its expenditures use capital funds to buy current output of investment goods. Net exports financed by borrowing from the United States is the second channel by which capital funds are used to buy current output.

It is sometimes difficult for students (and others) to accept this equality between saving and investment in view of the fact that saving and investment decisions are not made by the same persons or groups. But this constraint is enforced not by decisions of production units and demand sectors, but by accounting definitions. Producers can decide how much GNP to turn out in a given period, and they distribute incomes equal to the value of that production. The demand sectors spend most of their incomes to buy goods and services for current use ($C + G$), and save the rest of their incomes. The remainder of current output, GNP minus ($C + G$), must be purchased by the Business Capital Accounts or the Foreign Sector. (Remember that I_d includes inventory accumulation.) This *remainder of current output* also must equal total national saving. *Both investment and saving are equal to the value of current production less the value of goods currently consumed domestically.*

In equation form we may write:

Gross national saving = GNY − ($C + G$) = $S_p + S_g$ + GRE = S

Gross national investment = GNP − ($C + G$) = $I_d + I_f$ = I

---

† In this initial discussion, the presentation will be simplified a bit by assuming that transfer payments to foreigners (TR$_f$) are zero.

TABLE 4-6   Sources and Uses of Gross Saving, 1970

Item	Symbol	Billdar
Gross private saving		
Personal saving	S_p	54.1
Undistributed profits and IVA	S_b	11.7
Capital consumption allowances	D	87.6
Subtotal	S_{priv}	153.4
Government surplus or deficit ($-$) in national income and product accounts	S_g	-13.1
Statistical discrepancy	SD	-4.5
Gross national saving	S	135.8
Gross private domestic investment	I_d	135.3
Net foreign investment	I_f	0.4
Gross investment	I	135.8

Source: U.S. Department of Commerce, *Survey of Current Business,* July 1971, "U.S. National Income and Product Accounts," table 5-1.

Since GNP equals GNY, S and I must always be equal.† One can safely offer a large reward to the student who discovers an exception to this identity.

If saving and investment are both taken net of capital consumption allowances (D), a net identity remains, and may be interpreted in terms of changes in net worth. *For each sector, its net saving equals its change in wealth, if capital gains and losses are set aside.* The Business Capital Accounts Sector holds the real goods which constitute the increase in real wealth of the nation; the Personal and Government Sectors increase their wealth by acquiring financial claims against the Business Capital Accounts Sector, or against the Foreign Sector, or against each other. Thus possession and use of our stock of capital goods may be separated from its ownership. The flows of funds through capital markets (financial institutions) accomplish this separation and provide wealth owners with the types of financial claims they prefer in terms of liquidity, risk, and yield.

In the national economic accounts, the summary of the components of saving and investment for the year 1970 is shown in Table 4-6. The items included in saving have been described previously, as has the derivation of net foreign investment. Gross private domestic investment covers the following principal components.

† If net foreign transfers are included, we have:

Gross national saving $= \text{GNY} - (C + G + \text{TR}_f) = S_p + S_g + \text{GRE} = S$

and

Gross national investment $= \text{GNP} - (C + G + \text{TR}_f) = I_d + (X_n - \text{TR}_f) = I_d + I_f = I$

	Billdar in 1970
Nonresidential fixed investment	**102.1**
Structures	36.8
Producers' durable equipment	65.3
Residential structures	**30.4**
Nonfarm	29.8
Farm	0.6
Change in business inventories	**2.8**
Nonfarm	2.5
Farm	0.3
Gross private domestic investment	135.3

SAVING, INVESTMENT, AND WEALTH

Back in Chap. 2 we presented data on the wealth of the nation, excluding the value of human beings.   Now we can summarize how wealth originates and disappears. Recall that wealth is a *stock* of items which can produce a flow of utility, either directly from its services or indirectly by way of an income flow from using the item in production.   An individual or a firm may hold wealth items in many forms:

1 Human wealth: persons and their capacity to perform economically valuable work, the source of labor services
2 Nonhuman wealth:
   a Real tangible assets
       (1) Produced—capital goods (buildings, roads, improved farm land, machinery)
       (2) Nonproduced—natural resources (water power, unimproved farm land, harbors, natural forests, mineral deposits, pure air and water)
   b Rights to use real assets—leases, patents, copyrights
   c Money—cash and bank deposits (liquid claims)
   d Claims to receive future income and principal—loans, stocks, bonds, life insurance, retirement annuities, and other claims, as viewed by the owner or buyer of the claims
   e Obligations to pay future income and principal—items in 2*c* and 2*d* as viewed by the issuer or seller of the claims

As indicated by item 2*e*, wealth items can be negative as well as positive; there are liabilities as well as assets.   *The wealth of an economic unit is the sum of assets minus liabilities, or its net worth.*

These wealth items are relevant to macroeconomic analysis in several ways:
1 The real stocks of natural and human resources and of capital goods are important determinants of the production potential (or capacity output) of the

economy. They constitute the "land, labor, and capital" identified in classical economics as the ultimate sources of all productive services.

2   Since wealth items usually yield income, the ownership of wealth and financial claims in the economy is an important determinant of income distribution.

3   Wealth holdings and the desire to make changes in the portfolio of wealth items may influence the buying decisions of consumers and business units, and hence may affect aggregate demand.   Saving and investment flows are associated with changes in wealth, and the economy is sensitive to changes in the balance between desired saving and investment flows.

Let us make more explicit the relation between stocks of wealth and flows of saving and investment, for an individual unit and for the nation.   It is clear that an increase in the stock of human capital through improvements in personal health, education, motivation, and discipline is a most important element in determining the economic potential and growth of an economy.   In the absence of a human slave market, we do not capitalize persons and count them as items of wealth.   We regard human beings as ends in themselves, not as a means of production.

Human wealth aside, there are just four ways an individual economic unit can change its net worth.   It can (1) receive current income in return for services in production; (2) use up real assets in production or consumption; (3) pay or receive tax and transfer payments; and (4) incur capital gains or losses.   Economic units may also exchange one asset for another, or acquire an asset by increasing liabilities, but such exchanges do not alter the net worth of the transactors.   Capital transactions just change the composition of portfolios.

As applied to the Personal Sector, the rate of change of wealth (dW_p/dt, or $\dot{W}_p$) *equals* net disposable income *minus* consumption *plus* net capital gains (NCG_p).   Using the symbols for flows in the national income and product accounts, we have:

$$\dot{W}_p = Y'_d - C + NCG_p$$

or

$$\dot{W}_p = S_p + NCG_p$$

As we might have expected, the rate of change of wealth arises from saving out of current income and from capital gains minus losses.

The situation is similar in the Government Sector, where all purchases are counted as consumption-type expenditures which destroy wealth.

$$\dot{W}_g = T_n - G + NCG_g = S_g + NCG_g$$

If the Government Sector runs a deficit (S_g negative), the sector's wealth will be declining insofar as current-account flows are concerned.

The situation is a little more complicated in the Production Units and Business Capital Accounts Sectors.   In the former, payments of charges against production, including intermediate goods and services, involve decline in an asset (cash) offset by an equal rise in value of goods-in-process or of finished goods inventories (valued at cost). When finished goods are sold, the difference between receipts from sales and payments to others leaves undistributed profits and depreciation as the two flows retained by the business. Depreciation flows are offset by decreases in value of plant and equipment; so only the flow of undistributed profits results in a change in net worth arising from operations. In addition, businesses may experience capital gains or losses on their inventories and other assets. The total wealth change of the Business Sector becomes:

$$\dot{W}_b = \text{GRE} - D + \text{NCG}_b = S_b + \text{NCG}_b$$

where S_b equals undistributed profits corrected for inventory valuation adjustment.

By adding the results for all three domestic sectors, we obtain the formula for change in total national wealth (W):

$$\dot{W} = \dot{W}_p + \dot{W}_g + \dot{W}_b = (S_p + S_g + S_b) + (\text{NCG}_p + \text{NCG}_g + \text{NCG}_b)$$

or

$$\dot{W} = S_n + \text{NCG}$$

Apart from net capital gains, the rate of increase in national wealth equals the aggregate net saving flow. Since $S_n = I_n = I_{d,n} + I_f$ we could also write:

$$\dot{W} = I_n + \text{NCG} = I_{d,n} + I_f + \text{NCG}$$

Apart from net capital gains, the rate of increase in national wealth equals the net flows of domestic and foreign investment.

At first thought, something seems odd about this result. The wealth of any individual, business, or sector of the economy may include financial claims against other domestic units; that is, it may include money and nonmoney financial assets. But the rate of increase of total national wealth does not include the rate of increase in such financial assets, only the increase in domestically held real assets (stock of capital goods). The reason for this limitation becomes clear if we consider the process of aggregating wealth over all domestic units. For a single transactor we have:

> Wealth *equals* assets *minus* liabilities, that is, wealth *equals* real assets *plus* financial assets *minus* financial liabilities

If we aggregate over all units in the nation, financial claims held by one resident against another will appear both as assets and as liabilities, and hence will cancel out in total national wealth. Only net claims against foreigners will remain. Thus we find that

> National wealth *equals* the summation of real assets *plus* the summation of net financial claims against foreigners

where the summations are over all units in the national economy.

In summary, wealth of an economic unit is increased by current income from production, by receipt of tax and transfer payments, and by capital gains; wealth is reduced by consumption (including the using up of capital goods), by outflow of tax and transfer payments, and by capital losses. *Apart from capital gains and losses, the net change in wealth per unit of time must equal net income (after taxes and transfers) minus consumption.* But this difference is net saving.

For the economy as a whole, summation over all domestic units, with cancelation of financial claims and obligations between domestic units, yields:

$$\dot{W} = \text{GNY} - (C + G + D) = S_n = I_{d,n} + I_f$$

A physical analogy would be to compare wealth with a quantity (gallons) of water in a reservoir. Income is the rate of flow into the reservoir, say gallons per minute. Consumption is the rate of flow out of the reservoir in gallons per minute. Saving is analogous to a difference between rates of inflow and outflow and hence must equal the change per minute in the amount of water in the reservoir. This is a very general relation between stocks and their related flows:

Change per unit of time in a stock *equals* rate of inflow *minus* rate of outflow

Assuming no capital gains or losses, this relation applies to stocks of durable goods, to inventories, to money stocks, indeed to all balance-sheet items and to population and labor force as well.

It is intriguing to realize that aggregate investment and aggregate saving flows must always be equal to each other even though the saving from current income and the investment expenditures for real assets are made in large part by different decision makers with different motivations. Savers transfer funds to investors through markets for capital funds; the savers increase their wealth in the form of financial claims while businesses hold and use the capital goods in production.

Note, however, that we have said only that *actual* saving equals *actual* investment in real assets plus *actual* net change in foreign claims for the economy as a whole. We have not said that *desired* saving equals *desired* investment in real assets plus net change in foreign claims. *Desired* and *actual* flows may differ, and the analysis of how desired saving and desired investment are brought into equality occupies a large and important place in the analysis of dynamic adjustments and changes in equilibrium position for the economy.

HISTORICAL VARIATIONS IN SECTOR INCOMES, EXPENDITURES, AND SAVING

To illustrate the range of variation of the incomes, expenditures, and saving of the demand sectors, Table 4-7 summarizes the pertinent figures in selected years. To eliminate the scale factor arising from the long-term growth of GNP, all flows have been expressed as percentages of GNP.

TABLE 4-7   Sector Incomes, Expenditures, and Saving for Selected Years

	Percentages of Gross National Product						
	1929	1933	1940	1944	1948	1958	1969
Personal Sector							
Net disposable income	79.0	80.9	74.9	69.2	72.5	69.9	65.8
Personal consumption expenditures	74.9	82.5	71.1	51.6	67.4	64.9	61.9
Personal saving	4.1	−1.6	3.8	17.7	5.2	5.0	4.0
Business Capital Accounts							
Gross retained earnings	10.8	5.7	10.5	8.1	10.9	11.0	10.5
Gross private domestic investment	15.7	2.5	13.1	3.4	17.9	13.6	15.0
Business borrowing	4.9	−3.2	2.7	−4.8	7.0	2.6	4.4
Government Sector							
Net receipts	9.2	12.0	13.3	21.3	15.6	18.2	23.9
Purchases of goods and services	8.2	14.4	14.0	46.0	12.3	21.0	23.0
Government saving	1.0	−2.5	−0.7	−24.6	3.3	−2.8	0.9
Foreign Sector							
Net transfers to foreign recipients	0.4	0.4	0.2	0.1	1.7	0.5	0.3
Net exports	1.1	0.7	1.7	−0.9	2.5	0.5	0.2
Net foreign investment	0.8	0.4	1.5	−1.0	0.7	0.0	−0.1
Statistical discrepancy	0.7	1.1	1.0	1.2	−0.8	0.4	−0.6
	Percentages of Civilian Labor Force						
Unemployment rate	3.2	24.9	14.6	1.2	3.8	6.8	3.5

Sources: U.S. Department of Commerce, "The National Income and Product Accounts of the United States, 1929–1965," and "U.S. National Income and Product Accounts, 1964–69," table 7.1.

To trace the changes in years of high-level activity, we may compare the years 1929, 1948, and 1969, in which the unemployment rates were 3.2 percent, 3.8 percent, and 3.5 percent respectively.   The sector income distributions show that the percentages going to the Personal Sector declined from 79 percent in 1929 to 72.5 percent in 1948 and 65.8 percent in 1969; conversely, the net receipts of the Government Sector rose from 9.2 percent in 1929 to 15.6 percent in 1948 and 23.9 percent in 1969.   The shares taken by gross retained earnings and net foreign transfers changed little.   Consumer expenditures declined as a percentage of gross national product along with the decline in net disposable income.   Government purchases as a share of total output rose from 8.2 percent in 1929 to 23.0 percent in 1969.   Gross private domestic investment and net exports were both abnormally high in the early postwar period (1948), but in 1969 they were below their 1929 percentages of GNP.

The pattern in a deep depression is shown by the 1933 data.   Real GNP in that year was about 70 percent of its 1929 value, and unemployment was 24.9 percent of the civilian labor force! The income share of the Personal Sector rose moderately from 1929 at the expense of gross retained earnings of the Business Sector, but consumption increased more rapidly than disposable income as personal saving became negative. Net government receipts as a percentage of GNP actually rose, contrary to the later experience in mild recessions after World War II, but spending climbed even more than net tax receipts, so that governments were net borrowers of funds for current expenditures. Weakened demand for capital goods and demands by creditors for repayment of loans pulled gross private domestic investment spending below a replacement level. Businesses were repaying debt out of gross retained earnings, thus furnishing funds to capital markets to provide loans to consumers and government units! Such is the topsy-turvy world of a depression!

At the peak of the war effort in 1944, the income shares flowing to net disposable income and gross retained earnings were lower and government net receipts were higher than in 1929 because of wartime taxation. Private spending in both the Personal and Business Capital Accounts Sectors was restricted by rationing and controls. So personal saving was very high (about 25 percent of disposable income), and businesses were again supplying funds to capital markets. The Government Sector was purchasing some 46 percent of GNP, and more than half the government purchases were financed by borrowing the funds poured into capital markets by the private saving flows mentioned above.

The additional years tabulated in Table 4-7 illustrate the patterns of flows when the economy was emerging from the Great Depression of the 1930s, moving into the World War II boom (1940), and then slipping into one of the most serious postwar recessions (1958).

SPECIAL FEATURES OF THE NATIONAL INCOME AND PRODUCT ACCOUNTS

The preceding sections summarized the principal flows and general nature of national income and product accounts as a background for use in macroeconomic analysis. A few concluding comments on national economic accounting may be helpful in assessing the usefulness and limitations of these figures.

1 The overall measure of economic activity is one of value of production. The measure used (GNP) is the market value of the flow of current output of goods and services produced by currently rendered services of labor and property. This definition rules out increases in value arising from capital gains. In addition, some activities involving the use of labor and property services to produce useful goods are excluded.

a Hobby activities comparable to commercial production are not counted, for example, furniture repair in a home workshop, recreational fishing, dressmaking activities in the home. A shift of such activities from home to commercial purchases would raise GNP, and vice versa.

b Housewives' services are excluded, so that GNP will be lowered when a widower marries his housekeeper.

2 Some market transactions, notably illegal transactions, are excluded from GNP. Some flows of goods and services which are not exchanged through the marketplace are nevertheless included in GNP—the so-called *imputations*. These imputations bring into the accounts items of expenditure or income which are comparable to transactions by other economic units which do go through the marketplace.

a The treatment of owner-occupied houses has been noted. A fictitious rental payment is written into consumer expenditures as payment by the individual as a consumer to his or her unincorporated enterprise which owns the property. And a fictitious proprietor's income is imputed back to the homeowner from that enterprise.

b Food, lodging, and clothing supplied to domestic workers or to military personnel are counted as if purchased by the worker or military personnel, with an equal imputed income being paid from the Personal or Government Sector as compensation for the services rendered.

c Banks charge depositors less than cost for their services and cover the deficit out of interest received from lending their customers' deposits. The national income accountants impute full payment for those services in consumer expenditures and add equal payments of interest from the banks to their depositors.

d Food and fuel produced and consumed on farms are treated as if sold off the farm and purchased back in the marketplace.

e Imputed depreciation is charged on fixed investment held by nonprofit institutions in the Personal Sector.

3 There are knotty theoretical problems involved in assigning some purchases to the final goods category or to intermediate products. In one sense, all capital goods should be counted as intermediate products. They are used up in subsequent production of final goods, just as raw materials and supplies are. To count them in GNP when produced, and then to include them as depreciation charges in the value added for goods produced at a later date, is clearly double counting. Theoretically, it would be better to use net national product as our measure of national production activity—the value of current output less capital consumption allowances. However, as mentioned previously, the difficulty of measuring in current prices the value of capital goods being used up in production seems statistically insurmountable now. Since true capital consumption probably varies rather smoothly from year to year, the time pattern of net national product would probably closely resemble that of GNP in any event.

4 The failure to recognize stocks of capital goods, and hence of investment expenditures, in the Personal and Government Sectors is a more serious inadequacy of the accounts. To be sure, owner-occupied houses and the plant and equipment of nonprofit organizations in the Personal Sector are treated as invest-

ment expenditures, and depreciation charges are imputed to these stocks of capital goods. But automobiles, household furniture, and appliances in the Personal Sector and military hardware, roads and airports, government office buildings, public school buildings, sewer and water systems, and similar items in the Government Sector are not counted as capital goods. These items are often financed with borrowed funds, and are susceptible to waves of excess accumulation and postponement—just like business capital goods. Certainly GNP measures of investment understate our rate of accumulation of national wealth by excluding consumer and government accumulations of durable goods.

5   Another difficulty arises from the fact that many governmental services are utilized by businesses and are of a type which, if provided by private producers, would be counted as intermediate products. Think of roads, air and water navigation aids, police and fire-protection services, industry economic information, labor employment services, traffic-control systems, sewer systems. If we considered that business tax payments were purchases of these intermediate services from government enterprises, then GNP would be lower.

6   Finally, many critics have pointed to failures of GNP to account for the social costs and benefits associated with national productive activities. We do not value the increase in human capital arising from educational activities and health programs, or the deterioration in welfare associated with the growth of urban problems, depletion of soil fertility and mineral resources, and pollution of ground, air, and water. Some critics suggest that, in addition to measures of wealth and national product, the national accountants should develop measures of "illth" and "disproduct." Certainly these welfare and environmental factors are important in assessing our national living standards and progress. But the national accountants have no market measures for valuing most of them. Some private economists have worked on a "Measure of Economic Welfare" (MEW), however, and further attempts in this direction seem likely.

By understanding the concepts, coverage, and limitations of the national income and product accounts, we can use them intelligently in the analyses to which they are applicable. Other accounting frameworks have been developed to supplement the national income and product accounts in special areas. The *flow of funds accounts* trace money flows between sectors, not only within the income and product framework but through capital markets and financial institutions. The *interindustry (input-output) accounts* detail the flows of payments for intermediate goods and services among industries, as well as the value-added flows to demand sectors. The *balance-of-payments accounts* trace transactions between the United States and other countries, not only for exports, imports, and transfers on current account, but also for capital-account transactions in financial assets and liabilities. Finally, *accounts for nonhuman national wealth* held and owned by the major sectors of the economy are under development.

The foregoing presentation gives only the major highlights of the national income and product accounts as needed for general purposes of macroeconomic analysis.   Much more detailed statistics and explanations of data sources and estimation procedures are available in many publications of the U.S. Department of Commerce.   Principal Department sources are:

National Income, 1954 edition, original source for basic concepts and definitions and for detailed descriptions of sources and methods of estimation.

The National Income and Product Accounts of the United States, 1929–1965, and *U.S. National Income and Product Accounts, 1964–69*, which provide historical data.

"National Income" issues of the *Survey of Current Business*, published in July of each year, which contain more recent data.   The *Survey of Current Business* includes national income and product summary tables in each monthly issue.

APPENDIX   Distribution of Income from Production

The right-hand side of Table 4-2 describes the types of income payments made by production units out of the value of current production.   It does not indicate (1) how much income flows to each of the demand sectors, or (2) how these initial income payments are redistributed by various tax and transfer streams before becoming net current income in each sector, that is, income available for buying current output or for saving.

To trace the income flows in the economy, let us start with GNP of 974.1 billdar in 1970 and calculate its allocation as after-tax receipts to the four demand sectors.   The July 1971 "National Income Issue" of the *Survey of Current Business* provides the necessary data.   Tables 1.9, 1.10, and 2.1 are particularly useful.

Table 4-8 summarizes the allocation.   Total GNP is entered in a column for the Production Units Sector, and income streams are drawn from this to other columns to show payments made by the production units to the demand sectors.   By line 13 in the table all GNY has been distributed from the Production Units Sector.   Lines 14 through 27 show reallocations of income among the demand sectors.   Line 28 is the sum of income items down each column and equals the net or disposable income of each sector. Lines 29 and 30 show the sector allocation of this net income between GNP expenditures and saving or borrowing.

Starting from the top again, allocations of capital consumption allowances (D) to Business Capital Accounts leaves us with a subtotal called net national product (NNP). Theoretically this measures the flow of production available for current consumption or for increasing the wealth of the nation.   In practice, however, the reported values of D are a poor estimate of actual current consumption of the nation's capital stock at actual prices of the accounting period.

In lines 4 through 8, the remaining items of nonfactor costs are allocated to their proper sectors, with SD shifted to the bottom of the Production Units column to save adding a new column for this balancing item.   On line 8 we then have national income, the payments for currently rendered services of factors of production.   National income amounts to about 82 percent of GNP.

Lines 9 through 13 distribute national income to the Personal Sector and to Business Capital Accounts.   The Personal Sector receives both labor income (compensation of employees) and property income (rental income, net interest) as well as a stream (pro-

TABLE 4-8   Gross National Income Flows and Sector Accounts, 1970

(in billdar)

Line	Item	Symbol	Production Units Sector	Personal Sector	Business Capital Accounts	Government Sector	Foreign Sector	Source Table†
1.	Gross national product	GNP	974.1					1.9
2.	Capital consumption allowance	D	−87.6		+87.6			1.9
3.	Net national product	NNP	886.5					1.9
4.	Indirect business tax	TIND	−92.9			+92.9		1.9
5.	Business transfer payments (to 18)	TRBP	−3.9					1.9
6.	Subsidies less surplus of government enterprises	SUB	+1.7			−1.7		1.9
7.	Statistical discrepancy (to 28)	SD	+4.5					1.9
8.	National income	NI	795.9					1.9
9.	Compensation of employees	COMP	−601.9	+601.9				1.10
10.	Proprietors' income	PROP	−66.9	+66.9				1.10
11.	Rental income of persons	RENT	−23.3	+23.3				1.10
12.	Corporate-profits and inventory valuation adjustment	CPIVA	−70.8		+70.8			1.10
13.	Net interest	INTBP	−33.0	+33.0				1.10
14.	Dividends	DIV		+25.0	−25.0			1.10
15.	Contributions for social insurance	TSS		−57.6		+57.6		1.9
16.	Income from production			692.5	133.4	148.8		
17.	Corporate-profits tax liability	TCP			−34.1	+34.1		1.10
18.	Business transfer payments (from 5)	TRBP		+3.9				1.9
19.	Government transfers to persons	TRGP		+75.6		−75.6		1.9
20.	Government transfers to foreigners	TRGF				−2.2	+2.2	4.1
21.	Interest paid by government	INTGP		+14.7		−14.7		3.1, 3.3
22.	Interest paid by persons (from 26)	INTPP		+16.9				2.1
23.	Personal income	Yₚ		803.6				2.1
24.	Personal tax and nontax payments	TPI		−115.9		+115.9		2.1
25.	Disposable personal income	Yₐ		687.7				2.1
26.	Interest paid by persons (to 22)	INTPP		−16.9				2.1
27.	Personal transfers to foreigners	TRPF		−0.9			+0.9	2.1
28.	Net receipts		−4.5	669.9	99.3	206.3	3.1	7.1
29.	GNP expenditures			615.8	135.3	219.4	3.6	7.1, 1.1
30.	Excess of receipts		−4.5	+54.1	−36.0	−13.1	−0.5	7.1

† Tables in *Survey of Current Business,* July 1971, "U.S. National Income and Product Accounts, 1967–70."

prietors' income) which is partly wage and salary income and partly return on the proprietor's capital funds used in his or her business. The corporate profits flow in line 12 is corrected by the inventory valuation adjustment (IVA) to eliminate capital gains or losses arising from changes in prices of inventories. In line 14, dividends are brought back from Business Capital Accounts to the Personal Sector. In line 15, contributions for social insurance are allocated from the Personal to the Government Sector. This includes both employers' and employees' contributions, since the former are included in compensation of employees allocated to the Personal Sector in line 9.

Line 16 summarizes the distribution of gross national income before it is reallocated by transfer payments and income tax payments. At this point the Personal Sector has received 71 percent, Business Capital Accounts 14 percent, and the Government Sector 15 percent of the value of current production.

Line 17 accounts for the allocation of corporate profits tax liabilities to the government. Lines 18 through 20 show various transfers of income from one sector to another. These are gifts rather than market exchanges. Such nonmarket transactions include government and business gifts to nonprofit institutions, consumer bad debts, unemployment insurance benefits, old age and survivors benefits, family welfare and relief payments, government pension-system payments, military pensions, disability and retirement benefits, and United States government grants to foreigners for economic, welfare, and military aid.

Lines 21 and 22 record interest payments on government debt and on personal consumer debt. These items are treated similarly to transfers in the national economic accounts rather than as part of national income in line 13. It is argued that the capital funds borrowed by governments and individuals are used to buy goods and services which are currently consumed, not for the purchase of depreciable capital assets which yield their services in production. This treatment is consistent with the failure to recognize stocks of capital goods held by consumers and governments.

Lines 23 and 25 show two important subtotals in the Personal Sector. Personal income is the total flow of income to the Personal Sector before payment of income taxes. It includes labor and property incomes paid out from production less contributions for social insurance (line 16), plus receipts of transfer payments and government and personal interest.

After deduction of personal tax payments (net of refunds) from personal income, the resulting flow is called disposable personal income (Y_d). To achieve our goal of allocating total gross national income to demand sectors in a consistent manner, we need to make two additional adjustments which were introduced by the national income accountants in 1965. Interest payments on consumer debt (line 26) have already been discussed. In addition, individuals and institutions transfer part of their after-tax incomes to foreigners, for example, as contributions for relief work abroad or as remittances from immigrants to relatives back in their home country. Those flows are shown in line 27. (Gifts from individuals to other individuals or to nonprofit institutions within the United States are transactions between units in the Personal Sector and do not appear in national economic accounts at all.) After deducting from disposable personal income, the small payments of interest on consumer debt and transfers to foreigners, we obtain an item which may be called *net disposable personal income* (Y_d'). This is allocated either to consumer expenditures (C) or to personal saving (S_p).

$$Y_d' = C + S_p$$

All income allocations from production and all redistribution through taxes and transfers have now been shown. If we total each column, we obtain a net flow of current receipts by each sector (line 28). These receipts, totaled across line 28, are just equal to the GNP of 974.1 billdar with which we started at the top of the Production Units column. *All the value of production is distributed to the demand sectors as purchasing power available and adequate to purchase the national output.*

Units in some sectors save part of their current income. Units in other sectors spend more than their current income on purchases of currently produced goods and services, and borrow to do so. Line 29 shows aggregate GNP expenditures for each sector in 1970, and line 30 shows the difference between income and expenditures by sector. As usual, the Personal Sector spent less than its net disposable personal income in 1970, and hence saved 54.1 billdar. The Business Capital Accounts Sector borrowed 36 billdar to finance investment expenditures in excess of its net income from current production. The Government Sector ran a deficit, and the Foreign Sector also borrowed funds. Since both lines 28 and 29 total across to 974.1 billdar, the total across line 30 is zero. This does not mean that national saving and investment were zero, just that the flow of funds from current income to capital-funds markets equals the flow drawn from capital markets for expenditures on current production. We shall say more on this later.

PROBLEMS

1 Five types of transactions (flows of funds) in an economy are:

A Expenditures on final output, plus investment in business inventories

B Payments of income for current factor services (national income)

C Taxes, transfer payments, and government interest

D Financial transactions, including current saving

E Expenditures for intermediate goods and services, and capital consumption allowances

a Which of these flows constitute gross national product? Gross national income?

b Classify the following transactions into the five categories. (Some may involve components in more than one category.)

(1) Housewife buys housecleaning supplies, also paying sales tax.

(2) Family buys a new car for trade-in, some cash, and an installment loan.

(3) Son buys a used car.

(4) Housewife pays maid and also gives her lunch as part of compensation.

(5) Man pays income tax.

(6) Family contributes to Community Chest.

(7) Man pays rent on apartment to landlord.

(8) Homeowner makes payment on mortgage.

(9)   Farmer's family eats home-grown produce.

(10)   Man deposits his pay check in bank.

(11)   Man buys shares of stock.

(12)   Girl buys a postage stamp.

(13)   Man buys gasoline, using credit card.

(14)   Student pays tuition to private university.

(15)   Company pays its telephone bill.

(16)   Company pays its contributions into private health, unemployment, and pension funds.

2   The following tabulation summarizes the "Consolidated Statement of Current Earnings" for the General Electric Company in 1968.

	$ million/year
Income	
Sales of products and services	$8,382
Net earnings of General Electric Credit Corp.	15
Other income—dividends and interest on investments, royalties on patents	77
Total	$8,474
Costs	
Employee compensation, including benefits	$3,325
Materials, supplies, services, and other costs	4,062
Depreciation	300
Interest and other financial charges	70
Taxes, except those on income	78
Provision for income taxes	312
Increase (−) or decrease in inventories during year	−31
Total	$8,116
Net earnings applicable to common stock	$ 357
Dividends	235
Retained earnings	122

a   Insofar as possible, calculate the company's contribution in 1968 to (1) GNP; (2) national income; and (3) personal income.   Tables 1.9 and 1.10 in the *Survey of Current Business*, July 1974, "U.S. National Income and Product Accounts, 1970–1973," may be helpful.

b   What additional data would you need, if any, in order to make these estimates agree more closely with the concepts in the national income and product accounts?

c   Would any of the above estimates be changed if you knew how much of G.E.'s sales were to final users and how much to other businesses as materials, supplies, and services?   Explain.

3 The GNP, national income, and sector accounts for a three-sector economy, shown below, summarize the receipts and allocations of funds (billdar) for an economy in a given year.   There is no Foreign Sector.   Flows of funds are identified by appropriate symbols.

a Compute GNP by listing the symbols for the flows comprising GNP and adding the corresponding amounts.

b Calculate national income for the economy, indicating by their symbols which flows were combined to obtain the result.

c Calculate the distribution of GNY to the three sectors; that is, calculate the net disposable income of each sector.   Indicate by their symbols which flows were combined to obtain the total for each sector.

Income and Expenditure Flows in a Model Economy (billdar)

Business Sector

	Payments			Receipts	
		Business Production Units			
B1	Purchases of raw materials	200	B8	Sales of raw materials	200
B2	Indirect taxes	50	B9	Sales of capital goods	80
B3	Depreciation	50	B10	Sales to consumers	345
B4	Wages and salaries	270	B11	Sales to governments	65
B5	Rent and interest	40	B12	Change in business	
B6	Proprietors' income	50		inventories	10
B7	Corporate profits before tax	40			
		700			700
		Business Capital Accounts			
B13	Dividends	10	B17	Corporate profits before	
B14	Corporate-profits tax	20		tax	40
B15	Purchases of capital goods	80	B18	Depreciation	50
B16	Change in business		B19	Business borrowing	30
	inventories	10			
		120			120

Personal Sector

	Payments			Receipts	
				Wages and salaries:	
C1	Consumer expenditures	345	C4	from business	270
C2	Personal taxes	70	C5	from government	50
C3	Personal saving	35	C6	Rent and interest	40
			C7	Proprietors' income	50
			C8	Dividends	10
			C9	Transfer payments	
				from government	30
		450			450

Government Sector

	Payments			Receipts	
G1	Purchases from business	65	G4	Indirect taxes	50
G2	Wages and salaries of		G5	Corporate-profits tax	20
	government employees	50	G6	Personal taxes	75
G3	Transfer payments to		G7	Government borrowing	0
	persons	30			
		145			145

   d Complete the construction of a national economic budget by entering the GNP expenditures of each sector and then calculating the sector's excess of receipts over expenditures.

   e List by their symbols the flows comprising gross saving and gross investment. Total the amounts, and check to see that saving equals investment.

4 Historically there seems to have been a tendency for disposable personal income to rise as a percent of GNY during recessions and depressions.

   a Write the equation GNY equals the sum of its sectoral components, and solve for the ratio Y_d'/GNY.

   b In the light of the algebra of part **a**, explain what likely pattern of changes in income streams during a recession would lead to the observed increase in the ratio Y_d'/GNY.

   c Are there any likely changes which would tend to lower the ratio Y_d'/GNY?

   d Use data for 1969 and 1970 (recession year) to see if you can verify the patterns of change hypothesized in parts **b** and **c**.

5 List the principal types of activities or transactions which change the wealth (net worth) of an individual economic unit or of a sector.

   a Is production synonymous with wealth creation, and consumption synonymous with wealth destruction? Explain.

   b Distinguish between current-account activities or transactions and capital-account activities or transactions. Give examples of each type.

6 For the Personal, Business, and Government Sectors, separately define the concept of saving, indicating the principal flows which determine the amount of saving.

   a How do capital gains enter this calculation?

   b Would the sum of saving by the three sectors equal total national saving, hence total national investment?

7 Use data for the latest year from the current issue of the *Survey of Current Business* to construct a table similar to Table 4-8 in the appendix this chapter.

   a Note that the sums of items across rows 28 and 29 are equal. What relationship does this equality express? Write the relation in symbolic form.

   b What sectors were saving and what ones were dissaving in the latest year?

   c Is borrowing equivalent to dissaving? If not, what is the relation between the two concepts, if any?

   d Use the figures on lines 28 through 30 to demonstrate the equality of injections and withdrawals.

   e Use figures from those same lines to demonstrate the equality of saving and investment.

Part Two

Aggregate Supply and Demand for Products and Labor

Sector Decision Making, Macroeconomic Change, and Flow Equilibrium

INTRODUCTION

The conceptual and accounting framework which is used in macroeconomic analysis should be reasonably clear now. Next we shall proceed to analyze the decision-making process in each sector of the economy to understand what variables determine the real and money flows in the economy. Then we shall analyze the interaction of these decisions in various types of markets to see how changes are brought about and equilibrium states reached. From this demand-and-supply analysis we shall see that purchasing decisions by one sector will influence the production decisions of another sector. Changed production flows will then alter the inputs to decision making of the first sector and lead to changes in purchasing behavior. Thus interaction within the system leads to market adjustments and dynamic processes, probably with time lags between successive changes. We wish to trace these economic interactions and also to analyze the possibility and nature of equilibrium states, that is, conditions under which the flows, stocks, and prices in the system stay the same for several periods in a row. By way of biological analogy, we have studied the anatomy of the economic system; now we proceed to its physiology, the study of organic processes, change, and flow equilibrium.

SECTOR DECISION MAKING

Five types of decision-making sectors will be employed in this macroeconomic system: the Personal Sector, the Government Sector, Business Capital Accounts, the Production Units Sector, and Financial Institutions. The Foreign Sector units are similar to these domestic units and will not be discussed separately.

These sectors interact in four principal types of markets: *final product markets, markets for services of labor and property, markets for capital funds and other financial assets, and the market for the stock of money.* The first three involve demand and supply for flows, the fourth for a stock. Markets for intermediate products are not explicitly considered, but are subject to decision making comparable to that for factor inputs to production.

Perhaps a diagram will help clarify the framework of analysis we shall be using. Figure 5-1 shows the demand sectors, Production Units, and Financial

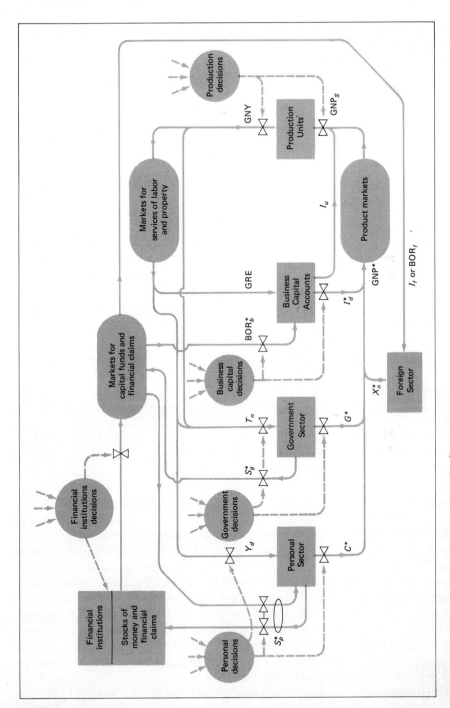

Figure 5-1 Sector decisions as controls of market flows.

Institutions as boxes comparable with those in previous diagrams. The GNP and GNY flows linking them are shown as solid lines rather than channels, and they pass through ovals which designate the markets for final output and for factor services. In this diagram, valves have been shown on some lines to indicate that the magnitude of flow through that channel is governed by current decisions of economic units in the sector whose decision-making circle is linked to the valve by a dashed line. Then, each of the decision-making circles has dashed lines flowing into it to represent the inputs of relevant information regarding current and past flows, stocks, and market conditions. Units in the given sector use the input information to generate forecasts and to make analyses relating income, prices, credit conditions, current stocks of goods and money, demographic factors, and similar data to their decision-making criteria (economic goals, utility, or profits). From these information inputs, forecasts, analyses, and weighing of costs and benefits of alternative lines of action come the decisions which express themselves as adjustments to the valves which control the flows for that sector. In later chapters we shall discuss the most important information inputs to each sector, that is, the most important variables controlling its flow decisions, and we shall attempt to derive functional relations between input information and output flow decisions. This process constitutes the economic behavioral analysis of macroeconomics.

PERSONAL SECTOR DECISIONS

The individuals and groups in the Personal Sector, and hence the sector as a whole, may be regarded as securing income flows in return for supplying labor and property services to the Production Units Sector. The amounts of services which persons desire to supply are determined by the information flowing to the Personal Sector and by the balance between expected benefits and costs from providing an additional (marginal) unit of services. The amounts of productive services actually provided, and hence the income flowing to the Personal Sector, depend also, of course, on the demand by Production Units for those services in factor markets.

Once their incomes have been determined, persons must decide how that income is to be allocated between desired consumption (C^*) and desired saving (S_p^*). (The asterisk will designate desired flows, as decided by the demand sectors.) Thus C^* indicates the market value of the flow of goods and services which units in the Personal Sector *would like* to purchase, given the present number and types of consumer units, their disposable personal income, market prices, interest rates and other credit conditions, consumer stocks of durable goods, money and other asset and debt items, and their expectations of these variables in future time periods. The flow C^* may be thought of as consumer orders for products they want delivered during the current time period. Desired net saving of the Personal Sector (S_p^*) is the portion of current income which consumer units *wish* to use for changes in their wealth, or net worth. Decisions to save or dissave involve acquisition or sale of stocks of real capital goods, money and other financial assets, and the borrowing of funds. They change the assets and liabilities on the balance sheet of the Personal Sector.

Note that *decisions regarding desired consumption and desired saving are interlocked because they are a joint allocation of current disposable income.*

$$C^* + S_p^* = Y_d$$

Once Y_d is given, a decision on C^* determines S_p^*, and vice versa.

Will the Personal Sector actually spend and save the *desired* flows out of a given disposable income? Actual C might fall short of C^* if an interruption of supply (for example, a strike) prevented delivery of desired goods or if C^* rose suddenly, as at the beginning of the Korean war when production plus inventories were inadequate to meet the demand. Then C was less than C^* and S_p was greater than S_p^*. Conversely, a consumer might place an order for future delivery of a car or an appliance and then have to take delivery, though he might not desire to do so in the later period. Then C would be greater than C^* and S_p would be smaller than S_p^*. However, almost always in this book we shall assume that actual, realized consumer spending and personal saving are equal to the desired values.

Equality of desired and actual flows describes a condition called *flow equilibrium* for the Personal Sector. If the units in the sector and their incomes do not change and if market conditions and expectations remain constant for many periods, desired and actual levels of consumption and personal saving will stay in balance at stable levels for many periods. This situation is described as a *static equilibrium.* However, desired and actual flows may stay equal to each other even if incomes and market conditions, and hence desired flows, change from period to period. Such a situation is described as a *dynamic equilibrium.* Both these situations will be discussed later, but more attention will be given to the former. The word "equilibrium" without a qualifier will refer to the state of *static equilibrium.*

GOVERNMENT SECTOR DECISIONS

The Government Sector has some control over its net income and saving through tax legislation and decisions on transfer payment, as shown in Fig. 5-1 by the interlocked controls on the valves governing T_n and desired surplus S_g^* (which may be negative). The government decision makers also control a valve regulating the Government Sector's desired purchases of current output of goods and services (G^*). These influences of the government on aggregate economic activity via their policies regarding purchases, taxes, and transfer payments are commonly referred to as *fiscal policy.* The criteria involved in governmental decision making reflect the economic goals of the nation, such as reasonably full employment of labor, satisfactory distribution of incomes, price stability, economic growth, balance-of-payments position, satisfactory balance between private and public consumption, and provisions for social welfare and the support of private economic activity. In furtherance of such objectives, the Government Sector's expenditures *need not be closely tied to its net receipts, but desired purchases and desired savings must be jointly determined so that they are consistent with the equality*

$$G^* + S_a^* = T_n$$

The flow S_g^* to or from markets for capital funds and financial claims indicates that the Government Sector influences the economy through capital markets as well as current-account flows.

Will the Government Sector actually purchase and save the *desired* flows out of given net receipts T_n? Again, we shall usually assume that actual flows of government purchases and saving (or borrowing) are equal to the desired flows. But there will be times when supply shortages make G less than G^*, and S_g greater than S_g^*. This might occur, for example, when the nation enters a war or an unexpected recession and when the time lags in drawing plans, in letting contracts, and in producing the buildings or equipment items ordered cause a shortfall of actual purchases. Conversely, the long lead times involved in government budgeting and procurement may give rise to situations when the government must buy goods or services which it would desire not to purchase at the time of delivery. Then G is greater than G^* and S_g is less than S_g^*. Normally, however, we shall assume that the government is in flow equilibrium, that is, G equals G^* and S_g equals S_g^*.

BUSINESS CAPITAL ACCOUNTS DECISIONS

In their capital-account decisions, managers determine how much they desire to spend for domestic investment (I_d^*), additions to fixed capital goods plus inventory accumulation. These decisions and the flow of internal funds (GRE) jointly determine the amount of borrowing desired from capital-funds markets (BOR$_b^*$). That is,

$$I_d^* - \text{BOR}_b^* = \text{GRE}$$

Presumably, these decisions depend on the schedule for anticipated rates of return for various investment projects and on the cost and availability of external capital funds as well as internal funds.

If the information inputs to the Business Capital Accounts decisions remain the same from period to period and if orders for investment goods are all filled on schedule, the Business Capital Accounts will be in flow equilibrium, that is, I_d will equal I_d^*. In a long-term stationary situation, the desired stocks of capital goods and inventories would normally be constant; so desired investment would simply be a replacement flow equal to current consumption of capital stock.

However, given the long lead times between investment decisions and delivery for some plant and equipment, and given the volatility of business expectations and credit conditions during business cycles, short-term divergences between actual and desired purchases are likely in this sector. If I_d is less than I_d^*, we may expect an increase in orders for investment goods in future time periods, a dynamic force for expansion. If I_d is greater than I_d^*, a contraction of investment spending will tend to depress future economic activity. In a later chapter we shall analyze in some detail how investment demand can cause cyclical movements in the economy. For the present, we shall assume that *actual fixed investment equals desired fixed investment but that actual inventory investment may differ from the desired change in* *business inventories.*   An inventory disequilibrium will lead to dynamic changes in aggregate demand and supply until a new static equilibrium is reached, in which I_d equals I_d^* and inventory change is zero.

PRODUCTION UNITS SECTOR DECISIONS

Inventory imbalances arise from poor sales forecasting and time lags in the Production Units Sector.   Managers in production units apply criteria of long-run profit maximization to knowledge of their production functions, of supply schedules of inputs, and of forecasts of demand for output; thus they arrive at decisions regarding the flow of production and the associated payments of costs and profits. See the linked decision arrows controlling the valves for flow of GNP_S and GNY at the Production Units Sector in Fig. 5-1.   The flows resulting from these decisions link the Production Units Sector with the Personal Sector in markets for services of labor and property and also with all demand sectors in the product markets. Here arises the possibility of mismatches.

At current prices for factor services, and given other current and expected market conditions, the Personal Sector might want to supply more labor and property services than the Production Units desire to employ.   Then there would be unemployed labor or property.   On the other hand, Production Units might desire more factor services than the Personal Sector wants to supply under existing market conditions.   Then there would be excess demand for factor services.   In either case, the factor markets would not be in equilibrium.   Decision makers on one or the other side of the market would be motivated to change their market offers to achieve their utility or profit-maximization objectives more fully.   Thus the market disequilibrium would lead to changes toward a desired equilibrium position, perhaps with some time lags, unless market restraints of some kind prevented the adjustments .   We shall discuss later such disequilibrium and adjustments in factor markets, primarily labor markets.

Now let us consider *possibilities of disequilibrium and change in product markets.*

The Production Units receive the flows of orders from the demand sectors. To meet those orders, producers have available current production (supply) plus inventories carried over from the previous period.   If current production, based on decisions made at the beginning of the current time period (or earlier), are inadequate to provide shipments on all orders, the producers may ship goods out of the beginning inventories or may build a backlog of unfilled orders for late delivery. On the other hand, if current production exceeds orders for current delivery, order backlogs may be run down (if customers are willing to take delivery in the current period) or an unplanned inventory buildup will occur.   The absorption of a discrepancy between flows of demand and supply by changes either in inventories or in backlogs of unfilled orders is illustrated in terms of the microeconomic demand-and-supply diagram in the appendix to this chapter.

Whether the mismatch between demand and supply is absorbed by an unplanned change in order backlogs or in inventories, the customers and producers are motivated to change their orders or production flows, and perhaps prices, in the next period. The economy is not in equilibrium. Flow equilibrium will be achieved for the Production Units Sector *only when receipts of orders for current delivery equal the planned production for the current period.* Theoretically, for full equilibrium, each production unit should be in this equilibrium state, but one could define a macroeconomic flow equilibrium as one in which individual producers experience offsetting imbalances, and the aggregate Production Units Sector continues in equilibrium.

In this text we shall assume that a mismatch between aggregate demand and supply is reflected in unplanned inventory changes rather than in changes in backlogs of unfilled orders. This situation is pictured in Fig. 5-1. Suppose the sum of orders by the three demand sectors (GNP*) is less than current production flow through the valve governed by producers' decisions (GNP_s). Then inventories accumulate. This is unplanned investment spending by Business Capital Accounts (I_u). The production units will be dissatisfied with the outcome of their production decisions for this period and will presumably change output flow or prices, or both, in succeeding periods. *The economy will be in equilibrium only if aggregate demand and supply are equal, so that unplanned inventory investment is zero.*

MACROECONOMIC DEMAND-AND-SUPPLY EQUILIBRIUM IN PRODUCT MARKETS

Let us express these conditions for flow equilibrium in equations and bring out their implications for the saving and investment balance. [An asterisk will be used to designate sector demands (or orders placed) as decided by the demand sectors. Subscript s will designate supply of goods and services, as decided by the Production Units Sector.] We may define aggregate demand (GNP*) and aggregate supply (GNP_s) as follows:

$$GNP^* = C^* + G^* + I_d^* + X_n^* \tag{5-1}$$

$$GNP_s = C_s + G_s + I_{d,s} + X_{n,s} \tag{5-2}$$

I_d^* and $I_{d,s}$ are considered to include flows of orders and production for *planned* inventory changes as well as for fixed capital goods. *Unplanned inventory investment (I_u) occurs when* GNP* *is not equal to* GNP_s. This can be shown by subtracting terms of Eq. (5-1) from corresponding terms of Eq. (5-2). We obtain

$$GNP_s - GNP^* = (C_s - C^*) + (G_s - G^*) + (I_{d,s} - I_d^*) + (X_{n,s} - X_n^*) \tag{5-3}$$

The excess of production over deliveries to meet customers' orders from the four demand sectors on the right side of Eq. (5-3) totals to the unplanned inventory accumulation in the economy. Hence we write

$$GNP_s - GNP^* = I_u \tag{5-4}$$

When the economy is in equilibrium, GNP_s *equals* GNP*, *unplanned inventory investment* I_u *will be zero, in total and also for goods of each of the four sectors.*

If GNP_s is greater than GNP* (or I_u is greater than zero), producers will wish to decrease production and/or lower prices in subsequent periods to eliminate the undesired accumulation of inventories. Indeed, they may wish to reduce production below orders temporarily, to cut the unwanted stocks accumulated in the first period. If GNP_s is less than GNP* (or I_u is less than zero), producers will wish to increase production and/or raise prices in subsequent periods, perhaps overshooting temporarily to rebuild stocks of inventories to desired levels.

If one could develop relations for calculating desired levels of inventory holdings in relation to GNP, it would be possible to estimate planned changes of inventories consistent with a given forecast of GNP*. Detection of deviations of actual inventory investment from the planned changes would give an indication of the probable direction of future changes in inventory investment. Despite difficulties in estimating planned changes in inventories, this is one factor used by forecasters of the national economy for predicting short-term fluctuations in economic activity.

EQUILIBRIUM BETWEEN DESIRED SAVING AND DESIRED INVESTMENT

Next let us explore the implications of the equilibrium equations for the balance between saving and investment flows. In the first place, note that *the value of current production* (GNP_s) *equals the value of aggregate income flows* (GNY); *that is, production units determine the levels of current output and incomes, except for profits.* Then, expressing aggregate income in terms of its four sector components, we have

$$GNP_s = GNY = Y_d + T_n + GRE + TR_f$$

Now subtract the aggregate demand Eq. (5-1) from the above equation, term by term, and note that in the Personal and Government Sectors the difference between net income and desired expenditures is equal to desired saving. We obtain

$$GNP_s - GNP^* = (Y_d - C^*) + (T_n - G^*) + GRE - I_d^* - (X_n^* - TR_f)$$
$$= (S_p^* + S_g^* + GRE) - (I_d^* + I_f^*)$$

Since the two bracketed sums on the right are just aggregate desired saving and aggregate desired investment, and since we know from Eq. (5-4) that the left side is unplanned investment in inventories, we may write

$$GNP_s - GNP^* = S^* - I^* = I_u \qquad (5\text{-}5)$$

This is a very important and revealing relation. *An excess of aggregate supply over aggregate demand is equivalent to an excess of desired saving over desired investment, and each equals the flow of unplanned inventory investment.* This seems reasonable. If the demand sectors desire to save more out of their income streams than the Business Capital Accounts and Foreign Sector want to borrow and spend for current output, aggregate demand will fall short of total output. We are assuming that desired purchases by consumers and governments are realized in the marketplace; so we must conclude that desired saving equals actual saving, that is, S^* equals S. However, since actual saving equals actual realized investment, that is, S equals I, we can substitute I for S^* in the preceding equation and find

$$I - I^* = I_u \qquad \text{or} \qquad I = I^* + I_u$$

Actual investment equals desired investment plus unplanned investment. Under the above assumption that producers fill all orders, it is investment spending that absorbs any mismatch between aggregate supply and aggregate demand, via unplanned inventory changes.

When the economy is in equilibrium, desired and actual purchases are equal for all sectors, and in the aggregate we have $\text{GNP}_s = \text{GNP}^*$ and $S^* = I^* = I$, with $I_u = 0$.

However, equilibrium in product markets does not guarantee that we have full employment of labor or that markets for capital funds and money are in balance.

The conditions for equilibrium in those additional markets, and the influence of those markets on aggregate demand and supply in product markets, will be the subjects of later chapters. Only when all markets are in balance can the economy be in a true state of equilibrium.

MACROECONOMIC EQUILIBRIUM IN A DYNAMIC ECONOMY

Implicitly, this discussion has dealt with what economists call short-run static equilibrium of the economy. If sectoral and aggregate output and demand flows come into balance and remain constant indefinitely thereafter, the economy must be one with no population growth, no changes in technology, capital stocks, or productivity, no shifts in income shares among the demand sectors, and no new products or changes in tastes.

Economists have approached the problem of macroeconomic analysis of a dynamic economy in two ways. First, dynamic changes are taken as changes in certain parameters or structural constants in the equations describing the economic system, for example, tax rates or money supply. Equilibrium solutions of the system are found first with the original constants and then with the new constants. The changes in output and income flows, in prices, interest rates, employment, and in other variables from the first short-term equilibrium state to the second are taken as the dynamic response of the system to the given changes in parameters. This approach is described as *comparative statics*.

A second approach involves use of a system of equations where each variable is dated and where some equations are added to the model relating changes or rates of change in endogenous variables to the time pattern of other variables. The solution of this system of equations yields the endogenous variables expressed as time series, which embody the response of the system to the impressed changes in exogenous variables. This may be called the *dynamic approach.* If the exogenous variables are relatively simple functions of time, usually constant rates of growth (exponential functions), then it becomes possible to derive steady growth patterns of the economy as a whole. This steady growth of the economy may be called a *dynamic equilibrium* because growth comes to be expected by the decision makers. Their anticipations are fulfilled and their decisions lead to further growth, as expected, in a stable, predictable manner.

APPENDIX   Imbalance between Supply and Demand as a Cause of Changes in Inventories or Backlogs of Unfilled Orders

In the familiar demand-supply diagram of microeconomics, let us identify *supply* as the *flow of current output of finished products* by a production unit during a specified time period. Identify *demand* as the *flow of orders received from customers for delivery during that time period.*

If orders received for current delivery fall short of current production, the producer will accumulate inventories or will ship some goods ordered for delivery in other time periods and thus run down the firm's backlog of orders. This situation is illustrated at price OP_1 in Fig. 5-2. At price OP_1, demand is $P_1 D_1$, evidenced by orders received by the producer. Supply is $P_1 S_1$, the flow of finished goods output of the producer. Actual deliveries to customers are $P_1 M_1$, indicating that shipments $P_1 M_1$ are cutting into the backlog of unfilled orders at a rate $\Delta(UO)/\Delta t = M_1 D_1$. Deliveries are still less than production by an amount $M_1 S_1$, so that inventories are accumulating at a rate $\Delta(INV)/\Delta t = M_1 S_1$. The deficit of orders received (demand) below flow of output (supply) will thus be reflected in a combination of declining order backlogs or a rise in inventories:

$$P_1 S_1 - P_1 D_1 = \frac{\Delta INV}{\Delta t} - \frac{\Delta(UO)}{\Delta t}$$

Such a situation, if expected to continue, will lead the producer to cut production or lower price, or both, in order to maximize long-term profits.

The reverse situation exists at price OP_2. Orders received (demand) exceed production (supply) by $S_2 D_2$. The excess of orders is met by deliveries out of inventories ($S_2 M_2$) and by increases in backlog of unfilled orders ($M_2 D_2$). The overall excess demand is

$$P_2 D_2 - P_2 S_2 = \frac{\Delta(UO)}{\Delta t} - \frac{\Delta(INV)}{\Delta t}$$

This situation, if expected to continue, would lead the producer to increase production or raise price, or both, in order to achieve a higher level of profit.

The equilibrium situation would exist with demand and supply equal at price OP_0. Then orders received, deliveries, and production would all be equal; inventories and unfilled orders would remain constant.

In the presentation of aggregate demand–supply disequilibrium situations in this book, it will be assumed that deliveries always equal orders received. This implies that any discrepancy between orders received and production is absorbed by inventory changes. The general relation becomes:

Production *minus* orders received *equals* change in inventories

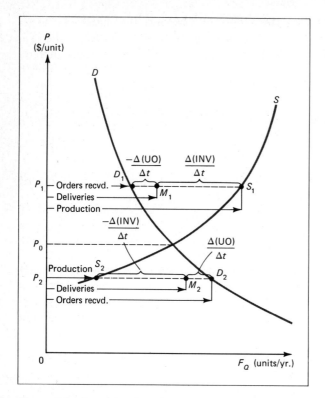

Figure 5-2   Demand and supply as determinants of changes in inventories and unfilled orders.

PROBLEMS

1 a   Does the value of current production of the economy (GNP_S) as determined by the Production Units Sector always equal aggregate incomes received by the four aggregate demand sectors (GNY)?   If not, under what conditions will they be equal?

b   Assume GNP_S equals GNY.   Subtract from GNY all the current-account expenditures (including transfers to foreigners) desired by the domestic demand sectors.   What remains on the right side of the above equation?

c   Subtract the same items of current-account expenditures from GNP_S.   What then remains on the left side of the above equation?

d   Explain the nature and economic meaning of the equality resulting after the subtractions in parts **b** and **c**.   Interpret the new equality in terms of rates of change of wealth in the four demand sectors.

e   If it be true that actual saving and investment flows are always equal, how can the economy ever be in disequilibrium?   Cite three practical situations that would lead to such a disequilibrium.

f   Show that if aggregate demand exceeds aggregate supply, desired investment must exceed desired saving by the same amount—under the assumptions used in this text.   (What assumptions are involved?)

2 The French economist Jean Baptiste Say, back in the early 1800s, noted that producers turn out goods in excess of their own needs only in order to exchange their excess output for goods they do not themselves produce.   He formulated Say's law that "supply creates its own demand."   This seems like the modern-day equality between the value of current production and the income payments resulting from that production.   It suggests that purchasing power always equals the value of output, so that aggregate demand and supply will stay in balance with only minor, temporary dislocations.

On the other hand, Marx noted that in a capitalistic economy part of the value of current output is distributed to property owners as rents, interest, and profits.   Hence workers do not receive the full value of what they produce. Thus, he stated, purchasing power would continually fall short of the value of output, and capitalist economies would inevitably be subject to periodic gluts of output and consequent business crises resulting from underconsumption.

Comment on these two points of view, using concepts of national income and product, aggregate demand and supply, desired and actual saving and investment, macroeconomic equilibrium, and other concepts.

3 A closed macroeconomic system (no Foreign Sector) is in equilibrium in period 1 with $GNP_S = GNP^* = 880$ billdar.   Out of the income stream, $D = 80$ billdar is withdrawn for capital consumption allowances.   Of the remaining stream of net national product, three-fourths flows to the Personal Sector as disposable income (Y_d), one-fourth to the Government Sector as net receipts (T_n), and nothing to Business Capital Accounts as undistributed profits.   Aggregate demand components are:

$C^* = 0.8 Y_d$
$G^* = 200$ billdar
$I^* = 200$ billdar

In period 2 government expenditures increase to 210 billdar, but desired investment stays at 200 billdar.   Producers did not expect the increase, and keep production at $GNP_S = 880$ billdar in period 2.   These data are summarized in the table on page 94.

a   Fill in the missing entries in the columns for period 1 and period 2.

b   In the column for period 3, indicate whether you would expect each of the flows to go up ($+$), down ($-$), or remain unchanged (0) from period 2.

c   Be prepared to identify the persons who made each decision to change, and to explain why they did so and what actions they took to carry out the decision.

	Period 1 (billdar)	Period 2 (billdar)	Period 3 (direction of change)
Aggregate supply (GNPs)	880	880	
Sector incomes:			
$Y_d = \frac{1}{4}(\text{GNP}_s - D)$	600		
$T_n = \frac{3}{4}(\text{GNP}_s - D)$	200		
GRE $= D = 80$	80	80	
GNY	880		
Sector demands:			
$C^* = 0.8\,Y_d$	480		
G^*	200	210	
I^*	200	200	
GNP*	880		
Sector saving:			
S_p^*			
S_g^*			
GRE			
S^*	—	—	
Indicators of disequilibrium:			
I			
I_u			
$\text{GNP}_s - \text{GNP}^*$			
$S^* - I^*$			

Chapter 6

Aggregate Supply and the Demand for Labor

DETERMINANTS OF GNP

Except for strikes and other calamities, it is the output decisions of managers of production units that determine the flow of *real* GNP. These output decisions simultaneously determine the real flows of labor and property services used in current production. Finally, given the prices of those factor services, income payments in the economy are determined by output levels, except for profits.

We assume that managers choose output levels which maximize *expected* profits. The estimates of expected profits are based on forecasts of demand (or revenue) curves and of the costs of producing various levels of outputs. Errors in forecasts of either element will cause profits to differ from expected levels. Suppose that output equals the forecast of demand, which is too high. Then, when the producers attempt to sell their output at the expected price, they either fail to realize profit on part of their output (which accumulates as unsold inventories) or they have to cut prices and realize a smaller profit margin on sales of their full output. Conversely, if demand exceeds their forecasts, they realize extra profit by selling down initial inventories or by raising prices and profit margins on their current output.

So we see that, *although producers can determine real GNP, demand factors enter into determining whether all the output will be purchased by final users and whether the price level expected by producers will be realized.* A mismatch between actual and expected demand will show up in unplanned inventory changes or in deviations of prices from expected levels. The discrepancy between forecast and realization will then lead to revised forecasts of demand in succeeding time periods and to changed output decisions in an attempt to maximize profits under the newly perceived market conditions. A similar line of reasoning holds with respect to errors in forecasts of costs; these errors may arise from misestimates of the prices or the quantities of inputs required for producing selected outputs.

SUPPLY CURVES FOR INDIVIDUAL FIRMS

To forecast or to control GNP, we need to understand the decision making of individual firms and then to aggregate this behavior to obtain a supply curve for the Production Units Sector as a whole. Consider a manager who is deciding on his firm's profit-maximizing output in the short run under the following conditions:

1 Prices of all inputs, notably money wage rates, have been forecast and are independent of the firm's level of output.
2 The firm's production function (that is, its inputs for various outputs) is known, so that total costs can be determined as a function of output level.
3 The level of GNP in the period ahead has been forecast, and the firm's demand curve (price versus unit sales) has been predicted as consistent with the GNP forecast.
4 The firm sells in a perfectly competitive market, so that its demand curve is a constant-price line for all levels of the firm's output.

Under these conditions the profit-maximizing output will be that level at which marginal costs (MC) equal marginal revenue (MR) in the short run.   In Fig. 6-1a, if the demand curve is $P_1 P_1$ at the forecast level of GNP, then the profit-maximizing output is y_{R_1} and point s_1 is one point on the supply curve of the firm.†

If we wish to find other points on the supply curve, we assume different demand curves and find the profit-maximizing level of output for each one.   For example, a higher GNP might yield a demand curve $P_2 P_2$.   This would lead the manager to schedule output y_{R_2} for profit maximization, and would determine point s_2 on the supply curve.   A still higher GNP would generate the demand curve

† Lower-case letters refer to individual firms and capital letters to aggregates covering all firms.

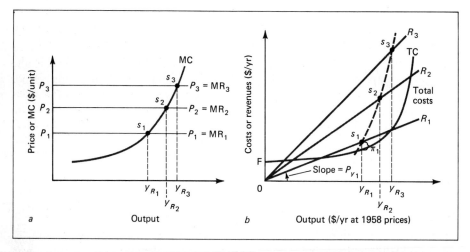

Figure 6-1 Supply curves for an individual firm.   (*a*) Price versus output.   (*b*) Revenue versus output.

$P_3 P_3$, output y_{R_3}, and point s_3 on the supply curve. In a perfectly competitive market, the supply curve coincides with the marginal-cost curve of the firm.

For aggregating the supply curves of all firms in the economy, we can add outputs (using dollar values at base-year prices as the common unit of measure), but it would be difficult to combine prices for various outputs in a meaningful way. To avoid this difficulty we convert the supply curves into relations between revenue and output for each firm. Both revenue (in current dollars per year) and output (in dollars per year at 1958 prices) can be added for all firms.

Derivation of the revenue type of supply curve is shown in Fig. 6-1b for the same firm as that in Fig. 6-1a. The firm's total-cost curve, labeled TC, starts from fixed costs OF. Corresponding to the three levels of GNP, we have the three revenue lines OR_1, OR_2, and OR_3. They are straight lines from the origin with slopes proportional to the three market prices P_1, P_2, and P_3 in the diagram in Fig. 6-1a. Given the total-cost curve and one of the revenue lines, the firm's management would choose the output at which the vertical distance from the TC curve to that revenue line is greatest, that is, the y_R for which profit $\pi = R - TC$ is greatest. The point on the revenue curve at this level of output is a point on the revenue-supply curve of the firm. In Fig. 6-1b, revenue lines OR_1, OR_2, and OR_3 determine profit-maximizing outputs y_{R_1}, y_{R_2}, and y_{R_3} respectively, and they in turn determine points s_1, s_2, and s_3 on the revenue-supply curve of the firm. The outputs thus determined are, of course, the same as those determined in Fig. 6-1a, since maximum profits in Fig. 6-1b occur at the output where slopes of revenue and total-cost curves are equal, corresponding to MR = MC in Fig. 6-1a. Also the revenues at points s_1, s_2, and s_3 in Fig. 6-1b equal the products Py_R at the corresponding points in Fig. 6-1a. The two forms of the supply curve provide essentially the same information, but the revenues shown in Fig. 6-1b at various points on the supply curve can be added up for all firms in the economy. (As is shown in the appendix to this chapter, similar revenue-supply curves can be derived for firms selling in markets which are not perfectly competitive. The difference is that the revenue lines OR_1, OR_2, and OR_3 are not straight lines but bend downward, indicating a lowering of price in order to sell more output.)

The supply curve of the firm shown in Fig. 6-1 indicates that as demand strengthens (higher GNP), output and price both rise. At low levels of output, where marginal costs remain nearly constant, a given increase in market price calls forth a relatively large increase in profit-maximizing output—movement from s_1 to s_2. At high levels of output, where marginal costs rise rapidly, the same increase in market price calls forth a much smaller increase in output—movement from s_2 to s_3.

THE AGGREGATE SUPPLY CURVE

If the outputs of all firms of the economy were measured in common units—dollars per year at 1958 prices—it would become possible to aggregate the revenue-supply curves of all these firms. Imagine that managers of all production units were using the same GNP forecast for the period ahead. Then each manager would select a profit-maximizing point on his firm's supply curve, say s_1. Summation of the outputs (y_{R_1}) at points s_1 for all firms would yield aggregate output; summation of the revenues (R_1) at points s_1 for all firms would yield aggregate revenues for the nation. From these aggregates, we would need to subtract the output and revenues for intermediate products. The result would be aggregate final output (Y_{R,s_1}) and aggregate revenues (Y_1) expected from sale of that output, as decided by production units consistent with a forecast GNP level Y_1. The point S_1 (with coordinates Y_{R,s_1} and Y_{s_1}) would be one point on the aggregate supply curve.

If all managers now adopted a second GNP forecast, they would shift their firms to new points s_2 on their individual supply curves. Aggregation of outputs and revenues at points s_2 for all firms—after exclusion of intermediate products—would lead to a second point on the aggregate supply curve, point S_2 with coordinates Y_{R,s_2} and Y_{s_2}. So one could develop points for the aggregate supply curve pictured in Fig. 6-2.

The aggregate supply curve shows what aggregate final sales (Y_s) *the Production Units Sector expects when it decides to produce each output* $Y_{R,s}$. The shape of the curve should be similar to that for the revenue-supply curves of individual firms. As we move outward along the curve, output and price both rise. However, as output rises, price holds nearly constant initially and then rises rapidly as capacity output is approached in the outer reaches of the aggregate supply curve. (Note that the GNP price index at a point S is $P_Y = Y_s/Y_{R,s} =$ the slope of a line from the origin to point S.)

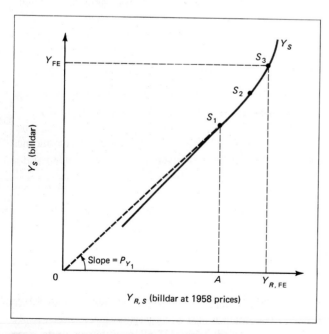

Figure 6-2 Aggregate supply curve of the economy.

For some purposes it is desirable to define a *full-employment level of output* somewhere in the upper reaches of the supply curve where prices start to rise rapidly as output increases. Currently the President's Council of Economic Advisers defines *potential GNP* as the real output which would be produced with 96 percent of the civilian labor force employed, that is, with 4 percent unemployment. This is not a measure based on capacity output in the sense of full utilization of the stock of capital goods. Rather, it is a measure based on a judgment as to the employment level which is consistent with reasonable price stability. In Fig. 6-2, such a full-employment level of output has been indicated as $Y_{R,\text{FE}}$. When the Production Units Sector decides on this level of output, it expects aggregate revenues from final sales equal to Y_{FE} and a GNP price index $P_Y = Y_{\text{FE}}/Y_{R,\text{FE}}$. which is the slope of a ray OS_3 in Fig. 6-2.

THE SUPPLY CURVE AND INCOME FLOWS

The revenue at each point on the aggregate supply curve equals the costs which the producers expect to pay out plus their expected profit. Except for the profit component, we may assume that these costs are met and become income receipts to the demand sectors of the economy. The profits flow will also be as expected if the demand forecast is correct.

The payments by the production units, excluding purchases of intermediate products, are just the components of GNY. To define them,

WL = aggregate compensation of labor, where W is the money wage rate and L the aggregate flow of labor services

$rP_K K$ = aggregate return to property including profits, where $P_K K$ is the money value of capital goods (K) used in production and r is its rate of return (percent per year)

D = capital consumption allowances

T_{ind} = indirect taxes

ONFC = other nonfactor charges

Then we may write

$$Y_S = (\text{WL} + rP_K K) + (D + T_{\text{ind}} + \text{ONFC})$$

= national income + nonfactor charges

The income components are marked off vertically in Fig. 6-3, where $AS = Y_S = \text{GNY}$. The income components would vary as output levels change, presumably in the same direction as output but with some changes in their relative proportions. These are income distributions before payment of income tax. The vertical distance AS could also be segmented to show the after-tax income available to each of the four final demand sectors of the economy, and we shall do this in some later analyses.

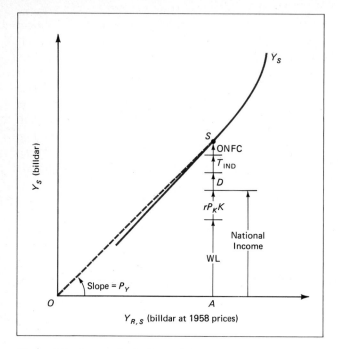

Figure 6-3   Income distribution and aggregate supply.

DEMAND FOR LABOR BY INDIVIDUAL FIRMS

Short-run changes in output are achieved primarily by changes in labor inputs, with stocks of capital goods and production technology held constant.   A typical short-run production function might be a curve like that in Fig. 6-4.   At zero output, there would be some employment of overhead labor—top management, accountants, clerical staff, research and engineering personnel, and certain others; so the curve would presumably cut the horizontal axis to the right of the origin. As output is increased, labor input must rise.   At first, the graph of output versus labor inputs may be nearly a straight line, that is, the marginal productivity of labor (MPL) may be nearly constant.   But as labor inputs increase further, with capital stock and techniques of production fixed in the short run, the increments to output resulting from uniform additions to labor inputs will fall, that is, MPL will decline.   Graphically, MPL is the slope of the production function, and a declining slope imparts a downward curvature to the graph of the function.

At what point on this production function will the firm earn maximum profits?   That depends.   It depends on the demand for output, on nonfactor charges, and on the wage rate.   Let us consider a firm with the same characteristics that were assumed in the discussion of the supply curve, that is, one with expected wage rate (W) and price of output (P) constant as the firm's output changes.   For such a firm, an increase of labor input by a small amount $d\ell$ will raise labor costs

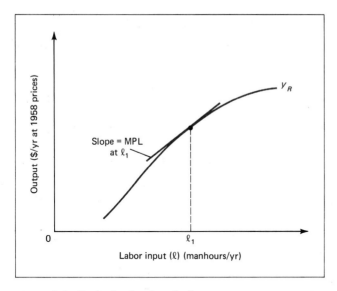

Figure 6-4   Production function of a firm.

by $W\,d\ell$.   *Revenues from sale of the added output must be enough to cover added labor costs and any other increase in variable costs if profits are to increase.*   Recall the equation for distribution of revenue:

$$R = Py_R = W\ell + rP_K k + \text{NFC} \qquad (6\text{-}1)$$

Let us assume that in the return-to-property term $(rP_K k)$, the rental and interest payments are fixed and only profits vary in the short run.   The nonfactor charges term (NFC) includes purchases of intermediate goods, capital consumption allowances, business property taxes, and indirect taxes (sales and excise levies).   The second and third items are nearly constant with respect to the firm's output, but costs of intermediate goods and indirect taxes may be roughly proportional to sales revenue.   So we may write:

$$\text{NFC} = \alpha + \beta R = \alpha + \beta P Y_R$$

where α and β are constants.

Given these assumptions regarding the terms in Eq. (6-1), and recalling our assumption that P and W are constant with respect to output or labor input, we may write the following equation for changes in the terms of Eq. (6-1) when a small increment in labor input $(d\ell)$ increases output by dy_R.

$$dR = P\,dy_R = W\,d\ell + d\pi + \beta P\,dy_R$$

The change in profits will be

$$d\pi = (1 - \beta)P\,dy_R - W\,d\ell \qquad (6\text{-}2)$$

The first term on the right equals the increment to revenue less the increment in nonfactor charges. This increment to net revenues must exceed the added labor cost ($W \, d\ell$) if profits are to increase. That is, for $d\pi$ is greater than zero,

$$(1 - \beta)P \, dy_R > W \, d\ell$$

or, in terms of derivatives,

$$(1 - \beta)P \frac{dy_R}{d\ell} = (1 - \beta)P(\text{MPL}) > W \tag{6-3}$$

Equation (6-3) provides a manager with a hiring rule for maximizing profits: *Increase employment of labor so long as the marginal net revenue of labor (MNRL) exceeds the wage rate, where*

$$\text{MNRL} = (1 - \beta)P(\text{MPL}) \tag{6-4}$$

and *stop when the two become equal.* We assume that MNRL exceeds W at low levels of labor input, because the marginal productivity of labor ($\text{MPL} = dy_R/d\ell$) will be high. As ℓ is increased, MPL declines and brings MNRL down to the level of W. Further increases in ℓ would lower profits. Thus *the criterion for profit maximization becomes*

$$\text{MNRL} = W \quad \text{or} \quad (1 - \beta)P(\text{MPL}) = W \tag{6-5}$$

Now we can answer the question about which point on the production function will yield maximum profits. Given P, β, and W, we can calculate the profit-maximizing value of MPL from Ep. (6-5). It will be

$$\text{MPL} = \frac{W}{P(1 - \beta)}$$

This value of MPL occurs at a particular point on the production function of Fig. 6-4; so that point determines the level of labor input (ℓ)—and the corresponding output (y_R)—which will maximize profits in the short run.

The *demand curve for labor* is a graph of profit-maximizing labor inputs at various levels of W. To derive it, we start from a given production function and find the relation between MPL and ℓ which that production curve implies. Given P and β, this permits use of Eq. (6-4) to express MNRL as a function of ℓ. *The MNRL curve is the demand curve for labor* since, for each specified wage rate (W), the profit-maximizing employment of labor will be determined at the point on the MNRL curve where MNRL = W, according to Eq. (6-5).

The preceding analysis is illustrated in Fig. 6-5. The top diagram is drawn for a given state of demand, yielding a market price P_1. Given P_1, the firm's revenue curve as a function of labor input (ℓ) may be derived from the firm's production function in Fig. 6-4 simply by multiplying all vertical distances by the constant P_1. By subtracting nonfactor charges from revenue, allowing for some variable component in NFC, we obtain the net revenue curve as a function of ℓ. Net revenue is allocated to labor cost and property incomes. Labor cost ($W\ell$)

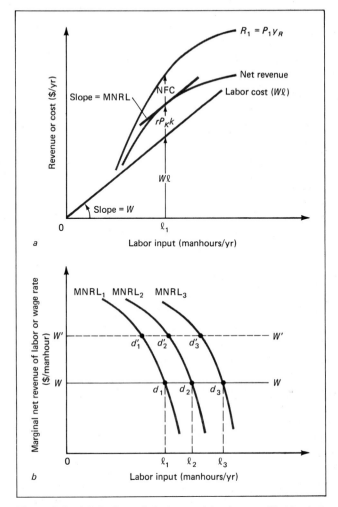

Figure 6-5 (*a*) Profit maximization on labor input. (*b*) Marginal net revenue of labor and the wage rate.

may be plotted as a straight-line function of ℓ, whose slope is W. The distance from the labor cost line up to the net revenue line measures property incomes at each level of labor input. The profit-maximizing employment of labor is ℓ_1, the value of ℓ for which the vertical distance from the labor cost line to the net revenue line is greatest.

Of course, that maximum vertical distance occurs at the value of ℓ for which the slopes of the labor cost curve and the net revenue curve are equal. So the profit-maximizing value of ℓ may also be found by marginal cost and revenue analysis. This approach is pictured in Fig. 6-5*b*. The MNRL_1 curve has ordinates equal, at each ℓ, to the slope of the net revenue curve above. It corresponds to a fixed price P_1, but values of $MNRL_1$ decrease as ℓ increases because the marginal productivity of labor declines along the production function at higher levels of ℓ. Recall that $MNRL_1$ equals $(1 - \beta)P_1(MPL)$. The profit maximizing employment of labor is determined at point d_1 where the horizontal line for the constant wage rate (W) intersects the $MNRL_1$ curve. The value ℓ_1 must, of course, be the same as that found in the total revenue and cost diagram in Fig. 6-5a.

What happens to labor demanded if the wage rate rises to W'? If markets for the firm's output (and hence P_1) remain unchanged, the $MNRL_1$ curve will be the same. The profit-maximizing output of labor is found at point d_1' where the new horizontal wage line W' intersects the $MNRL_1$ curve. The new employment of labor (ℓ_1') is less than the old (ℓ_1) because producers, faced by a higher wage rate, reduce employment and output so as to move leftward along the production function of Fig. 6-4 to a point where higher MPL will offset the higher wage rate. Recall that for profit maximization, we must have Eq. (6-5) satisfied.

$$MNRL = W \qquad \text{or} \qquad (1 - \beta)P(MPL) = W \tag{6-5}$$

For β and P_1 constant, MPL must change in proportion to W. In a similar fashion we could assume many different wage rates and determine other points such as d_1 and d_1'. All would lie on the $MNRL_1$ curve, which thus becomes the firm's demand curve for labor.

The effects of a change in the state of demand may be visualized from Fig. 6-5b also. A strengthening of demand that raises the price of the firm's output will shift its MNRL curve upward in proportion to the price rise. For example, a price increase from P_1 to P_2 will raise the marginal net revenue curve, or labor demand curve, from $MNRL_1$ to $MNRL_2$. Profit-maximizing employment of labor would increase to ℓ_2 if wage rate were W, or to the value of ℓ at point d_2' if wage rate were W'.

From the relation $MNRL = (1 - \beta)P(MPL)$, we note that the MNRL curve may be shifted up also by a decline in β (cut in sales tax) or by a rise in MPL (higher slope of production function at each value of ℓ). If either of these factors should improve, with market demand and price of output constant, the labor demand curve could shift upward further from $MNRL_2$ to $MNRL_3$. Then employment of labor by the firm would increase to point d_3 for wage rate W or to point d_3' for wage rate W'. Always, however, for a given demand curve for labor (MNRL curve fixed), a rise in wage rate will lead to a reduction of employment, and vice versa, if the producers are astute short-run profit maximizers.

In all this discussion, we have assumed that the firm is operating in a perfectly competitive product market, so that its product price remains constant for a given state of demand as the firm changes its output and employment of labor. For firms selling in imperfectly competitive markets, increased output can be sold only at a lower price. Their curves for MNRL will decline as output increases because of declining market prices as well as declining marginal productivity of labor. For such firms we may write

$$MNRL = (1 - \beta)(MR)(MRL)$$

where marginal revenue (MR) replaces the price factor (*P*) applicable to a firm in perfectly competitive markets. This MNRL curve will have the same general shape as that for the competitive firm and will shift upward with a decline in β, a strengthening of demand (rise in MR), or rise in marginal productivity of labor (MPL). So the labor demand curves for such firms are quite similar to those pictured by the MNRL curves in Fig. 6-5*b*.

AGGREGATE DEMAND FOR LABOR

If wage rates are the same for all firms in the economy, we can aggregate the curves of Fig. 6-5*b* over all production units fairly readily. Imagine that all firms forecast the same level of GNP as a basis for estimating their individual product demand curves. Each firm can then determine its labor demand curve (the MNRL curve) for that state of demand. Finally, we can imagine adding together the employment of labor over all firms at each of many specified levels of *W*. The result would be a curve of total employment at each wage rate, which would be the aggregate demand curve for labor at the forecast level of aggregate demand for output.

The aggregate demand-for-labor curve would have a shape similar to that for individual firms. It would drop off as employment increases because of declining MPL—and because of declining marginal revenue (MR) for firms not selling in competitive markets. The aggregate demand curve for labor would shift upward if demand strengthened, that is, if producers expected a higher Y^*. It would also shift up if variable nonfactor charges were reduced or if marginal productivity of labor increased for production units as a whole.

The aggregate curve for labor demand is graphed in Fig. 6-6, and its properties are discussed in the next section.

CORRESPONDENCE BETWEEN AGGREGATE SUPPLY CURVE AND AGGREGATE DEMAND FOR LABOR

The profit-maximizing levels of output and of labor inputs chosen by each firm must, of course, be consistent. That is, *the profit-maximizing level of employment must produce the profit-maximizing level of output consistent with the firm's production function.* Similarly, there must be a correspondence between points on the aggregate supply curve, the aggregate labor demand curve, and the aggregate production function for the Production Units Sector as a whole. Their mutual relationship is illustrated in Fig. 6-6. The aggregate supply curve is plotted in normal position in panel (*a*); the aggregate labor demand diagram is rotated 90 degrees counterclockwise in panel (*c*); the aggregate production function in panel (*b*) has its axes determined from the other two diagrams. In panel (*c*), the labor demand curves have been labeled L^* to emphasize the labor input variable. The heights of the curves are MNRL as before.

To trace the correspondence between the graphs, assume the following conditions:

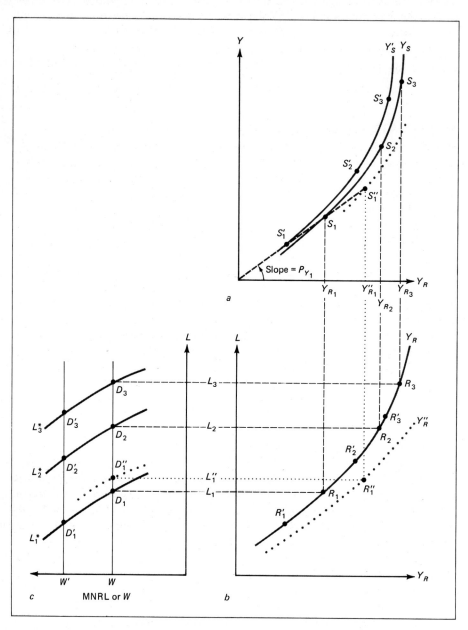

Figure 6-6 Relation between aggregate supply curve and aggregate labor demand. (*a*) Aggregate supply. (*b*) Production function. (*c*) Aggregate labor demand.

1   The state of aggregate demand is given, so that firms can derive their demand curves in a consistent manner, that is, based on the same GNP forecast.

2   Money wage rate and the prices of other inputs are assumed given.

3   Nonfactor charges are taken to be linearly related to total revenues.

4   Technology is assumed given, so that each production unit has a known production function.

For some initial set of these characteristics of the economy, each manager of a production unit determines a profit-maximizing point on the unit's supply curve, labor demand curve, and production function.   Aggregation of these points over all firms yields a point S_1 on the aggregate supply curve in panel (a), a point D_1 on the aggregate labor demand curve in panel (c), and a point R_1 on the aggregate production function in panel (b).   The GNP price index (P_{Y_1}) may be identified as the slope of a line from the origin to point S_1 in panel (a).

If demand for output becomes stronger, then aggregate supply moves up to a point such as S_2, involving a higher real output (Y_{R_2}) and higher price index (P_{Y_2}).   The higher price level shifts the labor demand curve up to L_2^* in panel (c), so that employment of labor increases to the point D_2, with wage rate assumed constant at W.   The higher output (Y_{R_2}) and employment of labor (L_2) are consistent at a new point R_2 on the aggregate production function.

A still stronger state of demand would lead to higher levels of output and price on the supply curve at S_3.   The higher price level would shift the labor demand curve up to L_3^* and result in employment L_3 at wage rate W.   On the production function, the new operating point is R_3 with output Y_{R_3} and employment L_3.

Suppose now that the money wage rate rises to W'.   In panel (c) the new profit-maximizing points for the three levels of aggregate demand are D_1', D_2', and D_3'.   In each case, employment of labor is less than when the wage rate was W, assuming that the labor demand curves stay fixed.   The lowered labor inputs imply lower outputs Y_R, as shown at the points R_1', R_2', and R_3' on the aggregate production function.   Assuming fixed price index P_Y for each of the three states of demand in panel (a), the points R_1', R_2', and R_3' project up to points S_1', S_2', and S_3' in panel (a).   These points trace out a new aggregate supply curve Y_S' lying to the left of the original supply curve.   This is what we might have expected.   A rise in wage rate shifts a firm's marginal cost curves up.   For unchanged price, or state of demand, each firm would maximize profits at a lower output.

Finally, suppose that the average and marginal productivity of labor were increased, perhaps by use of more capital equipment per man-hour or perhaps by improved managerial practices.   A rise in labor productivity would shift the aggregate production function outward, for example, to the curve Y_R'' in panel (b).   The increased MPL at each value of labor input would shift the labor demand curve up, since MNRL equals $(1 - \beta)P_Y(\text{MPL})$, and the new labor demand curve cuts the wage rate line W at point D_1''.   The employment thus determined (L_1'') may be projected across to the new production function at point R_1'', and this determines a new output level (Y_{R_1}'').   This output, together with the constant price level (P_{Y_1}), determines a point S_1'' on a new aggregate supply curve.   It would be shifted rightward in panel (a).   We might have expected this.   An increase in MPL shifts the marginal cost curve down so that the profit-maximizing level of output will be greater for any specified output price or state of demand.

SUMMARY

This chapter has discussed the behavior of the Production Units Sector, assuming that maximization of short-run profits is the managerial objective determining levels of output and employment of labor.   The principal variables which influence output and employment decisions were found to be:

1   Strength of demand, or market price of product (P_Y)

2   The money wage rate (W)

3   The marginal element in nonfactor charges (β)

4   The marginal productivity of labor (MPL), which corresponds to the slope of the production function and declines as employment increases

These variables in individual product markets determine supply curves for individual firms and, after aggregation, for the economy as a whole.   Each supply curve is drawn for particular values of W and β and for a given production function   As we move upward along a supply curve, successive points correspond to profit maximization, with demand stronger (price level higher) and MPL of labor lower as employment increases.

The same four variables just listed determine labor demand curves for individual firms and in the aggregate.   Each labor demand curve is drawn for particular values of P_Y and β and for a given production function.   As we move outward along a labor demand curve, successive points correspond to profit maximization at lower wage rates and lower MPL as employment increases.

For given production functions of firms and for given forecasts of market demand and factor prices, the equilibrium positions on supply curves are the outputs where MC equals MR for each firm; the equilibrium positions on labor demand curves are labor inputs for which MNRL equals W.   The output and employment levels thus determined are consistent, that is, they are coordinates of a point on each firm's production function—and on the aggregate production function for the economy.

This concludes our initial analysis of the decision making of production units. One crucial input to producer's decisions is their forecasts of GNP and of their market demand curves.   These expectations are based on the past and projected future behavior of the demand sectors of the economy.   So, in the next chapter, we shall turn to the analysis of sector demands and to the determination of aggregate demand for the economy as a whole.

APPENDIX   Derivation of Supply Curves for a Firm Whose Market Is Not Perfectly Competitive

Consider a firm whose production function and prices of inputs are known, so that its short-run total and marginal-cost curves can be determined.   Assume that the managers have forecast GNP in the period ahead and predicted the firm's demand curve consistent with that state of aggregate demand.   However, assume that this firm sells its output in markets that are not perfectly competitive, so that increasing output can be sold only at declining prices.

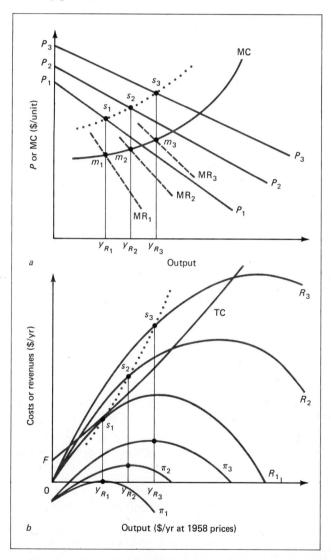

Figure 6-7   Firm's supply curves—imperfect competition.   (a) Price versus output.   (b) Revenue versus output.

For such firms, maximum profits occur at the output for which marginal revenue declines to equality with marginal costs, as was true for perfectly competitive firms. But here MR is less than price; so the supply curve lies above the marginal cost curve. The situation is illustrated in Fig. 6-7a.

The marginal cost curve rises as output increases. For a particular state of demand, we have the demand curve P_1P_1 and its associated marginal revenue curve MR_1. The marginal revenue and cost curves intersect at point m_1. This determines the profit-maximizing output y_{R_1}. Also by projection up to the P_1P_1 line, it locates point s_1 as one point on the supply curve of the firm.

Now assume a second level of GNP and the demand curve P_2P_2 and marginal revenue curve MR_2 associated with it. Profit-maximizing output is determined by the intersection of MR_2 and MC curves at m_2. This output (y_{R_2}) would be sold at a price corresponding to the point s_2 on the demand curve P_2P_2. A third state of demand would lead to a decision to produce output y_{R_3} at point s_3. The points s_1, s_2, and s_3 trace out the supply curve of the firm. The curve indicates that stronger states of demand lead to increases in both output and market price, as was true for the perfectly competitive firm.

Figure 6-7b shows the analogous determination of a revenue-output supply curve from total cost and revenue curves. The total cost curve is TC. For each of three revenue curves R_1, R_2, and R_3, the curve of profits as a function of output was calculated from the relation $\pi = R - TC$, and the three profits curves are plotted as π_1, π_2, and π_3. Maximum profits occur at y_{R_1}, y_{R_2}, and y_{R_3}, and the firm's supply curve is traced by points such as s_1, s_2, and s_3 on successive revenue curves where they rise farthest above the TC curve. The profit-maximizing levels of output and the prices and profits at points s_1, s_2, and s_3 are the same in the top and bottom panels of Fig. 6-7.

Note that the shape of the revenue-output supply curve for a firm not in perfect competition is similar to that for a firm selling in competitive markets (Fig. 6-1). So it should be possible to aggregate such supply curves for all firms. And the aggregate supply curve for the economy should have the same property that stronger demand leads to increases in both output and price level.

PROBLEMS

1 **a** Do the points forming the revenue-supply curve of a firm fall on total revenue curves, or on the total cost curve of the firm, or both?

  b Why will a firm selling in a perfectly competitive market increase output if market demand strengthens? How do managers of the firm know that demand has strengthened?

  c What are the principal assumptions underlying aggregation of the revenue-supply curves for firms?

  d Identify P_Y on a graph of the aggregate supply curve. What leads to the sharp rise in P_Y at high levels of output? (Remember that money wage rates are assumed constant.)

  e How is full-employment output of the United States economy defined? What is the basis for selecting a certain level of unemployment as consistent with full employment?

  f Does each point on the revenue-supply curve of a single firm correspond to a profit-maximizing choice of output by the firm's management? Explain your answer.

2 a Must GNP$_S$ always equal GNY? Explain why or why not.

b If producers revise their output decisions downward from one period to the next, describe what changes you would expect in the components of GNY, and how those changes would come about. For example, will all components change proportionally? Will income changes occur because of changes in real flows of inputs or in their prices? And so on.

c Explain why the marginal net revenue curves of Fig. 6-5*b* decline as labor inputs (ℓ) increase.

d Are all points on the aggregate demand-for-labor curves in Fig. 6-6*c* profit-maximizing points for individual firms?

3 a Sketch the Production Units Sector box, its input and output flows, and the associated decision-making circle. Label the input arrows to that circle to indicate the variables about which managers need information to arrive at decisions regarding their desired flows of output and of inputs of factors of production.

b Assume an exogenous increase for each input variable to the decision-making circle. Then indicate the direction of endogenous change which would occur in real output flow, employment of labor, price of output, and marginal productivity of labor as equilibrium is reestablished under the changed conditions. Assume in each case that all other input variables or functions are constant.

c The following data are reported for private nonfarm GNP and related variables in 1969 and 1970.

	1969	1970
Private nonfarm GNP (billdar)	800.3	835.4
Real private nonfarm GNP (billdar at 1958 prices)	642.8	640.3
	(Indexes: 1969 = 100)	
Private nonfarm GNP	100.0	104.4
Real private nonfarm GNP	100.0	99.6
Man-hours per year worked (L)	100.0	99.0
Output per man-hour (APL)	100.0	100.6
Compensation per man-hour (W)	100.0	106.8
Price index for private nonfarm GNP (P_Y)	100.0	104.8
Consumer price index (P_c)	100.0	105.8

Using these data, sketch curves like those of Fig. 6-6, and locate two points on your curves which represent the data for 1969 and 1970 as well as possible. (Assume the index for APL represents MPL also.)

d Explain the reasons for shifts in the three curves plotted in part (**c**). What was the percentage rise in unit labor costs? (It should be noted that unit nonlabor costs rose by $1\frac{1}{2}$ percent from 1969 to 1970.) What happened to real compensation per man-hour?

Chapter 7

Personal Consumption Expenditures and Saving

MICROECONOMIC ANALYSIS OF CONSUMPTION–SAVING DECISIONS

In this chapter we shall try to understand the economic decision making and behavior of the Personal Sector. In the next chapter we shall analyze how this behavior combines and interacts with the economic decision making and behavior of the other sectors in determining changes and equilibrium positions for the economy as a whole.

The consumer units in the Personal Sector consist of individuals, households, and nonprofit consumer organizations. Since we are all individuals and most of us are members not only of households but also of various nonprofit clubs and organizations, we can begin by recollecting what elements entered into our own buying decisions and those of our families and the organizations to which we belong.

First, of course, come *wants*. We are dissatisfied, in disequilibrium. We want a flow of services to "satisfy" our physical, psychological, and social needs. We reduce our tensions by securing a flow of services, or of goods that yield the services we desire. Not that we become fully content. There are constraints.

So we come to the second element, *costs*. The desired goods and services can usually be procured only by our work—work which yields either the goods and services directly or the income to purchase the goods and services in markets. Work is a cost, measured by the leisure forgone and the disutility of effort. But we do decide to work, at least up to the point where the gain in satisfactions secured from the work itself and from the products obtained by additional work remains adequate to balance the sacrifices involved in the extra work. How much we obtain for an increment of labor services depends on personal capabilities and training, and on the natural resources, capital stock, and production techniques with which our labor is associated.

Third, having obtained its flow of income, each consumer unit must decide first on the *allocation of that income between spending on current consumption and saving for future consumption*, and then on the allocation of current expenditures among all the possible goods and services that are available in the market. Rationally, these allocations will be made so that the expected satisfactions (marginal utility) from an added dollar allocated to any current or future consumption will be equal. When the allocation of income flow to these various uses has been increased up to that point, then no reallocation of one dollar per year of income from one use to another would increase prospective lifetime satisfactions of the decision maker.

So far microeconomic theory of consumption can carry us. Except that it seems to involve too thorough an exploration of alternatives, too clear and precise a calculation of benefits and costs, and perhaps too little regard for expectations and uncertainty in our decisions, the theory gives a reasonable description of the determination of consumer expenditures and saving in general terms. This economic theory of the consumer suggests that the important inputs to decision making which govern the short-run flows of income, consumption expenditures, and saving are the following:

1   Schedule of declining marginal utility of income and rising marginal disutility of work as a function of hours worked per year
2   Demand for labor and other factor inputs
3   Resulting current incomes and expected future incomes of the Personal Sector
4   Consumer "tastes," or the utility function relating satisfactions derived to the flows of various collections of goods and services consumed
5   The rate of discount consumers apply to future consumption
6   Market rates of interest as a factor for determining the relative costs of future versus present consumption and the present value of future income streams
7   Consumer wealth and its distribution among items of real goods, money, and securities
8   Market prices of each good and service, and expected future prices.

The nature and direction of influence of each of these variables will be discussed in relation to theories of aggregate consumption and saving functions described below.

FACTORS INFLUENCING AGGREGATE CONSUMPTION EXPENDITURES AND SAVING

The determinants of aggregate consumer spending and saving may be grouped under two main headings: characteristics of individual consumer units, and overall market and economic conditions.

CHARACTERISTICS OF INDIVIDUAL CONSUMER UNITS

Income is probably the attribute of a consumer unit which has most influence on its expenditure and saving flows. More than a century ago, students of consumer behavior found that budget-study data exhibited stable relations between household income and expenditures on various categories of goods and services—the so-called Engel curves. Income depends in part on characteristics of the consumer unit (education and training, motivation, age, real and financial assets owned) and in part on market conditions (employment levels, wage rates, interest rates). In modern times, a household's willingness and ability to draw on a line of consumer credit may provide a temporary source of additional purchasing power. And modern theories of rational consumer behavior suggest that utility maximization over the whole life cycle of a consumer unit will make current consumption depend on the expected future income stream and on current wealth in addition to current income level. For instance, Modigliani and Brumberg† analyzed the time pattern of consumption and saving for a representative consumer unit from the time it begins earning income to the end of its life. The expected income stream may well rise during the earning span, level off near the end, and then drop to low levels during retirement years. The desired consumption stream would presumably be more nearly level—perhaps rising during years when the consumption unit is augmented by dependent children. The different patterns of desired consumption and expected income can be reconciled by saving and borrowing at various stages of the life cycle. On balance, net saving will be required during the earning years to provide for consumption during retirement. This model implies that there will also be a planned life-cycle pattern of wealth for the consumer unit, for example, a pattern of net worth rising during the earning years and declining during retirement. This theory suggests that the ratio of current-period consumption or saving to income of a consumer unit will vary systematically during the life cycle because of long-term planning of the consumption stream. For an individual unit, current consumption will depend on expected income stream, current wealth, family composition, and age of the unit, as well as current income. And the lifetime pattern of allocation of income between current consumption and saving will be influenced also by such variables as uncertainty regarding future streams of income and consumption needs, desire to leave an estate for one's heirs, current versus expected market prices and interest rates, the consumer's rate of discount on future consumption, provisions of the tax and social security systems, prospects for capital gains, and similar conditions. "Tastes" may also be affected by such consumer characteristics as education and occupation of individuals, place of residence (rural or urban, city size, geographic area), home tenure (renter or owner, time in residence), and socioeconomic group with which the family is closely associated. Although these many family characteristics may be important in causing differences in expenditures or saving ratios between individual families in a cross section of the consumer population at a given time, they may have little influence on changes in aggregate consumption or saving ratios through time. Thus, if the distribution of consumer units among classes on each relevant characteristic is stable through time, progression of individual units through those classes will have little or no effect on aggregate consumption of the Personal Sector. To find the causes of changes in aggregate consumption or saving, we must look for market or economic conditions which alter the behavior of most consumer units in the same direction at the same time or which cause major shifts in the distribution of consumer units among classes on important characteristics.

---

† F. Modigliani and R. Brumberg. "Utility Analysis and the Consumption Function," in K. K. Kurihara (ed.), *Post Keynesian-Economics*, Rutgers University Press, New Brunswick, N.J., 1954.

OVERALL MARKET AND ECONOMIC CONDITIONS

In a growing economy, the increase in number of consumer units will normally raise aggregate consumption and saving proportionately, assuming that aggregate disposable income also rises at the same rate and that the population rise involves proportional increases in all categories of consumer units. Even if demographic developments are changing the proportions of consumer units in various classes by age, occupation, place of residence, marital status, and other characteristics, the "distributional effects" on the ratio of consumption to disposable income are likely to be of minor macroeconomic significance in the short run. It usually takes a large shift in distribution to change overall aggregates much, and the distributions of consumer units by demographic characteristics usually do not change rapidly.

However, macroeconomic changes in market and economic variables, such as employment, wage rates, incomes, taxes, price level, interest rates, capital gains or losses, stocks of consumer durables, and financial assets, can cause major short-run shifts in the distributions of consumer units experiencing various levels of these variables. And many households will be shifted in the same direction at the same time; so the changes in their consumption and saving flows will be additive rather than canceled out in the process of aggregation. Moreover, if the change in market and economic variables persists for some time in one level or direction, many consumer units will revise their *expectations* of future values of these variables in the same direction at the same time. Such learning and expectations, or even increased uncertainty about future market and economic conditions, can lead to short-run changes in consumers' allocation of their current income between consumption and saving.

It seems likely, then, that in attempting to explain short-run changes in aggregate consumption and saving flows, we must look to market and economic conditions for our explanatory variables. Even then, the list of possible candidates is large, and we must watch out for interrelations between macroeconomic variables whose counterparts may be assumed independent in microeconomic analysis. Thus analysis of the response of aggregate consumer expenditures to overall price levels cannot proceed on the assumption that disposable income or consumer wealth is unchanged when prices change. Also, previous business-cycle history and consumer expectations are not usually considered in microeconomic analysis of market equilibria.

Given all these complexities, how shall we explore the possible existence and stability of macroeconomic consumption and saving functions?

MACROECONOMIC CONSUMPTION FUNCTIONS

John Maynard Keynes, a famous British economist of the first half of this century, formulated a macroeconomic consumption function which has become the basis for economic models and econometric investigation since his time. Faced with all the complexities described here, he picked out real income as the most important influence on consumer decisions with respect to real consumption and saving. He hypothesized a basic relation between *real* consumption and *real* income as follows:† "The fundamental psychological law, upon which we are entitled to depend with great confidence both *a priori* from our knowledge of human nature and from the detailed facts of experience, is that men are disposed, as a rule and on the average, to increase their consumption as their incomes increase, but not by as much as the increase in their income." Keynes recognized and commented on the influence of most of the other factors which were mentioned earlier—distribution of consumer units by demographic characteristics, income distribution, capital gains or losses, time rate of discounting, interest rates, price level, expected future changes in prices and income, fiscal policy—but he insisted that in normal circumstances, their influence on aggregate real consumption would be small compared with the influence of real income. So he posited a stable relation between aggregate real consumption and aggregate real income, which he called the *propensity to consume*.

Letting C_R equal real consumption and $Y_{d,R}$ equal real disposable personal income, Keynes hypothesized that the relation between C_R and $Y_{d,R}$ was such that $\Delta C_R / \Delta Y_{d,R}$ is positive and less than unity.

This statement could be consistent with many types of functional relations. If we restrict ourselves to linear functions of the type $C_R = a + b Y_{d,R}$, we could have straight lines with the intercept a positive, negative, or zero. Since $b = \Delta C_R / \Delta Y_{d,R}$, its value must be between zero and 1. Three straight lines are graphed in Fig. 7-1 corresponding to the three possibilities for the intercept a. All have $b = 0.7$. The lower line can be ruled out, at least for low values of $Y_{d,R}$, for it implies that consumers would let real consumption drop to zero while they still had positive real income. Either of the top two lines would seem reasonable, and Keynes's comments indicate that he would favor the consumption function with a larger than zero, at least in the long run. He thought it likely that a greater proportion of income would be saved as real income increases. In the graph a 45-degree line is shown whose points will have ordinates equal to the $Y_{d,R}$ values, and whose slope is 1. The vertical distance between the consumption function line and this $Y_{d,R}$ line must equal real personal saving $S_{p,R}$.‡

A few frequently used definitions should be introduced at this point. The slope of the consumption function at any income is called the *marginal propensity to consume*, or MPC. Thus, MPC $= \Delta C_R / \Delta Y_{d,R} = b$. The ratio of total consumption to total income is, of course, called the *average propensity to consume*, or APC. Graphically, this is the slope of a ray from the origin to a point on the consumption function line. For a consumption function passing through the origin, $a = 0$ and APC $=$ MPC. For $a > 0$, we find MPC less than APC, or the slope of the consumption function is less than the slope of the ray from the origin to a point on the consumption function line.

† John M. Keynes, *General Theory of Employment, Interest, and Money*, p. 96, Harcourt Brace Jovanovich, Inc., New York, 1936.
‡ This assumes that $Y_{d,R} = Y'_{d,R}$, net disposable personal income.

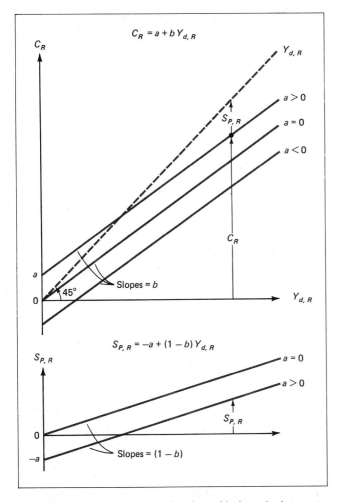

Figure 7-1 Linear consumption functions with alternative intercepts.

Comparable definitions are used for saving: for marginal propensity to save, $MPS = \Delta S_{p,R}/\Delta Y_{d,R} = 1 - b$; for average propensity to save, $APS = S_{p,R}/Y_{d,R}$. For the line with a equal to zero, MPS equals APS; for the line with a greater than zero, MPS is greater than APS.†

† Note that
$$S_{p,R} = Y_{d,R} - C_R = 1 - bY_{d,R} - a$$
So
$$\frac{S_{p,R}}{Y_{d,R}} = 1 - b - \frac{a}{Y_{d,R}}$$

Since $Y_{d,R} = C_R + S_{p,R}$, it follows that MPS + MPC = 1 and also that APS + APC = 1.   These two relations would be true even if the consumption and saving functions were curved lines, as they might well be.

EMPIRICAL CONSUMPTION FUNCTIONS

In the years since Keynes's hypothesis was developed, two types of data have become available to check on its accuracy and usefulness.   First are the budget-study or cross-section data gathered by the Bureau of Labor Statistics in order to revise the commodity weights used in computing its consumer price index.   Second are the time-series data of the national income accounts which show quarterly and annual aggregate consumer expenditures, saving, and disposable income for the years 1929 to date, either in current or in constant dollars.   Both sets of data may be expressed on a per capita basis to permit a more direct comparison. The definitions of after-tax income and of consumer expenditures are closely comparable in the two sets of data, except for the inclusion of imputed items in the national income and product accounts.   The Bureau of Labor Statistics data, however, relate only to urban families; the income and product data cover the entire United States population.

Figure 7-2 summarizes the findings for the 1960–1961 survey of consumer income and expenditures conducted by the Bureau of Labor Statistics (BLS). Data are reported for average consumer expenditures and average after-tax income for families in 10 income classes.   Since average family size is given within each income class, the income and expenditure data can be expressed on a per capita basis and plotted as shown.

The graph clearly supports Keynes's expectation.   In the year shown, there was some curvature to the relation and a clear tendency for families with incomes below $1,400 per year to spend more than their income, that is, they dissaved to maintain their standards of consumption.   A straight line averaged through the points might have an equation $c = 500 + 0.65y_d$.   Small letters will be used for per capita data and subscript R is omitted because money values may be taken as equal to real values in a base year (1960–1961).   According to the BLS, "On the average, families used only 91% of their after-tax income for current consumption in 1960–61, compared with 97% in 1950."

Figure 7-3a shows a similar chart for time-series data, for 1929 through 1973, for real per capita consumer expenditures and net disposable personal income, taken from the national income and product accounts.   Setting aside the years

---

(*Continued*)

If a equals zero, this savings ratio (or APS) will always equal $(1 - b)$, which must be positive.   If a is greater than zero, the savings ratio is negative at very low incomes, passes through zero, and then rises toward the limiting value $(1 - b)$ for very high values of $Y_{d,R}$.   See the lower graph in Fig. 7-1. Also, we have

$$\frac{C_R}{Y_{d,R}} = \frac{a}{Y_{d,R}} + b$$

If a equals zero, this average propensity to consume will be constant and equal to b; if a is greater than zero, the average propensity to consume will decline toward the value of b as real income rises.

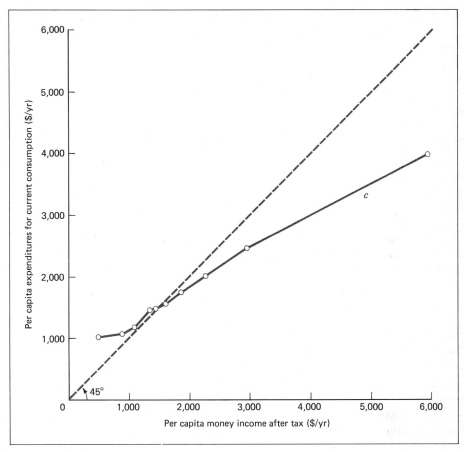

Figure 7-2 Per capita consumption expenditures and income for U.S. urban families, 1960 to 1961. (Adapted from "Contrasts in spending by urban families," BLS report no. 238-8, Feb., 1965.)

of rationing and shortages during World War II (1941 to 1946), the points lie mostly along a straight line with a very small intercept. The equation for the line might be $c = 25 + 0.93y_d$. In two years (1932 and 1933), reported personal saving was negative, so that consumption exceeded disposable income. There does appear to have been some tendency for consumers to maintain their standards of consumption during the income decline of the Great Depression in the 1930s, and during the mild postwar recessions in 1949, 1954, and 1960. This implies a decline in saving during those years.

Such movements are more clearly exhibited on the larger vertical scale in Fig. 7-3b, where real personal saving per capita is plotted versus real net disposable income per capita. A guideline indicates saving at a constant 6 percent of income. Note how in the depression years from 1930 to 1940, saving varied with income along a line with a very steep slope and did so again in the postwar recessions of 1948–1949, 1953–1954, and 1959–1960. These slopes would correspond to values

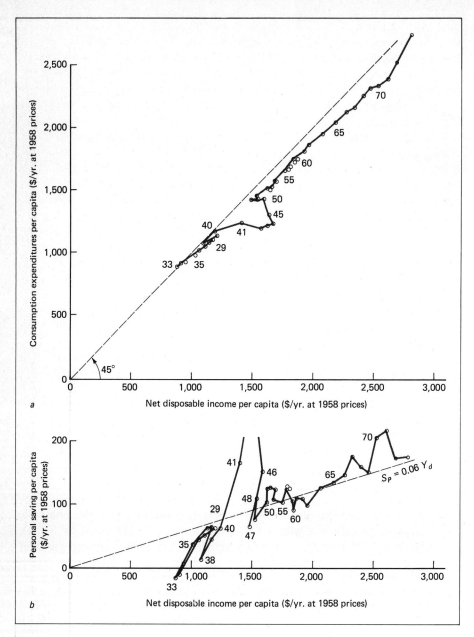

Figure 7-3   Real per capita consumption expenditures and saving from time-series data, 1929 to 1973.   (*National Income and Product Accounts of the U.S., 1929–1965,* and July issues of *Survey of Current Business.*)

of $(1 - b)$ equaling 0.15 to 0.20. It was, in part, the extrapolation of the 1929–1941 relation between saving and income in postwar forecasts which led to fears of low consumer spending and a recession in the early postwar years.

If we consider just the nonwar years of good business (low unemployment), the saving function would come close to going through the origin and would have a slope in the range of $(1 - b)$ equaling 0.055 to 0.075. In addition, research by Simon Kuznets and Raymond Goldsmith, going back to 1869, showed that the average saving ratio had remained nearly constant in aggregate time-series data covering that longer span of time and wider range of per capita income levels, down to one-third of the 1929 level. This long-term constancy of the saving ratio contradicts Keynes's expectation of a rise in the average propensity to save. Fortunately so, it may be argued. If rising real per capita income led to ever-larger saving ratios. then rising ratios of investment to national product would be required, or else the high levels of income would not be achieved. A rising proportion of investment expenditure to national product might be difficult to achieve in an economy with an ever-growing stock of capital goods, at least under normal private market mechanisms. Maintenance of the high income levels would then be possible only if the Government Sector absorbed the excess of private saving by borrowing the funds to finance a growing government deficit, or if the government managed to stimulate private investment by subsidies or controls, or if large-scale foreign investment were undertaken.

However that may be, we are faced with a problem of understanding the relation of consumption and saving to income in the Personal Sector. The long-term consumption and saving functions in years of high-level economic activity seem best described by straight lines through the origin, for which MPC equals APC and MPS equals APS. But during business depressions or recessions, consumption is maintained at a level higher than the long-term relation would suggest, at the expense of saving. That is, *in a short-term decline of income, the marginal propensity to consume is smaller than the long-term value, and the average propensity to consume rises.* In terms of the saving function, short-term declines in income lead to marginal propensities to save that are higher than the long-term value, with a corresponding decline in the average propensity to save. How can these data be reconciled?

THEORIES TO EXPLAIN LONG TERM CONSTANCY OF CONSUMPTION AND SAVING RATIOS

UPWARD-SHIFTING KEYNESIAN CONSUMPTION FUNCTION

If one accepts the consumption function with positive intercept as the true behavioral relation. the long-run constancy of the historical ratio of consumption to income might be explained by an upward shift of the consumption function as real per capita income increases. Thus, in the equation

$c = a + by_d$

the average propensity to consume will be constant if the intercept a rises in proportion to y_d. The historically observed points would fall on consumption lines with different intercepts, but would lie along a ray from the origin—so that observed c/y_d ratios would be constant.

What might cause increases in the intercept a? Possible factors include migration from farms to urban areas (where the consumption-to-income ratio is higher); upward shifts of consumption because of new products or advertising; and income redistribution via taxes and transfers in favor of consumer groups with higher average propensity to consume. It seems difficult to believe that the effect of these variables on the intercept a would have maintained the required close proportionality to advances in y_d. One other variable might do so. Wealth owned by the Personal Sector has advanced steadily in the process of economic growth. It might be that increasing real wealth per capita would cause an upward drift of the consumption function in step with rising real income per capita. Aggregate data do not permit analysts to disentangle the influences of these variables or to reject the alternative hypotheses decisively.

LIFE-CYCLE HYPOTHESIS

Other theorists reject the Keynesian consumption function and favor the hypothesis that consumer units at all income levels will desire to maintain the same average propensity to consume or to save, though the ratios are different for families at different ages or different stages of their life cycle. If families at various levels of income plan to maintain a steady level of consumption through their retirement years as well as in their earning years, the fraction of cumulative income to be saved during the earning years will be determined by the relative lengths of retirement period and earning span, regardless of income level. If the distribution of families by age and family type should remain stable through time, such consumption-saving behavior would lead to constant ratios C/Y_d and S_p/Y_d even as per capita incomes rise.

One may object that if each consumer unit consumes during retirement all the wealth it accumulates by saving during its years of employment, aggregation over all families should lead to zero saving, or an APC equal to 1. This assumes that consumer units do not leave any inheritance. In fact, they do, either because they plan to or because uncertainty over time of death leaves most individuals with assets to pass on to their heirs—and the government. (It must be assumed that receipt of an inheritance does not reduce saving *pari passu* in the younger generation.)

However, in a growing economy even the no-inheritance assumption can be consistent with net saving in the aggregate, as pointed out by Modigliani and Ando.†
In the first place, a growing population implies a larger proportion of working-age persons to retired persons than would a stationary population, hence an increased percentage of the population in the saver group, leading to positive saving in the

---

† F. Modigliani and A. Ando, "Test of the Life Cycle Hypothesis of Saving," *Bulletin of the Oxford University Institute of Statistics*, May 1957.

aggregate. Secondly, growth in per capita income will raise the incomes of current workers above the past incomes earned by currently retired persons during their working years, so that positive savers will be saving more (to support a higher level of consumption in their retirement years) than is being dissaved now by retired workers (whose level of consumption is less). Under simplifying assumptions, the investigators calculated that a 1 percent per year rise in aggregate real income, either from population growth or increase in per capita income, would lead to the saving of about 3 to 4 percent of aggregate income. A steady growth in aggregate income would lead to a stable saving ratio for the economy. The result is consistent with observed data, even though the level of the constant ratio might be altered by other influences, such as a desire to leave an inheritance or the existence of a social security system.

RELATIVE-INCOME HYPOTHESIS

Another theory of consumer behavior which would explain the long-run constancy of the consumption-income ratio in the face of rising per capita income is James Duesenberry's relative-income hypothesis.[†] Duesenberry suggested that the proportion of income which persons spend and save may be determined not by their absolute level of real income but by their relative position on the income scale among the persons and families with whom they associate or whom they desire to emulate. Thus, a person with a $5,000 annual income (at 1958 prices) in 1970 would spend a higher percentage of his or her income than would a person with the same real income in 1950, because the former is lower in the 1970 income distribution than was his or her counterpart in 1950. If the overall income distribution maintains nearly the same shape relative to the median income, and if the APC depends only on the percentile within which a consumer unit falls, the aggregate APC will stay constant as the median income rises. Statistical tests on cross-section data give support to this hypothesis, and it may well be part of the explanation for the long-run constancy of the aggregate APC.

SUMMARY

We see that there are several alternative hypotheses to explain why the aggregate APC and APS should stay constant in the long run. Historical data have not permitted decisive tests to choose among them.

The question now arises: Granted that these theories explain the long-run constancy of ratios C/Y_d and S_p/Y_d, can they also explain the undeniable evidence that in business recessions and in budget-study data the C/Y_d ratio rises at lower incomes? The hypothesis that the true consumption function has a positive intercept and shifts up through time is consistent with long-run, short-run, and budget-study data, but what about the other hypotheses?

[†] James S. Duesenberry, *Income, Saving and the Theory of Consumer Behavior*, Harvard University Press, Cambridge. Mass., 1949.

THEORIES TO EXPLAIN HIGHER APC AT LOWER INCOMES
RATCHET-TYPE CONSUMPTION FUNCTION

Duesenberry linked his relative-income hypothesis with a theory that when a person or family experiences a decline in income believed to be temporary, the consumer unit will try to maintain consumption standards near the previous high level, to which it expects to return. Saving will be reduced below the normal, long-term level which would be appropriate at the reduced level of income. Of course, if the lower income level persists for a long time and if the consumer loses faith that he or she will soon return to the former high level of income and consumption, the APC may drop and the APS rise to the levels appropriate to the reduced income; that is, the consumer moves back to his or her long-term consumption and saving lines.

This hypothesis implies that desired consumer spending depends both on current disposable income and on the previous higher level which may have been reached. Sometimes called the *ratchet consumption function*, it may be expressed by the equation

$$c^* = b_\ell y_0 - b_s(y_0 - y)$$

where c^* = desired per capita consumer expenditures
y = real per capita disposable income in the current period
y_0 = highest value of y reached to date
b_ℓ = long-run MPC
b_s = short-run MPC

If y rises in each period so that y equals y_0, then c^* equals $b_\ell y_0$, and consumption rises along a long-term consumption function line through the origin. If y drops below y_0, then c^* drops below its previous high level ($b_\ell y_0$) by an amount equal to the short-run MPC times the income decline, that is, it drops by $b_s(y_0 - y)$. An equivalent alternative expression is

$$c^* = (b_\ell - b_s)y_0 + b_s y$$

This brings out the fact that c^* depends both on current income and previous high level of income. In a recession, c^* varies with current income y along a short-run consumption function with slope b_s and intercept $(b_\ell - b_s)y_0$, an intercept which increases as the previous high income level y_0 rises. These relations are illustrated in the graph of Fig. 7-4, where b_ℓ is taken as 0.9 and b_s equals 0.7. The initial decline is shown starting from y_0 equaling $1,500 per year, and two other declines are indicated starting from income levels of $2,000 and $2,500 per year. During the recession periods, the short-run MPC is less than the long-term value, and APC rises as consumers try to maintain their consumption standards.

The corresponding saving function, where s_p^* equals desired real per capita personal saving, becomes

$$s_p^* = y - c^* = (1 - b_\ell)y_0 - (1 - b_s)(y_0 - y)$$

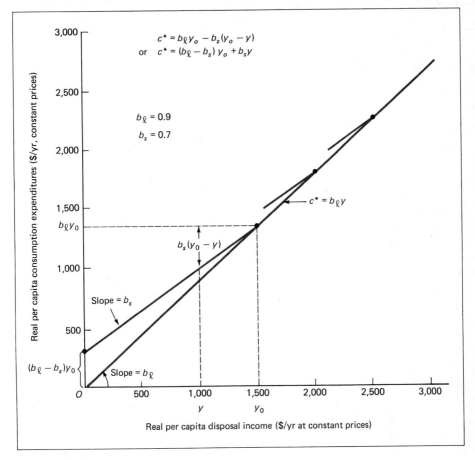

Figure 7-4 Ratchet consumption function.

or

$$s_p^* = -(b_\ell - b_s)y_0 + (1 - b_s)y$$

This relation is plotted in Fig. 7-5, where b_ℓ equals 0.9 and b_s equals 0.7 as before. The proportional difference between short- and long-run MPS, that is, 0.3 and 0.1, is greater than for the MPC, so statistical attempts to verify the ratchet-type behavior generally use saving data. Attempts to fit consumption or saving functions of this type to historical data have given statistically acceptable results. The hypothesis is given added validity by cross-section studies in which families whose income has declined from the previous year have lower saving ratios than families at the same current income level whose income has been stable, and conversely for families who have experienced increases in income.

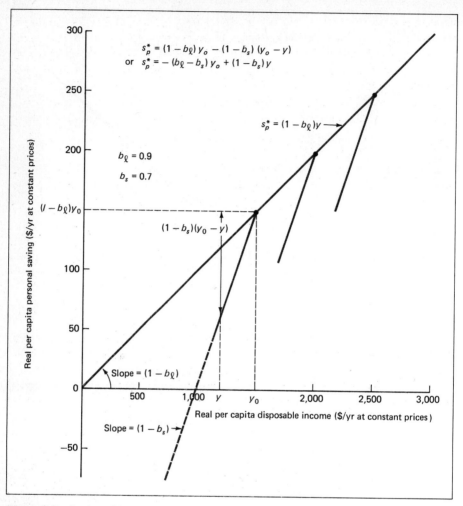

Figure 7-5 Ratchet saving function.

PERMANENT-INCOME HYPOTHESIS

The rise in APC at lower incomes in recessions or in budget-study data has also been explained by the hypothesis that current-period consumer expenditures are determined not by current-period disposable income but by long-term income expectations, or *permanent income*, as it was named by Milton Friedman,† whose theory rises from the foundations laid by Modigliani, Brumberg, and Ando.

---

† Milton Friedman, *A Theory of the Consumption Function*, Princeton University Press, Princeton, N. J., 1957.

Consumer units are assumed to engage in long-term (life-cycle) planning and make their planned consumption flows in each period consistent with their permanent incomes. The permanent income (y_p) for a consumer unit is defined as *a level income stream over its remaining expected lifetime whose present value would equal the sum of present net worth plus the present value of the actual (variable) income stream expected.* The actual income of the consumer unit in a given time period equals the sum of permanent income (y_p) and temporary or transitory deviations (y_t).

$$y = y_p + y_t$$

Similarly, actual consumption expenditures consist of permanent and transitory components.

$$c = c_p + c_t$$

The hypothesis goes on to state that permanent consumption is a multiple (k) of permanent income.

$$c_p = ky_p$$

The coefficient k varies with interest rate, with the ratio of net worth to total (human plus nonhuman) capital, and with age and tastes of individual units, but k is the same for consumer units at all levels of permanent income.

Finally, it is hypothesized that there is no time-series correlation between transitory and permanent components of either income or consumption, and none between transitory components of income and consumption. A change in actual income for a consumer unit will affect consumption in a systematic way (c_p) only to the extent that it alters permanent income (y_p). If the change is believed to be transitory, it will have very little effect on y_p and c_p; if the change is believed to be lasting and if it was not previously included in the expected future income stream, it will change y_p and c_p.

This response to actual income changes would explain the maintenance of consumer purchases during a recession. Actual income declines, but consumption is held in the normal relation to permanent income—which is altered very little. If lower incomes persist and lead to a downward revision of expected future incomes (hence y_p), c_p will be reduced. Consumer response to increases in actual incomes will depend similarly on whether the rise is expected to be transitory or lasting.

Measurement of permanent income is difficult; so it is difficult to make decisive tests of the hypothesis. Roughly, the effect is to introduce a time lag between unexpected income changes and resulting changes in consumer expenditures. In some econometric models, C is made a function of a moving average of disposable income over several prior periods. In others, a lagged consumption term is used to represent the influence of permanent income, and perhaps other slowly changing variables, on consumption.

$$C_t = b Y_{d,t} + \beta C_{t-1}$$

where t designates time periods. Such formulations seem to yield improved consumption functions, giving some evidence favorable to the hypothesis.

Other conclusions derivable from the permanent-income hypothesis seem to explain observed relations in budget-study, or cross-section, data. Thus, a young family expecting a rise in real income over the years ahead might spend a higher percentage of current income than would an older family at the same current-income level but expecting smaller future income. Also, entrepreneurs whose incomes fluctuate greatly from year to year might maintain a fairly stable time pattern of consumption and absorb the variation in current income by letting saving flows fluctuate.

The permanent-income hypothesis gives an interesting interpretation of the low slope and positive intercept of the consumption function derived from budget-survey data (Fig. 7-2). Families are classified, of course, according to reported current income. The reported consumer expenditures, however, reflect the *permanent income* of families in each classification. Each class, based on current income, will contain families for whom such income equals permanent income and also families whose permanent incomes are either above or below the current-income class limits but who have been shifted into that class by transitory components of income. For the middle-income classes (near the median), approximately equal numbers of "shifted" families will have higher and lower permanent incomes. For such classes, the average permanent income of families will be close to the average current income, and average consumer expenditures will reflect the normal relation between permanent consumption and permanent income. Income classes low on the scale will contain more families shifted down from higher permanent incomes than families shifted up from lower permanent incomes. So average observed consumer expenditures would be higher than would be predicted if the average current income in the class were taken to be average permanent income. Conversely, high-income classes will contain more families with positive than with negative transitory incomes. So average permanent income, and hence average consumer expenditures, would be lower than predicted by taking average class income to be average permanent income. The consequence is that the consumption function derived from budget surveys will be flatter than the true relation between c_p and y_p. The true relation would have smaller intercept, perhaps even go through the origin—implying constant ratio c_p/y_p at all income levels.

In summary, it would seem that the ratchet theory or permanent-income theory can go far toward reconciling the long-term constancy of the APC with the short-term and budget study data.

SHORT-RUN EFFECTS OF CHANGES IN POPULATION AND PRICES

Let us assume that real per capita consumption maintains a stable relation to real per capita disposable income for time-series data.

$$c = \alpha + \beta y_d$$

Expressed in terms of aggregate consumption and income, with N equaling the number of persons in the population and P_c equaling the consumer price index, the above relation becomes

$$\frac{C}{NP_c} = \alpha + \beta \frac{Y_d}{NP_c}$$

or

$$C = \alpha NP_c + \beta Y_d$$

Comparing this with the aggregate consumption function $C = a + b Y_d$, we note that the slopes of the aggregate current-dollar and the real per capita functions are the same, that is, b equals β. But the intercept of the aggregate current-dollar function rises as N and P_c increase. This has some interesting implications.

If Y_d and P_c stay constant but population rises, then C rises. This is so because the rise in N reduces real per capita income and raises the APC. Thus, a rise in population can have counter-cyclical effects in a recession. If Y_d should rise in step with N, then per capita real income and APC would stay constant, and C would increase in proportion to Y_d.

If N and Y_d were constant, an increase in price level would lower real per capita income and consumption, but it would raise aggregate current-dollar consumption C. Of course, Y_d will normally rise when P_c does. If the rise is proportional (with N constant), then real per capita income and consumption are unaffected, and C will increase in step with P_c and Y_d.

Of course, if the constant α in the real per capita consumption function were zero, then changes in N or P_c would not change the intercept of the aggregate current-dollar consumption function. Both the aggregate and the per capita consumption functions would pass through the origin; consumption would always be proportional to disposable income; and $APC = MPC = \beta$.

OTHER INFLUENCES ON CONSUMPTION AND SAVING DECISIONS

Although income, in some variant, still appears to be the dominant variable explaining levels and changes of aggregate consumption and saving flows, other variables have been found to be significant in econometric analyses of cross-section or time-series data. The role of population and price level for consumer goods has been noted above.

WEALTH

The stock of wealth owned by consumer units has long been regarded as a variable which should have a positive relation to the fraction of income spent for consumption. The underlying hypothesis is that consumers at various income levels will desire to own appropriate stocks of various wealth items as a part of their desired balance between current and future consumption. The wealth items may be financial assets (money or securities) which yield a future stream of income or of "liquidity services" to meet future payments or contingencies. Or the wealth items may be durable goods (houses, automobiles, furniture, and home appliances) which will yield a future stream of desired services. To acquire the desired real stock of wealth items, consumers will save part of their income during the accumulation stages of the family life cycle and may dissave during retirement. If events alter the desired or actual stock of wealth, consumers will deviate from normal consumption-saving flows to adjust their stocks of money, securities, and durable goods.

For example, changes in interest rates or a period of large capital gains or losses will alter the value of actual security holdings, and saving flows will be altered in the opposite direction to adjust actual stocks to desired levels. Or price rises may reduce the prospective real stock of wealth, and expected price rises can reduce the expected real income stream, resulting in added saving out of current income. Or a period of easy credit conditions may stimulate a build-up of stocks of consumer durables beyond the desired level and result in a subsequent reduction of durable goods buying. Or a prospective period of unemployment and financial uncertainty can stimulate saving flows to build up contingency reserves of assets. Econometric studies have exhibited significant effects of such variables, in cross-section and time-series data. Stable relations in aggregate data have been hard to establish, except for some stock-adjustment models for durable goods, presumably because wealth and permanent income tend to be highly correlated; or because expectational variables and desired stocks are hard to measure; or because periods when actual stocks deviate from desired levels significantly are disturbed periods exhibiting changes in many variables, with independent effects that are hard to disentangle.

HABITS

Another form of "stock effect" lies in the development of buying habits and customary levels of consumption. They promote persistence of buying patterns, and they apply to expenditures for nondurables and services as well as durables. In consumption-function equations, this effect shows up as a positive coefficient times the purchases in earlier periods. Such lagged effects (autocorrelation) were involved in the theories of Modigliani and Duesenberry and were examined in more detail later by Houthakker and Taylor.† The dynamic, lagged response of consumption to income changes in such models is similar to the response generated by permanent-income or ratchet-type models.

INTEREST RATES AND CREDIT

Since interest rates and other credit terms influence the trade-off ratios between current and future consumption, it seems reasonable that they should influence the allocation of aggregate incomes between consumption and saving. Investi-

---

† H. S. Houthakker and L. D. Taylor, *Consumer Demand in the United States*, 2d ed., Harvard University Press, Cambridge, Mass., 1970.

gations reveal negative correlations between interest rates and purchases of houses or consumer durables, a positive influence of interest rates on purchases of financial assets (though negative on demand for money), and a positive influence of easier credit terms on purchases of consumer durables. In the aggregate, the effects seem not to be large, perhaps because high interest rates have a positive income effect for owners of securities and occur at times when economic activity is strong, perhaps because the concomitant changes in other variables make it difficult to isolate the separate influence of interest rates and credit terms.

ATTITUDES AND EXPECTATIONS

Waves of optimism and pessimism have been taken as initiating or reinforcing variables in some business-cycle theories. It does seem plausible that expectations regarding future levels of employment, wages, prices, or interest rates might shift in the same direction at the same time for a majority of consumers, and might lead to temporary shifts of the consumption function. Uncertainty regarding future employment and income could motivate higher saving. Expectations of rising prices would logically stimulate current purchases, especially of durable goods. An unanticipated inflation accompanied by rising money incomes might well increase the average propensity to save because of a desire to restore the inflation-reduced stock of real wealth, and perhaps also because of an inflationary shift of income in favor of high-saving recipients of property incomes. Of course, definition and measurement of variables to represent consumer attitudes and expectations are difficult, and it is hard to separate the influence of such variables from the direct effects of the variables which generate the consumer attitudes and expectations.

OTHER INFLUENCES

Socioeconomic characteristics of consumer units, regional environments, advertising, and new-product introductions are other variables which have been found significant in markets for particular categories of consumer spending or saving. But they do not seem to influence the aggregate consumption function as much as the variables discussed above. A recent comprehensive survey and bibliography on consumption and saving has been prepared by Robert Ferber.†

Certainly there are sizable unexplained fluctuations of consumer expenditures in the historical data. They are large and frequent enough to keep economists humble and to cause problems both for the economic forecasters and the government policy makers. For example, the following table shows the changes in income, taxes, consumer spending, and saving which took place when the 10 percent surtax went into effect in mid-1968. Personal income rose steadily by 14 to 17 billdar each quarter. Taxes jumped in the third quarter by 10 billdar and by 4.4 billdar more in the fourth quarter. Net disposable personal income rose less than 7 billdar in the third quarter, but consumer spending jumped 15 billdar, aided by a 9-billdar decline in saving flows. Saving continued through the first half of 1969 at levels well below those of the first two quarters of 1968, despite higher disposable

† "Consumer Economics, a Survey," *Journal of Economic Literature*, vol. XI, no. 4, December 1973.

	1968				1969
	First Quarter	Second Quarter	Third Quarter	Fourth Quarter	First Quarter
Personal income	664.0	681.2	697.8	712.6	726.5
Less personal tax and nontax payment	89.0	92.9	102.7	107.1	114.0
Less consumer interest and transfers to foreigners	14.5	14.8	15.3	15.7	16.1
Equals net disposable personal income	560.5	573.6	579.9	589.8	596.4
Consumer expenditures	519.3	529.0	544.0	552.5	564.0
Personal saving	41.2	44.6	35.9	37.3	32.4

personal income, and thus offset much of the intended impact of higher tax on reducing consumer expenditures. Of the fluctuations in the personal saving rate before and after the tax rise, the President's Council of Economic Advisers wrote in its 1969 report: "There is no clear explanation for the extra saving in the period from late 1966 to mid-1968. . . . these developments meant that households had achieved relatively favorable financial position, which permitted them to increase expenditures markedly whenever they chose to do so. Why they happened to pick the third quarter of 1968 to exercise this option cannot be adequately explained."

Time lags in response to sudden changes in tax rates and income levels have been introduced in some models, but the length of the lag and the magnitude of the effect have not been reliable enough for practical use. Indeed, consumers seemed to have anticipated the tax cut of 1964 and increased buying in advance of the decline in tax rates.

DURABLE GOODS AND SAVING

Purchases of durable goods, those having an average life of three years or longer, should probably not be considered to reduce the wealth of the Personal Sector. A real asset is acquired which yields services over a long period of time, like investment goods bought by businesses on capital account. The component of consumption arising from durable goods should be the current-year services of those goods, analogous to capital consumption allowances in the Business Sector. If such consumption of consumer durables could be calculated, then current purchases of these durables could be treated as consumer gross investment. Like saving, it involves use of current income to provide for future consumption. That consumers do consider durables in some sense a substitute for saving is shown by the fact that in nonwar, nondepression years, there is a clear *negative* correlation between flows of expenditure for durable goods and of personal saving, which is net of increases in consumer debt.

In Fig. 7-6, the ratios of major components to net disposable personal income are plotted. The lower curves show spending for consumer durables and personal saving as percentages of net disposable personal income. Both ratios decline during the depression years, returning nearly to 1929 levels only in 1936–1937 and

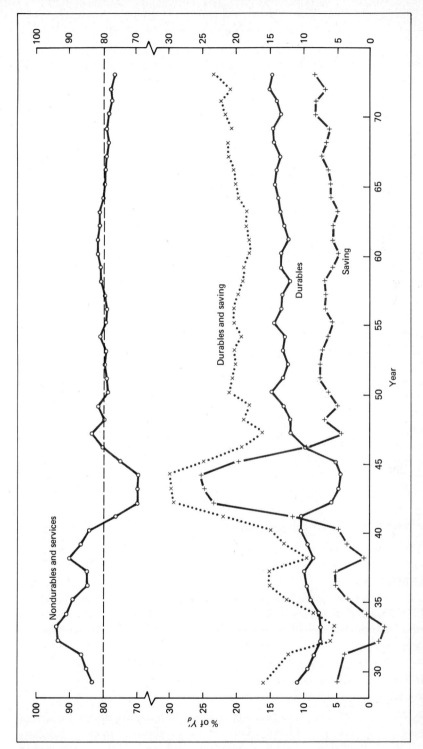

Figure 7-6 Consumer expenditure and saving components as a percent of net disposable income, 1929 to 1974. (*Survey of Current Business*, Tables 1.1 and 2.1, U.S. Department of Commerce, July 1974.)

1940. During World War II, production controls and rationing restricted purchases of durable goods (and nondurables and services also) so that saving rose sharply as a percentage of consumer income. Since World War II, durables and saving ratios have fluctuated in relatively narrow ranges, somewhat above their 1929 percentages, with no significant trend. The contrary fluctuations of durables spending and saving have led to greater stability for their combined ratio than for the individual series. The dotted line for "durables and saving" in Fig. 7-6 indicates that since 1948 the percentage of disposable personal income allocated to durables and saving has fluctuated between 18.1 percent (1960) and 22.5 percent (1971).

The sum of allocations to nondurables and services may be taken to represent true current consumption. Though services have risen and nondurables have fallen as a percentage of net disposable personal income in the postwar years, there has been a remarkable stability in their combined ratio. As shown in the top line of Fig. 7-6, their combined proportion of net disposable income has held close to 80 percent since World War II, a little higher in years of high unemployment and a little lower in years of low unemployment. The data for 1929 through 1947 showed a rise during the Depression of the 1930s and a decline during World War II, with its shortages and rationing.

Overall, the data of Fig. 7-6 reinforce the concept of long-term stability in the allocation of consumer income between consumer expenditures and saving flows. And it supports the notion that spending for durables is more volatile, less closely related to income, than are purchases of nondurables and services. Some of the variables, in addition to income, which have been shown to influence durable goods spending include existing stocks of durable goods owned by consumers (negative influence), new household formation or housing starts, consumer liquid assets, credit conditions, relative prices, and a measure of consumer attitudes or the unemployment rate. But compared with income, these other variables make relatively small contributions. Income is the dominant influence, perhaps modified by a ratchet term or by a lagged expenditure term as a proxy for permanent income.

PROBLEMS

1   a   List the macroeconomic variables which might influence the magnitude of desired aggregate personal consumption expenditures. (These would be the variables represented by the dotted arrows pointing into the Personal Decisions circle of Fig. 5-1 in Chap. 5.)

   b   For each of these variables, indicate whether its increase would tend to raise or lower consumer expenditures, assuming other input variables hold constant.

   c   Describe the way in which a general price rise, with relative prices of different goods unaffected, would change consumer expenditures—in current dollars and in constant dollars: (1) if Y_d were constant, and (2) if Y_d increased in proportion to prices. Assume population is constant.

d Why is the effect of a general price rise in macroeconomics different from the effect of a price rise for a single good in microeconomics? Can money incomes be assumed constant during a general price rise? Can real incomes?

2 Characteristics and comparisons of consumption functions

a Keynesian consumption function

(1) Write an equation for a Keynesian consumption function in linear form.

(2) Express in the symbols you used in this equation how you would calculate MPC and APC.

(3) Write the personal saving equation corresponding to this consumption equation, and express in symbols how you would calculate MPS and APS.

(4) If the Personal Sector decided to increase desired expenditures by a fixed amount at each level of Y_d, what changes would be needed in the equations for the consumption function and the saving function?

(5) What would be the corresponding graphical shifts for the two functions?

(6) What would be the effect of a rise in general price level on the Keynesian consumption function, as expressed in real units or as expressed in money units?

(7) What modifications of the Keynesian consumption function have been suggested by later research?

b Long-run versus short-run consumption functions

(1) What is the difference in the functional relation between real consumption and real disposable income (*a*) in the short term, and (*b*) in the long term?

(2) Suppose that real per capita disposable income rose to an historic high of $2,000 per person at 1958 prices in period 1 and then followed this pattern:

$t =$	1	2	3	4
$y_t =$	$2,000	$1,800	$2,000	$2,200

Assuming a long-run MPC of 0.9 and a short-run MPC of 0.75, write the equation for the ratchet-type consumption function, and calculate the values of real per capita consumer expenditures for periods 1 through 4.

(3) Write the equation for the corresponding real per capita saving function, and calculate values for s_p in periods 1 through 4.

c Permanent-income hypothesis

(1) How would consumer spending behavior, under this hypothesis, differ from the ratchet hypothesis when income fell from an historic high and when income rose above an historic high?

(2) If the permanent-income-type consumption function shows MPC as 0.9, that is,

$$c_t^* = 0.9 \; y_{p,t}$$

where $y_{p,t}$ is permanent income, and if $y_{p,t}$ is estimated as equal to an exponentially weighted moving average of past incomes (with smoothing constant of 0.4), then we find that

$$c_t^* = 0.36 \; y_t + 0.6 \; c_{t-1}$$

Given that c_0 is 1,800 in period 0 (equilibrium for $y_{d,0}$ equals $\bar{y}_{p,0}$, or $2,000), calculate c_t for the above equation in periods 1 through 4 with income fluctuating as follows:

$t =$	1	2	3	4
$y_t =$	$2,000	$1,800	$2,000	$2,200

(3) Comment on the comparison of these values of c_t with those obtained in part **b(2)** above with the same income pattern.

d Effects of population growth

Assume the per capita consumption function c equals $25 + 0.93 \; y_d$, in dollars per person at 1958 prices.

(1) With a population N_1 of 200 million persons, write the equation for the aggregate real consumption function $C_{R_1}^* = a + b Y_{d,R_1}$, where $C_{R_1}^*$ and Y_{d,R_1} are measured in billdar at 1958 prices and subscripts refer to year 1.

(2) If population rose to N_2 of 205 million persons in year 2, calculate the new consumption function $C_{R_2}^*$.

(3) Calculate the rise in C_R^* which accompanies a rise in $Y_{d,R}$ from 400 to 440 billdar: (*a*) if population stays constant at 200 million, and (*b*) if population rises from 200 million to 205 million as $Y_{d,R}$ increases.

e Effects of price changes

(1) If the general price level rises by x percent and money flow of disposable income stays constant, what can you say about the percentage changes in real consumption expenditures and in their money value? Assume the consumption function is stable when expressed in real units, that is, C_R versus $Y_{d,R}$.

(2) If the general price level and money flow of disposable income both increase by x percent, what can you say about the percentage changes in real consumption expenditures and in their money value? Assume that the consumption function is stable in real units.

(3) Based on your results, outline an empirical test which might show whether the consumption function is stable in real terms or in money terms. Would the test distinguish between the two hypotheses if the intercept a were zero?

Chapter 8

Interaction between Consumers and Other Sectors

INTRODUCTION

In view of the economic statistics and the econometric investigations that have become available since his day, it would seem that Keynes showed true genius in developing a simplified macroeconomic model of the complex economic system. He certainly picked out the most important input to consumer decision making when he made consumer spending and saving depend primarily on disposable income. We shall use his consumption function in linear form:

$$C^* = a + bY_d$$

This is the aggregate relation in current-dollar terms. The parameters a and b embody the influence of the many variables, other than current disposable income, which affect consumer spending. Changes in those variables will alter a and b, but such changes are judged to cause less short-run fluctuation in C^* than do changes in Y_d.

One linkage between consumer expenditures and the rest of the economy is obviously through disposable income. If the Production Units Sector decides to produce more, then more incomes will be paid out, and a major part of the increased income will flow to the Personal Sector as disposable personal income. The initial rise in output may well have been a response by the Production Units Sector to a rise in orders from the Government Sector or Business Capital Accounts. For the present, this step-up in demand will remain unexplained; call it an autonomous rise ("autonomous" meaning that it was caused by anything other than a rise in GNP). Initially, GNP will rise by the amount of this autonomous increment to demand. However, the consequent rise in disposable personal income will induce increased consumer spending, in accord with the consumption function. When the Production Units Sector receives increased orders for consumer goods, it will increase output to meet those orders and raise incomes again as it does so. This "second-round" increase in disposable income will cause the Personal Sector to increase consumer expenditures further, leading to additional rounds of increased output, incomes, and consumer expenditures. Clearly, we are here involved in a "feedback loop" in the economic system which can lead to a final response (rise in GNP) that will exceed the initial stimulus (rise in autonomous expenditures). How much the initial stimulus is amplified must be determined by working through the equations of the model of the economic system.

Before proceeding with that analysis, however, note that the Personal Sector may interact with the rest of the economy in another manner. The consumer units may provide the autonomous stimulus to the economy if they change their level of desired consumer expenditure (or saving) out of a given disposable income. Such a change in consumer spending-saving behavior could be represented analytically by changes in the parameters a or b, or both, in the consumption and saving functions. Graphically, this would imply a change in the level or the slope of the consumption function and an inverse change in the saving function. Such autonomous changes in consumption and saving must be kept logically separate from induced changes. *Induced changes arise from income changes* and appear graphically as movement along a fixed consumption or saving function. *Autonomous changes arise from changes in variables other than income* and appear graphically as a shift of the consumption or saving function.

PROPERTIES OF A SIMPLE FIVE-SECTOR ECONOMY†

For our initial analysis let us use what might be called a simple five-sector economy. It is characterized by the following properties.

1 Transfers to foreigners by the Government and Personal Sectors are assumed to be zero. This implies that net exports equal net foreign investment, which will be lumped with domestic investment in a gross investment term I in the model.

2 Interest paid by consumers and personal transfers to foreigners are zero. Then net disposable personal income is the same as disposable personal income, and the latter equals consumer expenditures plus personal saving.

3 The statistical discrepancy equals zero.

4 Government purchases and investment expenditures (domestic plus foreign) are all assumed to be autonomous, that is, not affected by the level of income flows resulting from producer's decisions on aggregate supply.

5 All prices remain constant, so that changes in money flows are equivalent to changes in real flows.

Our problem is to solve for GNP (or Y) when the economy is in flow equilibrium as determined by a set of values α for autonomous variables. The approach in formulating and solving the equations is the same as that described in Chap. 3.

---

† The symbols used in this chapter are listed, with their names, units, and reference pages, in an appendix at the end of the book.

Type of Equation	Equation	New Variables	
		Exogenous	Endogenous
1 Definition	$Y^* = C^* + G_\alpha^* + I_\alpha^*$	G_α^*, I_α^*	Y^*, C^*
2 Behavioral	$C^* = a + bY_d$		Y_d
3 Institutional	$Y_d = e + fY_s$		Y_s
4 Equilibrium condition	$Y_s = Y^*$		

SOLUTION OF THE MODEL

We have a system of four linear equations containing four endogenous variables. It should be solvable. We start with the aggregate demand definition [equation (1)] and use equations (2) and (3) to displace endogenous variables on the right side of equation (1) until only terms in Y_s and exogenous variables remain. The expression for aggregate demand becomes:

$$Y^* = (a + be + bfY_s) + I_\alpha^* + G_\alpha^*$$

We may simplify the notation by recalling that the last terms on the right are desired injections (INJ_α^*). The final solution for equilibrium state α is obtained by imposing the equilibrium condition $Y^* = Y_s$ and letting their equilibrium values be denoted by Y_α. We find

$$Y_\alpha = (a + be + bfY_\alpha) + \text{INJ}_\alpha^*$$

The solution is

$$Y_\alpha = \frac{1}{1 - bf}(a + be + \text{INJ}_\alpha^*)$$

or

$$Y_\alpha = k(a + be + \text{INJ}_\alpha^*) \tag{8-1}$$

where $k = 1/(1 - bf)$ is called *the autonomous expenditure multiplier of the economy*.

After Y_α has been solved for, we can calculate Y_d and C^* from equations (3) and (2) of the system. Explicitly,

$$Y_{d,\alpha} = e + fY_\alpha \tag{8-2}$$

and

$$C_\alpha^* = a + be + bfY_\alpha \tag{8-3}$$

PROPERTIES OF THE SOLUTION

1 Since b and f are normally each positive and less than 1, the reciprocal $1/(1 - bf)$ will be positive and greater than 1. Thus we have the multiplier $k = 1/(1 - bf) > 1$. As may be seen from its origins in equations (1) and (2) of the

model, the bf coefficient accounts for the impact of Y_s on C^* via the feedback path through Y_d.†

This feedback from GNP to one of its component inputs (C^*) leads to amplification of the influence of autonomous expenditures on GNP. If b and f were both near 1, the multiplier k could become enormous.

2   *The equilibrium level of* Y *will vary directly with changes in injections, though* δY *will exceed* δINJ^*. Consider two equilibrium states determined by different levels of injections, with k and $(a + be)$ constant.

$$Y_\alpha = k(a + be + \text{INJ}_\alpha^*)$$

$$Y_\beta = k(a + be + \text{INJ}_\beta^*)$$

Subtraction yields

$$Y_\beta - Y_\alpha = k(\text{INJ}_\beta^* - \text{INJ}_\alpha^*)$$

or, since the *symbol δ designates change of a variable from its initial to final equilibrium value,*

$$\delta Y = k(\delta \text{INJ}^*)\ddagger \tag{8-4}$$

3   *The equilibrium level of GNP will vary directly with the* autonomous *component of* C^*, that is, with the intercept $(a + be)$ of Eq. (8-3) specifying C^* as a function of Y_s. A change in position of this consumption function has a multiplied effect on GNP. From the solution equation [Eq. (8-1)], assuming a change in $(a + be)$, with INJ_α^* and k constant, we obtain directly

$$\delta Y = k\delta(a + be)$$

Thus, an autonomous change in consumer spending leads to an initial rise in output and incomes followed by *induced* changes in consumer spending, which constitute the multiplier effect. The mechanism is the same as for autonomous changes in injections. Shifts of the C^* line [Eq. (8-3)] may arise from an autonomous change in consumer behavior (a or b) or from a change in the function relating Y_d to Y_s, for example, a change in tax rates which alters e or f.§

---

† Note that
$$C^* = a + be + bf\, Y_s$$
and
$$\frac{dC^*}{dY_s} = bf = \left(\frac{dC^*}{dY_d}\right)\left(\frac{dY_d}{dY_s}\right)$$

‡ This is, of course, equivalent to the solution obtained by taking the derivative of the solution equation [Eq. (8-1)]:
$$\frac{dY_\alpha}{d(\text{INJ}_\alpha^*)} = k \quad \text{or} \quad dY_\alpha = k\, d(\text{INJ}_\alpha^*)$$

§ It should be noted that a change in b or in the coefficient $f(= dY_d/dY_s)$ will alter the multiplier $k = 1/(1 - bf)$.

4 *The equilibrium level of* Y *will be altered by a change in* k, *even if* autonomous expenditures $(a + be + \text{INJ*})$ remain constant. The value of the multiplier will change in the same direction as a change in either b or f. That is, k increases if consumers spend a larger share of an increment of Y_d or if a larger fraction of an increment in Y_s flows to Y_d.

$$\delta Y = (\delta k)(a + be + \text{INJ*})$$

AN ALTERNATIVE SOLUTION

Instead of obtaining our solution from the system of equations describing the equilibrium between Y^* and Y_s, we will sometimes find it convenient to use the system of equations describing equilibrium between desired withdrawals and desired injections.

Recall that desired withdrawals equal net income flows to the Government, Business Capital Accounts, and Foreign Sectors plus desired personal saving.

$$\text{WDL*} = T_n + \text{GRE} + \text{TR}_f + S_p^*$$

WDL* will be a function of Y_s, since sector income flows depend on GNY. Indeed, as shown in Chap. 4, withdrawals are equal to total incomes (GNY) less the flow of consumer expenditures. So we may write

$$\text{WDL*} = Y_s - C^*$$

	Type of Equation	Equation	New Variables	
			Exogenous	Endogenous
1	Definition	$\text{WDL*} = Y_s - C^*$		$\text{WDL*}, Y_s, C^*$
2	Behavioral	$C^* = a + bY_d$		Y_d
3	Institutional	$Y_d = e + fY_s$		
4	Definition	$\text{INJ}_\alpha^* = G_\alpha^* + I_\alpha^*$	G_α^*, I_α^*	INJ_α^*
5	Equilibrium condition	$\text{WDL*} = \text{INJ*}$		

To solve this set of five equations with five endogenous variables, we substitute equations (2) and (3) into equation (1), obtaining:

$$\text{WDL*} = Y_s - (a + be + bf\,Y_s) = -(a + be) + (1 - bf)Y_s$$

Then use the equilibrium condition [equation (5)] along with equation (4) to obtain

$$-(a + be) + (1 - bf)Y_\alpha = \text{INJ}_\alpha^*$$

The solution for Y_α is

$$Y_\alpha = \frac{1}{1 - bf}(a + be + \text{INJ}_\alpha^*)$$

Of course, the solution turns out to be the same as Eq. (8-1). The equilibrium conditions $Y_s = Y^*$ and WDL* = INJ* are equivalent. (The second may be derived by subtracting C^* from both sides of the equation $Y_s = Y^*$.)

A NUMERICAL EXAMPLE

Let us attempt to derive coefficients which will make the above equations describe the United States economy in the postwar years. Figure 8-1 displays the relation between C and Y_d' for the years 1929 through 1973 and has a dashed 45-degree line whose ordinates are Y_d'. Setting aside the depression years of the 1930s and the World War II years, we find that consumption is well represented by a line through the origin with slope MPC equaling 0.93. The consumption function for our model becomes

$$C^* = a + b Y_d' = 0.93 \, Y_d'$$

So we find that a equals 0 and b equals 0.93.

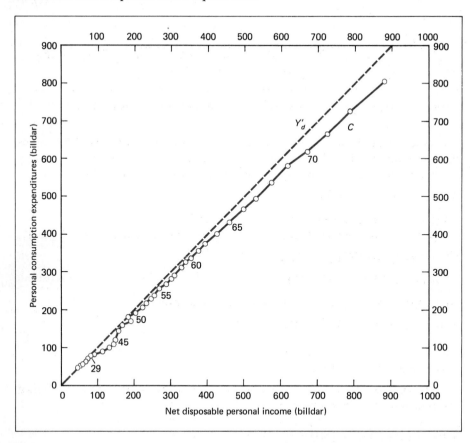

Figure 8-1 Consumption function for U.S. economy (at current prices).

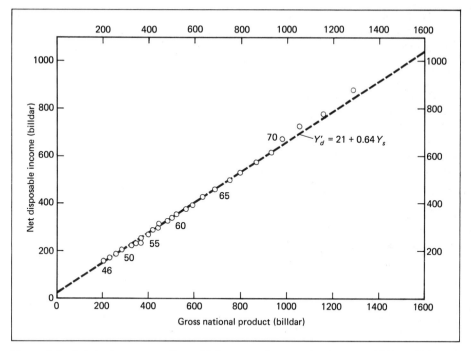

Figure 8-2 Relation between net disposable income and gross national product, 1946 to 1973.

Next we need the relation between net disposable income and GNP. The postwar data are plotted as a regression chart in Fig. 8-2. The relation is very close to linear, and the equation of a line through the points would be

$$Y_d' = e + f Y_s = 21 + 0.64 Y_s$$

Thus, in our model, e equals 21 and f equals 0.64.

We can combine these two equations to derive the relation between C^* and Y_s.

$$C^* = 0.93 Y_d' = 0.93(21 + 0.64 Y_s)$$

$$C^* = 19.5 + 0.595 Y_s = a' + b' Y_s$$

By comparison with previous equations we can identify

$$a' = a + be = 0 + 0.93(21) = 19.5$$

and

$$b' = bf = 0.93(0.64) = 0.595$$

We can now calculate the autonomous expenditure multiplier for the United States economy in the postwar years.

$$k = \frac{1}{1 - bf} = \frac{1}{1 - b'} = \frac{1}{1 - 0.595} = 2.47$$

Let us apply our equations to predict Y_s and δY_s for the years 1966 and 1967, when tax rates were constant and the consumption function stable. We assume that actual injections were equal to desired injections, that is, that there was no undesired inventory accumulation in those two years. This is probably a good assumption; despite the minirecession of 1967, the economy was operating at or above full employment during the two years.

Our solution equation becomes

$$Y = k(a' + \text{INJ*}) = 2.47(19.5 + \text{INJ*})$$

In the following table, historical values of INJ are used, and Y, Y_d', and C^* are calculated from the preceding equations.

Variable	Values in Billdar			
	1966		1967	
	Predicted	Reported	Predicted	Reported
G^*		156.8		180.1
I_d^*		121.4		116.6
X_n^*		5.3		5.2
INJ*		283.5		301.9
Y	748.5	749.9	793.8	793.9
C^*	464.9	466.3	491.8	492.1
Y_d'	500.0	498.9	528.0	532.4

For *changes* from 1966 to 1967 we have:

	Calculated	Reported
$\delta\text{INJ*} = 301.9 - 283.5 =$		18.4
$\delta Y = k(\delta\text{INJ*}) = 2.47(18.4) =$	45.4	44.0
$\delta C^* = b'(\delta Y) = 0.595(45.4) =$	27.0	25.8
$\delta Y_d' = f(\delta Y) = 0.64(45.4) =$	29.1	33.5

Injections rose from 1966 to 1967, since a strong increase in government purchases outweighed the decline in investment spending. GNP increased even more because consumer expenditures rose also. As a matter of fact, δC^* exceeded $\delta\text{INJ*}$. We would expect this on theoretical grounds if k is greater than 2.

Note that

$$\delta Y = k\delta\text{INJ*} = \delta C^* + \delta\text{INJ*}$$

Solving for δC^* in terms of $\delta\text{INJ*}$, we have

$$\delta C^* = (k - 1)\delta\text{INJ*}$$

So when k is greater than 2, we have δC^* greater than $\delta\text{INJ*}$, that is, the induced rise in consumer spending is larger than the initiating rise in autonomous expenditures.

The values predicted by our model are gratifyingly close to the reported values, indicating that the consumption function and allocation of GNY to Y_d' were stable for those two years and close to the postwar average relations represented by our lines.   For other years or pairs of years we could not expect such good results, because shifts in the consumption function would change the a and b parameters, or because changes in tax rates, transfer payments, or the profits share of income would alter the e and f parameters.

GRAPHICAL SOLUTION OF THE MODEL

The system of equations describing our simple model can be represented and solved very readily by a graphical approach.

1   Aggregate Demand-Supply Equilibrium

Equations (1) through (3) of our model permit determination of Y^* as a function of Y_s.

$$Y^* = C^* + \text{INJ}^*$$

or

$$Y_\alpha^* = (a' + b' Y_s) + \text{INJ}_\alpha^*$$

where $a' = a + be \simeq 20$ and $b' = bf \simeq 0.6$

This line can be plotted on a graph with Y_s plotted horizontally and Y^* vertically as in Fig. 8-3a.   Using rounded values for the constants determined for the United States economy postwar, we can plot the consumption function

$$C^* = 0.93 Y_d' = 20 + 0.60 Y_s$$

and then add vertically a distance representing INJ_α^*, here taken as 300 billdar. The resulting aggregate demand line

$$Y_\alpha^* = C^* + \text{INJ}_\alpha^* = 320 + 0.60 Y_s$$

is plotted in the top panel of Fig. 8-3.

To find the equilibrium value of GNP, we need to find the value of Y_s which generates aggregate demand Y^*, just equal to Y_s.   This situation may be discovered graphically by drawing a 45-degree line with ordinates Y_s and noting where it intersects the Y^* line.   The intersection falls at point α, for which $Y_\alpha = 800$ billdar, corresponding to the analytic solution.

$$Y_\alpha = k(a' + \text{INJ}_\alpha^*) = 2.5(20 + 300) = 800 \text{ billdar}$$

When $Y = 800$ billdar, we find $C^* = 20 + 0.6(800) = 500$ billdar, and this, added to $\text{INJ}^* = 3\,00$ billdar, makes $Y^* = 800$ billdar $= Y_s$.

If injections rise to $\text{INJ}_\beta^* = 350$ billdar, then the Y^* line is shifted vertically upward by 50 billdar to the position Y_β^* on the chart.   This aggregate demand line intersects the 45-degree line at point β, for which

$$Y_\beta = 925 \text{ billdar}$$

This is what we would expect analytically, since

$$Y_\beta = k(a' + INJ_\beta^*) = 2.5(20 + 350) = 925 \text{ billdar}$$

Note also that

$$\delta Y = k\delta INJ^* = 2.5(50) = 125 \text{ billdar}$$

as we would expect. Graphically, a vertical shift of the Y^* line has moved the intersection (equilibrium) point out along the 45-degree line far enough so that the increment (δY) exceeds the vertical shift (δINJ^*) by the amount of the induced rise δC^*.†

2. Equilibrium of Desired Withdrawals and Injections

The bottom section of Fig. 8-3 shows the graphical determination of the solution using the equality of desired injections and withdrawals as the equilibrium condition. (The analytical solution was described in the section "An Alternative Solution.") Withdrawals are a function of Y_s, since personal saving, net taxes, and gross retained earnings will all rise as GNY increases.

$$WDL^* = Y_s - C^* = -(a + be) + (1 - bf)Y_s$$

or

$$WDL^* = -a' + (1 - b')Y_s$$

For our numerical example, $a' = 20$ and $b' = 0.6$, so that

$$WDL^* = -20 + (1 - 0.6)Y_s = -20 + 0.4Y_s$$

This line is plotted in the bottom panel of Fig. 8-3. Since INJ* are the same for all values of Y_s, we plot INJ_α^* and INJ_β^* as horizontal lines at heights 300 and 350 billdar on the graph. The equilibrium levels of Y_α and Y_β are determined by the intersection of the fixed WDL* line with the two INJ* lines at points α and β respectively. As before, $Y_\alpha = 800$ billdar and $Y_\beta = 925$ billdar. The right triangle under the hypotenuse $\alpha\beta$ gives a good visual appreciation of the shift in equilibrium values. Its vertical side is δINJ^* and its horizontal side is δY_s. So we have $\delta INJ^*/\delta Y_s = 1/k$ is the slope of the WDL* line with respect to the horizontal.‡ It is clear that if k increases (b' increases), the WDL* line has lower slope and δY_s will be greater for any given δINJ^*.

The correspondence between this solution and the aggregate demand-supply solution in the top panel of Fig. 8-3 may be appreciated by noting that WDL* is the vertical distance from the C^* line up to the dashed Y_s line, while INJ* is the constant vertical distance from the C^* line up to the Y^* line. So the WDL* and INJ* lines in the bottom section are simply graphs of the Y_s and Y^* lines in

---

δY would equal δINJ^* only if the Y^* line were horizontal, implying that b' equals 0 and multiplier k equals 1. If the slope of the Y^* line increased nearly to 1, so that the Y^* line became nearly parallel to the 45-degree line, then a small upward shift of the Y^* line would lead to a huge rightward movement of the intersection point. This we might expect, since $b' \to 1$ implies $k \to \infty$ and $\delta Y = k\delta INJ^* \to \infty$.

‡ Algebraically, we find that $k = 1/(1 - b')$ yields $(1 - b') = 1/k$ for the slope of the WDL* line.

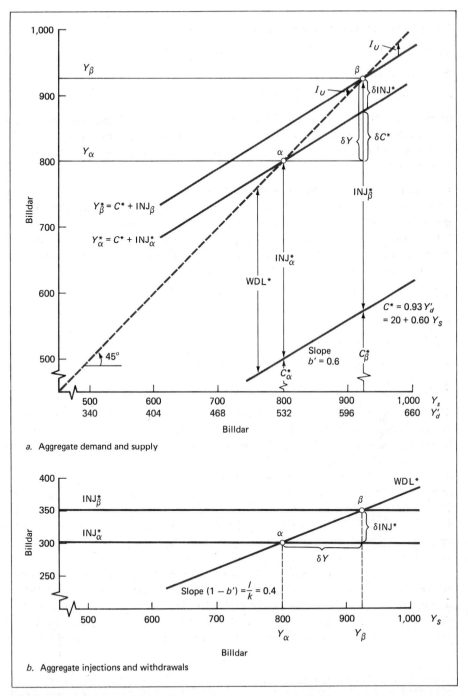

Figure 8-3   Macroeconomic equilibrium in a five-sector economy.   (a) Aggregate demand and supply.   (b) Aggregate injections and withdrawals.

the top section *as measured from the* C* *line as a base*. Obviously then, the two lines in the bottom section must intersect at the same value of Y_s as that at which the Y_s and Y^* lines intersect in the top section; that is, the equilibrium points α and β are aligned vertically in the two panels of Fig. 8-3.

STABILITY OF EQUILIBRIUM

What about the stability of the equilibrium, say at β? If producers underestimated demand and set production schedules to yield a flow of output (and hence incomes) less than 925 billdar, would the reduction of incomes lead to a reduction of aggregate demand and hence to a cumulative downward movement of the economy? Not so. Suppose the Production Units Sector decided to generate output, and incomes, at a rate of 900 billdar. Assuming that the aggregate demand line stayed fixed we would have

$$Y^* = C^* + INJ_\beta^* = 20 + 0.60(900) + 350 = 910 \text{ billdar}$$

Since Y^* would exceed output, undesired inventory reductions would occur.

$$I_u = Y_s - Y^* = 900 - 910 = -10 \text{ billdar}$$

See arrow for I_u on the chart. The undesired decline in inventories would lead producers to revise their demand forecasts upward and increase output in subsequent periods until the economy had moved back up to equilibrium at β.

What if producers increase output above the equilibrium level? Aggregate demand will rise by less than Y_s, since only consumer spending will increase and that by a fraction b' of the increase in Y_s. Thus, undesired inventory accumulation would occur.† In subsequent periods, producers would reduce output until the equilibrium point was reached, at which Y^* equals Y_s and I_u equals 0.

Of course, this analysis assumes that the aggregate demand line remains fixed. If it shifts down when producers lower output, or vice versa, we would have the possibility of dynamic, cumulative movements of the economy into a recession or a boom condition. But if the Y^* line stays fixed, the equilibrium point is stable against erratic fluctuations of supply.

THE PARADOX OF THRIFT

An interesting example of the interaction of consumer behavior with the rest of the economy is seen when the Personal Sector decides to make an autonomous increase in personal saving. It might be reasoned that the increase in desired saving will lead to an increase in investment (since $S = I$) and hence will stimulate the economy. In the words of an old musical hit: "It ain't necessarily so."

---

† Note that
$$\Delta Y^* = \Delta C^* = b' \Delta Y_s = 0.6 \Delta Y_s$$
Thus $1 - b' = 0.4$ of the added output would be undesired accumulation of inventories.
$$\Delta I_u = \Delta Y_s - \Delta Y^* = \Delta Y_s - 0.6 \Delta Y_s = 0.4 \Delta Y_s$$

An autonomous upshift of the S_p^* line, hence the WDL* line, implies a down-shift of the C^* line.   (Analytically, the value of a' has decreased.)   If INJ* remain constant, then the aggregate demand line Y^* shifts down also.   In the graphs of Fig. 8-3, we can see that a downshift of the Y^* line or a rise in the WDL* line will result in a *lower* equilibrium GNP.   Under our assumption of constant injec-tions, the autonomous decline in C^* (implied by the rise in S_p^*) leads producers to reduce output, and the consequent decline in incomes causes a further induced decline in C^* until a new equilibrium is reached.†

From another viewpoint, we may reason that with INJ* unchanged, the in-itial and final levels of WDL* must be the same.   So the initial *autonomous* rise in WDL* must be offset by an equal decline in *induced* WDL*, caused by a decline in incomes.

The increased propensity to save has just led to a decline in incomes sufficient to bring desired saving back down to the fixed level of investment spending, or withdrawals back down to the fixed level of injections.

Actually, the situation may be worse than this because the decline in consumer spending may lead producers to lower their desired investment for inventories or for plant and equipment.   This would make the equilibrium level of income lower still.

$$\delta Y = k(\delta a' + \delta \text{INJ}^*)$$

with both $\delta a'$ and δINJ^* negative.   There is one possible offset.   The increased saving flows might lower interest rates enough to give some support to consumer or investment spending.   More on this in later chapters.

DYNAMICS OF THE CHANGE TO A NEW EQUILIBRIUM
The derivation of equilibrium positions described above gave no indication of the intermediate steps by which an economy moves from an initial to a final equilib-rium position.   To study this process, we must calculate how each of the vari-ables describing the economic system changes from period to period.   This calcula-tion involves specifying the time lags with which the sectors respond to changed flows of orders, incomes, or other inputs to their decision making.

THE MODEL
For concreteness, let us assume that we are still in our simple five-sector economy and that:

---

† Analytically,
$$Y_\alpha = k(a_\alpha' + \text{INJ}_\alpha^*)$$
$$Y_\beta = k(a_\beta' + \text{INJ}_\alpha^*)$$
and
$$\delta Y = Y_\beta - Y_\alpha = k(a_\beta' - a_\alpha') = k(\delta a')$$
With $\delta a'$ negative, δY will be negative by k times as much.

1 Our time periods are quarter-years.

2 Consumers adjust expenditures quickly, so that desired and actual values of consumer spending are equal in each quarter.

3 Producers adjust output with a one-quarter lag, so that the flow of orders received (demand) in quarter t becomes the output flow in quarter $t + 1$. (This is equivalent to saying that production is set equal to the sales forecast for the period $t + 1$, which is taken equal to the actual demand of quarter t.)

4 Producers ship all the goods that are ordered in any period, absorbing the discrepancy between shipments and production by permitting inventories to change.

5 Producers let inventories change without any attempt to restore them to a desired level.

6 The structural constants of the system are fixed, that is, the values of a, b, e, and f do not change.

Let us analyze the change between the equilibrium positions described in Fig. 8-3. Initially, in period $t = 0$, the economy is in flow equilibrium, with INJ_α^* equaling 300 billdar, C_α^* equaling 500 billdar, and Y_α equaling 800 billdar. Desired injections rise to INJ_β^* of 350 billdar in period 1 and remain at that level thereafter. How do other variables in the system change through time as the economy moves to the final equilibrium with $Y_\beta = 925$ billdar?

GRAPHICAL ANALYSIS

First, let us analyze and visualize the adjustment on demand-supply and injections-withdrawal diagrams. Figure 8-4a presents the demand-supply analysis. Initially ($t = 0$), the economy is in flow equilibrium at point α, where the initial aggregate demand line (Y_α^*) intersects the 45-degree line (Y_s).

In period 1 the 50-billdar rise in injections lifts the aggregate demand line to its new position (Y_β^*). Since producers keep output in period 1 equal to demand in the preceding period, $Y_{s,1}$ equals 800 billdar. Given that income, the demand sectors order Y_1^* or 850 billdar of goods and services. Hence, 50 billdar of shipments must come from inventory. See the arrow on the chart for $I_{u,1} = Y_{s,1} - Y_1^* = 800 - 850 = -50$ billdar. The actual flow of current output and incomes in each quarter is at the circled point on the Y_s line.

In period 2, producers increase output to 850 billdar, the flow of orders received in the preceding period.

$$Y_{s,2} = Y_1^* = 850 \text{ billdar}$$

However, when incomes from production rise to 850 billdar, aggregate demand rises to 880 billdar because of the induced rise in consumer expenditures, $\Delta C_2^* = 0.6 \Delta Y_{s,2} = 30$ billdar. This rise in C^* was not forecast by producers and must be shipped from inventories.

$$I_{u,2} = Y_{s,2} - Y_2^* = 850 - 880 = -30 \text{ billdar}$$

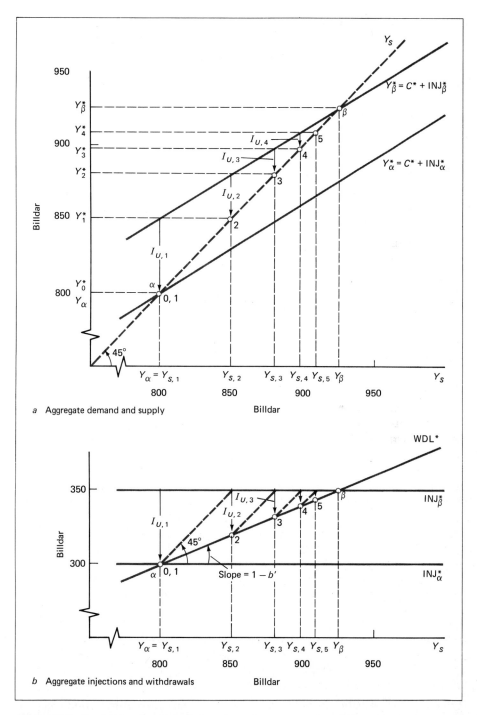

Figure 8-4 Graphs for dynamic adjustments to δINJ^*. (*a*) Aggregate demand and supply. (*b*) Aggregate injections and withdrawals.

The economy moves to point 2 on the Y_s line, but the excess of orders over production (with the concomitant unwanted decline in inventories) leads producers to increase output further in period 3.

So it goes, quarter after quarter. The producers' decision rule for determining output always leaves them short of meeting demand, in the amount of the unexpected rise $\Delta C_t^* = b' \Delta Y_{s,t}$. You might think the producers would learn the multiplier theory from sad experience, but they never do, in this model. Their error does become smaller and smaller, however, as may be noted on the graph from the narrowing of the gap between the Y^* and Y_s lines, hence the shortening of the arrows for $I_{u,t}$. Theoretically, it would take an infinite number of periods to reach the new equilibrium at β; practically, 99 percent of the adjustment is completed after 10 periods.

The bottom section of Fig. 8-4 pictures the adjustment on an injections-withdrawal diagram. The increase of INJ* in period 1 is met out of inventories $(I_{u,1})$. Producers respond by increasing Y_s in period 2 by

$$\Delta Y_{s,2} = -I_{u,1} = 50 \text{ billdar}$$

Graphically, we determine $\Delta Y_{s,2}$ by drawing a 45-degree line upward to the right from point 1 till it intersects the INJ_β^* line; the projection of this line segment on the horizontal axis is $\Delta Y_{s,2}$. The economy moves to point 2 on the WDL* line in period 2. As it does so, WDL* increases by the vertical distance from point 1 to point 2. The remainder of the vertical rise (point 2 up to the INJ_β^* line) is the induced rise in consumer spending since

$$\Delta Y_{s,t} = \Delta \text{WDL}_t^* + \Delta C_t^*$$

The increment $\Delta C_2^* = 0.6 \Delta Y_{s,2} = 30$ billdar, and it is drawn from inventories.

$$I_{u,2} = -\Delta C_2^* = -30 \text{ billdar}$$

Output is increased in period 3 by an amount $\Delta Y_{s,3} = -I_{u,2}$, determined graphically by a 45-degree line drawn northeast from point 2 to the INJ_β^* line. So the economy progresses upward by decreasing steps to the new equilibrium at β.

EQUATION SYSTEM—PERIOD ANALYSIS

The dynamic adjustment process may also be analyzed by using the system of equations for the model and solving one period at a time—a method called period analysis. The equations used are listed at the top of Table 8-1 and arranged in the order in which they enter the process of solution in each period.

The first equation describes the behavior of decision makers in the Production Units Sector; they set output in period t equal to orders received (aggregate demand) of the preceding period. Note that time subscripts must now be used on variables in the equations since the system is not in equilibrium in each period. This first equation replaces the equilibrium equation, since $Y_{s,t}$ is not equal to Y_t^* in the same period. Equations (2) through (4) specify the determination of aggregate demand. These first four equations would be adequate to yield a period-by-period solution for the adjustment of the system since they contain only four endogenous variables.

TABLE 8-1 Dynamic Adjustment to δINJ^* in a Simple Five-Sector Economy

Eq. (1): $Y_{s,t} = Y_{t-1}^*$

Eq. (2): $C_t^* = 20 + 0.60\ Y_{s,t}$

Eq. (3): $INJ_t^* = \begin{cases} 300 \text{ for } t \leq 0 \\ 350 \text{ for } t \geq 1 \end{cases}$

Eq. (4): $Y_t^* = C_t^* + INJ_t^*$

Eq. (5): $I_{u,t} = Y_{s,t} - Y_t^*$

Eq. (6): $INJ_t = INJ_t^* + I_{u,t}$

Eq. (7): $WDL_t^* = Y_{s,t} - C_t^*$

Time Period	(1) $Y_{s,t}$	(2) C_t^*	(3) INJ_t^*	(4) Y_t^*	(5) $I_{u,t}$	(6) INJ_t	(7) WDL_t^*
				billdar			
0	800	500	300	800	0	300	300
1	800	500	350	850	− 50	300	300
2	850	530	350	880	− 30	320	320
3	880	548	350	898	− 18	332	332
4	898	558.8	350	908.8	− 10.8	339.2	339.2
5	908.8	565.3	350	915.3	− 6.5	343.5	343.5
6	915.3	569.2	350	919.2	− 3.9	346.1	346.1
7	919.2	571.5	350 ·	921.5	− 2.3	347.7	347.7
8	921.5	572.9	350	922.9	− 1.4	348.6	348.6
9	922.9	573.7	350	923.7	− 0.8	349.2	349.2
10	923.7	574.2	350	924.2	− 0.5	349.5	349.5
Equilibrium	925	575	350	925	0	350	350

Equation (5) permits calculation of the degree of disequilibrium in each period as measured by the gap between supply and demand, which equals undesired inventory investment—in this case *dis*investment, since it is negative.

Equation (6) defines actual injections (INJ_t), which differ from desired injections by $I_{u,t}$. Equation (7) permits calculation of desired withdrawals. Since desired and actual withdrawals are always equal, according to the assumption of quick response by the Personal Sector, WDL_t^* must equal INJ_t from equation (6).

The columns of data in Table 8-1 show how the equations may be used to derive the complete dynamic solution numerically. In period 0 we have initial equilibrium values. In period 1, $Y_{s,1}$ and C_1^* are unchanged but the rise in injections leads to disinvestment in inventories, keeping actual injections unchanged along with WDL_1^*. The rise in $Y_{s,2}$ to equal Y_1^* leads to an induced rise in C_2^* and hence a further rise in Y_2^*. Inventory disinvestment continues, but at a smaller rate. Producers increase output in period 3 by the magnitude of $I_{u,2}$, and so it goes in decreasing increments toward equilibrium. By period 10, variables are quite close to their equilibrium values, shown on the bottom line of the table.

Although the system of seven dynamic equations in Table 8-1 seems to have only seven endogenous variables on their left sides, the system cannot be solved to determine one single value for each endogenous variable. Rather, as the period analysis shows, we obtain a time series of values for each variable. *Implicitly,*

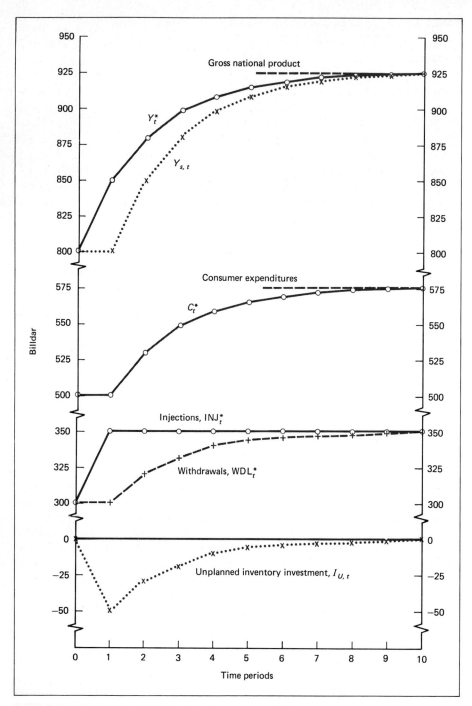

Figure 8-5   Adjustment to rise in injections, five-sector model.

the dynamic equations bring in time as an added variable, and the equations can be "solved" only to the extent of determining each of the endogenous variables as a function of time.

In Fig. 8-5, the solutions calculated in the preceding table have been plotted against time to display the dynamic adjustment as a set of time sequences for the system variables. The top panel shows GNP desired (Y_t^*) and actual ($Y_{s,t}$). Output lags behind orders by one period, and the difference between the two variables in each quarter equals $I_{u,t}$ (plotted in the bottom panel). Consumer expenditures (C_t^*) change in step with $Y_{s,t}$ because it determines $Y_{d,t}$. INJ$_t^*$ rise by a single step in period 1, but WDL$_t^*$ rises asymptotically to equality with the higher level of desired injections, keeping in step with rises in $Y_{s,t}$. Noting that desired withdrawals equal actual withdrawals and actual injections, we find

$$\text{WDL}_t^* - \text{INJ}_t^* = \text{INJ}_t - \text{INJ}_t^* = I_{u,t}$$

New equilibrium values are closely approximated after 10 periods of adjustment. Theoretically, equilibrium is never reached; practically, some other exogenous change will occur before the economy has fully adjusted to the original δINJ^*.

APPENDIX Analytic Solution to Equation System

A simple system of dynamic equations such as that in Table 8-1 can be solved analytically. The equations for aggregate supply and demand become:

$$Y_{s,t} = Y_{t-1}^*$$

$$Y_t^* = C_t^* + \text{INJ}_t^*$$

$$= a' + b'Y_{s,t} + \text{INJ}_t^*$$

For $t \geq 1$ we have desired injections constant at the new level INJ$_\beta^*$. Substituting the second equation into the first, with change in timing, we have

$$Y_{s,t} = a' + b'Y_{s,t-1} + \text{INJ}_\beta^*$$

This is an example of a difference equation. Its solution is

$$Y_{s,t} = k(a' + \text{INJ}_\beta^*) - k(\delta\text{INJ}^*)(b')^{t-1} \qquad \text{for} \quad t \geq 1$$

where $k = 1/(1 - b')$ and $\delta\text{INJ}^* = \text{INJ}_\beta^* - \text{INJ}_\alpha^*$.

Identifying $k(a' + \text{INJ}_\beta^*) = Y_\beta$ and $k(\delta\text{INJ}^*) = \delta Y$, we can write the solution as

$$Y_{s,t} = Y_\beta - (\delta Y)(b')^{t-1} \qquad \text{for} \quad t > 1$$

For our numerical example above,

$$Y_{s,t} = 925 - 125(0.6)^{t-1} \qquad \text{for} \quad t > 1$$

This would generate the values of $Y_{s,t}$ in column 1 of Table 8-1, and all other variables could be calculated from this by their respective equations.

An interesting property of the solution is the declining annual increments of $Y_{s,t}$, whose sum must equal $\delta Y = Y_\beta - Y_\alpha = k(\delta \text{INJ}^*)$. Let us check this out. From the first difference equation, we can obtain annual increments $(\Delta Y_{s,t})$ as follows.

$$Y_{s,t} = a' + b' Y_{s,t-1} + \text{INJ}_\beta^*$$

$$Y_{s,t-1} = a' + b' Y_{s,t-2} + \text{INJ}_\beta^*$$

Subtracting the second equation from the first, we find

$$Y_{s,t} - Y_{s,t-1} = b'(Y_{s,t-1} - Y_{s,t-2})$$

or

$$\Delta Y_{s,t} = b' \, \Delta Y_{s,t-1}$$

This tells us that each annual increment is a fraction b' of the preceding year's increment. This occurs because increments in $Y_{s,t}$ after period 1 arise from previous-year rises in induced consumption, and these are a fraction b' of the increment in ΔY_s in period $t-1$. In tabular form we have:

Period (t)	$\Delta Y_{s,t}$ (algebraic)	$\Delta Y_{s,t}$ (numerical)
1	0	0
2	δINJ^*	50
3	$b'(\delta \text{INJ}^*)$	30
4	$(b')^2(\delta \text{INJ}^*)$	18
5	$(b')^3(\delta \text{INJ}^*)$	10.8
6	$(b')^4(\delta \text{INJ}^*)$	6.5
7	$(b')^5(\delta \text{INJ}^*)$	3.9
8	$(b')^6(\delta \text{INJ}^*)$	2.3
9	$(b')^7(\delta \text{INJ}^*)$	1.4
10	$(b')^8(\delta \text{INJ}^*)$	0.8

The initial rise is in injections; all subsequent increments are in consumer expenditures. The sum of this infinite geometric series is

$$\delta Y_s = \sum_{t=2}^{\infty} \Delta Y_{s,t} = \sum_{t=2}^{\infty} (\delta \text{INJ}^*)(b')^{t-2} = \frac{\delta \text{INJ}^*}{1 - b'}$$

$$= k(\delta \text{INJ}^*)$$

$$= 2.5 \,(50) = 125 \text{ billdar}$$

This is as it should be. The multiplier effects on C^* have just been stretched out through time because of the one-period lag in response of decision makers in the Production Units Sector.

PROBLEMS

1   Consider a five-sector economy such as that described early in this chapter. The income distribution is such that $Y_d' = 21 + 0.64 Y_s$.
Assume a consumption function $C^* = 0.94 Y_d'$, and $G_\alpha^* = 219.5$ billdar. Let $I^* = 140.0$ billdar.

   a   Write the system of demand-and-supply equations for this economy.

b Obtain an expression for Y^* as a function of Y_s.

c Solve for equilibrium values for Y, Y'_d, C^*, S^*_p, INJ*, and WDL*.

d Compare these with actual values for 1970 as reported: $Y = 977.1$, $Y'_d = 673.8$, $C = 617.6$, $S_p = 56.2$, INJ $= 359.5 =$ WDL. Suggest possible reasons for discrepancies.

e Plot graphs for this problem comparable to those in Fig. 8-3. Now assume that exogenous demands have been forecast: $G^*_\beta = 234.2$ billdar and $I^*_\beta = 153.5$ billdar for 1971. Assume the income distribution and consumption function as above.

f Solve for equilibrium values for Y, Y'_d, C^*, S^*_p, INJ*, and WDL* in 1971.

g Compare your calculated values with the reported values for 1971: $Y = 1,054.9$, $Y'_d = 727.6$, $C = 667.1$, $S_p = 60.5$, INJ $= 387.8 =$ WDL. Suggest possible reasons for the differences between calculated and observed values.

h What is the value of the exogenous expenditure multiplier in this model economy? Does k equal $1/(1 - b)$, where $b =$ MPC $= 0.94$? If not, explain why not.

2 Use numerical values for 1970 as given in the first part of Prob. 1.

a Derive an equation for WDL* as a function of Y_s. (Note that WDL* $= S^*_p + T_n +$ GRE and that $T_n +$ GRE $= Y_s - Y_d$. Alternatively, WDL* $= Y_s - C^*$.)

b Solve for the equilibrium value of Y by setting WDL* $=$ INJ*.

c From the equation in part **(b)**, calculate δY as a function of δINJ*.

d From the equations for Y_d as a function of Y_s and for C^* as a function of Y_d, derive the equation for δC^* as a function of δY_s, applying to shifts between equilibrium positions. Then find the relative sizes of an exogenous δINJ* and the resulting δC^* which is induced by the subsequent δY_s.

e How would you obtain the value of the multiplier if you know the equation for WDL* in part **(a)** above?

3 To illustrate the "paradox of thrift," start with income distribution and injections the same as those given at the begining of prob. 1, but assume the Personal Sector decides to save 10 billdar more at every level of Y_d.

a Solve for the new equilibrium values of Y, Y_d, C^*, S^*_p, INJ*, and WDL*.

b Is the multiplier useful in analyzing the change in equilibrium values from those in prob. 1? Explain.

c What happened to S^*_p and WDL* from the equilibrium in prob. 1 to that in this problem? Explain.

d In a real-world economy, do you think such a shift in saving behavior would raise investment in the new equilibrium? Explain.

Chapter 9

Demand for Fixed Investment and Inventories

DETERMINANTS OF INVESTMENT EXPENDITURES

In early 1969 when surveys of business firms' plans to invest in plant and equipment showed very high estimates for the year, an article in *The Morgan Guaranty Survey* attempted to evaluate these plans on the basis of underlying causal factors which are believed to influence investment spending. Also, the McGraw-Hill Economics Department surveyed United States businesses to learn the reasons for their strong expenditure plans. Some of the relevant factors brought out in these studies included the following:

1　In the long run, sales were expected to rise strongly, leading to need for increased productive capacity.
2　New facilities were needed to reduce costs
   a　By reducing unit labor requirements, especially for scarce, skilled manpower, and
   b　By installation of new machinery embodying improved technology to replace obsolete equipment.
3　Wage rates had risen relative to the prices of capital goods, making substitution of capital for labor profitable.
4　Capacity was needed for new products.
5　Prices of capital equipment and plant construction were rising and expected to continue increasing.
6　Recent profits had been higher than anticipated, which
   a　Raised estimates of the profitability of new investment, and
   b　Provided internal funds for spending on capital goods.
7　Increases in interest rates were expected, along with shortages of available external funds.
8　Expiration or reduction in the rate of surtax on profits was anticipated.
9　Prices of output were expected to continue rising.

All these factors were counted on to support capital expenditures. A few restraining forces were noted, however.

10   In the year ahead, aggregate demand was expected to weaken somewhat as the surtax of 1968 and tighter monetary policy slowed the economy.

11   High rates of capital investment in recent years would tend to reduce investment because
  a   The most lucrative projects had already been undertaken, and
  b   In many manufacturing industries, the capacity output from current plant and equipment might be adequate to meet expected demands some distance into the future, as evidenced by low operating rates.

12   Current high interest rates or unavailability of capital funds might cause cancellation or postponement of some investment projects.

On the whole, the positive factors clearly seemed to dominate in early 1969, and investment was, in fact, $11\frac{1}{2}$ percent greater in 1969 than in 1968.

A government survey in January–February, 1970, showed that businessmen were planning a further 10 percent rise in plant and equipment expenditures from 1969 to 1970.   However these plans proved too optimistic,   In the event, capital spending rose only $5\frac{1}{2}$ percent, presumably because restrictive fiscal and monetary policies led to a decline in real aggregate demand in 1970.   Also, interest rates rose to postwar highs, corporate profits after taxes fell by 12 percent for 1970, and operating rates fell far below capacity in many manufacturing industries.

As was the case with analysis of consumer spending and saving, it would seem that a very large number of factors are involved in the investment decision. Also, investment decisions have their payoffs well into the future, so that a large element of forecasting or expectation is involved.   Is any simplification possible, or any selection of a few dominant causal variables?

MICROECONOMIC THEORY OF CAPITAL INVESTMENT

Classical economic theory includes an analysis of investment decision making which implicitly takes into account the determinants listed here.   It makes the investment decision depend on a comparison of the marginal efficiency of investment, or rate of return, with the market rate of interest.   This theory has been carried over into the business world in comparisons between the rate of return and the cost of funds for a proposed capital-spending project, or between the present value of the returns from an investment and the initial outlays on the project.

Investment decisions are assumed made in order to maximize long-term profits, or rate of growth of wealth, for owners of the business.   Since the profits from a capital investment are spread over the lifetime of the capital good, the profit-maximization principle comes to involve estimation of the flow of revenues and costs many years ahead, perhaps fluctuating from year to year.   Some summary measure of the estimated future stream of profits is needed to permit evaluation of a given investment project or a comparison between alternative projects. One way of evaluating the project is to calculate the present value of the forecast stream of profits and to compare that with the current outlays required.   An alternative approach stems from the fact that each year's profit flow equals the increase in the firm's net worth (wealth) before dividend payments. So projects can be evaluated on the basis of the rate of growth of net worth which they are expected to yield. These alternative approaches lead to the same decision in most cases but may differ in unusual instances, as will be described below.

PRESENT VALUE OF NET RETURNS COMPARED WITH CURRENT OUTLAYS

Businessmen must make allowance for the fact that $1 in the present is worth more than $1 in the future, because a "present dollar," loaned now, will accumulate interest between now and a future date. At a time t years in the future, a present amount A_0, accumulated at a rate i percent per year compounded annually, would amount to

$$A_t = A_0(1 + i)^t$$

Conversely, the amount one would have to lend now to achieve wealth A_t at a time t years in the future would be

$$\mathrm{PV}_{A_t} = \frac{A_t}{(1 + i)^t} = A_0$$

That is, A_0 is the present value of the amount A_t payable t years in the future.

Now consider a proposed investment project which involves an increment to real capital stock (δK) for which the current outlay required (or cost) is $P_K\, \delta K$, where P_K is the price of the capital goods. Forecasts by the marketing and production departments indicate that undertaking this project will lead to future increments in the streams of revenue and of cash operating costs (excluding depreciation and cost of capital funds) for the firm which may be designated δR_t and δCC_t extending from $t = 1$ to $t = L$, the estimated lifetime of capital goods. Thus the incremental stream of net cash receipts associated with the project becomes, for the year t,

$$\delta NR_t = \delta R_t - \delta CC_t$$

We also assume that the scrap value of the capital good L years ahead is estimated as δSV_L.†

To evaluate the proposed project we pose the question: How much money would we need to lend now at interest rate i to obtain future payouts equal to this prospective stream of net revenues, and would that amount lent be greater or less than the outlays required for the proposed investment project? *If the required lending exceeds the outlay for capital goods which yield the same net returns, we should prefer the real investment to the financial investment, and vice versa.* The required lending would be the sum of the present values of the items in the future stream of net receipts.

---

† δ's refer to deviations of variables from their values if the investment project were not undertaken.

$$PV = \frac{\delta NR_1}{1+i} + \frac{\delta NR_2}{(1+i)^2} + \frac{\delta NR_3}{(1+i)^3} + \cdots + \frac{\delta SV_L'}{(1+i)^L} \tag{9-1}$$

$$= \sum_{t=1}^{L} \frac{\delta NR_t}{(1+i)^t} + \frac{\delta SV_L}{(1+i)^L}$$

In this calculation, the future receipts δNR_t include the return of principal for the amounts loaned now, as well as the interest paid on them. By analogy, the net revenue stream from the investment in capital goods should include the return of the original outlays and the interest return on them, that is, the cash operating costs subtracted from revenues should not include depreciation or interest charges on the investment outlays.

The real investment will increase the wealth of the business faster than the alternative financial investment having the same net revenues when $P_K \delta K$ is less than PV. The investment in capital goods should not be undertaken if $P_K \delta K$ is greater than PV. This comparison assumes equal risk or uncertainty in the estimate of future net revenues for the alternative investments.

If various capital spending projects were to be compared, one might rank them by the ratios of $PV/P_K \delta K$ and choose the projects with higher values of this ratio.

RATE OF RETURN AND COST OF CAPITAL

The second way of evaluating a capital expenditure against the alternative of lending the funds is to calculate the *rate of return* or *marginal efficiency of investment* (MEI) for the project. *This is the rate of interest which, applied to the initial capital outlays, would be sufficient to provide the net returns which are forecast for the project. If this rate of return is greater than the rate of interest which could be obtained by lending the funds, investment of the funds in real capital assets is to be preferred.*

The rate of return r is calculated by using it to discount the forecast stream of returns back to the present and by setting the sum of those present values equal to the current outlay on the investment project.

$$P_K \delta K = \frac{\delta NR_1}{1+r} + \frac{\delta NR_2}{(1+r)^2} + \frac{\delta NR_3}{(1+r)^3} + \cdots + \frac{\delta SV_L}{(1+r)^L}$$

$$= \sum_{t=1}^{L} \frac{\delta NR_t}{(1+r)^t} + \frac{\delta SV_L}{(1+r)^L} \tag{9-2}$$

The net returns for each period, the scrap value, and the initial investment outlays are assumed known in this equation, and r is to be determined. (Usually various r's are assumed, the sum of the present values of the stream of incremental net revenues is calculated for each r, and by interpolation the r is estimated for which the present value of the stream equals the current outlay.)

The value of r may be compared with the rate of interest (i) obtainable by lending funds. (The lending rate i is assumed the same as the interest rate at which funds could be borrowed.)   *If $r > i$*, the investment project should be undertaken, since it yields a higher rate of growth of wealth than would the alternative investment in financial claims (loans).   *If $r < i$*, the investment project should not be undertaken.   Sometimes the rates of return on projects are compared not to the market rate of interest (i) but to a "cutoff rate of return" set by management as the minimum acceptable rate.   If real investments were considered riskier than lending, the estimated r value for the project, or the incremental net revenue stream, could be lowered to allow for risk.

Alternative capital investment projects might be ranked by their rates of return, and the ones with higher r values might be chosen first.

CORRESPONDENCE BETWEEN THE TWO METHODS

For reasonably smooth future streams of net returns, the two approaches would lead to the same "go—no go" decision on a given project.   The logic is shown graphically in Fig. 9-1a where interest rates are plotted vertically and present value of streams of future net receipts are plotted horizontally.   The two investment projects δK_1 and δK_2 each involve current outlays of $P_K \delta K$.   For the first, the sum of its net stream of returns plus scrap value totals to OA.   This is the present value of that stream of net receipts when discounted at a zero interest rate.   As the value of i is increased, the present value of the stream is reduced, because the denominators become larger in Eq. (9-1).   So the present value curve for investment project 1 slopes upward to the left from point A.   At some interest rate, the present value of the future stream of net revenues becomes equal to $P_K \delta K$, and this interest rate by definition is the rate of return r_1 for that project.   According to our first criterion for evaluating an investment project, we would undertake the project at market rates of interest below r_1, because $PV_1 > P_{K1} \delta K_1$ at such levels of interest rates; conversely, we would reject the project when $i > r_1$ because $PV_1 < P_{K1} \delta K_1$.   The present value criterion leads to the same decisions as does

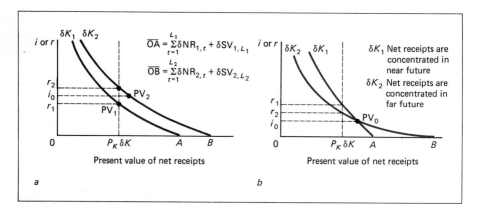

Figure 9-1   Present value and rates of return for investment projects.

the criterion "go" for $i < r_1$ and "no go" for $i > r_1$. Note also that we would not undertake a project for which $OA \le P_K \delta K$. The total net receipts from such projects would equal, or fall short of, initial outlays, and the rate of return would be zero or negative.

Now imagine that a second project δK_2 is to be compared with the first one. It involves the same current outlays but yields a larger total stream of net revenues OB. Its present value curve at each interest rate is greater than the present value for project 1, and its rate of return r_2 is greater than r_1. So project 2 would be preferred on either criterion at any market rate of interest. However, at interest rates above r_2, neither project would be undertaken; at interest rates below r_1, both would be more profitable than lending. At an interest i_0 between r_1 and t_2, the present value of project 2 exceeds its current outlays so that it would be undertaken; but the present value of project 1 falls short of its current outlays so that it would not be carried through.

In some cases, the choice between two investment projects might change with the market rate of interest. Fig. 9-1b, the present value graphs for two such projects are plotted. Project 1 has a smaller total of net receipts than does project 2, but its net receipts are assumed concentrated in the near future and those for project 2 in the distant future. Consequently, an increase in the interest rate reduces the present value of the stream of net receipts from project 2 more rapidly than it does the present value of the stream for project 1. At values of i above i_0, the present value of project 1 exceeds that for project 2, so that it would be preferred. As drawn, r_1 is greater than r_2, but an increase in $P_K \delta K$ could have made them equal (at PV_0) or reverse the inequality. In such cases as this, the criterion involving comparison of current outlays to the present value of future net receipts at the current market rate of interest would be preferred to the use of comparative rates of return r_1 and r_2.

Whichever criterion is used for evaluating and comparing investment projects, we need only assume that, for various market rates of interest, it is possible for business managers to rank prospective projects and to decide which are more profitable than alternative real and financial investment opportunities.

THE INVESTMENT DEMAND CURVE AND THE SUPPLY OF FUNDS

Given the ranking of investment projects according to their profitability, we can imagine each producer constructing a schedule showing what projects would be undertaken at various levels of the market rate of interest. Such a schedule, which is called the *marginal-efficiency-of-investment schedule*, is illustrated in Fig. 9-2a. The height of each bar corresponds to the rate of return for a project and the width of each bar measures the current outlay for each project in the year for which plans are being made and financing is being arranged. At a market rate of interest i_0, projects 1 through 5 would be undertaken, involving investment expenditures in year t of $I_t^* = \sum_{j=1}^{5} P_{Kj} \delta K_j$. Project 6 would increase net worth

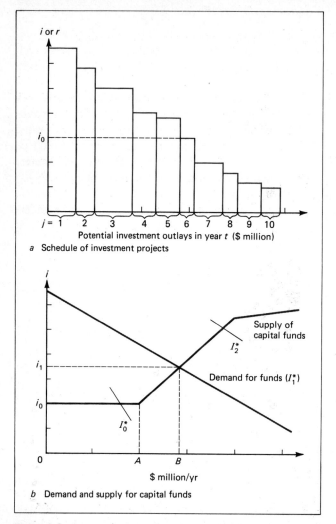

Figure 9-2 Investment demand and supply of capital funds
(a) Schedule of investment projects. (b) Demand and supply for
capital funds.

but only as fast as would the acquisition of financial claims by lending the funds
$P_{K6} \delta K_6$; or if the funds were borrowed for project 6, it would not do more than
cover the interest cost on the borrowed capital funds plus depreciation.

This schedule is shown as a continuous straight-line function in Fig. 9-2b;
investment spending increases as interest rate decreases because more projects
have rates of return in excess of the lower market rate of interest. The line may
be called the *investment demand curve* or the marginal-efficiency-of-investment
(MEI) curve. It may be thought of as:

1 The *desired dollar expenditures for capital goods* at various rates of interest, or

2 The *flow of capital funds demanded* for vesting into real goods.

Capital funds might come either from internal sources (gross retained earnings or cash flow) or from external sources (sale of debt instruments or new stock issues). Even internally generated funds should be counted as having a cost, since they could earn interest outside if they were not spent for capital goods by the firm itself. Sometimes the supply-of-funds curve is shown with three sections:

1 A horizontal section for internally generated funds, that is, depreciation allowances and undistributed profits;

2 A rising section indicating borrowing at rising interest rates; and

3 A more slowly rising segment showing the rising costs of obtaining capital funds by issuing new shares of stock.

The level i_0 applicable to internal funds may be taken as the market rate which could be earned if those funds were invested outside the firm in financial assets of comparable risk to the firm's real investment. It is the opportunity cost of using the funds within the firm. The rate of interest for borrowed funds may rise as the amount borrowed increases because of the increasing risk to lenders associated with increase in a firm's debt in relation to its net worth or its annual flow of profits. The risk borne by equity capital is greater still, and the rate of return paid for it is shown as highest.

Given the curves for demand for capital funds (investment demand schedule) and for the supply of funds, the equilibrium point at their intersection jointly determines the flow of funds for investment and the interest rate at which they are obtained. The rational, profit-maximizing manager would increase his investment in capital goods to OB, financing the expenditure by using his gross retained earnings OA and borrowing AB additional funds at interest rate i_1. For some firms the current outlays required for projects with rates of return in excess of i_0 may be less than the firm's cash flow OA, in which case (I_0^*), the supply of funds, will be entirely internal. For others the investment demand line may be shifted far to the right (I_2^*) so that considerable borrowing or even equity financing is profitable.

THE INVESTMENT DEMAND CURVE RELATED TO EMPIRICAL DETERMINANTS OF INVESTMENT EXPENDITURES

How does this theory, expressed in terms of present value of expected stream of net returns, rate of return (or marginal efficiency of investment), and the investment demand curve, correspond with the many factors affecting investment expenditures which were listed in the opening section of this chapter? Despite its seeming simplicity, the theory does incorporate the effects of most of those empirical factors.

Consider *long-run sales expectations.* If producers come to believe that future sales will increase, particularly if the expected level will be maintained in excess of current productive capacity, investment projects involving an expansion of current facilities will come to have present values of expected streams of net revenues in excess of current outlays for the projects, or rates of return in excess of current interest rate. The increase in projects with r greater than i will shift the investment demand curve rightward by adding new projects to the schedule, or upward by increasing the estimated rate of return on projects previously under consideration. The result is that investment spending will increase at any given market rate of interest. Conversely, of course, an expected decline in sales will shift the investment demand curve downward or to the left and reduce the number of projects which meet investment criteria.

Technological advances in production methods or *development of new products* will bring into existence new projects with high prospective rates of return, either from an addition to the stream of revenues (new products) or from a reduction in costs of production (new machines). Again the investment demand curve is shifted to the right.

Expected price changes may also shift the investment demand curve. If output prices are expected to rise and input prices remain relatively stable, the net revenue stream for some projects will rise enough to make them profitable at the current market rate of interest. An expected rise in input prices, with output prices relatively stable, would have opposite effects. The expected rise in costs would reduce the stream of net revenues, lower the rate of return, and tend to decrease investment expenditures. But if the rise were in labor wage rates, substitution of capital for labor will be stimulated, tending to increase investment expenditures.

If output prices and unit costs for labor and nonfactor charges are both expected to rise at a steady growth rate (g), net cash receipts will be inflated by the same multiplier. If we let subscript 0 denote constant-price measures (no inflation) and if r stands for the rate of return under inflationary conditions, then r is determined from

$$P_K \delta K = \sum_{t=1}^{L} \frac{\delta \text{NR}_{o,t}(1+g)^t}{(1+r)^t} + \frac{\text{SV}_{0,L}(1+g)^L}{(1+r)^L}$$

The comparable noninflationary equation is

$$P_K \delta K = \sum_{t=1}^{L} \frac{\text{NR}_{0,t}}{(1+r_0)^t} + \frac{\text{SV}_{0,L}}{(1+r_0)^L}$$

If $P_K \delta K$ has the same value in the two situations, then we can identify

$$1 + r_0 = \frac{1+r}{1+g}$$

or approximately

$$r = r_0 + g$$

The rate of return on investment projects is increased by the amount of the expected steady growth in prices.

However, at times when a steady inflation is expected, market rates of interest will tend to rise by the amount of the growth rate of prices also. If they do so, only the investment projects which were profitable in the noninflationary situation will be so under conditions of a steady expected rate of inflation. Actual desired investment in real units remains unchanged when r and i rise by equal amounts; the change in current dollar units would be proportional to the change in prices of capital goods (P_K).

Of course, in an inflationary period, prices of capital goods will probably rise also, increasing $P_K \, \delta K$ for each and every project. When P_K rises, the rate of return on a given project declines because its current outlay rises. The number of profitable projects will be reduced, at a given market rate of interest, but the effect on the aggregate dollar value of current outlays depends on the elasticity of real investment with respect to P_K for marginal projects in the neighborhood of the equilibrium point.

The retarding influence of a rise in P_K on real investment may be temporarily offset if prices of capital goods are expected to rise continuously. Then businesses may shift some investment to the present, at the expense of future investment, in order to purchase the capital goods at current prices. This shift would be weakened if prices of capital goods should rise faster than wage rates, so that labor would be substituted for capital.

Government actions may also influence the net returns on investment projects, and hence the position of the investment demand curve. Changes in tax rates on profits or in allowable depreciation schedules would change expected streams of net returns. An investment tax credit or a tax on investment spending would alter the effective price of capital goods and shift the investment demand schedule.

Finally, the *current stock of capital goods* resulting from earlier flows of investment expenditures affects the number of projects with high rates of return. Presumably each year the highest-yielding investment projects are undertaken. This will leave only projects with lower rates of return available in later years, at least in a *static economy* where the principle of declining marginal productivity of capital holds true as the stock of capital goods increases. In a *dynamic economy*, the number of potential investment projects with high yields is replenished by advances in productive technology, by new product opportunities, by growth in population, and by rises in standards of living. Any stability in the position of the investment demand curve over time must arise from offsetting effects of these countervailing forces.

The effects of the market variables just discussed can thus be brought into the framework of analysis underlying the MEI schedule. It is convenient to plot the investment demand schedule as a function of interest rate because i is compared with rates of return on projects in capital budgeting decisions and because the supply-of-funds curve can naturally be plotted against interest rate. The effects of other variables appear in the standard investment demand diagram as shifts in the level or slope of the investment demand curve plotted versus i. If, in fact, investment spending is not sensitive to interest rates or if interest rates do not change much from year to year, most of the variability in actual investment spending may be caused by other variables as they shift the investment demand curve and hence its intersection with the supply-of-funds schedule. Since many of these other variables involve expectations, which are highly uncertain and which may change in the same direction for many producers at the same time, it seems likely that shifts of the investment demand schedule may well be relatively large and frequent.

Some market forces will affect the level of investment through the *supply-of-funds schedule*. Changes in the level of business activity will alter cash flows of businesses and thus the amount of investment that can be financed from internal funds. A rise in internal funds will normally increase investment expenditures. Changes in personal saving flows would affect the position of the supply-of-funds curve also, with an increase in saving lowering the interest rate at which a given flow of loans would be supplied and perhaps also lowering the cost of equity funds. Finally, actions by the authorities in their use of monetary policy to control economic activity could lead to changes in overall levels of interest rates and in availability of capital funds; for example, the whole supply-of-funds schedule could shift up and the segment representing borrowing of funds become nearly vertical. This would curtail investment expenditures, especially by firms dependent on external financing.

INVESTMENT DEMAND IN THE NATIONAL PRODUCT ACCOUNTS

In the national product accounts, the four principal components of gross private domestic investment are: *nonresidential structures, producers' durable equipment, residential structures*, and *change in business inventories*. The first two correspond closely to what we usually think of as plant and equipment expenditures by business, but it does include also such expenditures by nonprofit organizations in the Personal Sector.

The first three components comprise "fixed investment" for the economy.

In nonrecession, nonwar years, total gross private domestic investment normally ranges between $14\frac{1}{2}$ percent and $16\frac{1}{4}$ percent of GNP. It stood at 15.7 percent in 1929, dropped to less than 2 percent in 1932, and recovered to a high of 14.4 percent in 1941. During World War II, private investment was held down to 6 percent or less of GNP. After the war, the ratio rose to 18 percent in 1948, and after the Korean war started, investment amounted to 19 percent and 18 percent of GNP in 1950 and 1951. The proportion dropped in postwar recession years to the $13\frac{1}{2}$ percent to 14 percent range, but in more normal years it has ranged between 14.5 percent and 16.2 percent.

In Fig. 9-3 the various components of investment are plotted as a percentage

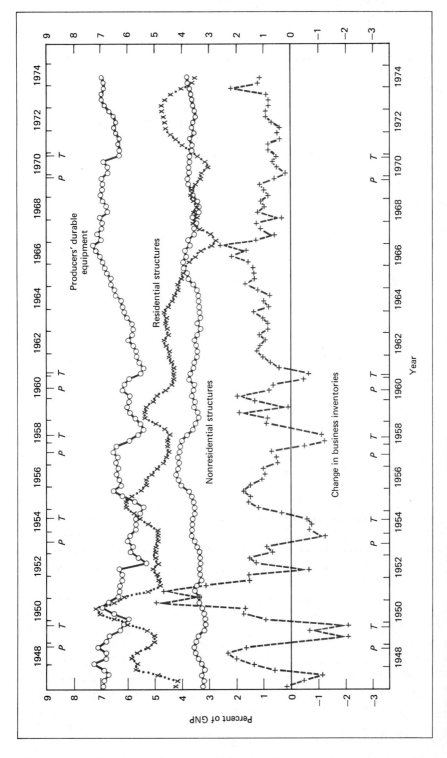

Figure 9-3 Components of gross investment as percent of GNP.

of GNP.   They display different time patterns which suggest that different causal
factors dominate the decision making in major areas of investment.   The chart
clearly indicates that inventory investment, though the smallest of the four components, often dominates the quarter-to-quarter movement in the total.   Except
for the unplanned decline in inventories during 1947 and perhaps one quarter of
1952, the period of negative inventory investment serves to identify the four mild
postwar recessions of 1949, 1954, 1958, and 1960–1961.   Indeed, the National
Bureau of Economic Research finds that inventory investment does reverse
direction somewhat earlier in time than the peaks and troughs of general business.

The National Bureau of Economic Research chronology for postwar business
cycles in the United States is summarized in the accompanying table.   The peak
months are indicated by P on the time scales of Fig. 9-3 and the troughs by T.

Trough	Peak	Months of Expansion	Months of Contraction
October 1945			
	November 1948	37	
October 1949			11
	July 1953	45	
August 1954			13
	July 1957	35	
April 1958			9
	May 1960	25	
February 1961			9
	November 1969	105	
November 1970			12

The two solid lines with circles are the equipment and the plant portion of
nonresidential fixed investment.   The top line, for producers' durable equipment,
shows that machinery buying drops off during recessions and recovers fairly quickly,
but after the business cycle hits the trough.   It is classed as a lagging series with
respect to business-cycle turning points.   (See 1950, 1955, 1958, 1962.)   Spending
was higher during the reequipment boom in the early postwar years and rose again
to about 7 percent of GNP during the full-employment years from 1966 to 1969,
marked by the Vietnam war.   Investment in nonresidential structures (plant) is
less volatile and shows longer lags at peaks and troughs of business cycles, as would
be expected for building projects requiring more time for planning and construction.

Despite the fact that plant and equipment expenditures lag behind fluctuations
in general business, investment decisions may still be an important causal factor
in generating fluctuations in overall activity.   In the first place, appropriations of
funds, orders placed for equipment, and construction contracts are series which
tend to lead at peaks and troughs; their effects may show up in the GNP accounts
first as inventory accumulation of raw materials and work in process while production is started on the investment projects.   In the second place, the data in Fig. 9-3
show that plant and equipment spending usually declines as a percentage of GNP

during periods when GNP declines and that it rises as a percent of GNP during business expansions.   So it would appear to be an active element, if not in causing, then in reinforcing and intensifying, the fluctuations in general business.

Investment in residential structures shows a different time pattern, tending to be inversely related to the plant and equipment component just discussed. Apart from the early postwar "catch-up" peak in 1948, residential construction tends to start rising during the contraction phase of business fluctuations, to reach a peak early in the expansion phase, and then to decline as the expansion continues.   It is, in general, countercyclical with a lag of two or three quarters. During the strong expansion of plant and equipment spending from 1965 on, the contrary movement between it and residential construction activity is clearly evident, including some recovery in housing activity when business investment weakened temporarily in 1967 and 1968.   The generally accepted explanation for this opposite movement in residential and nonresidential fixed investment is that they compete for capital funds and for construction labor and facilities.   Since plant and equipment projects are more profitable and are backed by firms with better access to capital markets, business investment tends to dominate the competition and squeeze residential construction during the latter part of an economic expansion.   In a recession when demand for funds and production inputs for nonresidential fixed investment slacks off, residential construction booms.

STATISTICAL ANALYSES OF THE DETERMINANTS OF INVESTMENT EXPENDITURES
The microeconomic theory discussed above suggests that higher plant and equipment expenditures will be associated with lower long-term interest rates.   We do not obtain this result from the historical data.   Long-term interest rates are roughly coincident with cycles in general business activity; so *high interest rates tend to coincide with high levels of plant and equipment spending at the end of business expansions*, and *low interest rates occur at the end of business contractions along with low capital spending by producers.*

NONRESIDENTIAL FIXED INVESTMENT
In Fig. 9-4, nonresidential fixed investment, expressed as a percentage of GNP to remove the scale effect in a growing economy, is related to a long-term interest rate (Moody's Baa bond yield).   We do not get the standard form of the investment demand curve.   Points for the first postwar decade (1947 through 1956) show large fluctuations of business fixed investment between 9.1 and 10.5 percent of GNP within a narrow band of interest rate (3.2 to 3.9 percent per year).   For the period 1957 through 1965, plant and equipment spending varied from 9.0 to 10.5 percent of GNP again, but within a higher narrow band of interest rate (4.7 to 5.2 percent per year).   During the prosperous years from 1966 to 1969, interest rates rose steadily and plant and equipment spending ranged from 10.3 to 10.9

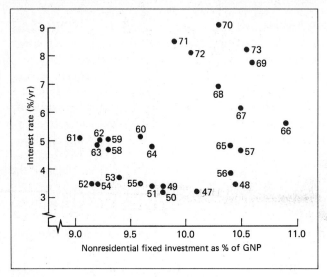

Figure 9-4   Relation between nonresidential fixed investment and long-term interest rate (Moody's Baa).

percent of GNP.   If we consider the point for each year to be at the intersection of aggregate demand and supply curves for capital funds, as in Fig. 9-2, it is clear that the two curves must have shifted left to right and right to left many times, and that since 1965, they seem to have shifted upward from their level in the first two postwar decades.

On the basis of this and other evidence, the relevance of the classical demand curve for investment as a function of interest rate has been questioned.   It must be granted that other causal variables are more important than interest rate at times.   Modern econometric analyses bring in such variables as orders or expected sales in relation to current capacity, technological advances and new products, recent cash flow and profit expectations, prices and expected prices of capital goods, of output, and of labor, investment tax credit, and pollution-control requirements. Also, recent models have made allowance for the time lags in the investment process—lags in awareness of investment opportunity, in preparation of proposals, in management reviews and appropriation of funds, in drawing of plans, in placement of orders and contracts, in production, delivery, and installation.   Time lags between initial decision and final installation may range from 1 or 2 up to 12 quarters.   So investment expenditures at period t should be correlated with values of decision variables as they existed one-half to three years earlier.

With these refinements in statistical techniques, it has been found that interest rates (lagged) do contribute significantly to the explanation of investment expenditures, especially for residential and commercial construction, but also for manufacturing plant and equipment. The small historical variations in interest rate accompanying large changes in investment may be accounted for by simultaneous shifts in demand for capital funds (investment) and in the supply of funds. As economic activity expands, internal funds increase as profits rise, and external sources expand as personal saving rises and government borrowing falls. In a recession, internal and external sources of funds are reduced, along with the demand for funds to finance investment.

RESIDENTIAL STRUCTURES

Residential construction expenditures are determined in the long run principally by the *growth in population and households*, in addition to demand for *replacement of dwelling units* that become uninhabitable or are destroyed by fire, urban reconstruction, natural disasters, and other events. Given the population and income factors, there will be a desired stock of dwelling units, and a flow of new construction will be induced to adjust the existing stock toward the desired level. In the short run, as was noted from the time series for components of gross private domestic investment, investment in residential structures competes with business fixed investment for available capital funds and production resources in the construction industry—and comes out second best. Thus, *availability and cost of long-term credit* are major determinants of housing activity; they affect the decisions of both the speculative builders of houses and the prospective purchasers of new homes. In addition to credit conditions, other short-run factors related to housing activity include construction costs relative to levels of rent and perhaps income levels as an influence on the average value of units constructed. Again the explanatory variables must be timed properly, for there appears to be an interval of some two or three quarters between changes in the causal variables and the resulting change in investment in residential structures.

CHANGE IN BUSINESS INVENTORIES

Inventory investment involves the short-term adjustment of stocks of raw materials, work in process, and finished goods toward desired levels. *The desired levels are related to expected levels of sales or production, and the investment flow (rate of change of stocks) would be related to the discrepancy between desired and actual inventories at the beginning of the decision period.* This would be the *planned* investment in inventories. As we noted in Chap. 5, *unplanned* inventory changes will also occur if forecasts of sales are incorrect. Since it takes time to obtain sales data, to determine whether any deviations from sales forecasts are likely to continue, and then to change production schedules or placement of orders so that the inflow of goods into stocks can be altered, there may be an appreciable time lag in planned inventory investment behind the changes in causal variables. The lags may vary among manufacturing and wholesale and retail trade, and may be

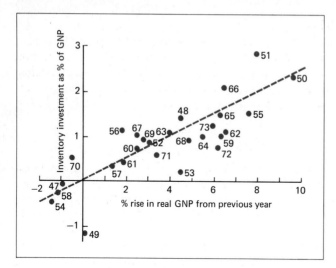

Figure 9-5 Inventory investment versus rise in GNP.

different for nondurables and for durable goods. Statistical research indicates that lags of up to two quarters are involved.

Over a year's time, inventory investment should reflect largely the planned adjustment of stocks in line with sales. Figure 9-5 shows, for annual sales data in the postwar period, the result of relating inventory investment as a percentage of GNP to the rate of rise of GNP from the previous year. A straight line through the origin would imply that *the rate of increase in stocks of goods is proportional to the rate of rise of real GNP*, or that inventories are linearly related to output.† The diagram shows that a relation of this type is reasonable, with a 4 percent per year rise in GNP being associated with inventory investment equal to 1 percent of GNP. This implies that the incremental stock to GNP ratio is 0.25 year.

Statistical investigations for industries, manufacturing and trade separately, and for nondurables and durable goods strongly support the stock-adjustment theory presented above. In some models for manufacturing inventory investment, a variable is added for *change in unfilled orders*, to account for accumulation of materials and work in process on orders not yet reflected in sales. In the non-durable goods area, a term for *price rise* has sometimes been found significant, representing a speculative or capital gains motive for investment in inventories. Interest rate does not appear to be an influential decision variable, presumably because of the short-run nature of the investment commitment and the over-riding importance of technological and marketing considerations in requiring that inventories be adjusted to production and sales levels.

---

† Let Q_R be the real quantity of inventory holdings. If $Q_R = \alpha Y_R$, where α equals a desired stock-to-sales ratio, then $dQ_R/dt = \alpha(dY_R/dt)$, or $I_{Q,R} = \alpha(dY_R/dt)$, where $I_{Q,R}$ is the net real investment in inventories. Division by Y_R yields $I_{Q,R}/Y_R = \alpha(dY_R/dt)/Y_R$, the equation of the dashed line in Fig. 9-5, where $\alpha = 0.25$.

INVESTMENT MODELS

This discussion should make it clear that no single investment demand function can be constructed which is as satisfactory and simple as the consumption function relating consumer expenditures to disposable personal income. Three principal investment-expenditure models are prominent in theoretical literature: *interest rate models, level-of-output models, and stock-adjustment or accelerator models.*

INTEREST RATE MODELS

Historically, the investment demand curve used most often in macroeconomic models was the MEI curve showing real investment spending increasing as interest rate decreased. Investment spending could be changed by fiscal or monetary policy which altered the supply-of-funds schedule, and hence the rate of interest. Business managers responded by changing aggregate investment spending *along* the stable demand curve.

It was recognized that the causal factors underlying the MEI schedule could change and shift the investment demand curve, but such shifts were considered exogenous to the model and were not analyzed. Moreover, the primary interest of early theorists was in describing the economy in a state of flow equilibrium and in explaining the shifts of equilibrium values of the variables in the economy resulting from exogenous shifts in some variable or parameter of the system. Thus, they did not bring dynamic elements into the model explicitly—population growth, technological change, stock-adjustment processes, formation and changes of expectations, or time lags in decision making and production. *In flow equilibrium, tomorrow is the same as yesterday, stocks are completely adjusted to flows, forecasts are perfect, and time lags are irrelevant.* There was some logical inconsistency here. The mere existence of investment flows, if they differ from pure replacement investment, will change the stock of capital goods and thus shift the investment demand schedule.

LEVEL-OF-OUTPUT MODELS

The second type of model recognizes the influence of the level of output on capital expenditures. High levels of economic activity normally lead to high levels of personal saving and of profits and cash flow, thus shifting the supply-of-funds curve to the right. Simultaneously, high GNP may lead to expectations of increased sales and output in the future. This will raise the estimated rate of return on investment projects, and shift the investment demand curve upward. Both these effects will increase the amount of investment spending as determined at the intersection of the curves for investment demand and supply of capital funds. Interest rates may remain constant or change somewhat, but they are not considered to have an appreciable influence on investment decisions in this model.

Such a model clearly has dynamic implications. An increase in output would raise investment spending, and the increase in investment spending plus the multiplier effects on consumer expenditures will increase aggregate demand and output further. Time lags should be built into such a model. The course of adjustment and the nature of equilibrium for level-of-output models are taken up in the next chapter.

STOCK-ADJUSTMENT OR ACCELERATOR MODELS

The preceding discussion regarding plant and equipment spending, residential construction, and inventory investment referred to the possibility that decision makers may wish to keep *stocks* of capital goods or of inventories in some desired relation to the expected *flow* of output or sales. Changes in either the desired or the actual level of stocks will lead to disequilibrium. Then the *decision makers are motivated to alter investment flows so as to bring the actual stock toward the desired level.* If they try to retire the discrepancy in stocks very quickly, very large changes in investment flows are required, larger in percentage than the initial change in desired stock. In the real world, inventory stocks will be adjusted more quickly than stocks of fixed capital goods, and equipment stocks more quickly than structures. Complicated " distributed lags " are needed to account for the timing in relations between causal variables and the resulting flow of investment expenditures.

Since desired investment flows are dependent on changes in desired stocks, and since the latter are proportional to changes in output or sales, we have

$$I^*_{Q,R} = \text{const} \times \frac{dY_R}{dt}$$

This dependence of investment flow on the rate of change of output flow led to the designation of this theory as an *accelerator* model.

APPENDIX Relation of Classical Marginal Productivity of Capital to Investment Criterion Based on Present Value

Another approach to capital-investment decisions which the classical economists developed was the analysis in which capital stock was changed, *with labor inputs held constant.* Marginal-revenue versus marginal-cost analysis was applied for profit maximization. It is assumed that the consequence of an increase in capital stock was an increased flow of output, continuing indefinitely into the future and sold at a constant price P_Y. For an increase δK, we have

Marginal revenue $= P_Y \delta Y_R = P_Y(\text{MPK})\delta K$

and

Marginal cost $= iP_k \delta K$

where $\text{MPK} = dY_R/dK$ is the marginal productivity of capital.

Capital stock should be increased so long as marginal revenue exceeds marginal cost, that is,

$P_Y(\text{MPK})\delta K > iP_K \delta K$

or

$$MPK > \frac{iP_k}{P_Y}$$

If we recognize the existence of nonfactor marginal costs equal to $\beta P_Y \delta Y_R$, the annual marginal net revenue attributable to the increase in capital stock becomes

$$(1 - \beta)P_Y \delta Y_R = (1 - \beta)P_Y(MPK)\delta K,$$

and capital stock should be increased so long as

$$(1 - \beta)P_Y(MPK)\delta K > iP_k \delta K$$

or

$$(1 - \beta)MPK > \frac{iP_K}{P_Y} \tag{9-2}$$

The equivalence between this solution and the present value criterion described in this chapter may be established if we identify the annual marginal net revenue from adding δK units of capital stock with the annual net revenue (δNR) in a constant stream of returns extending far into the future. Thus,

$$\delta NR_t = (1 - \beta)P_Y(MPK)\delta K$$

$$PV = \sum_{t=1}^{\infty} \frac{\delta NR_t}{(1 + i)^t} = (1 - \beta)P_Y(MPK)\frac{\delta K}{i}$$

According to the present value criterion, we should make the addition to capital stock if $PV > P_K \delta K$, which becomes equivalent to

$$(1 - \beta)P_Y(MPK)\frac{\delta K}{i} > P_K \delta K$$

$$(1 - \beta)(MPK) > \frac{iP_K}{P_Y} \tag{9-3}$$

This is the same test as that obtained from the marginal-productivity-of-capital approach in Eq. (9-2). So the various approaches lead to the same criterion in this simple case where the increase in net revenues arises from a constant annual stream of incremental product sold at constant prices.

It should be noted that the present value and rate-of-return approaches are more general than the marginal-productivity-of-capital approach. The former may be applied to investments which cut costs rather than increase output and revenue. They may also be applied to research or market development projects which do not involve increases in capital stocks. And they are applicable to cases in which the incremental stream of net revenues may fluctuate from period to period.

PROBLEMS

1 An investment project costing $P_K \delta K$ is expected to yield an incremental variable stream of net cash flow δNR_t in years $t = 1, 2, 3, \ldots, L$ where L is the life of the investment.

a **(1)** Explain how you would calculate the rate of return for such an investment.

   (2) Explain how you would calculate the present value of the future stream of net cash flow for this investment.

   (3) If present value of the stream of net cash flow equals present cost of the investment project at interest rate i, what must be the relation between interest rate i and rate of return r?

b Assume that this project increases annual output flow by $\delta Y_{R,S}$ taken as constant over an infinitely long life of the investment, and assume the output is sold at a constant price P_Y.

   (1) Write the equation for calculating r, using the above quantities.

   (2) Derive the relation between r and the marginal productivity of capital.

   (3) Discuss which of the two criteria, r or MPK, is more useful in choosing among alternative investment projects.

2 **a** List the variables which are inputs to the decision-making circle for the Business Capital Accounts as they decide on I_d^* and on how much to borrow (BOR_b^*).

b For each of five important decision variables, indicate whether an *increase* in that variable, with other variables assumed constant, would increase ($+$), decrease ($-$), or leave unchanged (0) the equilibrium flow of desired investment expenditures I_d^*.

c Where possible, indicate whether the increase of each causal variable in part **(b)** affects I_d^* by:

   (1) Shifting the I_d^* schedule

   (2) Shifting the schedule for supply of capital funds

   (3) Some other causal linkage (explain it)

d Sketch the marginal-efficiency-of-capital (or investment demand) curve and briefly explain the rationale underlying it. Label axes clearly.

e For the economy as a whole, describe the sources for the aggregate flows of funds into capital markets. Can they be identified, in whole or in part, with any flows reported in the national income and product accounts? Explain.

3 **a** Theoretical economic models often indicate that a flow of net investment since it increases the stock of capital goods, will shift the MEI curve (I_d^* schedule) progressively downward. Historical data show a rising flow of net investment along with an increasing capital stock. What assumptions underlying the theoretical model are not fulfilled in the historical situation? Can historical data be reconciled with the theoretical model? Explain.

b The investment demand curve shows I_d^* decreasing as interest rates rise, but historical data indicate some positive correlation between investment expenditures and market rates of interest. (See Fig. 9-4.) Explain this seeming discrepancy.

   c For movement of the economy through a complete business cycle, from trough to peak and then back to trough again, describe how you think the two curves of Fig. 9-2*b* would change. What would these movements of demand and supply for capital funds suggest regarding the cyclical patterns of fluctuation for interest rates and investment expenditures?

4 For each of the following theories of investment demand, indicate the important causal variable that theory emphasizes, and explain why business firms alter their investment decisions in response to that variable.

   a Marginal-efficiency-of-investment theory

   b Level-of-output investment theory

   c Accelerator principle

Chapter 10
Interaction of Investment Spending with the Rest of the Economy

INVESTMENT DECISION MAKING AND FEEDBACK EFFECTS

To the business executive, "investment" involves the exchange of financial assets for real assets. Liquid assets are "vested" in stocks of goods which yield services in future time periods. The purchaser of capital goods does not experience a decline of net worth as a result of the exchange, but places himself in a less liquid, more risky position with the expectation of earning a rate of return greater than that available from the acquisition of financial claims. Over time businesses "liquefy" investments by accumulating depreciation allowances for fixed capital goods, or by using up inventories of raw materials and goods in process during production, and by selling inventories of finished goods.

Investment activity interacts with other economic activities in several ways.

1 Since investment involves expenditure for current output, it affects aggregate demand in the economy, directly and via multiplier repercussions on consumer incomes and demand.

2 If current purchases of capital goods and inventories exceed the using up of stocks of those goods which were brought into the current period, then stocks of fixed capital goods and inventories are increased. This increase, if planned, will tend to bring actual stocks closer to desired levels and to reduce investment flows in succeeding periods.

3 On the other hand, the rise in aggregate demand accompanying increased investment expenditures may raise expectations of future sales and profits and lead to higher desired stocks of fixed capital goods and inventories. This will tend to stimulate investment flows further.

4 The demand for capital funds which accompanies investment spending will affect markets for capital funds and financial claims, and hence, interest rates.

5 The rise in aggregate incomes stemming from increased investment expenditures will tend to increase both the flow of internally generated funds (especially undistributed profits) and the flow of personal saving to capital-funds markets, hence to lower interest rates.

6 If investment spending is enough to bring the economy to full or overfull employment, rises in current and expected wage rates and in prices of other inputs will affect expected rates of return and hence shift the investment demand function.

Some of these channels of interdependence and feedback between investment decisions and the rest of the economy are pictured in Fig. 10-1. The circle for business investment decisions in the center of the diagram shows that inputs to these decisions include information on current stocks of inventories and fixed capital goods; current gross retained earnings; forecasts of sales, prices, and availability of inputs; new products, profits, and desired future stocks of fixed capital goods and inventories; current information and expectations regarding interest rates and availability of external funds. The investment decisions which emerge from these inputs control the flows of business borrowing (BOR_b^*) and of investment spending (I_d^*). These flows affect orders, production decisions, and gross national income payments, as well as capital-funds markets. Changes in these flows will alter flows and prices in markets for services of labor and property and in markets for capital funds and financial claims. These changes in turn affect personal decisions with respect to consumption and saving, which in turn affect later business investment decisions.

However, there will be time lags in the transmission of these influences around the feedback paths, so that current investment decisions will not simultaneously affect the input information to those decisions. If we use periods up to quarter-years in length, fixed capital investment expenditures are largely exogenous, in the sense that the variables governing investment levels in this quarter are the levels of the causal variables in previous quarters. Inventory investment may be affected by current-quarter flows of sales and orders.

These considerations indicate that *net investment expenditures are a dynamic element in the economy*. For a stationary economy in flow equilibrium, inventory investment would be zero and fixed capital investment would be limited to replacement flows. If gross investment exceeds capital consumption, then stocks of capital goods must be rising, and the MEI curve will tend to shift downward as stocks grow. Any flow equilibrium must be short term unless a growing population, technological advance, changes in fiscal or monetary policy, or other exogenous forces generate new investment projects with rates of return equal to those completed in prior time periods. Investment is a dynamic element also because it depends so strongly on forecasts of future demand-and-supply conditions in several markets and because of the time lag between investment decisions and their pay-offs. Expectations of economic conditions may change appreciably in one quarter-year, and actual market conditions may deviate significantly from those expected during the lifetime of an investment project. In the following sections, interaction between investment spending and the economy will be displayed, both as shifts between short-term flow equilibrium positions and as a dynamic adjustment process between equilibria.

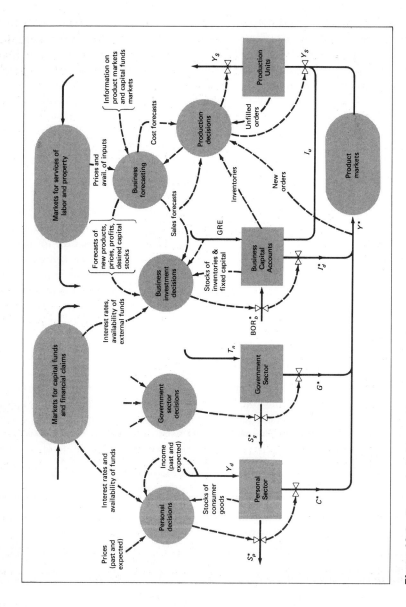

Figure 10-1 Economic decision making and feedback loops.

INTEREST RATE MODEL AND THE IS LINE

In this section, we deal with shifts between equilibrium states of the economy, so-called *comparative statics.*

In interest rate models, desired investment (I_d^*) *is taken to be a declining function of a long-term interest rate* i. If the function is linear, we have

$$I_d^* = g - hi \tag{10-1}$$

where g and h are positive constants.†

Let us see how this investment demand function fits into the equilibrium solution for the simple five-sector economy used in Chap 8. Now we change the earlier assumption that investment is exogenous and replace it with the dual assumptions that I_d^* depends on i and that i is exogenous. The equation system determining equilibrium is as follows.

			New Variables	
Type of Equation	Equation	Exogenous	Endogenous	
1 Definition	$Y^* = C^* + I_d^* + G^* + X_n^*$	G^*, X_n^*	Y^*, C^*, I_d^*	
2 Behavioral‡	$C^* = a' + b' Y_s$		Y_s	
3 Behavioral	$I_d^* = g - hi$	i		
4 Equilibrium condition	$Y^* = Y_s$			

‡ This equation is a combination of the consumption function $C^* = a + bY_d$ and the institutional relation $Y_d = e + f Y_s$, implying $a' = a + be$ and $b' = bf$.

This system of four equations has four endogenous variables; it should be solvable. Replacing endogenous variables in equation (1) by the right-hand sides of equations (2) through (4), and letting Y stand for either Y^* or Y_s in equilibrium, we obtain

$$Y = a' + b' Y + g - hi + G^* + X_n^*$$

The solution for Y is

$$Y = k(a' + g - hi + G^* + X_n^*) \tag{10-2}$$

where $k = 1/(1 - b')$ is our autonomous expenditure multiplier.

According to this solution equation, an increase in interest rate will reduce equilibrium GNP because it reduces investment expenditures. Let us visualize

† The true behavioral equation, allowing for a four-quarter lag of investment spending behind interest rate, would be

$$I_{d,t}^* = g - hi_{t-4}$$

This equation is used in the analysis of the dynamic disequilibrium process in Appendix A of this chapter.

When the economy is in equilibrium, interest rate is constant, so that $i_t = i_{t-4}$. Thus in comparative statics we may suppress time subscripts and use the behavioral equation (Eq. 10-1).

this by using the aggregate demand-supply and aggregate injections-withdrawals diagrams from Chap. 8. See Fig. 10-2.

In Fig. 10-2a, the aggregate demand line Y_α^* is drawn for a specified interest rate i_α. It is the graph of the equation

$$Y_\alpha^* = (a' + b'Y_s) + (g - hi_\alpha) + G^* + X_n^*$$

Aggregate demand may be thought of as the sum of an *autonomous* part Y_{AUT}^*, which is independent of Y_s, and an *induced* part Y_{IND}^*, which consists of terms involving Y_s. Thus, we may write

$$Y_\alpha^* = Y_{\text{AUT}}^* + Y_{\text{IND}}^* = (a' + g - hi + G^* + X_n^*) + b'Y_s$$

The autonomous part becomes the intercept on the vertical axis for the graph of Y^* versus Y_s, and the induced part is the component of Y^* that increases in proportion to Y_s, with slope b'. At point α the Y^* line intersects the 45-degree line Y_s, and we can determine the equilibrium value Y_α.

Now an increase of interest rate will reduce I_d^*, and hence Y_{AUT}^*. The aggregate demand line shifts down, and so must equilibrium GNP. In Fig. 10-2 it is assumed that interest rate rises from i_α to i_β and then to i_γ, reducing equilibrium output from Y_α to Y_β and then to Y_γ. If we know that full-employment GNP is Y_{FE}, it would be necessary to lower interest rate below i_α in order to raise I_d^* and the Y^* line to levels which would yield equilibrium at the point labeled FE.

Equivalent conclusions may be drawn from the diagram for aggregate injections and withdrawals in Fig. 10-2b. Increases in i lower the horizontal INJ* line and determine equilibrium points at progressively lower values of Y along the fixed WDL* line. (Desired withdrawals equal $Y_s - C^*$, and the consumption function is assumed fixed.)

A graph of the relation between interest rates and resulting equilibrium values of GNP would be a line with Y decreasing as i increases. Actually, it is just the graph of the solution equation [Eq. (10-2)], which is a straight line.

$$Y = k(a' + g - hi + G^* + X_n^*) = k(a' + g + G^* X + {}_n^*) - khi \qquad (10-2)$$

The intercept on the Y_s-axis will be $k(a' + g + G^* + X_n^*)$ and the slope with respect to the i-axis will be $-kh$.

This relation is interpreted graphically in Fig. 10-3. Figure 10-3a shows investment demand as a linear function of interest rate, with the constant term g determining the intercept on the horizontal axis and with $-h = \Delta I_d^*/\Delta i$ as the slope of the line with respect to the i-axis. The value of g accounts for the influence on the level of desired investment from all variables other than i, and changes in those other variables would shift the investment demand line horizontally by changing the value of g.

Figure 10-3b gives a graphical interpretation of the solution equation for this model. To obtain the quantity in parentheses in Eq. (10-2), we start with the I_d^* line from Fig. 10-3a and add horizontally the other components of autonomous expenditures, $(G^* + X_n^*)$ and a'. This gives us the line VH, whose horizontal

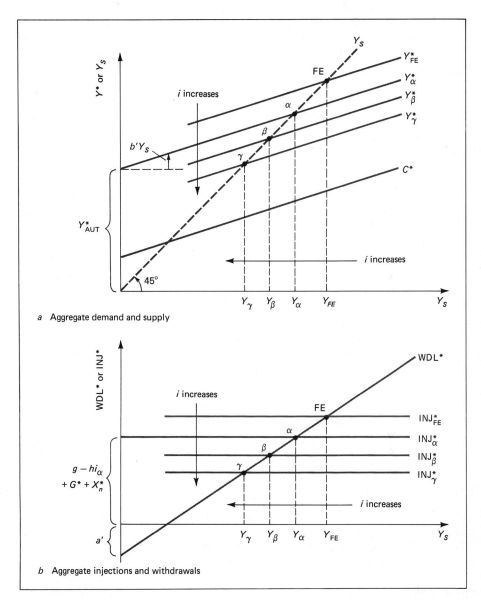

Figure 10-2   Effect of interest rate on equilibrium GNP.   (*a*) Aggregate demand and supply. (*b*) Aggregate injections and withdrawals.

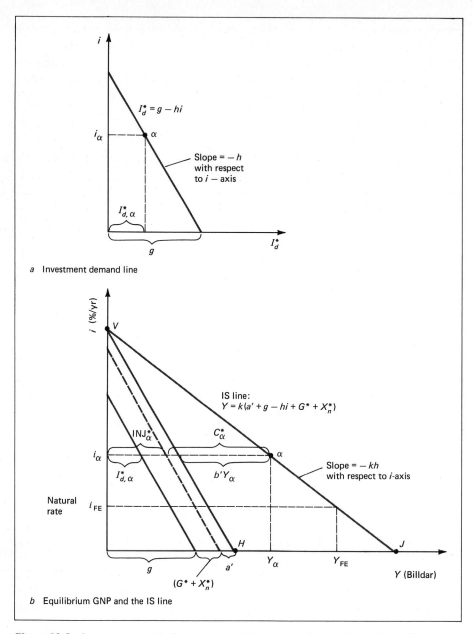

a Investment demand line

b Equilibrium GNP and the IS line

Figure 10-3 Interest rate model of investment. (*a*) Investment demand line. (*b*) Equilibrium GNP and the IS line.

distances display $(a' + g - hi + G^* + X_n^*) = Y_{AUT}^*$ as a function of i. To obtain Y, we need only multiply those horizontal distances by k, obtaining the line VJ. At any interest rate (i_α), the horizontal distance out to the line VJ equals the equilibrium value of gross national product (Y_α). Along the horizontal line to the equilibrium point α, we can identify segments corresponding to INJ_α^* and C_α^*, and to their components $I_{d,\alpha}^*$, $G^* + X_n^*$, a' and $b' Y_\alpha$, as labeled in Fig. 10-3b.†

The VJ line thus pictures the solution equation [Eq. (10-2)] expressing equilibrium values of Y as a function of i. In classical economics the equilibrium solution was derived by equating desired investment with desired saving; hence the name "IS line" (for investment-saving). Today, we might call it the "DS line," for equilibrium of aggregate demand and supply, or even the "IW line," for equality between desired injections and withdrawals. The important point to remember is that *the GNP values on the IS lines are equilibrium values corresponding to given levels of interest rates.*

If we mark off on the horizontal axis the particular distance Y_{FE} which corresponds to full-employment output, we can project upward to the IS line and across to the i-axis to find the interest rate i_{FE} which would bring the economy into equilibrium at full employment. This particular rate of interest has historically been called the *natural rate*, though "full-employment rate" would be more descriptive. For, given investment and consumer demand functions and given values of G^* and X_n^*, an interest rate above this "natural rate" would lead to an equilibrium below full-employment output; an interest rate below the equilibrium "natural rate" would lead to equilibrium only if output were above full employment, indicating that the economy would be in an inflationary situation. The important concept involved is that *achievement of full employment in this model requires an appropriate rate of interest.* Monetary policy contributes to this goal by influencing interest rates, and hence I_d^* and Y.

How sensitive will equilibrium GNP be to changes in interest rate? From the equilibrium equation [Eq. (10-2)], we can obtain immediately, for increments between equilibrium positions,

$$\delta Y = -kh(\delta i)$$

We might call the product $(-kh)$ the *interest rate multiplier.* It is the slope of the VJ line with respect to the i-axis.

The interest rate multiplier depends, first, on the (negative) sensitivity of investment demand to interest rate. A large value of h implies that a given decrease in i will lead to a large increase in I_d^*. Graphically, increased h implies a larger

---

† The horizontal segment between the VH and VJ lines is correctly labeled $b' Y_\alpha$ because:

The horizontal distance to the VH line is

$$(a' + g - hi + G^* + X_n^*) = \frac{1}{k} Y_\alpha$$

The horizontal distance to the VJ line is Y_α

$$Y_\alpha - \frac{1}{k} Y_\alpha = \left(1 - \frac{1}{k}\right) Y_\alpha = b' Y_\alpha$$

angle between the I_d^* line and the i-axis, hence between the VH line and the i-axis. Secondly, the interest rate multiplier depends on the responsiveness of consumer demand to an increase in GNP, since $k = 1/(1 - b')$ and $b' = \Delta C^*/\Delta Y$. An increase in strength of this feedback, for example by a decline in personal tax rates or a rise in marginal propensity to consume, would raise k. Graphically, larger k leads to a greater angle between the VH and VJ lines. These two influences may be summarized by noting that

$$\frac{\Delta Y}{\Delta i} = \left(\frac{\Delta Y}{\Delta I_d^*}\right)\left(\frac{\Delta I_d^*}{\Delta i}\right) = k(-h) = -kh$$

In the extreme situation when investment demand is not affected by i, then $h = 0$ and the I_d^* line would be vertical. So would the IS line. Then lower interest rates would not raise equilibrium Y by raising investment. Control of the economy would need to be exerted by other factors which affect the position of the IS line. The solution equation [Eq. (10-2)] or the graph of Fig. 10-3b makes clear what the possibilities are; namely, changes in g, G^*, or X_n^*, or in the parameters of the consumption function a' or b'. Increases in the first four of these items raise autonomous demand and increase equilibrium Y by a multiple k of the autonomous shift. When b' increases, induced consumer expenditures rise at each value of Y, which is equivalent to an increased value of k.†

As a numerical example, consider a model with the following equations.

$$Y^* = C^* + I_d^* + G^* + X_n^*$$

$$C^* = 20 + 0.60\,Y_s \text{ billdar}$$

$$I_d^* = 240 - 25i \text{ billdar}$$

$$G^* + X_n^* = 210 \text{ billdar}$$

$$i_\alpha = 5 \text{ percent per year}$$

The equilibrium solution becomes

$$Y = 2.5\,(20 + 240 - 25i + 210)$$

$$Y = 2.5\,(470 - 25i)$$

This is the *equation of the IS line*. From it we find that when i_α is 5 percent per year, $Y_\alpha = 2.5\,[470 - 25(5)] = 862.5$ billdar, with components

$$C^* = 537.5 \text{ billdar}$$

$$I_d^* = 115 \text{ billdar}$$

$$G^* + X^* = 210 \text{ billdar}$$

---

† Graphically, increased autonomous expenditures translate the VH line rightward by δY_{AUT}^* and move point V up and point J outward proportionately. An increase in b' leaves the point V unchanged but rotates the VJ line outward from the i-axis.

If interest rate were lowered to $i_\beta = 4$ percent per year, we calculate $Y_\beta =$ 2.5 $[470 - 25(4)] = 925$ billdar.

The increment to GNP between equilibrium positions is $\delta Y = 62.5$ billdar, as we might have expected from

$$\delta Y = -kh(\delta i) = -2.5\,(25)(-1) = 62.5 \text{ billdar}$$

The interest rate multiplier is 62.5 billdar per percentage point change in i.

In Appendix A to this chapter the solutions for the period-by-period adjustment of the variables in response to step-function changes in interest rates are calculated and graphed as a time series.

LEVEL-OF-OUTPUT MODELS

In level-of-output models *desired investment is positively related to actual levels of output or sales*, perhaps because of the influence of sales volume on cash flow, or on expected sales and rates of return. Assuming a linear relation, we write

$$I_d^* = g' + h' Y_s \tag{10-3}$$

The constant term g' reflects the influence of variables other than Y_s. The coefficient h' measures the sensitivity of investment demand to GNP.

In a dynamic analysis, we should need to allow for some time lag of I_d^* behind Y_s. Here we are adopting a comparative statics approach and use Eq. (10-3) without time subscripts in solving for variables when the economy is in a state of flow equilibrium. The relevant set of equations is as follows:

$$Y^* = C^* + I_d^* + G^* + X_n^*$$

$$C^* = a' + b' Y_s$$

$$I_d^* = g' + h' Y_s$$

$$Y^* = Y_s$$

The solution equation, obtained by substitutions into the first equation and by setting $Y = Y_s = Y^*$ in equilibrium, is

$$Y = a' + b'Y + g' + h'Y + G^* + X_n^*$$

Solving for Y, we obtain

$$Y = k'(a' + g' + G^* + X_n^*) = k' Y_{AUT}^*$$

where $k' = 1/(1 - b' - h')$, and Y_{AUT}^* is the part of Y^* that does not depend on Y_s.

In this model we have two feedback paths from Y_s to aggregate demand, one involving consumer expenditures induced by income changes and a new one involving investment spending induced by changes in GNP. As we might expect, the added feedback via I_d^* increases the value of the exogenous expenditure multiplier k'. The larger the induced investment coefficient h', the greater k' will be.

In fact, if h' is so large that $(h' + b')$ equals 1, then k' becomes indefinitely large.†
Mathematically, any step-function rise (or fall) in autonomous spending would
lead to an unending round of equal increases (or decreases) in induced consumption and investment, causing GNP to expand (or contract) without limit. In
actuality, the expansion (contraction) would be limited by shifts of the behavioral
functions, nonlinearities in the system, or by policy actions aimed at correcting
extreme cyclical fluctuations.

The graphical solution for equilibrium is shown in Fig. 10-4 for a model with

$C^* = 20 + 0.6\, Y_s$ billdar

$I_d^* = -70 + 0.2\, Y_s$ billdar

$G^* + X_n^* = 210$ billdar in state α and 230 billdar in state β

Equilibrium solutions, in billdar, are:

	α	β
Y	800	900
C^*	500	560
I_d^*	90	110
$G^* + X_n^*$	210	230

The multiplier in this model is

$$k' = \frac{1}{1 - b' - h'} = \frac{1}{1 - 0.6 - 0.2} = 5$$

Thus,

$$\delta Y = k'(\delta G^* + \delta X_n^*) = 5(20) = 100 \text{ billdar}$$

Note in Fig. 10-4a that the aggregate demand line Y^* is not parallel to the C^*
line, because I_d^* increases with Y_s. The slope of the Y^* line is $b' + h' = 0.8$. If
$b' + h' = 1$, then the Y^* line is parallel to the dashed 45-degree line for Y_s; and
changes in autonomous expenditures, which shift the Y^* line vertically, will cause
the intersection point to move indefinitely far along the Y_s line.

The sensitivity of I_d^* to Y_s shows up in Fig. 10-4b as a rising injections line,
with slope of $h' = 0.2$. If this slope increases to $h' = 1 - b' = 0.4$, the INJ* line
becomes parallel to the WDL* line, and we have the instability of the economy
noted previously. The solutions in panels (a) and (b) of Fig. 10-4 are, or course,
equivalent.

In summary, the level-of-activity model involves greater response of the
economy to a change in exogenous expenditures because of the added feedback
via induced investment expenditures. The multiplier is greater, but analytically
the solution is similar.

---

† An equivalent statement is the following: If h' is so large as to equal the marginal rate of
withdrawals, that is if $h' = 1 - b' = \Delta(\text{WDL}^*)/\Delta Y_s$, then k becomes indefinitely large. The INJ*
and WDL* lines plotted as functions of Y_s are parallel under this condition.

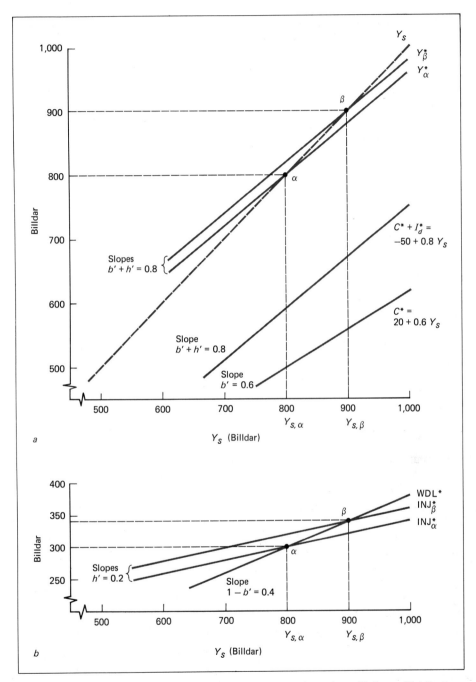

Figure 10-4   Equilibria for level-of-output model.   (*a*) Demand-supply equilibrium.   (*b*) Injections-withdrawals equilibrium.

In practice it may be desirable to combine the two preceding models and write

$$I_d^* = g'' - hi + h' Y_s$$

The equilibrium solution would be

$$Y = k'(a' + g'' - hi + G^* + X_d^*) = k' Y_{AUT}^*$$

where $k' = 1/(1 - b' - h')$

$$Y_{AUT}^* = a' + g'' - hi + G^* + X_n^*$$

In terms of the graphical solutions;

1 In Fig. 10-3 for the interest rate model, g'' would replace g in autonomous expenditures and the VJ line would be rotated further out from the i-axis to reflect the larger value of k' arising from induced investment; or

2 In Fig. 10-4 for the level-of-output model, the Y^* line and the INJ* line would shift vertically an amount $-h(\delta i)$ when interest rate changes by δi.

STOCK-ADJUSTMENT OR ACCELERATOR MODELS

In accelerator models, it is hypothesized that managers attempt to adjust stocks of fixed capital goods or of inventories to a desired level, which is related to expected demand for output. The flow of investment spending is then determined so that it would adjust stocks toward the desired level at some optimum rate of approach.

Analytically, the model may be represented by linear equations as follows.

$$K_t^* = \alpha + \beta \hat{Y}_t^*$$

where K_t^* is the desired real stock at end of period t, and $\hat{Y}_t^*$ is the forecast of demand (orders flow) during period t.

$$I_{K,t}^* = \mu(K_t^* - K_{t-1})$$

where $I_{K,t}^*$ is the desired flow of net investment,† $(K_t^* - K_{t-1})$ is the discrepancy between desired stock at end of period t and actual stock at end of period $t - 1$, and μ is the rate of adjustment which measures the fraction of the discrepancy in stocks which would be eliminated in one period by the accelerator investment.

Note that *when the economy is in equilibrium, desired and actual stocks must be equal; so accelerator investment will be zero.* Thus the accelerator model deals with dynamic adjustments when the economy is in disequilibrium. It has no

---

† Since capital stock K is a real variable, measured in constant-dollar units, the flows of aggregate demand and of investment spending in the two equations should be in real units also. Until price changes are analyzed, it will be assumed that the price factor is absorbed in the coefficients β and μ. That is, β is inversely, and μ is directly, proportional to prices of investment goods.

relevance to comparative statics. To elucidate the accelerator theory, we must analyze the period-by-period response of the economy to some exogenous shock. The stock-adjustment process is a feedback effect which influences the time path of economic variables during the movement of the economy from one equilibrium toward another.

For fixed capital stock (K) the accelerator theory implies that desired investment in a given period will equal: (1) a replacement component I_ρ^* to offset current consumption of capital goods, and (2) a net investment (accelerator) component I_k^* to adjust the existing capital stock toward the desired level. In equation form

$$I_{d,t}^* = I_{\rho,t}^* + I_{K,t}^* = \rho K_{t-1} + \mu(K_t^* - K_{t-1})$$

where $K_t^* = \alpha + \beta \hat{Y}_t^*$.

The investment decision is made at the beginning of period t regarding investment spending during period t. The rate of capital consumption is ρ per year. K_{t-1} is the actual stock of fixed capital at the end of period $t-1$. K_t^* is a capital stock which would provide optimum capacity at end-period t from a profit maximizing criterion, given the forecast of demand $\hat{Y}_t^*$ for period t. The higher the value of μ, the more quickly is actual stock adjusted to the desired level, and the higher the flow of net investment required. If μ equals 1, the adjustment is completed in one period. If μ is less than 1, we have introduced a time lag into the adjustment of actual to desired stock.

To complete the model we need a forecasting function to define $\hat{Y}_t^*$ in terms of past levels and trends in sales, orders, or other variables. Not knowing just how business analysts forecast sales, we shall use the naïve assumption that sales in period t are forecast to be the same as sales in period $t-1$, that is, $\hat{Y}_t^* = Y_{t-1}^*$.

The equations for the dynamic model become the following.

		New Variables	
Equation	Exogenous	Endogenous	
$Y_t^* = C_t^* + I_{\rho,t}^* + I_{K,t}^* + G_t^* + X_{n,t}^*$	$G_t^*, X_{n,t}^*$	$Y_t^*, C_t^*, I_{\rho,t}^*, I_{K,t}^*$	
$C_t^* = a' + b' Y_{s,t}$		$Y_{s,t}$	
$I_{\rho,t}^* = \rho K_{t-1}$		K_t	
$I_{K,t}^* = \mu(K_t^* - K_{t-1})$		K_t^*	
$K_t^* = \alpha + \beta Y_{t-1}^*$			
$K_t = K_{t-1} + I_{K,t}^* \Delta t$			
$Y_{s,t} = Y_{t-1}^*$			

We have seven linear equations and seven endogenous variables; a solution should be obtainable. In a sense, time is an eighth variable implicit in the set of equations, so the solution actually expresses Y^* or any other variable as a function of time. Substituting other equations into the first one so as to progressively eliminate all endogenous variables except Y_t^*, we find

$$Y_t^* = a' + b' Y_{t-1}^* + \rho K_{t-1} + \mu(\alpha + \beta Y_{t-1}^* - K_{t-1}) + G_t^* + X_{n,t}^* \tag{10-4}$$

To obtain the solution to the equation when the economy has reached equilibrium, we note that (1) the investment flow $I_{K,t}^*$, undertaken to adjust capital stock to the desired level [bracketed term in Eq. (10-4)], will be zero in equilibrium; and (2) time subscripts may be omitted in equilibrium because values are the same in period t as in period $t - 1$.   Thus we find

$$Y = a' + b'Y + \rho(\alpha + \beta Y) + G^* + X_n^*$$

which yields

$$Y = \frac{a' + \rho\alpha + G^* + X_n^*}{1 - b' - \rho\beta} = k(a' + \rho\alpha + G^* + X_n^*)$$

The multiplier $k = 1/(1 - b' - \rho\beta)$ is higher than $1/(1 - b')$ to allow for the fact that higher levels of Y induce higher replacement investment.

Equation (10-4) embodies what is called *interaction between the multiplier and the accelerator*.   An autonomous rise in expenditures has several effects in this model: (1) It increases aggregate demand, and hence, output and income in its own right.   (2) It engenders a follow-on sequence of rises in consumer income and demand (multiplier effect).   (3) It increases desired capital stock and thus stimulates net investment expenditures (accelerator effect), with its consequent multiplier effects on consumer demand.   (4) The net investment increases capital stock and thus ultimately reduces the motivation for further additions to capital stock (capacity effect).   This sequence of reactions raises the possibility of an initial multiplier-accelerator response which is stronger than that arising from the multiplier alone, followed by a downward reaction on investment spending as capital stock (or production capacity) is built up to desired levels.   In the early part of the adjustment, since net investment exceeds its equilibrium level of zero, it follows that aggregate demand and desired capital stock will exceed their final equilibrium levels.

This opens the possibility that the model may exhibit oscillations above and below the equilibrium level instead of a steady approach to the new flow equilibrium.   The discovery of this response in a multiplier-accelerator model of the economy was an exciting one in the annals of business-cycle analysis, and this mechanism remains a central feature of most modern theories of business fluctuations.   It opened the possibility that business cycles may occur because of the normal behavior of decision makers in the economic system rather than because of exogenous events such as harvest cycles, wars, and changes in fiscal and monetary policies.   That is, the possibility of an endogenous business cycle was demonstrated.

The calculations and graphs in Table 10-1 and Fig. 10-5 present an example of a fixed capital stock-adjustment model with an oscillatory solution.   The coefficients in the model are reasonable ones for the United States economy.   We have used $a' = 20$ and $b' = 0.6$ in previous consumption functions.   Also, $\rho = 0.075$ implies an average life of 13.3 years for items of equipment and plant in the fixed capital stock.   And $\mu = 0.25$ indicates that a discrepancy between desired

TABLE 10-1 Dynamics of Fixed Capital Stock Adjustment: Calculations

Eq. 1:	$Y_{s,t} = \hat{Y}_t^* = Y_{t-1}^*$
Eq. 2:	$C_t^* = 20 + 0.6\,Y_{s,t}$
Eq. 3:	$I_{\rho,t}^* = 0.075\,K_{t-1}$
Eq. 4:	$I_{K,t}^* = 0.25(K_t^* - K_{t-1})$
Eq 5:	$G_t^* + X_{n,t}^* = \begin{cases} 210 \text{ for } t \leq 0 \\ 220 \text{ for } t \geq 1 \end{cases}$
Eq. 6:	$Y_t^* = \text{col. 2} + \text{col. 3} + \text{col. 4} + \text{col. 5}$
Eq. 7:	$I_{u,t} = Y_{s,t} - Y_t^*$
Eq. 8:	$K_t^* = 1.5\,\hat{Y}_t = 1.5\,Y_{t-1}^*$
Eq. 9:	$K_t = K_{t-1} + 0.25\,I_{K,t}^*$

Period	(1) $Y_{s,t}$	(2) C_t^*	(3) $I_{\rho,t}^*$	(4) $I_{K,t}^*$	(5) $G_t^* + X_{n,t}^*$	(6) Y_t^*	(7) $I_{u,t}$	(8) K_t^* (end-period)	(9) K_t
0	800	500	90	0	210	800	0	1,200	1,200
1	800	500	90	0	220	810	−10	1,200	1,200
2	810	506	90	4	220	820	−10	1,215	1,201
3	820	512	90	7	220	829	− 9	1,230	1,202.8
4	829	517	90	10	220	837	− 8	1,243.5	1,205.3
5	837	522	90	12.5	220	845	− 8	1,255.5	1,208.4
6	845	527	91	14.8	220	852.5	− 7.5	1,267.5	1,212.1
7	852.5	531.5	91	16.6	220	859	− 6.5	1,278.7	1,216.2
8	859	535.5	91	18.1	220	865	− 6	1,288.5	1,220.7
9	865	539	91.5	19.2	220	869.5	− 4.5	1,297.5	1,225.5
10	869.5	542	92	19.6	220	873	− 3.5	1,304	1,230.4
11	873	544	92	19.8	220	876	− 3	1,309.4	1,235.4
12	876	546	93	19.6	220	878	− 2	1,314	1,240.3
13	878	547	93	19.2	220	879	− 1	1,317	1,245.1
14	879	547	93	18.4	220	879	0	1,318.5	1,249.7
15	879	547	94	17.2	220	878	1	1,318.5	1,254.0
Equilibrium	835	521	94.0	0	220	835	0	1,252.5	1,252.5

and actual capital stock would lead managers to make net investment adequate to retire one-fourth of the discrepancy in the first year, one-fourth of the remaining discrepancy in the second year, and so on. This problem uses $\alpha = 0$ and $\beta = 1.5$, indicating that in the United States economy the total stock of fixed capital is about 1.5 times the annual flow of GNP. In the equation for K_t [Eq. (9)], $\Delta t = 0.25$ because the period analysis will be worked out for quarter-years, whereas the flows are expressed as annual rates; thus the change in capital stock caused by a net investment flow of 20 billdar will be only \$5 billion in a quarter-year's time.

From Eq. (10-4) we find the long-term multiplier to be

$$k = \frac{1}{1 - b' - \rho\beta} = \frac{1}{1 - 0.6 - 0.075\,(1.5)} = \frac{1}{0.2875} = 3.48$$

For the initial and final equilibrium states we calculate the following:

	Initial Equilibrium	Final Equilibrium
$G^* + X_n^*$ (exogenous)	210	220
Y	800	835
C^*	500	521
I_ρ^*	90	94
I_K^*	0	0

The dynamic adjustment of the system to the rise in exogenous expenditures by 10 billdar in period 1 is calculated quarter by quarter in Table 10-1 and is graphed in Fig. 10-5. Since producers are assumed not to have forecast the rise in exogenous expenditures, output in period 1 remains at the original equilibrium level of 800 billdar, and the rise in shipments is met out of inventories ($I_{u,1} = -10$ billdar). Output increases by 10 billdar in period 2, but aggregate demand goes up by 10 billdar more, a 6-billdar rise in consumer expenditures, and a 4-billdar rise in net fixed investment to adjust the fixed capital stock toward the desired level of $1,215 billion, consistent with the increased orders in period 1 (which becomes the forecast for orders in period 2). The rise of orders in period 2 leads to an increase in sales forecast in period 3. This stimulates further increases in consumption and in investment spending. So it goes.

However, successive increments in aggregate output and aggregate demand become smaller because the marginal propensity to consume and the net investment coefficient sum to less than 1. Also, increments to fixed capital stock tend to narrow the gap between desired and actual capital stock and hence to reduce demand for net investment. Eventually, net investment spending turns down (period 12) because desired capital stock rises less than does actual capital stock in the previous period. Aggregate output, and hence consumer expenditures, continues rising two more quarters because of the rise in replacement investment and the one-quarter lag in output behind demand. At this point, flows and capital stocks in the economy are well above the final equilibrium values, which are indicated by horizontal dashed lines in Fig. 10-5. Activity declines, though capital stock continues rising. Net investment remains positive until period 25, when desired and actual capital stocks have reached equality. Then net investment swings negative and the economy continues to decline below its equilibrium level. The downward movement is eventually reversed by a mechanism which is the opposite of that on the upward cycle, and the variables oscillate around their final equilibrium levels in waves of decreasing amplitude.

Higher values of b', μ, and β, would make the system even more responsive, or less stable. The economy could respond to a rise in exogenous spending by waves of increasing amplitude and never settle down at the new calculated equilibrium values, or it could rise continuously at an increasing rate of change until some limiting constraints stop the expansion—perhaps full employment of labor, or fiscal and monetary policy actions. On the other hand, smaller values of b', μ, and β would decrease the sensitivity of the economy, so that it could respond

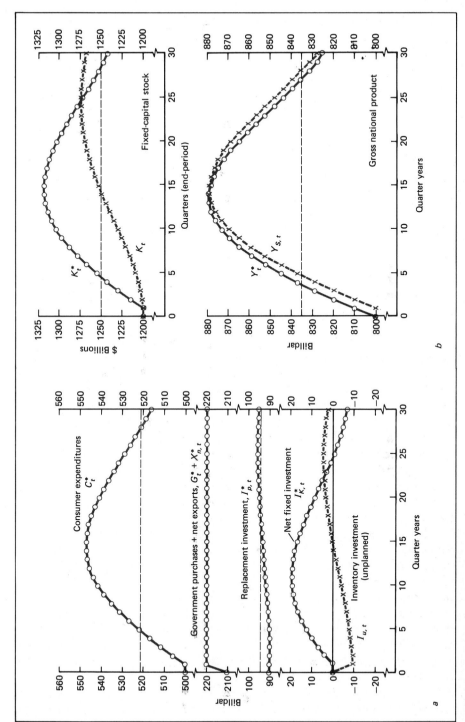

Figure 10-5 Dynamics of fixed capital stock adjustment.

to a rise in exogenous demand by a smaller cycle that damped down to equilibrium values quickly, or by a smooth movement to the new equilibrium without over-shooting.   The richness of the possible behavior for this stock-adjustment model makes it possible to explain historical economic fluctuations of many types, given alternative values of the constants in the model at various past times in the history of the economy.

The above model dealt with the adjustment of stocks of plant and equipment. Fixed capital investment cycles are fairly long; for example, in our illustrative model the economy reached a peak about three and a half years after the initial stimulus and took another three years to return to the final equilibrium level.   What about shorter cycles in the economy?   Could they arise from a quicker adjustment of inventory stocks toward desired levels?

The answer is "Yes."   Inventories are adjusted toward desired levels more quickly than fixed capital goods, that is, the μ coefficient is larger for inventory swings.   And the length of cycle for an oscillatory solution varies inversely with μ.   Illustrative calculations and graphs for an oscillatory inventory model are presented in Appendix B to this chapter.

Accelerator terms may be combined with interest rate and level-of-activity terms in a composite model for investment demand.   It may well be that different causal factors dominate decision making for different categories of investment—plant and equipment, residential construction, or inventories—or for different industries, such as manufacturing, public utilities, farming, or trade.

The stock-adjustment or accelerator mechanism adds a feedback path in which investment depends on the rate of change of demand.   In conjunction with multiplier effects, this opens up rich possibilities of endogenous cyclical responses of an economy to step-function shocks.   Stock-adjustment terms are included in behavioral equations in most modern econometric models of the economy and help to explain business fluctuations.

SUMMARY ON INVESTMENT DEMAND

The present value of net cash receipts or the rate-of-return schedule, when combined with a schedule for the cost of capital funds, seems to provide an adequate theoretical base for developing a realistic picture of the determinants of investment spending.   However, the fact that those schedules for demand and supply of capital funds are usually plotted against long-term interest rate does not imply that i is the most important variable in explaining short-term fluctuations in investment. Changes in levels of economic activity cause shifts in both schedules and need to be brought into a realistic theory of investment spending.   Economic activity affects the supply-of-funds schedule via corporate profits and cash flow, personal saving flows, government borrowing or repayment of debt, and demand for money stocks.   These influences will be discussed in following chapters.

In this chapter, it has been emphazised that fluctuations in the level of econom-ic activity change managers' sales expectations, and hence shift the MEI schedule by making more or fewer investment projects profitable at the current market rate of interest.   If a concept of desired stock of fixed capital and of inventories in relation to expected sales is introduced, we find that the MEI curve shifts upward and downward with business activity.   These shifts, combined with the multiplier effect of induced consumer expenditures, can lead to an eminently reasonable explanation of cyclical movements in the economy.

There remains the task of relating these fluctuations in desired stocks and in investment flows to shifts in the supply-of-funds curve and to changes in fiscal and monetary policy actions.   These topics will be discussed in later chapters.

APPENDIX A   Interest Rate Model: Adjustment Process following Change in Interest Rate

To calculate how the economy of the interest rate model moves from one equilibrium position toward another following a change in interest rates, we must write the equations of the model with variables designated by periods and with time lags made explicit.   Then we must solve the equations numerically, period by period, starting from an initial equilibrium disturbed by a change in interest rates in period 1.

Assumptions, in addition to those of the simple five-sector model, are these:

1   Assume that consumers adjust C^* to disposable income of the same period, that is, $C_t^* = a' + b' Y_{s,t}$.

2   Assume that domestic investment responds to interest rates with a four-quarter lag, that is, $I_{d,t}^* = g - h i_{t-4}$.

3   Take G_t^* and $X_{n,t}^*$ to be exogenous and constant.

4   Assume that producers adjust output of consumer goods with a one-quarter lag but match production to orders for other goods, that is,

$$Y_{s,t} = C_{t-1}^* + I_{d,t}^* + G_t^* + X_{n,t}^*$$

The equations of the model are as follows.

		New Variables	
Equation		Exogenous	Endogenous
1.	$Y_t^* = C_t^* + I_{d,t}^* + G_t^* + X_{n,t}^*$	$G_t^*,\ X_{n,t}^*$	$Y_t^*,\ C_t^*,\ I_{d,t}^*$
2.	$C_t^* = a' + b' Y_{s,t}$		$Y_{s,t}$
3.	$I_{d,t}^* = g - h i_{t-4}$	i_{t-4}	
4.	$Y_{s,t} = C_{t-1}^* + I_{d,t}^* + G_t^* + X_{n,t}^*$	C_{t-1}^*	

Equations (2) and (3) may be substituted into equation (4) to yield an expression for aggregate supply as a function of exogenous variables and aggregate supply (incomes) in the preceding period.

$$Y_{s,t} = a' + b' Y_{s,t-1} + g - h i_{t-4} + G_t^* + X_{n,t}^* \tag{10-5}$$

Once $Y_{s,t}$ is determined for any period, C_t^* can be calculated, and Y_t^* can be totaled. The calculations can be made period by period to trace out the time path of each of the variables in the economy.

The ultimate equilibrium solution can be derived without going through the period analysis. Note that in equilibrium, all variables will be the same, period after period, so that time subscripts become irrelevant. If we drop them from Eq. (10-5) and solve for Y_s, we obtain

$$Y_s = \frac{a' + g - hi + G^* + X_n^*}{1 - b'} = k(a' + g - hi + G^* + X_n^*) \tag{10-6}$$

This is the same as Eq. (10-2) in the text.

As a specific example, consider a model with $C_t^* = 20 + 0.6 Y_{s,t}$ and $I_{d,t}^* = 240 - 25i_{t-4}$. Assume $G_t^* + X_{n,t}^* = 210$ billdar. Let the interest rate be 5 percent per year initially and the economy be in flow equilibrium. Then assume that interest rate changes to $i_\alpha = 4$ percent per year for periods 1 through 6 before reverting back to $i_\beta = 5$ percent per year for periods 7 through 15. Corresponding to these interest rates, the levels of desired investment are $I_{d,\alpha}^* = 140$ billdar and $I_{d,\beta}^* = 115$ billdar. The equilibrium levels of GNP, calculated from Eq. (10-6), become $Y_\alpha = 925$ billdar and $Y_\beta = 862$ billdar.

In Table 10-2, the calculations are worked out for the period-by-period adjustment of the economy to the exogenous changes in interest rate. The results are plotted as

TABLE 10-2 Interest Rate Model: Calculations for Dynamic Adjustment

Eq. 1:	$Y_{s,t} = C_{t-1}^* + I_{d,t}^* + G_t^* + X_{n,t}^*$
Eq. 2:	$C_t^* = 20 + 0.60 Y_{s,t}$
Eq. 3:	$I_{d,t}^* = 240 - 25i_{t-4}$
Eq. 4:	$G_t^* + X_{n,t}^* = 210$ billdar (exogenous)
Eq. 5:	$Y_t^* = \text{col. 2} + \text{col. 3} + \text{col. 4}$
Eq. 6:	$i_t = 5$ in $t = 0$, 4 in $t = 1$ through 6, 5 in $t = 7$ through 15
Eq. 7:	$INJ_t^* = \text{col. (3)} + \text{col. (4)}$
Eq. 8:	$I_{u,t} = Y_{s,t} - Y_t^*$
Eq. 9:	$WDL_t^* = Y_{s,t} - C_t^*$

	(1)	(2)	(3)	(4)	(5)	(6)	(7)	(8)	(9)
Period	$Y_{s,t}$	C_t^*	$I_{d,t}^*$	$G_t^* + X_{n,t}^*$ (exog.)	Y_t^*	i_t (exog.)	INJ_t^*	I_t^*	WDL_t^*
0	862	537	115	210	862	5	325	0	325
1	862	537	115	210	862	4	325	0	325
2	862	537	115	210	862	4	325	0	325
3	862	537	115	210	862	4	325	0	325
4	862	537	115	210	862	4	325	0	325
5	887	552	140	210	902	4	350	−15	335
6	902	561	140	210	911	4	350	− 9	341
7	911	566	140	210	916	5	350	− 5	345
8	916	569	140	210	919	5	350	− 3	347
9	919	571	140	210	921	5	350	− 2	348
10	921	572	140	210	922	5	350	− 1	349
11	897	559	115	210	884	5	325	+13	338
12	884	550	115	210	875	5	325	+ 9	334
13	875	545	115	210	870	5	325	+ 5	330
14	870	542	115	210	867	5	325	+ 3	328
15	867	540	115	210	865	5	325	+ 2	327

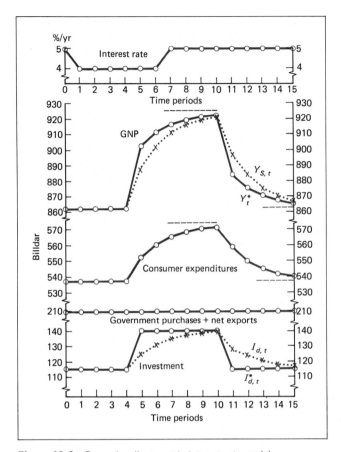

Figure 10-6 Dynamic adjustment in interest rate model.

time series in Fig. 10-6. Although the interest rate drops to 4 percent in period 1, the decision and production time lags in the investment function postpone any response in the economy until period 5. Then *desired* investment rises, along the investment demand curve $I_{d,t}^* = 240 - 25i_{t-4}$. Aggregate supply rises by the same amount as investment in period 5, but aggregate demand rises more because of the rise in consumer expenditures induced by the income increase. Since producers did not forecast the rise in consumer spending, aggregate supply is less than aggregate demand; thus in the GNP panel the small x's (output) fall below the solid line representing aggregate demand, and inventories are reduced. The unplanned reduction of inventories shows up also in the investment panel, where the x's show actual investment falling below the desired investment (solid line). In periods 6 through 10, the economy is completing its multiplier adjustment toward the new equilibrium levels consistent with the higher level of investment. The equilibrium levels, shown by horizontal lines, are $Y_\alpha = 925$ billdar, $C_\alpha^* = 575$ billdar, $INJ_\alpha^* = I_\alpha^* + G^* + X_n^* = 350$ billdar.

However, in period 7 the interest rate returns to its original 5 percent per year level, perhaps in response to the increased demand for funds accompanying economic expansion.

This causes a decline in *desired* investment in period 11, back to the original level of 115 billdar. Aggregate demand drops in period 11 because of declines in investment and consumer spending, and then continues downward toward the original equilibrium level. Aggregate supply lags on the decline, so that unplanned inventories accumulate. Actual investment exceeds desired investment by the undesired flow of goods into inventory.

This model shows that changes in interest rates, perhaps planned and executed by the monetary authorities, can lead to cyclical fluctuations in the economy if desired investment is sensitive to interest rates, that is, if h is positive. In the real world, the response to changes in interest rate would come somewhat sooner as inventories are built up by capital goods producers after their receipt of orders for plant and equipment. Also, the time lags would differ for various items of machinery and construction, so that the rise in investment following a decline in interest rate would be spread out more in time instead of occurring in a single jump. Finally, the changes in interest rates would probably be gradual rather than the step-function assumed here. All these modifications would tend to "round the corners" on the sharp changes plotted in Fig. 10-6 and make the response of the economy more like that observed in historical data.

APPENDIX B   Stock-Adjustment or Accelerator Model: Inventory Adjustment following Change in Exogenous Expenditures

The equations for an inventory adjustment model are much like those for the fixed capital accelerator model, except that (1) it must be recognized that the *undesired* inventory investment affects the level of inventories as well as does the flow of *desired* net investment in inventories; (2) the desired stock to sales ratio will be lower; and (3) managers will want to retire a gap between desired and actual inventories more quickly than for fixed capital stocks. Fixed capital investment ($I_{f,t}^*$) is here taken as exogenous. The naïve forecast is used again, that is, the forecast of orders in period t is taken equal to actual orders in period $t-1$. Thus, $Y_{s,t} = \hat{Y}_t^* = Y_{t-1}^*$ for equations (4) and (7) below. Letting the stock of inventories at end-period t be designated by Q_t, we obtain the following equation system.

| | | New Variables | |
Equation		Exogenous	Endogenous
1	$Y_t^* = C_t^* + I_{f,t}^* + I_{Q,t}^* + G_t^* + X_{n,t}^*$	$I_{f,t}^*, G_t^*, X_{n,t}$	$Y_t^*, C_t^*, I_{Q,t}^*$
2	$C_t^* = a' + b' Y_{s,t}$		$Y_{s,t}$
3	$I_{Q,t}^* = \mu'(Q_t^* - Q_{t-1}^*)$		Q_t^*, Q_t
4	$Q_t^* = \alpha' + \beta' Y_{t-1}$		
5	$Q_t = Q_{t-1} + (I_{Q,t}^* + I_{u,t}) \Delta t$		$I_{u,t}$
6	$I_{u,t} = Y_{s,t} - Y_t^*$		
7	$Y_{s,t} = Y_{t-1}^*$		

We have seven linear equations with seven endogenous variables; so it should normally be possible to solve for any of the endogenous variables as a function of time. The solution for aggregate demand becomes

$$Y_t^* = a' + b' Y_{t-1}^* + I_{f,t}^* + \mu'(\alpha' + \beta' Y_{t-1}^* - Q_{t-1}) + G_t^* + X_{n,t}^* \tag{10-7}$$

The equilibrium solution, obtained by setting inventory investment to zero and neglecting time subscripts, becomes the ordinary multiplier solution.

$$Y = k(a' + I_f^* + G^* + X_n^*)$$

where $k = 1/(1 - b')$.

Equation (10-7) embodies multiplier and accelerator effects, that is, an autonomous rise in aggregate demand can induce follow-on rises both in consumer demand $(b' Y_{t-1}^*)$ and in investment demand $(\mu' \beta' Y_{t-1}^*)$. So we have the possibility of overshooting the new equilibrium and oscillating around it. As an illustrative example, consider a model in which (1) we have the same consumption function as previously, $C_t^* = 20 + 0.6 Y_{s,t}$; (2) fixed investment, government demand, and net exports are exogenous and rise suddenly in period 1 from a total of 300 billdar to 310 billdar; (3) business firms desire inventory stocks (Q^*) equal to one-fourth of Y_α $(\alpha' = 0$ and $\beta' = 0.25)$, and (4) managers try to retire a gap between desired and actual inventories in one year's time $(\mu' = 1)$.

For such a model, the long-term multiplier is $1/(1 - b') = 2.5$ and the solutions in the initial and the final equilibrium positions are as follows:

	Initial Equilibrium	Initial Equilibrium
$I_f^* + G^* + X_n^*$ (exogenous)	300	310
Y	800	825
C^*	500	515
I_Q^*	0	0

The calculations for the dynamic response of this model to the 10-billdar increment in the exogenous expenditures in period 1 are tabulated in Table 10-3 and are plotted in Fig. 10-7. Again, the solution is oscillatory, but cycles are shorter and the amplitude of the waves is somewhat smaller than for the fixed capital investment cycles in Fig. 10-5. The shorter fluctuations reflect the fact that inventories are brought to desired levels faster than are stocks of fixed capital goods. The comparative amplitudes of waves of induced investment reflect two offsetting factors: (1) the ratio of inventories to output is much smaller than the comparable ratio for fixed capital, 0.25 compared with 1.5; but (2) the accelerator or rate-of-adjustment coefficient is larger for inventory, 1 as against 0.25.

Note that induced investment $I_{Q,t}^*$ returns to zero first in period 11, because desired stocks have fallen to equality with actual stocks at end of the preceding period. So long as GNP is rising, desired inventories are rising, and induced inventory investment is positive. Desired inventory investment becomes negative after period 11 because desired inventories continue to fall with GNP, but actual inventories rise for a while because positive unplanned inventory change, caused by the production lag, exceeds the negative stock-adjustment component. After a lag of three quarters, actual inventories decline as desired, and eventually the gap between desired and actual stocks becomes smaller. Then desired inventory investment $(I_{Q,t}^*)$ turns up, and GNP follows suit with a short lag. The economy is in an expansion phase once more, and a cyclical movement continues, gradually settling down to the new equilibrium level.

The constants chosen in this example make the system very oscillatory, but not unstable. Higher values for b', μ', or β' could yield an unstable system that would

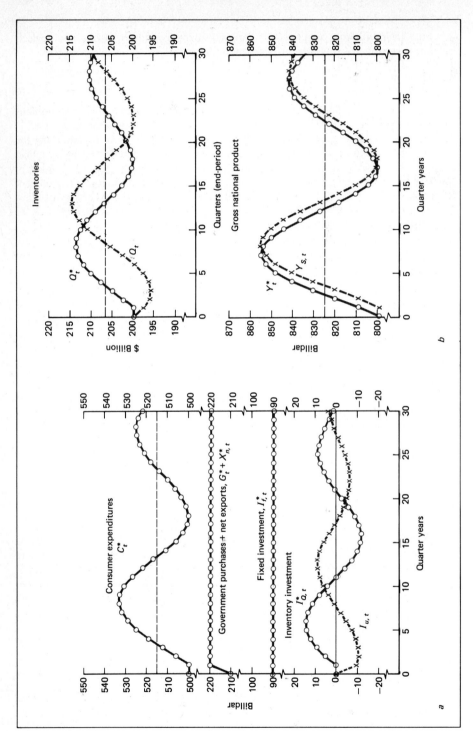

Figure 10-7 Dynamics of inventory adjustment.

TABLE 10-3   Dynamics of Inventory Adjustment: Calculations for Illustrative Model

Eq. 1:	$Y_{s,t} = Y^*_{t-1}$
Eq. 2:	$C^*_t = 20 + 0.6\,Y_{s,t}$
Eq. 3:	$I^*_{f,t} = 90$
Eq. 4:	$I^*_{Q,t} = Q^*_t - Q_{t-1}$
Eq. 5:	$G^*_t + X^*_{n,t} = \begin{cases} 210 \text{ for } t \le 0 \\ 220 \text{ for } t \ge 1 \end{cases}$
Eq. 6:	$Y^*_t = $ col. 2 + col. 3 + col. 4 + col. 5
Eq. 7:	$I_{u,t} = Y_{s,t} - Y^*_t = $ col. 1 − col. 6
Eq. 8:	$Q_t = Q_{t-1} + 0.25(I^*_{Q,t} + I_{u,t})$
Eq. 9:	$Q^*_t = 0.25\,\hat{Y}^*_t = 0.25\,Y^*_{t-1}$

	(1)	(2)	(3)	(4)	(5)	(6)	(7)	(8)	(9)
Period	$Y_{s,t}$	C^*_t	$I^*_{f,t}$	$I^*_{Q,t}$	$G^*_t + X^*_{n,t}$	Y^*_t	$I_{u,t}$	Q_t (end-period)	Q^*_t
0	800	500	90	0	210	800	0	200	200
1	800	500	90	0	220	810	−10	197.5	200
2	810	506	90	5	220	821	−11	196.0	202.5
3	821	512	90	9.2	220	831.5	−10.5	195.7	205.2
4	831.5	519	90	12.2	220	841.2	− 9.7	196.3	207.9
5	841.2	525	90	14.0	220	849	− 7.8	197.8	210.3
6	849	529	90	14.4	220	853.5	− 4.5	199.8	212.2
7	853.5	532	90	13.6	220	855.7	− 2.2	202.6	213.4
8	855.7	533	90	11.3	220	854.7	+ 1.0	205.7	213.9
9	954.7	533	90	8.0	220	851.0	+ 3.7	208.6	213.7
10	851.0	531	90	4.2	220	845.0	+ 6.0	211.1	212.8
11	845.0	527	90	0.1	220	837.0	+ 8.0	213.1	211.2
12	837.0	522	90	− 3.9	220	828.0	+ 9.0	214.4	209.2
13	828.0	517	90	− 7.4	220	819.0	+ 9.0	214.8	207.0
14	819.0	511	90	−10.0	220	811.0	+ 8.0	214.3	204.8
15	811.0	506	90	−11.5	220	805.0	+ 6.0	212.9	202.8
Equilibrium	825	515	90	0	220	825	0	206.5	206.5

expand without limit or exhibit oscillations with increasing amplitude.   Lower values of these coefficients would yield stable solutions with moderate fluctuations or a smooth approach to final equilibrium values.

PROBLEMS

1   Consider a closed economy for which the consumption and investment demand schedules (in billdar) are:

$$C^* = a' + b'\,Y_s = 20 + 0.6\,Y_s$$

$$I^* = g - hi = 240 - 25i$$

Time lags are disregarded here because, in equilibrium, all variables keep the same values for many time periods.

a Find equilibrium values of Y, C, I, and S when i_α is 5 percent per year and G_α equals 200 billdar.

b By how much will equilibrium values of Y, C, I, and S change if the interest rate changes so that i_β becomes 4 percent per year? (Use relations between δ's.)

c Write an equation which relates equilibrium values of Y to interest rate i. (This is the equation of the IS line.) Express the dependence of equilibrium Y on i in terms of both symbols and numerical values.

d Explain how the interest rate multiplier $\delta Y/\delta i$ is determined, what changes would influence its magnitude, and what the behavioral reasons are for the change in Y when i changes.

e Are all points on the IS line equilibrium points? Justify your answer.

f What happens to the IS line and to equilibrium value of Y when government expenditures change by a given amount? What if i changes by a given amount? What if both changes occur simultaneously?

2 Consider a model economy in which investment demand is a function of interest rate and level of output. Disregard time lags, because we shall calculate only equilibrium solutions.

$$C^* = a' + b'Y_s = 20 + 0.6\,Y_s$$

$$I_d^* = g'' - hi + h'Y_s = 80 - 20i + 0.15\,Y_s$$

$$G^* + X_n^* = 200 \text{ billdar}$$

a Write the solution equation for equilibrium Y in symbols. Describe how the addition of the $h'Y_s$ term affects the IS line for this economy.

b Assuming that i is 5 percent per year, calculate the initial equilibrium values for Y, C, and I_d.

c Now assume that G^* plus X_n^* increases to 210 billdar in period 1 and thereafter. If i stays at 5 percent per year, calculate the new equilibrium values of Y, C, and I_d.

d Write an equation for the autonomous expenditure multiplier for this economy — in symbols and in numbers. Is it true now that $\delta Y = k\,\delta \text{INJ}^*$? If not, what relation replaces it?

e Finally, assume that the economic expansion following the increase in G^* plus X_n^* leads to a rise in interest rate to 5.25 percent per year. Now solve for equilibrium values of Y, C, and I_d.

f Explain why δY is different in parts **c** and **e** above. Is the concept of the autonomous expenditure multiplier applicable in part **(e)**? If so, calculate its value.

3 a Express the concept of accelerator investment in one or two sentences. Explain the rationale for business investment decisions which underlies this behavior. Is the rate-of-return concept applicable?

b Describe all the feedback paths linking production-income flows to aggregate demand in an economy characterized by accelerator investment.

c Describe what is meant by interaction of the multiplier and accelerator. What observable behavior of the macroeconomic system can be accounted for by this interaction?

4 Consider a dynamic economy similar to that described in the text. The behavioral equations for the economy are:

Consumption: $\qquad\qquad C_t^* = a' + b' Y_{S,t} = 20 + 0.6 Y_{S,t}$

Replacement investment: $\qquad I_{\rho,t}^* = \rho K_{t-1} = 0.1 K_{t-1}$

Net investment: $\qquad\qquad I_{K,t}^* = \mu(K_t^* - K_{t-1}) = 0.2(K_t^* - K_{t-1})$

Desired capital stock: $\qquad K_t^* = \beta Y_{t-1}^* = 1.5 Y_{t-1}^*$

For quarterly time periods, noting that investment flows are at annual rates, we find:

$$K_t = K_{t-1} + 0.25 I_{K,t}^*$$

As for output decisions, let us assume, more realistically than the text example, that production units schedule production for period t equal to orders received in period $t-1$ less one-half of the undesired inventory change in the period $t-1$. That is,

$$Y_{S,t} = Y_{t-1}^* - 0.5 I_{u,t-1}$$

The economy is in an initial equilibrium with

$$Y_0 = 600 \qquad C_0 = 370 \qquad I_{\rho 0} = 90 \qquad K_0 = 900 \qquad \text{and} \qquad G_0 = 130$$

with all values in billdar. Then G^* increases to 140 for period 1 and later, that is, $G^* = 140$ for $t \geq 1$.

a Write the equations describing the dynamic behavior of the economy.

b Calculate the new equilibrium values of Y_S, Y^*, C^*, I_ρ^*, I_K^*, K^*, and K.

c Calculate values for the above variables and for $I_{u,t}$ for periods 1 through 4. (The values for periods 5 through 20 are given on p. 208.)

d Plot the above variables for periods 0 through 20 on graphs similar to those of Fig. 10-5.

Solution for Periods 5 through 20

t	$Y_{s,t}$	C_t^*	$I_{\rho,t}^*$	$I_{K,t}^*$	G_t^*	Y_t^*	$I_{u,t}$	End-Period K_t^*	K_t	
0										0
1										1
2										2
3										3
4										4
5	647.2	408.3	90.6	12.99	140	651.9	−4.7	970.82	909.12	5
6	654.3	412.6	90.9	14.45	140	657.9	−3.6	981.39	912.73	6
7	659.7	415.8	91.3	15.38	140	662.5	−2.8	989.62	916.58	7
8	663.8	418.3	91.7	15.85	140	665.8	−2.0	995.82	920.54	8
9	666.8	420.1	92.1	15.94	140	668.1	−1.3	1,000.23	924.52	9
10	668.7	421.2	92.5	15.71	140	669.4	−0.7	1,003.07	928.45	10
11	669.7	421.6	92.8	15.23	140	669.9	−0.2	1,004.60	932.26	11
12	670.0	422.0	93.2	14.54	140	669.8	+0.2	1,004.97	935.90	12
13	669.6	421.8	93.6	13.71	140	669.1	+0.5	1,004.47	939.32	13
14	668.8	421.3	93.9	12.78	140	668.0	+0.8	1,003.21	942.46	14
15	667.6	420.6	94.2	11.79	140	666.6	+1.0	1,001.40	945.41	15
16	666.1	419.7	94.5	10.75	140	665.0	+1.1	999.15	948.10	16
17	664.4	418.6	94.8	9.69	140	663.1	+1.3	996.57	950.52	17
18	662.5	417.5	95.1	8.65	140	661.2	+1.3	993.77	952.68	18
19	660.6	416.3	95.3	7.63	140	659.2	+1.3	990.84	954.59	19
20	658.6	415.1	95.5	6.66	140	657.3	+1.3	987.87	956.25	20

Chapter 11

The Government Sector and Fiscal Policy

THE ROLE OF GOVERNMENT IN ECONOMIC ACTIVITY

"The Congress hereby declares that it is the continuing policy and responsibility of the Federal government to use all practicable means consistent with its needs and obligations and other essential considerations of national policy... to promote maximum employment, production, and purchasing power." This quotation from the Employment Act of 1946 indicates that *decision makers in the federal government have some power to control the economy, do not simply adapt to an uncontrollable economic environment.* Here we are moving into a strange, new world. Decision makers can affect macroeconomic activity, and they can commandeer income rather than earn it.

"Maximum employment, production, and purchasing power" are *national economic goals* for which the federal government has responsibility; usually we abbreviate these objectives to the goal of "full employment" or maintenance of output close to the "potential GNP." Other economic goals include: (1) reasonable overall stability of prices; (2) rising productivity and standard of living; (3) equitable distribution of incomes and consumption; (4) support for a competitive, efficient, growing private market economy; (5) maintenance and promotion of a desirable human environment or "quality of life"; and (6) reasonable balance of international payments.

The power of governments to achieve these targets is far from complete. The 1970 *Economic Report of the President* says: "The basic full-employment growth path of an economy is not readily raised by any of the policy instruments that we now know about." Also, full achievement of one goal may interfere with achievement of another goal. For example, really full employment may lead to unacceptable price rises, or correction of deficits in the balance of payments may require a reduction of domestic consumption and employment. Also, private decision makers are extremely ingenious in finding ways to circumvent government taxes and controls which interfere with maximization of their private benefits. So government policy makers need to weigh decisions on the basis of social costs versus benefits of alternative possible actions, from local school construction to national income tax rates.

POLICIES AND INSTRUMENTS OF THE GOVERNMENT SECTOR

Broadly conceived, the areas for government economic policy making include:

1 Decisions to *support the efficient operation of the private economy* in production, consumption, and exchange through markets (including financial markets).

2 Decisions which *determine how our limited national output shall be divided between private and governmental uses,* and what assortment of goods and services shall be purchased by governments as collective consumption units.

3 Decisions which *determine who shall surrender income, or claims on output, to make output available to the Government Sector.* This involves income redistribution via tax policy and also transfer, interest, and subsidy payments, all of which redistribute gross national income differently from the allocation arrived at in markets for services of labor and property.

4 Decisions which *help maintain aggregate demand and supply in balance near full-employment levels of output.*

5 Decisions regarding *economic relations with foreign countries.*

The actions which governments may take to achieve policy objectives in these areas are numerous and varied. Some of the principal instruments are classified below in connection with the types of decisions just outlined.

EFFICIENT OPERATION OF THE PRIVATE ECONOMY

Governments are responsible for formulating and enforcing the "rules of the game" with respect to contracts, property rights, human rights, fraud, and methods of competition in the conduct of private business. So we have laws on contracts, police forces, child labor and other labor statutes; industrial safety regulations; antidiscrimination statutes; the Pure Food and Drug Act; antitrust laws; regulations on security markets; insurance and banking laws; consumer-protection legislation; pollution and safety controls; regulation of prices and standards of service for public utilities and transportation industries. In some cases, the government intervenes directly in setting market prices, for example, agricultural support prices, minimum-wage laws, maximum interest rates, gas and electric utility rates, interest-equalization taxes on foreign securities.

Governments also promote private production and consumption by providing necessary support facilities and services. Included here are streets and highways, air and water navigation aids, flood control, water supply and waste removal, fire and police departments, post offices, the central banking system, national defense, public health facilities and research, recreational facilities, agricultural and industrial research, employment offices, vocational training programs, mediation services, gathering and publication of statistics on demographic and economic conditions (domestic and foreign), and automobile and driver registration—the list is very long. Sometimes governments undertake needed research and development on projects too costly or risky for private enterprise, for example, development and uses of nuclear energy, the supersonic transport plane, or space travel, or they provide insurance or loans in case of natural catastrophes. Subsidies for soil conservation, shipbuilding, and operation of local airlines support private activities considered socially desirable, and tax advantages encourage prospecting for oil. Government lending and loan-guarantee programs supplement private capital markets, for example, through crop loans, Federal Housing (FHA) and Veterans Administration (VA) mortgage lending, Federal National Mortgage Association, Small Business Investment Corporation (SBIC), home loan banks, and the Export-Import Bank. Tariffs, voluntary quotas, and other restrictions on imports may also be regarded as aids to domestic business.

ALLOCATION OF OUTPUT TO PRIVATE AND GOVERNMENT USES

Governments are agents for purchasing goods and services which government legislators decide can better be procured collectively than by individual persons or businesses. The Government Sector is a final demand sector of the economy. In 1973 the federal government bought over 8 percent of GNP, and state and local governments bought over 13 percent. (These are purchases by the general Government Sector, exclusive of current-account purchases by government enterprises.) In that year about 50 percent of federal and 56 percent of state and local purchases consisted of compensation of employees (GNP originating in government); the remainder were expenditures to buy products of private business, some 10 percent of GNP in total. It is clear that spending by the Government Sector is a major factor in labor markets and in markets for the output of private businesses.

In some cases, cost-benefit ratios or rates of return can be estimated for various projects as a basis for rational choice. But it seems clear that government spending decisions are often made on the basis of pressures from persons or groups who would benefit or taxpayers who would be hurt. Or the calculus may be the political advantage to the legislators voting the expenditures.

Because noneconomic criteria loom large, it is difficult to specify the variables most influential in the decision making which governs the valve on the flow of government expenditures. Tax income is *not* a decisive determinant, because units in the Government Sector have the power, within the limits set by taxpayer resistance, to set government income by fiat. Income may be more important in determining state and local spending, but rising federal grants are loosening that constraint. Credit conditions do seem to influence state and local bond issues, and hence expenditures, after a time lag. Probably population should be taken into account as a decision variable in the case of state and local expenditures, where education, highways, health, civilian safety, and welfare services are major elements. But national defense expenditures dominate federal spending, and its economic determinants are not at all clear.

The forecaster of government expenditures can, however, use budgets or noncausal methods of projection. The federal budget is prepared and publicized well in advance of expenditures; so it provides a lead-decision indicator in that area. State and local government expenditures have been rising so steadily in the postwar years that a simple trend extrapolation usually is eminently satisfactory for the aggregate figure.

TAXATION AND INCOME REDISTRIBUTION

Taxation diverts income from the private sectors of the economy to the Government Sector on a nonmarket basis, that is, taxes paid are not closely related to the government services provided to a taxpayer. So taxation (1) reduces private spending and permits the Government Sector to claim part of current output, and (2) changes the distribution of income (and hence purchases of GNP) among persons and businesses in the private sectors. The income redistribution is carried further by government payments of transfers, interest, and subsidies, which in 1973 amounted to 32 percent of total government tax and nontax receipts in the national economic accounts. These payments include social security benefits, welfare payments, unemployment insurance, veterans' benefits, interest on government debt, and business subsidies, less net income of government enterprises. The first three of these items are countercyclical components of private incomes. Together with our progressive income tax schedule, they tend to raise personal income as a percent of GNP during recessions and lower the ratio in expansion periods. Their countercyclical effect does not depend on current action by Congress; hence they are called *automatic stabilizers*. As we shall see, their effect is to reduce the multiplier in the economy. Consumer incomes and expenditures are stabilized. Government income is destabilized, but that does not affect government spending much. So aggregate demand is stabilized overall.

Taxes and transfer payments can also be changed on a *discretionary* basis by legislative action to help achieve economic policy targets. Thus income taxes were cut in 1964–1965 to stimulate aggregate demand, and a surtax was imposed in 1968 to prevent inflationary aggregate demand during the period of high military expenditures for the Vietnam war. Tax changes can aim to affect particular groups or activities, for example, investment tax credit; the interest-equalization tax on purchase of foreign securities; sales taxes on gasoline, liquor, and tobacco; tariffs on particular commodities. Transfer payments and subsidies can also be specific in their impact; among them are increases in social security benefit payments, occupational retraining allowances, federal grants to inner-city educational programs, interest subsidies on mortgage loans to low-income families, subsidies for shipbuilding in the United States, or grants to cities for urban transportation programs.

One difference in the impact of direct (income) versus indirect (sales and business property) taxes should be noted. *Indirect taxes* levied on businesses are a nonfactor charge against the value of output and will normally be passed on to customers in increased prices. The effect on private purchases comes about through increased prices of goods without a compensating rise in private money income, so that private real purchasing power has been cut and that of the Government Sector maintained or increased. Sales taxes also change the relative prices and profitability of taxed and nontaxed products. *Direct taxes* (corporate profit taxes and levies on personal income and property) are assumed not to be reflected quickly or completely in product price changes; they reduce the flow of money incomes to private sectors directly.

An important difference between the federal, and the state and local, governments with respect to their revenue situation may be noted. Federal tax revenues come primarily from personal and corporate income taxes, excise taxes, and contributions for social insurance. With tax rates constant, receipts from these sources rise and fall with levels of economic activity, and they increase as wage and price levels rise. State and local revenues are more heavily dependent on the property taxes. With constant tax rates, their receipts will not show as great cyclical fluctuations or as great an increase as the economy grows. Also, they do not grow in proportion to increases in wage rates and prices. It follows that the federal government revenues decline more in a recession, but that they grow faster during periods of economic growth and inflation. Consequently, state and local governments have been squeezed between rapidly growing demand for and costs of the services they provide, and their less buoyant revenues. This has lead to a marked rise in federal grants-in-aid to state and local governments, a source of funds which in 1969 amounted to over one-fifth of total state and local government receipts in the national economic accounts.

MAINTENANCE OF FULL EMPLOYMENT

As was noted above, *the federal government has little control in the short run over aggregate supply*, except as its mediation services, persuasive powers, or legal actions prevent or terminate strikes. Another exception is government control over production during wartime via price and wage controls, allocation of materials, production quotas, and direction of labor force into preferred occupations and industries. In the longer run, governments may affect the growth rate of the economy to some extent by influencing:

1 The size of the labor force—via actions affecting immigration, birth and death rates, compulsory school attendance, retirement age, unemployment benefits and welfare provisions, programs to encourage and train persons for employment, prevention of discrimination, and other programs

2 The average hours worked per year—via laws on the standard workweek

3 Output per man-hour (productivity)—via government research and development programs in basic science, agriculture, public health, atomic energy, and military and space technology; or through education and retraining of workers in needed occupations and skills; or by encouraging investment in new technology by means of investment tax credits, the patent system, and government funding of development contracts; or by maintaining market competition (except in some monopolized labor markets and regulated industries)

In the short run, most of the government's efforts at maintaining full employment involve attempts to keep aggregate demand closely in line with potential GNP. The policy instruments available include direct government control over their own expenditures and some influence on expenditure decisions of all the other final demand sectors.

Government purchases of current output include purchases from private businesses and compensation of government employees. The former results in increased private output and consequently increases income payments for services of labor and property, whereas the latter generates employment and personal labor incomes directly and immediately in the Government Sector. It is sometimes argued that wage and salary incomes are likely to be respent more quickly and more completely (higher marginal propensity to consume) than are property incomes. Also, payments made directly to the Personal Sector yield a greater multiplier effect. On the other hand, if government purchases from private business lead to the need for more capacity or for different plant and equipment to produce the government output, the economy will be stimulated by induced investment as well as by the consumption multiplier. There are difficulties in timing the changes in government purchases in a countercyclical pattern, however. Programs cannot be cut off quickly enough during the business expansion and cannot be started quickly enough in a recession. For longer-term adjustments of aggregate demand to the trend of potential GNP, the method has more promise but should be evaluated in terms of the desired division of total output between government and private uses.

The *government influences consumer expenditures* not only through the income effect from government spending but also (1) by changing the ratio of disposable income to GNP, and (2) by changing the propensity to consume out of a given disposable income. Changes in tax rates and transfer payments are the principal policy tools for altering the proportion of gross national income which becomes available to the Personal Sector. It is usually assumed that such changes are independent of government purchases of goods and services. If so, the shift of income streams between the Government and Personal Sectors will affect consumer spending without altering government purchases, and thus will change aggregate demand. Different types of taxes or transfer payments may have different impacts on consumer spending per dollar shift in the government budget; that is, changes in tax rates for different income brackets, sales taxes, property taxes, increases in social security taxes or benefits, or veterans bonuses may affect persons with different marginal propensities to consume. But the differences are probably not great enough to be a major factor in selecting the policy tools to be used.

Given the ratio of disposable personal income to GNY and its distribution among income recipients, the second type of government influence on consumer spending is via the allocation of disposable personal income between saving and spending. Money and capital-market conditions influence consumer decisions, for example, through the cost and availability of consumer credit, interest rates on savings, perhaps rate of change in money supply. The impact of government actions and publicity on consumer expectations regarding future employment, incomes, and prices may be another influence. And, in the long run, provision for old age and survivors insurance, health insurance, and unemployment benefits may affect the saving ratio.

As for the *influence of government actions on investment demand*, there is first the stock-adjustment or accelerator effect of government purchases on inventories and fixed capital investment. Taxes and subsidies which change the estimated rate of return on investment projects should also be effective, for example, changes in corporate profits tax rates, investment tax credits, subsidies for projects of given types, or long-term purchasing contracts for output. The availability and cost of funds will be a factor also, and the government can influence capital flows through general monetary policy or through more specific programs to make loans to small businesses, channel funds into mortgage markets, or guarantee loans for certain types of projects. Government economic policies and public statements regarding them may also influence business expectations and hence plans for capital spending. In the longer run, government provision of supporting "social capital," support for research and innovation, provision of marketing information, or requirements of investment in pollution control would influence the investment component of aggregate demand.

Finally, *government actions can influence levels of exports and imports*, directly by government purchases and sales abroad or indirectly by their impact on private marketing activity. Some of the policy instruments available to the government in this area are tariffs, quotas on imports, subsidies for exporters, provision of credit and marketing information for exporters and importers, trade fairs, promotion or restriction of tourism, and regulations on movements and uses of private capital funds abroad. Foreign aid is often tied to purchases from the United States, as are purchases for military establishments abroad. The ultimate policy tool would be a change in the exchange rate of the United States dollar compared with foreign currencies, which would alter relative prices of exports and imports.

Thus, the government decision makers, primarily in the federal government, have considerable influence on the level of aggregate demand, know how to exercise it, and are committed to use that influence to keep the economy operating near full-employment levels. Any future administration which permits the United States economy to fall into a depression will quite surely be committing political suicide. The penalties for permitting aggregate demand to exceed potential GNP and engender continuing inflation are probably not so quick and drastic, but would quite surely be levied in time. Moderate economic fluctuations are still to be expected; in a private enterprise economy, the government cannot control the direction, magnitude, and timing of private economic behavior closely enough to keep aggregate demand equal to full-employment output in each quarter or year. But, given modern understanding of macroeconomics plus our automatic stabilizers, and assuming economic statistics are quickly available from computers, it would seem that the federal government can influence economic behavior enough to avoid major departures from the path of full-employment growth for the economy.

THE GOVERNMENT SECTOR IN THE NATIONAL INCOME AND PRODUCT ACCOUNTS

As was explained in Chap. 4, the Government Sector combines federal, state, and local government units of all kinds, except for government enterprises.†

TYPES OF BUDGETS

Government Sector receipts and expenditures in national income accounts are chosen to measure the impact of government activities on current production and income flows in private sectors of the economy. Capital transactions involving secondhand real assets or financial claims are excluded. But social insurance contributions and benefits are included, since they are counted as part of the current income and payment streams which change the net worth of the paying and receiving sectors. Interest payments on government debt are counted, like transfer payments, as expenditures not related to current services rendered in production. Only compensation of general government employees is counted as value added, or GNP originating, in the Government Sector. Finally, the timing of government receipts and expenditures is chosen to reflect the estimated economic impact and to synchronize with the accounting of the private sectors of the economy. Thus personal taxes are entered on a cash basis, but corporate profits, social security contributions, and sales taxes are recorded on an accrual basis. Government purchases are entered in the national economic accounts at time of acquisition rather than when payment is made.

The national income accounts closely match the *unified budget* categories of the federal budget except for (1) the timing of personal tax payments, (2) the timing of expenditures, and (3) the exclusion of financial transactions.

One other type of budget should be mentioned at this point—the *high-employment* or *full-employment* budget. This is a modified version of government receipts and expenditures in the national income accounts in which receipts and expenditures are adjusted to the levels they would reach if GNP were at the full-employment level. The calculation of this budget involves three steps: (1) Full employment is defined (usually as 96 percent employment of the civilian labor force) and the real GNP consistent with it is calculated. (2) Income streams which provide tax bases for government receipts are calculated as they would be if the economy were operating at its full-employment level; that is, wages and salaries, corporate profits, sales subject to sales tax, and personal income are estimated consistent with production equal to potential GNP. (3) Full-employment government receipts are calculated by applying high-employment tax rates to these derived full-employment income components. In addition, one adjustment is made to government expenditures, a calculation of unemployment benefits payments consistent with full employment of the labor force. Other expenditures are assumed not to vary with the level of employment.

---

† Government enterprises, "agencies of government whose operating costs are at least to a substantial extent covered by the sale of goods and services," have their operating statements consolidated with those of private businesses.

TABLE 11-1   Major Components of Government Receipts and Expenditures

		Percent of GNP in 1973	
		Federal	State and Local
1	Gross receipts:	**20.0**	**14.9**
2	Personal tax and nontax receipts	8.8	2.9
3	Corporate profits tax accruals	3.4	0.5
4	Indirect business tax and nontax accruals	1.6	7.6
5	Contributions for social insurance	6.1	0.9
6	Grants-in-aid to state and local governments	—	3.1
7	Non-GNP expenditures:	**12.2**	**1.1**
8	Transfer payments	7.4	1.6
9	Net interest paid	1.3	−0.1
10	Subsidies less current surplus of government enterprises	0.4	−0.4
11	Grants-in-aid to state and local governments	3.1	—
12	Net receipts (line 1 minus line 7)	7.8	13.8
13	Purchases of goods and services:	**8.2**	**13.1**
14	Compensation of employees	4.1	7.4
15	Structures	0.4	2.2
16	Other purchases	3.8	3.5
17	Surplus (or deficit): (line 12 minus line 13)	**−0.4**	**0.7**

Source: U.S. Department of Commerce, *Survey of Current Business,* July 1974, "U.S. National Income and Product Accounts," tables 3-1 and 3-3.

   The full-employment budget figures provide a measure of the government surplus, deficit, or balance that would exist if the economy were at full employment.   This can yield a guide to government decision makers in choosing policies to hold economic activity near potential output.   More on this later.

MAJOR COMPONENTS IN NATIONAL INCOME AND PRODUCT ACCOUNTS

Table 11-1 summarizes the major components of government receipts and expenditures for both the federal and the state and local sectors, and gives their relative magnitudes in 1973 as percentages of GNP.   Gross receipts of the Government Sector totaled 31.8 percent of GNP in 1973 (eliminating grants-in-aid), with the federal government taking 20.0 percent and state and local governments 11.8 percent of GNP.   In 1929, the comparable figures were 3.7 percent federal and 7.2 percent state and local, totaling about 11 percent of GNP.   The federal government's heavy reliance on taxes related to income flows is noteworthy, compared with state and local government reliance on indirect taxes (largely levies on property and sales).   The federal government accounts for the preponderance of non-GNP payments, including grants-in-aid to state and local units.   Net receipts and GNP purchases of state and local governments are much higher than those of the federal government.   As shown in Table 11-2, about 70 percent of federal

TABLE 11-2   Major Components of Government Purchases of Goods and Services

		1973 Percentages	
		Federal	State and Local
1	National defense	69.8	0.3
2	Space research and technology	2.8	—
3	General government	7.7	10.1
4	International affairs and finance	0.8	—
5	Education	0.9	42.0
6	Health, labor, welfare	5.5	28.4
7	Veterans' benefits and services	3.1	0.0
8	Commerce, transportation, and housing	6.2	15.7
9	Agriculture and agricultural resources	0.3	0.8
10	Natural resources	2.9	2.5
	Total (percent)	100.0	100.0
	Total (billdar)	106.6	169.8

Source: U.S. Department of Commerce, *Survey of Current Business*, July 1974, "U.S. National Income and Product Accounts," table 3.10.

government purchases were for national defense.   The largest components in state and local spending were education; health, labor and welfare (including health and hospital services, police and fire protection, public assistance and relief); and commerce, transportation, and housing (largely highway spending).

HISTORICAL PERSPECTIVES

Figures 11-1 and 11-2 portray graphically the trends and fluctuations in government receipts and expenditures.   In the Great Depression from 1929 to 1940, *federal receipts* at first declined, then rose faster than GNP from 1932 to 1940; but sizable deficits were incurred in most years, that is, the total expenditure line lies above the gross receipts line.   During World War II, federal gross receipts exceeded 20 percent of GNP for the first time, but expenditures reached 45 percent of GNP, with the deficit being financed by heavy wartime borrowing.   Postwar fluctuations in federal receipts as a proportion of GNP are accounted for by tax cuts (early postwar years, 1954, 1958, 1964–1965, 1970–1971), tax rises (in 1950–1951, 1968), and by recessions (in 1949, 1954, 1958, 1960–1961, 1970).   In the absence of rate changes, federal revenues tend to rise as a percentage of GNP in prosperous periods and decline in recession periods because of the progressivity of our tax structure.   In 1969, gross receipts were larger in relation to GNP than at their peak in World War II.

Federal purchases of goods and services rose in the depression years, skyrocketed in World War II, dropped in the early postwar years when defense spending was curtailed, peaked again during the Korean war (1950–1953), then held fairly steady as defense spending was maintained at a high level (8 to 10 percent of GNP).   The vertical distance between the purchases line and the total expenditure

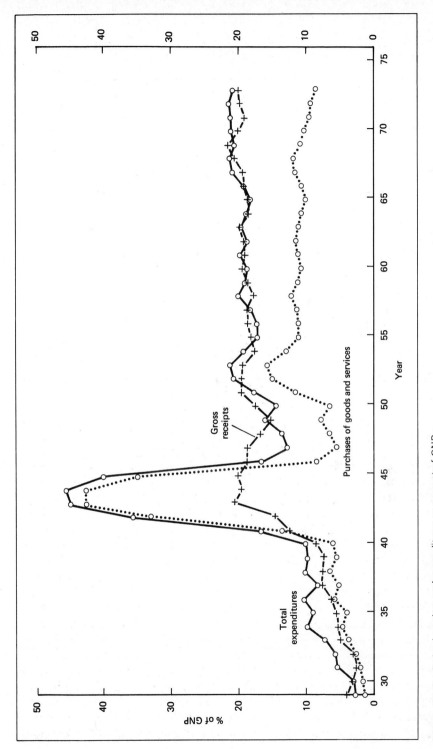

Figure 11-1 Federal receipts and expenditures as percent of GNP.

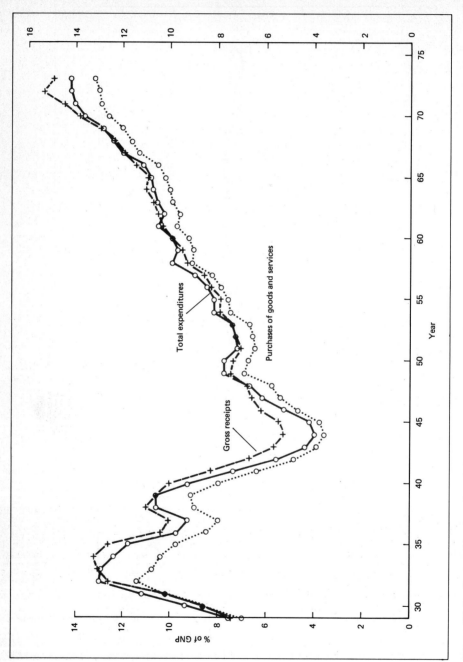

Figure 11-2 State and local receipts and expenditures as percent of GNP.

line shows non-GNP expenditures, which have recently ranged between 9 and 12 percent of GNP.

The pattern of *state and local receipts and expenditures* is rather different (Fig. 11-2). Gross receipts held nearly steady in dollar amounts during the Depression of the 1930s; so the percentage fluctuated inversely with GNP levels. Expenditure percentages showed a similar pattern to receipts, except that they exceeded receipts from 1929 to 1934 and fell short in later years. Receipts and expenditures were both depressed during World War II, but sizable surpluses were generated. After the war, expenditures showed a continuing rise as pressures for state and community services, educational buildings, public works, and highways increased. Expenditures outran gross receipts (including federal grants-in-aid) between 1949 and 1961, and fell a little short in 1962 through 1966, and again from 1970 through 1973. Expenditure increases required continual upward adjustment in sales and property tax rates and introduction of more income taxes at the state and local levels. Note that the relative stability in state and local budgets gives the ratios to GNP a countercyclical movement in the recession years of 1949, 1954, 1958, and 1960–1961.

INTERACTION OF THE GOVERNMENT SECTOR WITH THE REST OF THE ECONOMY

GOVERNMENT PURCHASES OF GOODS AND SERVICES

Government purchases of GNP are taken to be autonomous in short-term models of economic fluctuations. As we know, sustained changes in exogenous spending will have a multiplier effect on GNP because of the follow-on rise in consumer spending which it generates and because it may shift the investment demand function also. (1) If the rise in government purchases is for employee compensation, the multiplier action should come quickly and have maximum impact, because all the funds go to consumers. (2) If the increased government spending is for business services or " off-the-shelf" goods, the impact may also be rather quick, but it will be reduced in magnitude by the diversion of part of the business receipts to income streams other than personal income. In addition, some of the government purchases of goods may be initially drawn from inventory; so there will be a delay in added generation of disposable income. (3) If the increased government spending is for construction or major equipment items, orders will presumably have been placed well in advance of time the expenditures occur, and private income generation will have preceded the government purchases, appearing in GNP as investment in business inventories related to government contracts.

Accelerator effects of government purchases on investment spending will depend on many factors: (1) whether the government orders are for new types of goods requiring new production facilities, such as new military planes and space vehicles; (2) whether business managers judge that the increased government spending will last long enough to justify capital investment; and (3) whether the government orders are large enough to require output beyond current capacity.

Government spending may also feed back on investment spending via *level-of-output effects* and influences on *interest rates*, depending on methods used to finance the government spending.

GOVERNMENT REALLOCATION OF INCOMES VIA TAXES AND TRANSFER PAYMENTS
Government taxation and transfer payments reallocate gross national income between Private and Government Sectors and among taxpayers and recipients of government transfers, interest, and subsidies in the Private Sector. The effect on aggregate demand depends on the relative marginal propensities to spend for the persons or groups who lose versus those who gain income by the reallocation.

The income reallocation is exogenous in the sense that tax rates and transfer benefits schedules are set by legislative action and are not a market-determined payment for services rendered or goods exchanged. Also, they may be determined from policy considerations in order to influence the aggregate level of economic activity rather than be set to provide revenues to cover desired government expenditures. However, once tax rates and benefit schedules have been set, government receipts are endogenous in the sense that they depend on levels of other money flows in the private economy. *The discretionary (exogenous) and passive (induced) aspects of government receipts and transfer payments must be kept conceptually distinct in analyzing the impact of government receipts and transfers on the economy.*

Figure 11-3 shows how the Government Sector's gross and net receipts and its non-GNP expenditures have fluctuated in relation to GNP. Dotted lines are shown before the points for years in which legislated federal tax rates were significantly different from rates in the preceding year. The discretionary changes in tax rates seem to cause vertical displacement of the receipts lines, that is, changes in their intercept values. Between such changes, gross and net receipts show fairly good straight-line relations to GNP, with dips in recession years. The equations for these lines in the most recent period of near-stability for tax rates (1964 through 1967) are given in Fig. 11-3.

$$T = 0.364\,Y_s - 60 \text{ billdar}$$

$$T_n = 0.244\,Y_s - 27 \text{ billdar}$$

Note that, in both cases, *the intercepts are negative, indicating a progressive tax structure, such that the percentage rise in tax receipts exceeds the percentage rise in GNP.* The slopes of the functions (coefficients of Y_s) indicate that about 36 percent of an increment in GNP flows to the Government Sector on a gross basis and over 24 percent on a net basis, after allowing for transfers, interest payments, and net subsidies. Actually, tax rates are not exactly constant within the subperiods indicated because changes in state and local rates have been continuous during the postwar period. Also social security tax and benefit rates and other components in the non-GNP expenditures have been adjusted fairly frequently, so that the curves reflect these upward adjustments of rates through time as well

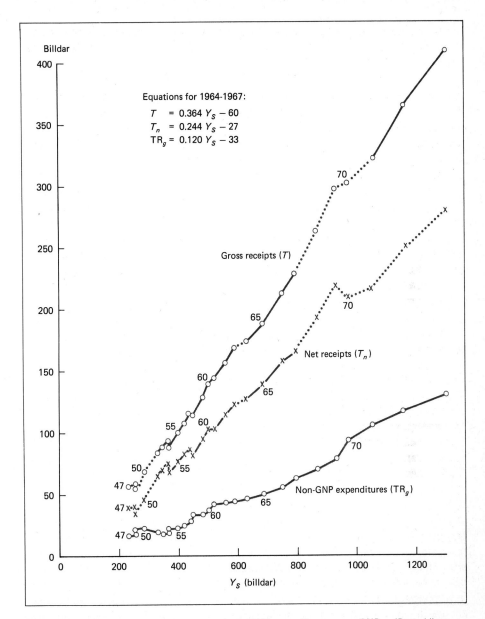

Figure 11-3 Government sector receipts and non-GNP expenditures versus GNP. (Dotted lines indicate change in income tax rates.)

as the effects of rising GNP. Since GNP and "time" are closely correlated variables, it is practically impossible to isolate their separate influences on tax receipts and transfers.

EFFECTS OF CHANGES IN NET TAX FUNCTIONS

Now let us analyze the influence on economic activity of exogenous shifts in government tax rates and transfers—"transfers" here referring to all government non-GNP payments. The analysis will be expressed algebraically in symbols first; then a numerical example will be worked out.

We begin with linear equations relating sector income flows to GNP (or Y_s).

$$Y_d' = e + f Y_s \tag{11-1}$$

$$T_n = \tau_1 + \tau_2 Y_s \tag{11-2}$$

$$\text{GRE} = \gamma_1 + \gamma_2 Y_s \tag{11-3}$$

$$\text{TR}_f = \phi_1 \text{ (assumed independent of GNP)} \tag{11-4}$$

Since the sum of these income flows equals Y_s, we have

$$Y_s = (e + \tau_1 + \gamma_1 + \phi_1) + (f + \tau_2 + \gamma_2) Y_s \tag{11-5}$$

For the two sides of Eq. 11-5 to be equal, we must have:

$$e + \tau_1 + \gamma_1 + \phi_1 = 0 \tag{11-6}$$

and

$$f + \tau_2 + \gamma_2 = 1 \tag{11-7}$$

Changes in tax rates or transfers imply changes in τ_1 and/or τ_2 in the equation for T_n, that is, changes in the intercept or slope of the line for net tax receipts in Fig. 11-3. Because of the constraints of Eqs. (11-6) and (11-7), a change in τ_1 will cause an equal and opposite change in the sum $(e + \gamma_1 + \phi_1)$, and a change in τ_2 will cause an equal and opposite change in the sum $(f + \gamma_2)$. After all, at a given value of Y_s, if *net taxes flowing to the Government Sector are changed exogenously, incomes flowing to the other demand sectors must change exogenously in the opposite direction. The impact on aggregate demand will depend on how the private demand sectors alter their GNP purchases when their incomes change.* Since private persons and companies probably alter their purchases by different amounts in response to a given change in income, the impact on GNP of a given exogenous change in net taxes will depend on the type of tax and on who pays it.

1 *Assume an exogenous change in net taxes by* $\delta\tau_1$, that is, a "lump-sum" change, and assume that it is *levied on persons*, perhaps an increase in property taxes or a rise in social security benefits. The resulting exogenous change in disposable income will cause an *exogenous* shift in C^*, and this will have further multiplier repercussions on Y_s and incomes, including an induced countereffect on net taxes.

For analytical purposes, assume $INJ^* = (G^* + I_d^* + X_n^*)$ are independent of sector incomes, and assume the usual consumption function

$$C^* = a + bY_d' = (a + be) + bfY_s = a' + b'Y_s$$

in view of Eq. (11-1).

Under the above assumptions, the change in "lump-sum taxes" by $\delta\tau_1$, will lead to an equal but opposite change in e, the intercept of the Y_d' function of Eq. (11-1). Thus,

$$\delta a' = b\delta e = -b\delta\tau_1$$

The effect of the exogenous tax change is to cause an opposite exogenous vertical shift of the disposable income line by $\delta e = -\delta\tau_1$, and of the consumption and aggregate demand lines by $\delta a' = b\delta e = -b\delta\tau$.

We know from solving previous models that in equilibrium, $Y = k(a' + INJ^*)$. Hence, the change in equilibrium GNP, caused by $\delta\tau_1$ becomes

$$\delta Y = k\delta a' = -kb\delta\tau_1 \tag{11-8}$$

(Note that the coefficient b is involved here, not b'.) Thus *a change in personal lump-sum taxes alters equilibrium GNP in the opposite direction*, with a coefficient $-kb$.

Since injections are constant, all the change in GNP must occur in C^*.†

In Eq. (11-8) the coefficient $(-kb)$ is sometimes called the *lump-sum tax multiplier* for the economy. As would be expected, it is negative. Note also that its magnitude is less than k, since $b < 1$. This implies that *an exogenous change in taxes causes a smaller change in equilibrium GNP than does an exogenous change in G* of equal magnitude*. This indicates that a larger change in taxes than in government purchases would be needed to cause a specific change in GNP. Note that a given change in G^* is itself the initial, exogenous change in aggregate demand which sets the multiplier process in operation. But an equal exogenous change in T_n affects private *incomes* in the first instance and leads to a smaller initial, exogenous change in aggregate demand, depending on how much the private sectors want to change their GNP purchases as a result of the income change.

The different multipliers for government expenditures and taxes imply that *equal increases in G* and T_n may stimulate aggregate demand while leaving the balance of the government budget unaltered*. Under the assumption that all changes in T_n are oppositely reflected in Y_d', it can be shown that *the "balanced-budget multiplier" is 1*. That is, equal increases in G^* and T_n will lead to a rise in equilibrium GNP of $\delta Y = \delta G^*$. The mathematical derivation of this result is given in Appendix A of this chapter, but the logic is simple. If government pur-

---

† The change is the sum of autonomous and induced components.
$$\delta C^* = \delta a' + b'\delta Y = -b\delta\tau_1 + b'(-kb\,\delta\tau_1) = -b\,\delta\tau_1(1 + b'k) = -bk\,\delta\tau_1 = \delta Y$$
since $b' = 1 - 1/k$.

chases rise but taxes drain off additional incomes equal to this rise in GNP, total private incomes will be unchanged. Assuming no shift in sectoral distribution of private incomes, private demand for GNP will be unchanged, and the rise in aggregate demand will be equal to δG^*.

2   A second way of altering net taxes is to change the marginal rate of taxation on personal incomes, that is, to alter τ_2 in Eq. (11-2). If business taxes are unchanged (γ_2 constant), then, according to Eq. (11-7), the change in τ_2 implies an equal and opposite change in f, the marginal coefficient relating disposable personal income to GNP.

Noting that b' equals bf, we obtain

$$\delta b' = b\delta f = -b\delta\tau_2$$

So the marginal propensity to consume out of GNP has been altered by the change in income tax rates. Graphically, the lines relating Y_d' and C^* to Y_s have both been rotated about their intercept points as pivots, and the direction of rotation is opposite to that for the T_n line.

The effect on the equilibrium level of GNP can be calculated from the solution equation

$$Y = k(a' + \text{INJ}^*) \tag{11-9}$$

where $k = 1/(1 - b')$

The change in income tax rates alters equilibrium GNP by changing the multiplier k *in the opposite direction to the change in tax rates.*†

Let the multiplier before tax change be $k_0 = 1/(1 - b_0')$, and after tax change be $k_\beta = 1/(1 - b_\beta')$. Then we may write

$$\delta Y = (k_\beta - k_0)(a' + \text{INJ}^*) = \delta k(a' + \text{INJ}^*) \tag{11-10}$$

Since the tax change affected only consumer demand, the entire change in GNP shows up in autonomous plus induced changes in consumer expenditures.‡

Thus, lower marginal tax rates imply higher multipliers and a more responsive, less stable economy; higher marginal tax rates imply lower multipliers and more automatic stabilization against shifts of exogenous expenditures.

---

† Some authors, assuming $\gamma_2 = 0$, use the relation $f = 1 - \tau_2$ [from Eq. (11-7)] and write the multiplier as

$$k = \frac{1}{1 - b'} = \frac{1}{1 - b(1 - \tau_2)} = \frac{1}{1 - b + b\tau_2}$$

This brings out the inverse relation between τ_2 and k.

‡ The proof that $\delta C^* = \delta Y$ is as follows:

$$\delta C^* = \delta(b'Y) = b_\beta' Y_\beta - b_0' Y_0$$

$$= \left(1 - \frac{1}{k_\beta}\right)[k_\beta(a' + \text{INJ}^*)] - \left(1 - \frac{1}{k_0}\right)[k_0(a' + \text{INJ}^*)]$$

$$= (k_\beta - 1)(a' + \text{INJ}^*) - (k_0 - 1)(a' + \text{INJ}^*)$$

$$\delta C^* = (k_\beta - k_0)(a' + \text{INJ}^*) = \delta Y$$

3   Changes may be made in taxes and transfers other than those relating to the Personal Sector, for example, tariffs, sales taxes, corporate profits taxes, and business property taxes.   The impacts of these taxes on aggregate demand and on equilibrium GNP depend on their effects on incomes and prices and on the responsiveness of the demand sectors to such price and income changes.   The analysis will not be pushed further at this place, but it seems clear that taxes on businesses which reduce GRE may not have so large an impact on aggregate demand in the short run as do personal income taxes.   And taxes which alter prices operate more by changing real demand than by changing money flows.

A NUMERICAL EXAMPLE

As an illustrative example, let us calculate the predicted impact of a 10-billdar rise in personal income taxes in 1966.   That was the year when the inflationary impact of Vietnam war expenditures began to show up.   Fiscal restraint would have been desirable to restrict aggregate demand to the level of full-employment GNP.   (Historically, taxes were not raised until mid-1968.)

1   Model with No Tax Rise

For our equations to represent the actual economy of 1966, we may use the relations between consumption function and disposable income, and GNP from Chap. 8 (Figs. 8-1 and 8-2).   To these we add the net tax receipts function graphed in this chapter (Fig. 11-3) and note that transfers to foreigners (plus statistical discrepancy) were constant at about 2 billdar.   The equation for gross retained earnings can then be derived to make all income equations add to Y_s.   The results, expressed in billdar, are:

$$Y_d' = 21 + 0.640\,Y_s$$

$$T_n = -27 + 0.244\,Y_s$$

$$\mathrm{GRE} = 4 + 0.116\,Y_s$$

$$\mathrm{TR}_f + \mathrm{SD} = 2$$

The consumer expenditure functions become

$$C^* = 0.93\,Y_d' = 0.93(21 + 0.640\,Y_s) = 19.5 + 0.595\,Y_s$$

In 1966, reported injections were

$$\mathrm{INJ}^* = G^* + I_\alpha^* + X_n^* = 156.8 + 121.4 + 5.3 = 283.5 \text{ billdar}$$

Using this in our equation model [Eq. (11-9)], we calculate equilibrium GNP as

$$Y = 2.47(19.5 + 283.5) = 748.5 \text{ billdar}$$

where $k = 1/(1 - b') = 1/(1 - 0.595) = 2.47$

This calculated value of Y compares closely with the reported value of 749.9 billdar for 1966.

TABLE 11-3 Comparison of 1966 Sector Flows with and without 10-Billdar Rise in Personal Taxes (billdar—all at current-year prices)

		Actual Flows	Values Predicted by Model		
			(1) No Tax Rise	(2) Lump-Sum Tax Rise	(3) Rise in Marginal Rate
				$(\delta\tau_1 = 10)$	$(\delta\tau_2 = 0.0134)$
Gross national product	(Y)	749.9†	748.4	725.4	726.0
Sector net incomes					
Net disposable income	(Y'_d)	498.9	500.0	475.3	476.0
Net tax receipts	(T_n)	157.9	155.6	160.0	159.8
Gross retained earnings	(GRE)	91.3	90.8	88.1	88.2
Foreign transfers and statistical discrepancy	$(TR_f + SD)$	1.8	2.0	2.0	2.0
Sector purchases of GNP					
Consumer expenditures	(C^*)	466.3	464.9	441.9	442.5
Government purchases	(G^*)	156.8	156.8	156.8	156.8
Gross domestic investment	(I_d^*)	121.4	121.4	121.4	121.4
Net exports	(X_n^*)	5.3	5.3	5.3	5.3
Saving or borrowing					
Personal saving	(S_p)	32.5	35.1	33.4	33.5
Government surplus	(S_g)	1.1	−1.2	3.2	3.0
Business borrowing	(BOR_b)	30.1	30.6	33.3	33.2
Foreign borrowing and statistical discrepancy	$(BOR_f$ and SD)	3.5	3.3	3.3	3.3
Autonomous expenditures multiplier	(k)		2.470	2.470	2.396

† Full-employment GNP was 737 billdar in 1966.

Once GNP is known, the above equations can be used to calculate sector incomes, and C^* can be derived. Then saving and borrowing can be calculated by sectors. The results of these computations are presented in column 1 of Table 11-3, where they are compared with actual 1966 data and with the figures calculated from the models with changes in taxes.

2 Model with 10-Billdar Lump-Sum Rise in Personal Taxes

A lump-sum rise in personal taxes by 10 billdar will raise τ_1 and lower e by that amount, so that the sector income equations become:

$$Y'_d = 11 + 0.640 Y_s$$

$$T_n = -17 + 0.244 Y_s$$

$$GRE = 4 + 0.116 Y_s$$

$$TR_f + SD = 2$$

The consumer expenditure equation becomes

$$C^* = 0.93(11 + 0.640\,Y_s) = 10.2 + 0.595\,Y_s$$

Assuming that INJ* would be unchanged by the rise in personal taxes, we calculate equilibrium GNP as:

$$Y = 2.47(10.2 + 283.5) = 725.4 \text{ billdar}$$

where $k = 1/(1 - 0.595) = 2.47$, as before the tax change.

The tax rise shifted the consumer expenditure and aggregate demand functions by

$$\delta a' = -b(\delta\tau_1) = -(0.93)(10) = -9.3 \text{ billdar}$$

The resulting decline in GNP was

$$\delta Y = -kb(\delta\tau_1) = -2.47(0.93)(10) = -23.0 \text{ billdar}$$

In Table 11-3 the complete breakdown of sector incomes, expenditures, and saving or borrowing is shown in column 2. We note that, compared with values in column 1, GNP is lower by 23 billdar, with all the decline in aggregate demand occurring in C^*. Changes in sector incomes are the result of the exogenous tax change plus changes induced by the change in GNP.

$$\delta Y_d' = \delta e + f\delta Y = -10 + 0.640(-23) = -24.7 \text{ billdar}$$

$$\delta T_n = \delta\tau_1 + \tau_2\delta Y = +10 + 0.244(-23) = 4.4 \text{ billdar}$$

$$\delta\text{GRE} = \delta\gamma_1 + \gamma_2\delta Y = 0 + 0.116(-23) = -2.7 \text{ billdar}$$

The rise in government income increases its surplus. The decline in private incomes reduces personal saving and increases business borrowing.

3 Model with Rise in Marginal Tax Rates

As an alternative tax policy, consider a rise in marginal tax rates on personal income sufficient to yield an added 10 billdar at the original levels of income. In the income equations, this implies a rise in τ_2 by

$$\delta\tau_2 = \frac{10}{Y_0} = \frac{10}{748.4} = 0.0134$$

and an equal decline in f. The sector income equations become:

$$Y_d' = 21 + 0.6266\,Y_s$$

$$T_n = -27 + 0.2574\,Y_s$$

$$\text{GRE} = 4 + 0.116\,Y_s$$

$$\text{TR}_f + \text{SD} = 2$$

The consumer expenditure equation becomes

$$C^* = 0.93(21 + 0.6266\,Y_s) = 19.5 + 0.5827\,Y_s$$

Assuming INJ* equal 283.5 billdar as before, we find:

$Y = 2.396 (19.5 + 283.5) = 726.0$ billdar

where $k = 1/(1 - 0.5827) = 2.396$

The rise in marginal tax rates has lowered the multiplier k from 2.470 to 2.396 because of the decline in b'.

$\delta b' = b\delta f = -b\delta\tau_2 = -0.93(0.0134) = -0.0124$

The resulting decline in GNP from the situation before the tax rise is:

$\delta Y = (k_3 - k_1)(a' + \text{INJ}*)$

$\qquad = (2.396 - 2.470)(19.5 + 283.5) = -22.4$ billdar

In Table 11-3, the sector accounts for this case are shown in column 3. The 21-billdar decline in GNP is all in C^*.

$\delta C^* = \delta a' + \delta(b' Y) = 0 + (\delta b') Y_0 + b_3'(\delta Y)$

$\qquad = (-0.0124)748.4 + 0.5827(-22.4)$

$\qquad = -9.3 - 13.1 = -22.4$ billdar

The changes in income flows, as compared to the situation before the tax rise (column 1), involve exogenous and induced components as follows:

$\delta Y_d' = \delta e + \delta(fY) = 0 + (\delta f) Y_0 + f_3(\delta Y)$

$\qquad\qquad = -0.0134(748.4) + 0.6266(-22.4)$

$\qquad\qquad = -10.0 - 14.0 = -24.0$ billdar

$\delta T_n = \delta\tau_1 + \delta(\tau_2 Y) = 0 + (\delta\tau_2) Y_0 + \tau_{2,3}(\delta Y)$

$\qquad\qquad = +0.0134(748.4) + 0.2574(-22.4)$

$\qquad\qquad = +10 - 5.8 = 4.2$ billdar

$\delta\text{GRE} = \delta\gamma_1 + \gamma_2 \delta Y = 0 + 0.116(-22.4) = -2.6$ billdar

A comparison of the figures in columns 2 and 3 shows that the impacts of the two tax increases are very similar. GNP and C^* do not decline quite so much when marginal tax rates change, because tax revenues bear more of the burden of an income decline and private incomes less when the tax schedule is more progressive.

The effects of the two policies on equilibrium GNP are illustrated in Fig. 11-4, both for the aggregate demand-supply and for the injections-withdrawal equilibria. (The lines for case 3 are distorted a little to separate the points 2 and 3 enough to be seen.) In both cases the movement from point 1 consists of a vertical autonomous shift to point 1' and then an induced movement along the new Y^* or WDL* line to the new equilibrium point. The autonomous shift in both cases

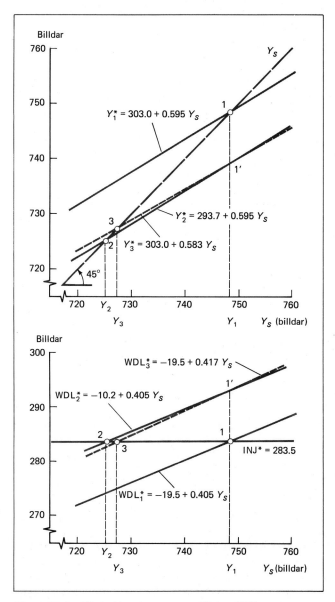

Figure 11-4   Effect of tax change on equilibrium GNP.

amounts to b (autonomous tax shift) $= 0.93(10) = 9.3$ billdar.   The induced shift leftward along the Y^* or WDL* line then leads to a decline in equilibrium GNP by $\delta Y = k(9.3)$.   This differs for cases 2 and 3 because of the different values for k.

Either of these tax increases would have held GNP below the full-employment level of 737 billdar in 1966, on the model's assumptions that the tax rise would not have led to shifts of the relation between C^* and Y'_d nor to changes in G^*, I^*_d, or X^*_n.   The effects of autonomous shifts in WDL* or INJ* are worked out in Appendix B.

GOVERNMENT INFLUENCE ON CAPITAL MARKETS

Changes in government purchases, taxes, and transfer payments are instruments of fiscal policy.   In the previous discussion, it was assumed that investment spending remained constant.   This implied that interest rates and the supply-of-funds schedule were unchanged.   But exogenous changes in government tax receipts or expenditures will lead to changes in the government surplus (or deficit) and to compensating changes in personal saving and gross retained earnings flows.   So shifts occur in flows of funds through capital markets which may well affect interest rates, and hence those investment expenditures which are sensitive to interest rates. These effects will depend on how the Government Sector uses its surplus funds or how it finances a deficit.   Consideration of government capital-funds transactions brings us into the area of monetary policy, which will be discussed in later chapters.

In addition to financing its current-account surplus or deficit, governments engage directly in capital-funds transactions which influence the direction and extent of private economic activity.   Governments borrow funds for public enterprises and for private companies which agree to locate plants in a state; they lend to small businesses and channel funds into the mortgage market; they provide crop loans to farmers and loans to finance foreign trade.

In the complete picture of government influence on economic activity, the effects of monetary policy and capital-market transactions must be taken into account as well as the effects of fiscal policy.

FISCAL POLICY TO MAINTAIN FULL EMPLOYMENT

What fiscal policy should the government adopt to bring the economy to equilibrium at full-employment output?

Since the government cannot change the aggregate supply curve much in the short run, it must try to ensure that aggregate demand will equal full-employment output when the demand sectors receive the incomes generated at full employment. An equivalent statement is that desired injections must equal desired withdrawals at full employment.

Think of the situation this way: For a given tax structure, total incomes paid out by Production Units will vary with the level of output (Y_S) they choose and will be divided among demand sectors according to a stable pattern of distribution.

If, in addition, desired consumption expenditures are related to disposable income by a stable functional relation, then we conclude that consumer spending will be related to Y_S in a stable fashion. We have used a linear relation for this consumption function.

$$C^* = a' + b'Y_S$$

Desired withdrawals have been defined as GNY less consumer expenditures; so they will also exhibit a stable relation to Y_S.

$$WDL^* = Y_S - C^* = -a' + (1 - b')Y_S$$

Withdrawals consist of income flows to Government, Business Capital Account, and Foreign Sectors, plus personal saving.

$$WDL^* = Y_S - C^* = S_p^* + T_n + GRE + T R_f$$

It seems reasonable that these components of income and saving should rise and fall in a stable, positive relation to Y_S; that is, they have induced components. As may be seen in the functional relation above, WDL* will change by a fraction $(1 - b')$ of a change in Y_S, where $(1 - b')$ is the reciprocal of the autonomous expenditure multiplier and is about 0.4 for the U.S. economy.

Now we know that economic equilibrium is reached at a level of Y for which desired injections equals desired withdrawls. Injections include government purchases, gross private domestic investment, and net exports; and these components of aggregate demand are assumed to be largely autonomous, that is, not dependent on Y_S in the short run. The equilibrium equation becomes†

$$INJ^* = WDL^*$$

or

$$G^* + I_d^* + X_n^* = -a' + (1 - b')Y$$

The solution is

$$Y = \frac{1}{1 - b'}(a' + G^* + I_d^* + X_n^*) \tag{11-11}$$

If the government wishes equilibrium Y to be the full-employment level (Y_{FE}), then it may:

1 Adopt tax, transfer, and monetary policies which lead to values of a' and b' consistent with a solution value Y_{FE} for a given level of INJ* in Eq. (11-11),

2 Adjust government expenditures so that the solution of Eq. (11-11) becomes Y_{FE}, for given values of a', b', I_d^*, and X_n^*,

3 Adopt fiscal and monetary policies which lead to private injections $(I_d^* \times X_n^*)$ adequate to yield a solution Y_{FE}, for given values of a', b', and G^*, or

4 Use a combination of the above policy actions.

---

† Adding C^* to both sides would convert this into the equilibrium condition $Y^* = Y_S$.

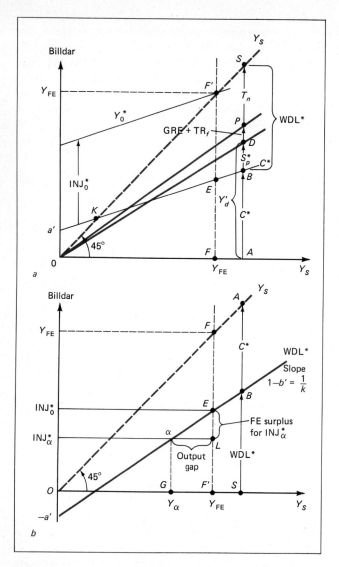

Figure 11-5 Fiscal policy and full employment. (a) Aggregate demand and supply. (b) Aggregate injections and withdrawals.

If we assume a' and b' fixed, the achievement of equilibrium at full employment is illustrated in the diagrams of Fig. 11-5. In Fig. 11-5a, autonomous desired injections INJ_0^* are added to the consumption-function line to yield an aggregate demand line Y_0^* which intersects the 45-degree line at an equilibrium point corresponding to full-employment output ($Y_{FE} = OF$). Lines have been added to the diagram to illustrate the concept of sector incomes and withdrawals as a function of Y_S. The vertical distance from the Y_S axis up to the line OD measures net disposable income (Y_d'); so the vertical distance from the C^* line to the line OD measures personal saving. The vertical distance between lines OD and OP measures $GRE + TR_f$ as a function of Y_S; and the distance from the OP line up to the 45-degree line must then equal T_n. Total withdrawals are measured by the vertical distance from the C^* line up to the 45-degree line (Y_S), which agrees with the definition

$$WDL^* = Y_S - C^*$$

WDL^* increases as Y_S increases and equals INJ_0^* when $Y_S = Y_{FE}$.

In Fig. 11-5*b* this WDL^* distance has been plotted versus Y_S. The equation of the line is

$$WDL^* = -a' + (1 - b')Y_S$$

A horizontal line at height INJ_0^* intersects the WDL^* line at the equilibrium level of GNP, in this case the full-employment level Y_{FE}.

If injections are less than the level INJ_0^* which leads to equilibrium at full employment, then the economy will come into equilibrium with GNP less than potential output. For example, if injections are INJ_α^*, then equilibrium GNP will be $Y_\alpha(=OG)$, at which $WDL_\alpha^* = INJ_\alpha^*$. The shortfall of output below the full-employment level is called the *output gap* (GF'). It arises because desired injections (INJ_α^*) fall short of the full-employment level of desired withdrawals (F'E). If producers mistakenly produced output Y_{FE}, there would be a surplus of WDL_{FE}^* over (autonomous) desired injections INJ_α^*. The surplus would equal undesired inventory accumulation (I_u) and would lead to a subsequent decline in Y_S toward the equilibrium level Y_α. This full-employment surplus ($WDL_{FE}^* - INJ_\alpha^*$) is another measure of the policy problem faced by the government. To bring the economy to equilibrium at full employment, with a' and b' fixed, the government must reduce this full employment surplus to zero; that is, it must raise desired injections to the level of desired withdrawals at full employment, where $WDL_{FE}^* = -a' + (1 - b')Y_{FE}$.†

The implications of this analysis for fiscal policy aimed at maintaining full employment are clear. If aggregate demand is less than Y_{FE}, then: (1) government purchases should be increased, or (2) taxes and transfers should be changed so as to stimulate more private injections or to shift the consumption function upward. (In the withdrawals-injections diagram, the line for desired withdrawals shifts down by the same amount that the consumption line shifts up, leading

---

† The output gap and the full-employment surplus are related via the multiplier, since a rise in injections by the amount of the full-employment surplus would raise equilibrium Y by the amount of the output gap. That is,

output gap $= k \times$ (full-employment surplus)

or

$$\frac{\text{full-employment surplus}}{\text{output gap}} = \frac{1}{k} = 1 - b' = \text{slope of WDL* line}$$

to higher equilibrium Y for given injections.)   The magnitude of needed shifts in injections and withdrawals lines combined equals the full-employment surplus corresponding to actual desired injections.   Usually the slope of the WDL* line is known, so that one can project upward along it from point α to the point E where the WDL* line intersects the vertical F'F erected at full-employment GNY. The vertical distance of point E above the level of actual injections (INJ_α^*) gives the full-employment surplus $(=\text{LE})$.   Refer to the analysis in Appendix B.

It is possible to push the analysis a little further in answer to the question: Is it the Government or the Private Sector that would have a surplus of withdrawals over injections at full employment?   The full-employment surplus can be broken into government and private components as follows.

$$\text{FE surplus} = \text{WDL}_{\text{FE}}^* - \text{INJ}_\alpha^*$$

$$= (T_{n,\text{FE}} - G_\alpha^*) + [(S_p^* + \text{GRE} + \text{TR}_f)_{\text{FE}} - (I_{d,\alpha}^* + X_{n,\alpha}^*)]$$

$$= \text{government FE surplus} + \text{private FE surplus}$$

In order to bring the economy to equilibrium at full-employment GNP the Government Sector must adjust its spending and its tax structure to make the government FE surplus offset any FE surplus of the Private Sectors.   Government economists would have to predict the influence of monetary policy and market variables on S_p^*, $I_{d,\alpha}^*$, and $X_{n,\alpha}^*$ to arrive at an estimate of the private FE surplus with existing tax and transfer policies.   Then fiscal policies would need to be adopted to yield a FE government surplus equal and opposite to the estimated private surplus at full employment.†   Different combinations of changes in G^*, taxes, and transfers could achieve the desired government FE surplus.   So a policy mix might be chosen to achieve some other goal in addition to full employment, for example, a desirable allocation of potential GNP between private and public uses or stimulation of investment to provide for economic growth.   Though the logic is reasonably clear, its application is complicated by changes in private behavior patterns, by legislative inertia or opposition to required policies, by the possibilities of using monetary policy to affect private injections and withdrawals, and by the fact that full-employment GNP is a moving target, changing because of changes in labor force, productivity, and prices.

As an example of the use of withdrawals-injections analysis in evaluating fiscal policy, consider Fig. 11-6, which shows withdrawals and injections in relation to full-employment levels for the years 1964 through 1967, a period when federal income tax rates were essentially unchanged while Vietnam war expenditures were rising.   GNP is plotted along the horizontal scale, and three vertical

---

† Allowance would need to be made for the effect of tax and transfer changes on private withdrawals at full employment.   As shown earlier in this chapter, an autonomous downshift of the T_n line by $\delta\tau_1$, assuming all the impact on personal disposable income, will cause an upshift of the C^* line by $b\delta\tau_1$. Hence the private WDL* line will shift up by $(1-b)\delta\tau_1$ because of the rise in S_p^*.   The combined shift of government and private WDL* lines will be $-\delta\tau_1 + (1-b)\delta\tau_1 = -b\delta\tau_1$, equal and opposite to the shift of the C^* line, as it must.

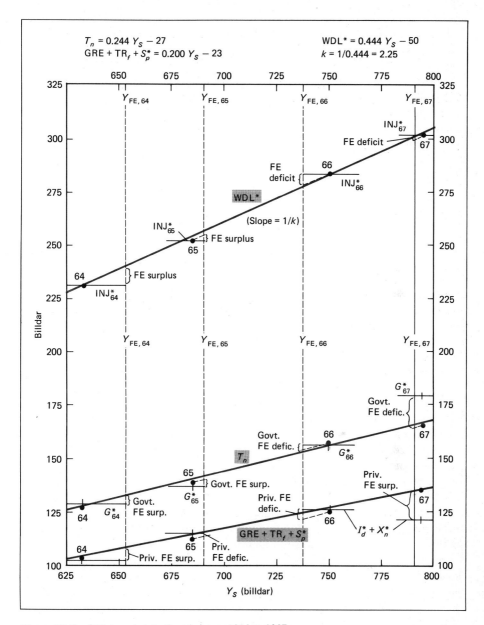

Figure 11-6 Withdrawals-injections balance, 1964 to 1967.

dashed lines indicate the current-dollar values of potential GNP (full-employment output) in the four years. (The increasing increments in full-employment GNP reflect the more rapid rises in GNP price index in the later years.) The circles labeled by years locate the actual values of withdrawals versus GNP for each year, and the three sloping straight lines are averaged through the points to illustrate the way in which private withdrawals, net government receipts, and total withdrawals varied with GNP during this four-year period. The ordinates of the total withdrawal line (WDL) are the sum of the ordinates for the two lower lines, and its equation is the sum of the other two equations. Thus,

$$T_n = 0.244 Y_s - 27 \qquad \text{for government withdrawals}$$

$$\text{GRE} + \text{TR}_f + S_p^* = 0.200 Y_s - 23 \qquad \text{for private withdrawals}$$

$$\text{WDL}^* = 0.444 Y_s - 50 \qquad \text{for total withdrawals}$$

Recall that *the slope of the WDL line is always* $(1 - b') = 1/k$. Hence $k = 1/0.444 = 2.25$.

The total full-employment surplus (or deficit), and that for the Government and Private Sectors separately, can be estimated graphically for each year by drawing a line from each withdrawals circle parallel to the appropriate WDL guideline to an intersection with the vertical Y_{FE} line for that year. A horizontal INJ* line, also projected to intersect the proper Y_{FE} line, will cut off on it a segment equal to the FE surplus $= \text{WDL}^*_{FE} - \text{INJ}^*_\alpha$.†

In 1964 the total FE surplus was about 9.3 billdar, with Government and Private Sectors contributing about equal amounts. Consequently, the economy came into equilibrium at a GNP some 21 billdar below the full-employment level of 653 billdar. [Note that k times the FE surplus equals 2.25 (9.3), or 21 billdar.] An increase in government purchases by 9.3 billdar would theoretically have stimulated the economy to full employment, assuming that private injections and the government and private withdrawals lines were unchanged.

Alternatively, if it were decided to keep government purchases constant, the economy might be brought to full employment in 1964 by lowering the WDL line, keeping injections constant. A cut in taxes (downward shift in the T_n line) would do this. However, a tax reduction (assume the lump-sum type $\delta\tau_1$) greater than 9.3 billdar would be needed, because private withdrawals would shift up as private income flows increase at each level of GNP. If we assume that the tax cut applies only to personal income so that the upshift of private withdrawals occurs because of a rise in personal saving, then the exogenous shift in the total withdrawal line

---

† The assumption that the injections would not change if Y_s were different in a given year is probably not strictly true. Government purchases and domestic investment are largely determined by decisions made on the basis of the previous year's economic conditions, but there may be some short-term alterations of plans. And inventory investment and imports (hence X_n) will depend to some extent on current-year GNP. If the variation of injections versus GNP can be quantified, the injections lines should be expressed as equations with terms involving Y_s and should be drawn with appropriate slopes on the graph instead of being taken as horizontal.

becomes autonomous $\delta\text{WDL}^* = b(\delta\tau_1)$. Since this must equal the FE surplus of 9.3 billdar, we find the required

$$\delta\tau_1 = \frac{9.3}{b} = \frac{9.3}{0.93} = 10 \text{ billdar}$$

Actually, taxes had been reduced about 8 billdar in early 1964. This stimulus, plus rising expenditures for the Vietnam war, raised aggregate demand above Y_{FE} by 1966, leading to a full-employment deficit, or excess of injections over desired withdrawals at Y_{FE}. As shown by the points for 1966 in Fig. 11-6, in both the Government and Private Sectors, injections had risen above the level of full-employment withdrawals. Actual GNP at 750 billdar was some 12.5 billdar above the full-employment level. Hence a decline in G^* by

$$\delta G^* = \frac{1}{k}(Y_{FE} - Y_{66}) = 0.444\,(-12.5) = -5.6 \text{ billdar}$$

would have been required to bring the economy down to Y_{FE}, or a rise in net taxes by

$$\delta T_n = \frac{1}{kb}(Y_{FE} - Y_{66}) = -\frac{-12.5}{(2.25)(0.93)} = 6.0 \text{ billdar}$$

Even though the Government Sector had a small actual surplus, that surplus would have become a deficit at Y_{FE}, whereas a government FE surplus was needed to offset the Private Sector's deficit at full employment.

In 1967, fiscal policy was out of control. Government expenditures rose rapidly, and despite the rise in net revenues, a deficit of 14 billdar opened up in the government accounts. At full employment the deficit would have been about the same. Fiscal restraint was clearly in order. Tax rises were requested by President Johnson, but Congress did not vote them until early 1968. Fortunately the Private Sectors were running a large surplus at full employment, presumably because the tight money policy in 1966 had reduced private injections.

This review of fiscal policy in relation to full-employment output during the mid-1960s brings out the dynamic nature of the problem. *Since the full-employment GNP increases each year because of the increase in real potential output and also because of price rises, therefore full-employment government net receipts will increase each year if the tax structure is kept constant.* This rise in full-employment revenues has been called a *fiscal drag* or a *fiscal dividend*, depending on the point of view of the commentator. If government purchases are held constant, the rise in full-employment net receipts would generate a rising full-employment surplus and tend to prevent aggregate demand from rising in step with full-employment output. The economy would be continually dragged down below full-employment levels. On the other hand, the rise in full-employment net receipts with no discretionary changes in tax rates does give the Government Sector added leeway in its policy making. The gains may be used either to expand government purchases to levels consistent with full-employment receipts, or to stimulate private spending via added transfer payments or reduction in taxes.

Sometimes the rule for fiscal policy action is stated as follows: *Adjust government purchases and tax rates so that the government budget will be in balance when the economy is operating at full employment.* If government purchases are fixed at this level (the FE value of T_n), the variation of government net receipts with GNP would lead to a government deficit if the economy were in recession, and to a government surplus if the output exceeded the full-employment level. The net effect of Government Sector injections and withdrawals would thus be stimulating when aggregate demand was weak, restraining when aggregate demand was excessively high, and neutral when the economy was just at full employment.

Such a policy of budget balance at full employment is better than a policy of balancing the budget year by year, in good times and bad. However, as shown in Fig. 11-6, under some market conditions private desired injections may not equal the full-employment level of desired private withdrawals. Then, if the economy is to operate at its potential, it will be necessary for the Government Sector to maintain a FE surplus which is the negative of the private FE surplus. Of course, the government, by its monetary and fiscal policy actions, has some influence on the magnitude of the full-employment surplus or deficit of the Private Sectors. More on this as we analyze money and credit markets in the next few chapters. The underlying principle is that the economy will come to equilibrium at a level for which INJ* = WDL*, equivalent to $Y^* = Y_s$. For equilbrium at *full employment*, INJ* = WDL^*_{FE}, or $Y^* = Y_{\text{FE}}$ when Y_s is at the full-employment level.

APPENDIX A.   The Balanced-Budget Multiplier

Assume that government expenditures and personal taxes are increased so as to leave the budget balance unchanged after the new equilibrium is reached.

$$\delta G^* = \delta T_n \tag{11-12}$$

Find the change in equilibrium GNP, assuming I_d^* and X_n^* constant.

Assume also that GRE and TR_f are constant, that is, γ_2 equals zero in Eq. (11-3) of the text. Then

$$\delta Y_d = \delta Y - \delta T_n \tag{11-13}$$

and

$$\delta \text{WDL}^* = \delta T_n + \delta S_p^* \tag{11-14}$$

1.   Solution Using Aggregate Demand-Supply Equilibrium

$$\delta Y = \delta C^* + \delta G^*$$

$$\delta Y = b\, \delta Y_d + \delta G^* \tag{11-15}$$

Substituting Eq. (11-13) into Eq. (11-15), we obtain

$$\delta Y = b(\delta Y - \delta T_n) + \delta G^*$$

Combining terms in δY and noting that δT_n equals δG^*, we obtain

$$(1 - b)\delta Y = (1 - b)\delta G^*$$

Thus, δY equals δG^* and the balanced-budget multiplier $\delta Y/\delta G^*$ equals 1.

2. Solution Using Aggregate Withdrawals-Injections Equilibrium

$$\delta INJ^* = \delta G^*$$

$$\delta WDL^* = \delta T_n + \delta S_p^*$$

Equating δINJ^* and δWDL^* in equilibrium and noting that δT_n equals δG^*, we find

$$\delta G^* = \delta G^* + \delta S_p^*$$

It follows that

$$\delta S_p^* = 0 = (1 - b)\delta Y_d$$

which requires $\delta Y_d = 0$.

From Eqs. (11-12) and (11-13), this yields

$$\delta Y_d = \delta Y - \delta G^* = 0$$

or

$$\delta Y = \delta G^*$$

Once again, the balanced-budget multiplier is 1, given the initial assumptions.

APPENDIX B Effects on Equilibrium GNP from Autonomous Shifts in Injections and Withdrawals

From the basic definition we have

$$WDL^* = (T_n + GRE + TR_f) + S_p^* = (Y_s - Y_d) + [-a + (1 - b)Y_d] \qquad (11\text{-}16)$$

If we assume Y_d to be a linear function of Y_s, then WDL* becomes a linear function of Y_s also. Substitution of $Y_d = e + fY_s$ into Eq. (11-16) leads to

$$WDL^* = -(a + be) + (1 - bf)Y_s = -a' + (1 - b')Y_s$$

(This could also be derived from the relations $WDL^* = Y_s - C^*$, where $C^* = a' + b'Y_s$.)

The autonomous part of WDL* is $WDL_{AUT}^* = -a'$; the induced part is $(1 - b')Y_s = (1/k)Y_s$, since $k = 1/(1 - b')$. Thus we have

$$WDL^* = WDL_{AUT}^* + \frac{1}{k}Y_s$$

In equilibrium, of course, WDL* equals INJ*, and we are assuming INJ* to be autonomous here.

1 *Assume an autonomous shift in INJ*, with the WDL* line fixed.* To find δY, we note $\delta INJ^* = \delta WDL^* = (1/k)\delta Y$, or

$$\delta Y = k(\delta INJ^*)$$

This is a familiar result, assuming no *shift* of consumption as a function of Y_s.

2 *Assume a change in the autonomous part of WDL*, with INJ* constant.* In equilibrium:

$$\delta INJ^* = \delta WDL^*$$

$$0 = \delta WDL_{AUT}^* + \frac{1}{k}\delta Y$$

$$\delta Y = -k(\delta WDL_{AUT}^*) = k(\delta a')$$

If the autonomous shift in the WDL* function is caused by an autonomous shift in personal taxes $\delta\tau_1$, then $\delta\text{WDL}^*_{\text{AUT}} = -b\delta e = b(\delta\tau_1)$, and

$$\delta Y = -kb(\delta\tau_1)$$

3 *Assume autonomous shifts in both δINJ^* and WDL**.   In equilibrium

$$\delta\text{INJ}^* = \delta\text{WDL}^*$$

$$\delta\text{INJ}^* = \delta\text{WDL}^*_{\text{AUT}} + \frac{1}{k}\,\delta Y$$

$$\delta Y = k(\delta\text{INJ}^* - \delta\text{WDL}^*_{\text{AUT}})$$

$$\delta Y = k(\delta\text{INJ}^* + \delta a')$$

As might be expected, *the change in Y resulting from autonomous shifts in both injections and withdrawals is the sum of the changes in Y caused by each shift separately*.   The result is the same as that obtained from the aggregate demand-supply approach where we found

$$Y = k(a' + \text{INJ}^*)$$

and

$$\delta Y = k(\delta a' + \delta\text{INJ}^*)$$

PROBLEMS

1 **a** Given a projection of full-employment real GNP for the year ahead and given government budgets determining how much of the year's output will be purchased by the Public Sector, what principal fiscal policy instruments do federal decision makers have for ensuring that private demands will be adequate to bring aggregate demand to the full-employment level, neither too little nor too much?

   b What are some reasons for possible inadequacy of fiscal policies to lead to equilibrium at full employment?

   c Since labor force and average productivity of labor normally increase from year to year, full-employment GNP, in real or money measure, will rise through time.   What changes, if any, will need to be made in fiscal policies in order to keep aggregate demand near to the value of rising full-employment output?

2 Developing countries frequently find that attempts to increase investment lead to inflation.   Apparently, the high MPC, and hence high expenditure multiplier, leads to a large rise in C^* when I_d^* is increased, making aggregate demand exceed the value of full-employment output at current price levels. What fiscal policies would be helpful in permitting large domestic investment without serious inflation?

3 Evaluate the statement: "Government deficits cause inflation by causing aggregate demand to exceed the full-employment level of output."

4 In 1968 the U.S. economy experienced aggregate demand above the full-employment level, largely because of rising military purchases related to the Vietnam war. The federal government adopted a fiscal policy to restrain aggregate demand. A 10 percent federal surtax on personal incomes was levied as a counterinflationary measure, effective at mid-1968. However, in the second half of 1968 the average propensity to save declined, raising the consumption function just as the higher taxes took effect.

Let us calculate equilibrium values of Y and of sector incomes and purchases in the second half of 1968 under three sets of assumptions.

A Tax and consumption functions the same as in the first half of 1968

B Federal tax rates on personal incomes up 10 percent, but consumption unchanged as a function of Y'_d

C Federal tax rates up by 10 percent and marginal propensity to consume up by 0.013

The approximate equations describing each set of assumptions are tabulated below.

	State A	State B	State C
Y'_d =	$25 + 0.640Y_s$	$25 + 0.633Y_s$	$25 + 0.633Y_s$
T_n =	$-21 + 0.244Y_s$	$-21 + 0.251Y_s$	$-21 + 0.251Y_s$
GRE =	$-4 + 0.116Y_s$	$-4 + 0.116Y_s$	$-4 + 0.116Y_s$
$TR_f + SD$ =	0	0	0
C^* =	$0.924Y'_d$	$0.924Y'_d$	$0.937Y'_d$
INJ * =	334.5	334.5	334.5

a Calculate equilibrium values to fill in the following table. (Dollar flows are in billdar.)

	State A	State B	State C	Actual 2H, 1968
Y				882.7
Y'_d				584.8
T_n				200.6
GRE				97.2
$TR_f + SD$				0
C^*				548.2
INJ *				334.5
S^*_p				36.6
S^*_g				-3.0
k				—
Y_{FE}	873	873	873	873

b Describe briefly what the above calculations reveal about the attempt to moderate inflation during 1968. At the actual level of Y for the second half of 1968, by how much was the increase in tax rates depressing C^*, and by how much was the decline in propensity to save raising C^*?

5 To make an injections and withdrawals analysis:

a Use information from the following Economic Data Sheet to plot an injections and withdrawals chart similar to Fig. 11-6 but covering 1967 through 1972. On it plot the withdrawals lines from Fig. 11-6, to serve as a guideline for estimating slopes.

Economic Data Sheet, 1966–1972

	1967	1968	1969	1970	1971	1972
			(in billdar)			
Withdrawals : Total	$301.9	$328.1	$350.8	$359.4	$387.7	$429.0
S_p	40.4	39.8	38.2	56.2	60.5	52.6
GRE	93.0	95.4	97.0	97.0	110.2	125.9
$TR_{f,p} + SD$	0.1	−1.9	−5.3	−5.4	−1.2	−2.6
Private WDL's	$133.5	$133.3	$129.9	$147.8	$169.4	$175.9
Government WDL's						
$(T_n + TR_{g,f})$	168.4	194.8	220.9	211.6	218.3	253.1
Injections : Total	$301.9	$328.1	$350.9	$359.4	$387.7	$429.0
I_d	116.6	126.0	139.0	136.3	153.7	179.3
X_n	5.2	2.5	1.9	3.6	−0.2	−6.0
Private INJ's	$121.8	$128.5	$140.9	$139.9	$153.5	$173.3
Government INJ's						
(G)	180.1	199.6	210.0	219.5	234.2	255.7
Full-employment GNP	$791.3	$856.0	$933.2	$1,023.8	$1,112.8	$1,196.4
$Y_{R,FE}$ (1958 prices)	673.0	699.9	727.9	757.0	787.3	818.8
GNP price index	117.59	122.30	128.20	135.24	141.35	146.12
Actual GNP	$793.9	$864.2	$930.3	$977.1	$1,054.9	$1,158.0
Percent of full-employment GNP	100.3%	101.0%	99.3%	95.4%	94.7%	96.6%
Civilian labor force, percent unemployed	3.8%	3.6%	3.5%	4.9%	5.9%	5.6%
Federal full-employment budget, surplus or deficit	$−11.8	$−7.1	$11.8	$7.6	$0.6	$−7.4
Percent of full-employment GNP	−1.5%	−0.8%	1.3%	0.7%	0.1%	−0.6%
Money supply (daily average, in $ billion)	$181.6	$194.3	$206.5	$215.7	$230.7	$245.6
Interest rate (Aaa bonds, Moody's)	5.51%	6.18%	7.03%	8.04%	7.39%	7.21%
			Percent Change from Previous Year			
GNP price index	3.2%	4.0%	4.8%	5.5%	4.5%	3.4%
Money supply	3.9%	7.0%	6.3%	4.5%	7.0%	6.4%
Current-dollar GNP	5.9%	8.9%	7.6%	5.0%	8.0%	9.8%
Constant-dollar GNP	2.6%	4.7%	2.7%	−0.4%	3.3%	6.2%

b Calculate full-employment surpluses or deficits year by year, in total and for the Government and Private Sectors separately.

c Discuss what the chart and the Economic Data Sheet reveal regarding reasons for weakness of the economy in 1970 through 1972. Comment on fiscal policy during this time. Income taxes were raised in mid-1968 and then lowered again in 1970 and 1971. Social security tax rates were increased at the beginning of 1969, 1971, and 1972; social security benefits were raised in the second quarter of both 1970 and 1971, and in the fourth quarter of 1972. A price-and-wage freeze and controls began in August 1971, and continued throughout 1972.

Part Three

Money, Capital Markets, and General Equilibrium

Chapter 12
The Demand for Money Stocks

THE IMPORTANCE OF STOCKS OF MONEY IN MACROECONOMIC ANALYSIS

Money is purchasing power. We receive money incomes for providing services of labor or capital funds; we pay out money for the goods and services that yield us satisfactions and also for taxes and for financial investments that we hope will provide future benefits. Money is exchanged against real goods and services in GNP flows; it is the medium by which income, transfer, and tax payments are made; and it is used in myriad other exchanges of real and financial assets among individual economic units in this country and abroad.

But income and expenditures are *flows*. Why is the size of the *stock* of money in an economy important for determining and analyzing the *flows* of gross national income or GNP? Streetcorner economists suggest that money is important. Inflation is described as "too much money chasing too few goods," whatever that means. The Federal Reserve System is viewed as having the power to make money "tight" or "easy," whatever that means. And bankers are regarded as having arbitrary power to create or destroy money, and to determine the amount of credit extended and the interest rates charged.

Money does matter. The size of the stock of money in an economy, and its rate of change, influence the level and rate of change of economic activity and of prices. The causal chains linking money stock to GNP flows are not entirely clear yet, and the monetary reaction coefficients in econometric models are not agreed on, but we can push deeper than the streetcorner economist's concepts just cited.

DEFINITION OF MONEY

Money is a generally accepted medium of exchange, acceptable in payment for goods and services rendered to a buyer and acceptable for settlement of debts. Its units of measure are the currency units of the nation involved—dollar, pound, franc, mark, ruble, krona, lira, yen, rupee, peso, and many others. Its physical embodiment nowadays is in coins, paper currency, and bank deposits (usually exchanged in the form of checks), but at various times in history, money has taken such forms as cattle, gold, shells, beads, cigarettes, and beaver skins.

The current Federal Reserve Board definition for money of the United States is this: "Averages of daily figures for (1) demand deposits of commercial banks other than domestic interbank and U.S. government, less cash items in process of collection and F.R. float; (2) foreign demand balances at F.R. Banks; and (3) currency outside the Treasury, F.R. Banks, and vaults of commercial banks."†
This measure is chosen as the best estimate of the stock of circulating medium being used by economic units to carry on economic transactions.

Demand deposits counted in the money supply include those owned by the Personal, Business, and Foreign Sectors, state and local governments, and financial institutions except commercial banks. Deposits of commercial banks with other commercial banks and with the Federal Reserve banks are excluded because they are related to operation of the monetary system itself rather than to normal economic transactions. Federal government deposits are also excluded because they reflect fiscal and monetary policy decisions rather than normal economic activity; for example, they may consist of the proceeds of government bond sales to the nonbank public, temporary deposits of taxes with commercial banks, and federal government deposits with its banker, the Federal Reserve System. Cash items in process of collection and Federal Reserve float are payments in transit from buyer to seller. During the period after a check has been added to the recipient's account but has not yet been subtracted from the payer's bank account, the amount of money involved will be reported by the banks in both accounts. The money supply would be overstated unless the amount of these payments in process of collection and the Federal Reserve float of uncleared checks were subtracted from the deposit totals as reported by the banks. Foreign demand deposits at the Federal Reserve banks are included in the money supply because they are balances held by foreigners to conduct economic transactions with United States buyers and sellers.

Currency consists of gold and gold certificates, coins issued by the U.S. Treasury, and "paper money" (notes issued by the Federal Reserve System plus a small amount issued by the U.S. Treasury). Monetary gold and gold certificates are not held outside the Treasury and the Federal Reserve banks; so the currency component of the money supply consists of the coins and notes "in circulation," that is, held by the general public outside the Treasury, Federal Reserve banks, and vaults of commercial banks. Note that state and local governments are considered part of the "general public," as are financial institutions other than commercial banks, for example, savings banks, savings and loan associations, security dealers, and insurance companies.

An item of money, then, is a claim against the U.S. Treasury (coins and U.S. notes), or against the Federal Reserve System (Federal Reserve bank notes), or against a commercial bank (demand deposits). The institutions against which

---

† Board of Governors of the Federal Reserve System, *Federal Reserve Bulletin*, August 1974, p.A14, footnote to table, "Measures of the Money Stock."

these claims are held are regarded as so reliable that their promises to pay may become generally acceptable as a medium of payment for goods and services and debts. The distinguishing characteristics of these claims are their *liquidity* (general and immediate acceptability) and the fact that *they yield no interest.*

Some other claims are so widely accepted or so readily converted into money that they are close substitutes for items of money as just defined. One widely used broad definition of the money supply does include savings deposits in commercial banks in addition to the items of the preceding definition. But no further stretching of the definition has gained wide acceptance; that is, deposits in savings banks, in credit unions, or in savings and loan associations are not counted as money, nor are letters of credit, credit cards, or U.S. savings bonds.

BENEFITS AND COSTS OF MONEY

Money serves as (1) *a unit of account or of market value*; (2) *a means of payment*; and (3) *a store of purchasing power.*

In a money economy the market value of any item of wealth or any service may be expressed in terms of a single *money price*, which *is the number of units of money which exchanges for one unit of physical goods or services in the marketplace.* The use of a *single unit of account* provides an enormous simplification in conducting economic transactions as compared with a barter economy. In barter, each good or service would have many prices, expressed as the number of physical units of each other item which exchange for one physical unit of the good in question. The monetary unit also provides a convenient, single measure in terms of which assets and debts may be valued and compared at different times. Modern business accounting, with its operating statement and balance sheet, would be well-nigh impossible without a monetary system.

The utility of having one or two types of assets which serve as a universally acceptable means of payment is enormous, both for the individual and for society as a whole. It is hard to imagine the conduct of trade or the taxing and spending by a central government in a barter economy. Perhaps the closest comparable situation in modern times is that existing in an economy subject to hyperinflation: confidence in the currency breaks down and workers take time off from their jobs to rush to markets with their rapidly depreciating day's pay or travel to the countryside to barter cigarettes or other wealth items for food.

Even under normal conditions, a stock of money provides a flow of services to its holder by reducing the costs of making market transactions. Suppose that an individual decides to keep his or her money assets to a minimum. When a man, say, receives his paycheck, he immediately buys securities or a savings deposit in order to receive interest income on his funds. Then each day he sells part of the securities or withdraws from the savings account to obtain funds to meet that day's expenditures. Over the period until his next paycheck arrives, he earns interest on funds averaging half his paycheck. But he has incurred costs of his time plus brokerage fees for all the securities transactions and withdrawals he has made. Holding his paycheck in the form of money and paying out checks or currency day by day saves all those transactions costs. This flow of services derived from a money stock needs to be weighed against the loss of interest income in determining the optimum average money balances to hold. Some compromises would be possible, for example, by devoting half the paycheck to interest-bearing assets for the first half of the month and then cashing them to meet expenditures for the second-half month, or by paying credit charges to postpone payments until the next paycheck is due.

In addition to its services in reducing transactions costs, a money stock may reduce risks of capital losses or increase prospects of capital gains. Securities held by savers fluctuate in dollar value and may be "down" at the time when they must be sold to raise funds. Also, an available stock of money enables the holder to make advantageous purchases of other assets at times when their prices are relatively low. Of course, there is some small risk in holding money—the risk of theft of currency or of bank failure. And there is the risk of loss of purchasing power during an inflation. The latter risk is shared by savings deposits and bonds, but might be avoided by holding real assets or corporate stock.

On balance, proportions of money and other assets held by wealth owners in the nation will be determined by a balancing at the margin of the costs and benefits of holding various assets. One can imagine developing a marginal net-return curve for each alternative asset, based on estimated values of marginal streams of future services or income derivable from increments to the stock of each asset—just as was done for capital goods in Chap. 9. These would be the demand curves for stocks of alternative assets. Presumably, marginal net revenue for each asset, and hence its rate of return, will decline as the stock of a given asset is increased. The stock of each asset held would be that at which the rate-of-return (demand) curve declines to the level of the appropriate interest rate—rate of return on substitutable assets. In equilibrium, the rate of return on increments of all assets would be equal for each wealth owner; one could not improve one's own welfare by any portfolio shifts between various assets.

For the economy as a whole, the cost of producing money used to be large when it involved mining precious metals with market value equal to the face value of the coin, or when it required export of goods with value equal to the gold or silver needed for the domestic money system. Fractional-value coins, fiat paper money, and credit instruments have greatly reduced the cost of producing money domestically, and credit-based money (special drawing rights) has emerged in the international money markets. The central government now has a monopoly of legal power over the creation of money, and it bears the cost of protecting the public against counterfeiting, theft, and bank failures (via regulations and deposit insurance). But these costs are minimal as compared with the *social costs* which occur if the size of the money stock is inadequately controlled by the authorities with resulting deflation and unemployment, or inflation and the loss of real value of accumulated savings.

THE TIME PATTERN OF DEMAND FOR MONEY STOCKS

Demand for money is based on the desire to hold a stock of a uniquely liquid asset in the portfolio of wealth items of a person, group, business, or government unit. It may well be compared to the demand for inventories of finished goods by a production unit or the demand for a stock of durable goods by a household. Such stocks provide a reservoir in which large, perhaps irregular, inflows of goods may be stored until required to maintain future needed outflows—outflows which are to some extent unpredictable. The situation is analogous to a retailer's accumulation of inventories when large shipments are received and then drawing stocks down later to meet customer demand as it develops. Just so an individual, a business, or a government unit receives income on a different time schedule from its pattern of payments, and a fluctuating stock of money is held to bridge the gap between income receipts and payments, and to ensure that required payments can be met safely and conveniently when due.

In each of these cases, if inflows occur in sudden spurts (receipt of a shipment of goods or of a paycheck) and if outflows are at a more steady rate (retail sales, or flow of expenditures), then the graph of the stock of goods or money in the reservoir as a function of time will be a sawtooth pattern. The spurts of inflow yield a vertical rise in stock, and a constant rate of outflow leads to a steady decline along a straight line whose (negative) slope is proportional to the rate of outflow from the reservoir. See Fig. 12-1, drawn for an economy with Personal and Business Sectors only.

The height of the solid line indicates the money holdings of the Personal Sector of the economy (M_p). In panel (a) it is assumed that all persons are paid their income ($150 billion each month) at the first of the month. They make expenditures at a constant rate ($150 billion per month) and reduce their stocks of money to zero just before payday. This pattern of money holdings has an average value of $75 billion. If the Business Sector is the only other sector in the economy, its pattern of money stocks (M_b) varies inversely to that of the Personal Sector, and the holdings of the two sectors add up to a constant money stock. Business payments of income to persons shift money stocks from the business to the Personal Sector at the beginning of each month, and consumer purchases from business shift money stocks back to the Business Sector gradually during the course of the month. *The total money stock for the economy is constant* at $150 billion, as shown by the horizontal dashed line in panel (a). Money stocks of business (M_b) can be measured downward from this dashed line to the solid line, and its average is also $75 billion if M_b reaches zero at the beginning of each month. The sum of M_p and M_b in Fig. 12-1a might be called the *transactions balances* for the economy, since they are the amounts held to facilitate the flow of transactions in the course of production and consumption.

Probably neither persons nor businesses will want their money stocks to run down to zero at any time. They will wish to maintain some *minimum* stock to permit meeting unexpected fluctuations in flows of income or payments. And they may want a *cash reserve* of funds available to take advantage of favorable

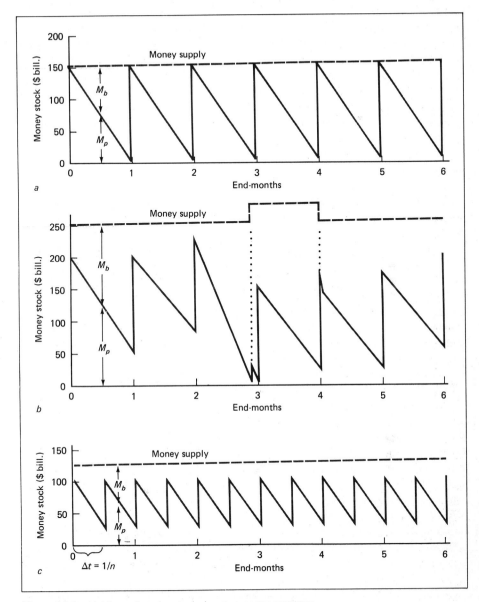

Figure 12-1 Holdings of money stocks in a two-sector economy.

buying opportunities for financial or real assets. Such precautionary and speculative balances, which we shall call *asset balances*, are pictured in panel (*b*) of
Fig. 12-1. Each of the Personal and Business Sectors is assumed to desire an
average holding of $50 billion for these asset balances. Maintenance of such
stocks and of transactions balances as pictured in panel (*a*) would lead to a repeti-

tion each month of the pattern shown for the first month in panel (*b*). Total money stock in the economy would be $250 billion, with $150 billion providing transactions balances and $100 billion providing asset balances. Actual monthly fluctuations in money holdings will vary from this desired pattern, and such deviations are pictured in an extreme form in panel (*b*). In month 2, consumer spending is at a slower rate than in month 1 ($120 billion per month), so that at the end of month 2, Personal Sector balances are larger and Business Sector balances smaller than normal by $30 billion.

In month 3, personal spending rises fantastically to a $260 billion per month rate, which would pull Personal Sector balances $30 billion below zero by the end of month 3. It is assumed that when their money stocks drop to zero near the end of period 3, persons borrow $30 billion from commercial banks to maintain their expenditure rate and keep their money stock above zero until payday. Note that this bank borrowing increases the money supply by $30 billion over the time the loans are outstanding, as shown by the step-up in the money supply line when the loans are made. The additional money is held initially by the Personal Sector but is transferred to the Business Sector by consumer spending in the closing days of month 3. During month 4, persons cut down their expenditure rate to only $130 billion per month, and thus are able to rebuild their money stocks by $20 billion. After receiving their $150 billion of paychecks at the beginning of month 5, persons pay off their bank loans ($30 billion), reducing the money supply of the economy by this amount. By holding spending down to a $120 billion per month rate in months 5 and 6, the Personal Sector is able to restore its money holdings to the desired asset level of $50 billion at the end of month 6.

Through all these changes, the money stock held by the Business Sector is the vertical distance downward from the dashed money supply line to the solid line representing money held by the Personal Sector. During period 3, when consumers were spending at such a phenomenal rate, business holdings of money rose well above the normal expected or desired levels. Under such circumstances, it might well be that business units would have been glad to use some of their excess cash to acquire earning assets. So they might have provided the funds which consumers wanted to borrow, either directly by extending credit to customers or indirectly by way of time deposits in banks or purchases of credit-market instruments from financial institutions, which then provide loans to customers. If the money needs of the Personal Sector were met in this way, the money supply would not have been increased by personal borrowing from commercial banks.

This example illustrates that markets for capital funds can permit exchanges of money stocks between units with more and those with less cash balances than they desire, thus obviating increases in money supply via bank borrowing. *Nonbank credit extension is one way to permit financing a given flow of transactions with less money supply than would otherwise be needed.* It permits a better synchronization of payments and receipts, including the credit transactions, than would otherwise be possible. As an extreme example, we can imagine an economy in which businesses extend credit to customers at the same time and in the same amount as their purchases of goods and services. Then, on the national payday, businesses pay persons their business incomes from production and simultaneously receive repayment of a like amount for credit extended to persons during the preceding pay period. Average transactions balances are practically zero!

A similar result can be achieved by making paydays more frequent. Panel (*c*) in Fig. 12-1 illustrates the situation with semimonthly pay periods but with monthly income and expenditures at a rate of $150 billion per month, as in panel (*a*). The average of transactions balances for each sector is $37.5 billion, one-half the $75 billion semimonthly income payments. Hence, total transactions balances are $75 billion, only half as much as in panel (*a*) with its monthly income payments. In general, under the assumption of a steady expenditure rate and income payments of Y_m per month, with n pay periods per month the required transactions balances would be equal to Y_m/n. In panel (*c*), Y_m is $150 billion per month, n equals two pay periods per month, and M_T equals $75 billion. Note that if n equals infinity, so that income payments and expenditures are exactly synchronized, average transactions balances would be zero. Total money supply in panel (*c*) is shown as $125 billion, which allows for $25 billion of asset holdings of money in each sector. Even then the total money supply is less than in panel (*a*), which involves the same monthly income and expenditure flows and no asset stocks of money.

The same pattern of fluctuations in transactions balances as that shown in panel (*c*) would be obtained if persons received their total monthly income of $150 billion at the beginning of each month but used half of it to pay off indebtedness to businesses. At midmonth, persons borrow $75 billion from business to finance their purchases for the second half of the month. Repetition of this cycle would yield a pattern of transactions balances equivalent to that for semimonthly pay periods. Carried to its extreme, the condition of continuous extension of credit described two paragraphs back becomes equivalent to the condition of an infinite number of pay periods per month, in that each situation would require practically no average transactions.

FACTORS RELATED TO THE DEMAND FOR MONEY STOCKS

This discussion suggests several conditions which influence the demand for money stocks in an economy: (1) the *magnitude of the income and expenditure flows*; (2) the *length of the average payroll period* in the economy, as affecting the degree of synchronization of income and expenditure flows; (3) the *efficiency of the capital-funds markets* in transferring money stocks from units with more to those with less than they desire, in the short run; and (4) *uncertainty regarding the future course of the economy and expectations regarding future changes in prices and interest rates*. These conditions will affect the magnitude of the demand for money stocks to hold as a liquid asset in portfolios—to meet unexpected fluctuations in income and expenditures, or to make profitable purchases and sales of real or financial assets.

Conditions (2) and (3) involve conventional practices, institutions, and market structures which will normally change very slowly. So short-run changes in the demand for money stocks will reflect primarily changes in conditions (1) and (4).

If we wish to write an equation explaining the demand for money (and we do), some measure of aggregate income and expenditure flows should be included, say gross national product (Y). *Transactions demand for money should certainly vary with the level of GNP. The amount of money held for precautionary purposes might well rise with GNP* also, because the doller value of unexpected fluctuations in income or expenditures may well be estimated to maintain a fairly stable percentage relation to GNP. *Desired precautionary holdings* will probably also *increase with the degree of uncertainty regarding the business outlook*, for example, when consumers foresee rising unemployment or when business owners experience sales declines or difficulty in obtaining credit.

The amount of money held as an asset in portfolios depends on total wealth owned and on the relative attractiveness of holding money rather than other assets, as measured by the attributes of liquidity, yield, and risk. Money stocks possess liquidity par excellence, no yield, and little risk except when inflation cuts their purchasing power. Time deposits, savings accounts, and U.S. government savings bonds are very liquid, have moderate yield, and little risk except for loss by inflation. Other government securities, corporate bonds, mortgages, and other credit-market instruments have liquidity depending on their marketability and have higher yield; but they are more risky than the previously mentioned assets, the degree depending on the term to maturity and the riskiness of the agency or assets behind the given debt instrument. To the risk of loss from inflation, there are added varying degrees of risk of capital loss from price fluctuations of the security if sold before maturity. Then there are preferred and common stocks of corporations, with high liquidity if actively traded and with greater risk of fluctuation in market value but greater opportunity for gain from capital appreciation as well as dividend yield. Life insurance and pension funds are illiquid asset holdings with relatively low yield and low risk. And finally, wealth may be held in real assets—land, buildings, art objects, jewelry, and other tangible property.

We see that money is just one asset in a continuous spectrum of types of assets by which wealth may be carried through time. Shifts among holdings of these types of assets occur as owners' preferences and economic conditions change their relative attractiveness. Since a rise in yields (interest rates) on savings accounts and securities would raise the opportunity cost of holding wealth in the form of money, it is logical, and empirically true, that *money holdings tend to be low when interest rates are high, and vice versa.* We usually think of interest rates as affecting asset holdings of money, but it is possible that transactions balances would be reduced also. At very high rates of interest, individuals and businesses with fluctuating transactions balances would find it worth their time and transactions costs to lend funds which would be idle even for a short period, thus achieving better synchronization of inflows and outflows of funds, as illustrated in Fig. 12-1c.

Expected changes in interest rates would affect the composition of portfolios also. Since the market value of fixed-interest securities (bonds) varies inversely with interest rates, it is reasonable that wealth owners will shift out of bonds and hold money at a time when they expect interest rates to rise, that is, bond prices to fall. Conversely, when interest rates are high and expected to fall (when bond prices are low and expected to rise), wealth holders will have an incentive to shift out of money and buy bonds to realize capital gains on their expected appreciation.

The impact of *inflation* on the demand for money is mixed. Higher prices raise the money value of a given flow of goods and services. And rises in prices of labor and capital services increase income flows. So the demand for transactions balances would increase. But if inflation were expected to continue, money stocks would be an asset whose real purchasing power would decline, and there would be an incentive to shift portfolios into corporate stocks or real assets. (The purchasing power of savings deposits, bonds, and mortgage principal, life insurance, and pension funds would decline also since they are denominated in dollars, but during a continuing inflation, interest rates would rise to offset at least part of the prospective capital loss from inflation.)

On the other hand, it is sometimes believed that consumers will wish to maintain *precautionary stocks of money at a given level of real purchasing power;* to the extent that this is true, money stocks will need to be built up in an inflationary period. The possibility of measuring these effects empirically is slight because of the concurrent changes in interest rates, prices, and unmeasurable expectations regarding each of those factors.

EQUATIONS FOR MONEY DEMAND

The net result of this analysis is that the demand for money, in the short run, probably varies directly with aggregate real income and product flows, directly with level of prices, inversely with rate of rise of prices (or expectation of future inflation), inversely with level of interest rate, and directly with rate of rise of interest rates (or expectations of decline in bond prices). The dependence on aggregate wealth and on institutional factors is dropped in most short-run analyses. Thus we may write, for the demand for money stocks (M^*),

$$M^* = f\left(Y_R, P_Y, i, \exp \frac{dP_Y}{dt}, \exp \frac{di}{dt} \right)$$

If we restrict our attention to equilibrium states of the economy when prices and interest rates are constant, and hence expected changes are zero, we have

$$M^* = f(Y_R, P_Y, i)$$

If we further assume that the demand for money is separable into two parts, one linearly dependent on aggregate money GNP and one linearly dependent on interest rates, we have the simple equation

$$M^* = \mu_1 P_Y Y_R + (M_{2,0} - \mu_2 i) = \mu_1 Y + (M_{2,0} - \mu_2 i) \tag{12-1}$$

where $M_{2,0}$ would be calculated asset demand for money at zero interest rate, and where μ_1 and μ_2 are assumed to be positive coefficients. Usually the first term on the right is identified as transactions plus precautionary balances (M_1^*) and the second term as asset holdings (M_2^*). This is a convenient way of thinking. But it should be kept in mind that transactions and precautionary holdings may be affected by very high interest rates. Also precautionary stocks and asset holdings may be related to national wealth or long-term income expectations (permanent income) as well as to current income and interest rates; so the intercept term $M_{2,0}$ may be constant in the short run but shifting up in the long run. Note, too, that $M_{2,0}$ and μ_2 will presumably vary in direct proportion to the general price level, since we hypothesize that holders of asset balances will want to keep their *real* purchasing power constant. Finally, we express desired asset holdings M_2^* as a linear function of interest rate for ease of exposition, but we take this to be an approximation to the true curve over a limited range of values.

The first component of the desired money stock, which we here call *transactions balances*, is kept in a fixed ratio to money GNP and will fluctuate in proportion to GNP, whether the GNP change arises from variations in real output or in the aggregate price level. The coefficient μ_1 has the dimensions of money stock to money flow, hence has units of time, and expresses the number of years of GNP flow which is held in the stocks of transactions balances. The inverse of this ratio can be identified as the income velocity of transactions balances, comparable to the turnover rate of a retailer's inventory.

$$V_1 = \frac{Y}{M_1} = \frac{1}{\mu_1} \quad \text{times per year}$$

Historically, classical economists analyzed the demand for money on the assumption that total demand for money was directly proportional to GNP.

$$M^* = kY = kP_Y\,Y_R$$

or, using the velocity factor $V = 1/k$,

$$M^*V = Y = P_Y\,Y_R \tag{12-2}$$

Equation (12-2) expresses the famous *quantity theory of money* used to analyze the relation of money stocks to economic activity. On the assumption that V is a constant—determined by payroll periods, capital-funds markets and institutions, and individuals' desires for precautionary balances—it is clear that the quantity of money demanded is directly proportional to GNP. The governmental authorities can control the money supply and hence, via market mechanisms which will equate the demand to supply of money stocks, can control the level of GNP. Constancy of V may have been a good assumption a few centuries ago when a larger proportion of money payments was related to income and product transactions, when there were fewer credit-market instruments available as close substitutes for money, when markets for capital funds were less developed, and perhaps when interest rates were more stable. In the United States during the postwar years, more of the quarter-to-quarter change in GNP, and just as much of the rise over the whole period also, have arisen from the changes in V than from those in M. From 1947 to 1973, GNP rose 457 percent while money supply and income velocity each rose by 136 percent.

The introduction of a component of desired *money stocks held as an asset*, related to variables other than GNP, could help account for variations in the velocity (turnover rate) of total money stock. Since the asset holdings M_2^* are assumed to have a turnover rate near zero, the overall income velocity will decrease as the proportion of asset to transactions holdings of money increases. Mathematically, we find

$$V = \frac{Y}{M^*} = \frac{Y}{M_1^* + M_2^*} = \frac{Y}{M_1^*(1 + M_2^*/M_1^*)}$$

$$V = \frac{V_1}{1 + M_2^*/M_1^*} = \frac{V_1}{1 + (M_{2,0} - \mu_2 i)/\mu_1 Y}$$

Since we are assuming V_1 to be constant, the income velocity V will decline as M_2^*/M_1^* increases, and vice versa. Since M_2^* is related negatively to interest rates, this is equivalent to saying that V is low when interest rates are low, and vice versa. It would be expected that desired stocks of money will be higher (V lower) when interest rates are low, and that high interest rates will lead individuals and companies to maintain lower cash balances, increasing the turnover rate V. The United States data since 1929 seem consistent with this interpretation. Velocity declined from 1929 to 1946, a period when interest rates dropped; velocity rose from 1946 to the 1970s while interest rates were rising. But it is likely that institutional developments in security markets, changes in pay periods, and rising per capita wealth may have influenced these trends also.

GRAPHS FOR DEMAND FOR MONEY

Figure 12-2 illustrates these components of demand for money stock graphically. In panel (*a*) asset demand for money as one item in the national portfolio of wealth is shown. The horizontal distance from the left-hand axis to the slanted line is the *demand for money as an asset at various rates of interest*. Such demand increases as interest rates are reduced. The horizontal distance out to the "total wealth" line at the far right represents the total wealth held in the portfolio of economic units of the nation. Money is just one of the many forms of holding wealth, and the slanting line shows how this total stock is shifted between holdings of money and nonmoney assets at various levels of interest rates. As drawn in panel (*a*), a 12 percent per year interest rate is assumed to induce holders to shift all idle cash into nonmoney assets to earn that high yield.

An added feature, called the *liquidity trap*, is shown in the graph by making the asset demand for money stock infinitely elastic at a very low interest rate,

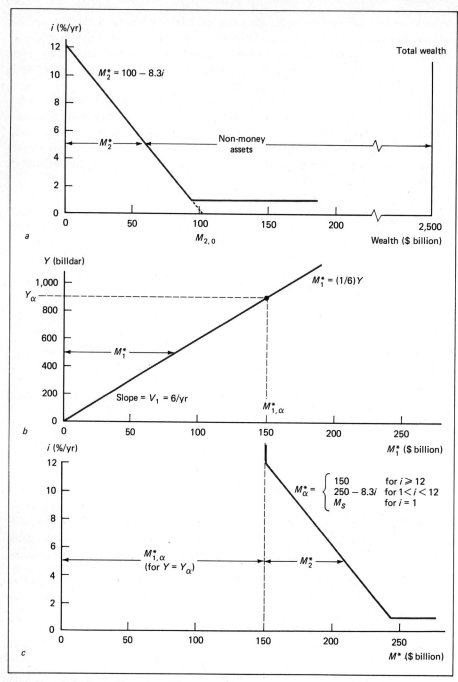

Figure 12-2 Demand for money stock. (a) Asset demand. (b) Transactions demand.
(c) Total demand.

here taken as 1 percent.   The logic of this construction is that very low rates of interest do not cover transactions costs of shifting excess money holdings into securities and out again later, and do not provide sufficient inducement to offset the marginal utility of holding even very large stocks of idle cash.

The potential existence and shape of this liquidity-trap portion of the money demand curve is a topic of contention among economists.   We may well consider it to be a straight-line approximation to the true curve relating M_2^* to i at low interest rates.

Panel (*b*) illustrates the transactions demand for money stock, directly proportional to GNP.   Here transactions velocity is taken as 6; so the horizontal distance M_1^* is one-sixth of the value of GNP, plotted on the vertical axis.

In panel (*c*), total demand for money stock is plotted as a function of i, assuming a fixed GNP of 900 billdar. $M_1^* = \frac{1}{6}(900) = \$150$ billions is added horizontally to the asset demand in panel (*a*), yielding the downward sloping line in panel (*c*).   It is just the M_2^* line of panel (*a*) shifted \$150 billion to the right. For smaller GNP, the shift would be less; for greater GNP, the shift would be greater.   The amount of the shift can be determined graphically by projecting down to panel (*c*) the value of M_1^* determined in panel (*b*) by the given value of GNP.   Note that at an interest rate i of 12 percent per year, asset demand for money is reduced to zero, so that the entire money stock is held as transactions balances.

Now let us see how the supply of money stock is determined, so that we can proceed to analyze the interaction of demand and supply for money and the effects of money markets on the rest of the economy.

PROBLEMS

1   Holdings of money stocks yield no interest income.
   a   Why do persons want to hold stocks of money?
   b   Why do businesses or government units want to hold stocks of money?
   c   Does a money stock yield a flow of services comparable to the flow from a stock of capital goods held by producers or from a stock of durable goods held by consumers?
   d   Would you say that a stock of money has a rate of return?   Explain.

2   In choosing whether to hold one unit more or less of money, a decision maker presumably balances marginal costs and benefits against one another.
   a   What are the elements of marginal costs and benefits?
   b   What are the substitutes to holding one less unit of money?
   c   By what types of transactions might those who wish to hold more money obtain it from others?
   d   What would motivate some holders of money stocks to part with some of their money?

   e What would you regard as the price of money?

   f What is the price of the flow of services yielded by a money stock?

3 Compare and contrast the equation for aggregate demand for money with the equation for aggregate demand for stocks of capital goods. Which economic variables, by their increase, would shift the aggregate demand for money up? Which would shift it down?

4 Suppose that the supply of money is constant.

   a Explain why an increase in GNY would raise interest rates.

   b Using the symbols in the text, show what the ratio $\delta i / \delta Y$ would be.

   c What happens to the two components of demand for money, that is, M_1^* and M_2^*?

   d Using the data in Fig. 12-2, calculate the elasticity of demand for money with respect to Y when i is 6 percent per year.

5 **a** Explain the concept of income velocity of money (V) and of income velocity of transactions balances (V_1).

   b What factors might cause changes in the desired levels of these velocity ratios?

   c In 1973, reported data show M equaled \$263.6 billion and Y equaled 1,295 billdar. The rate of turnover was 102.6 times per year for demand deposits averaging \$167 billion at banks in 233 metropolitan areas reporting to the Federal Reserve System. Compare and comment on the values of income velocity and of turnover rate for demand deposits.

Chapter 13
Supply of Money

TYPES AND OWNERSHIP OF MONEY STOCKS

Money stock has been defined as liquid claims against commercial banks and the monetary authority (the U.S. Treasury and the Federal Reserve System) held by the nonbank public (excluding the U.S. government). The principal types of money are currency (coins and "paper money") and demand deposits. Table 13-1*a* shows the amounts of these claims outstanding at the end of 1972, and Table 13-1*b* shows who held that money stock. About 22 percent of the money supply was in currency and 78 percent was in demand deposits. About 62 percent of the

TABLE 13-1   Money Stock by Types and Ownership (year-end level, 1972)

a   Money supply by types

	($ billions)
Liabilities of monetary authority	
Currency held by nonbank public, domestic	57.8
Foreign demand deposits	0.4
Liabilities of commercial banks	
Domestic demand deposits, except	
U.S. government and interbank	202.5
Foreign demand deposits	7.6
Total U.S. money supply (including mail float)	268.3

b   Money supply by ownership

		($ billions)
Households		156.5
Nonfinancial businesses		55.2
Corporate businesses	36.0	
Nonfarm noncorporate businesses	12.5	
Farm businesses	6.8	
State and local governments		14.6
Financial sectors		17.0
Rest of the world		8.0
Mail float		17.0
Total U.S. money supply (including mail float)		268.3

Source: Data from Board of Governors of the Federal Reserve System, *Flow of Funds Accounts* 1945–1972, August 1973, p. 113.

money supply was held by the household sector, close to the percent of GNP purchased by consumers. Nonfinancial businesses held about 22 percent of the money stock, and financial businesses about 7 percent. State and local governments held 6 percent, and foreigners the remaining 3 percent of the United States money stock.

BALANCE SHEETS OF THE MONETARY AUTHORITY AND COMMERCIAL BANKS

Since money stocks are claims against the commercial banks and the monetary authority, we must analyze the balance sheets of those institutions in order to discover what assets "lie behind" our money stocks and how the money supply of a country may be changed.

COMMERCIAL BANKS

From Table 13-1*a* we note that about 78 percent of the U.S. money supply consists of demand-deposit liabilities of commercial banks. The balance sheet for the commercial bank sector should reveal what assets are back of that component of the money supply.

The principal types of items on the balance sheets of commercial banks are shown in Table 13-2. The description of each item is followed by its symbol. Usually an item is designated by letters providing a reminder of the type of asset or liability involved, and it is followed, where needed, by a subscript indicating the sector that owns the item. Among the symbols the letter D is used for bank deposits within the Monetary Sector, and A is used for nonmoney financial claims (liabilities) of other types (loans, bonds, mortgages, notes, deposits in savings banks, and so on). The value of shares of stock in corporations is indicated by NW, meaning net worth. Letter prefixes indicate the sector issuing a claim, except that DD and TD denote demand and time deposits issued by commercial banks. The sectors used in the discussion of money and capital markets, and their symbols, used as subscripts and prefixes, are:

C for commercial banks
FR   for Federal Reserve banks, the monetary authority

TABLE 13-2   Commercial Banks Balance Sheet

Assets	Liabilities and Net Worth
Deposits at Federal Reserve banks (FRD_C)	Loans from Federal Reserve banks—borrowed reserves (BR)
Vault cash, or currency (CU_C)	Demand deposits owned by nonbank public (DD_{NB})
	Demand deposits owned by U.S. government (DD_{US})
U.S. government securities (USA_C)	Time deposits owned by nonbank public (TD_{NB})
Other earning assets (NBA_C)	Bank debt securities owned by nonbank public (CA_{NB})
Property ($PROP_C$)	Net worth (CNW_{NB})

NB for the nonbank public, including the Personal Sector, state and local governments, Business Capital Accounts (including financial intermediaries), and similar Foreign Sector accounts

US for the federal government

Demand deposits (checking accounts) held at commercial banks are liquid claims against commercial banks held by the nonbank public. The nonbank public acquires these claims by providing the banks with earning assets, such as U.S. government securities, mortgages, bonds, loan contracts, and promissory notes. Since banks receive interest on these earning assets and do not pay interest on demand deposits (in the United States), it is usually profitable for the commercial banks to provide the nonbank public with money (liquid assets) in exchange for the public's illiquid promises to pay future interest and principal.

REQUIRED RESERVES

What limits commercial banks in the creation of money? One constraint is business prudence—the credit-worthiness of the persons, businesses, and government units whose promises the banks accept. A second constraint is the legal requirement to hold "reserves" at or above specified levels in relation to demand and time deposits. These required reserves (RR) serve the dual purpose of (1) ensuring liquidity of the bank in the event of large-scale withdrawals of deposits, and (2) ensuring the power of the Federal Reserve System to control the money supply of the economy.

$$RR = r_D(DD_{NB} + DD_{US}) + r_T(TD_{NB}) \tag{13-1}$$

where

r_D = reserve ratio required for demand deposits (normally 10 to 20 percent)
r_T = reserve ratio required for time deposits (normally 3 to 6 percent)

To meet their reserve requirements, the commercial banks may hold deposits at Federal Reserve banks (FRD_C) or currency in their vaults (CU_C).

$$R = FRD_C + CU_C \tag{13-2}$$

Since these reserve assets do not earn interest, bankers try to keep their ratio to earning assets low. As we shall see, the Federal Reserve System has the power to control the amount of commercial bank reserves. Hence they can limit the creation of demand deposits (money) by commercial banks.

The currency component of the money supply consists of U.S. Treasury currency (coins) and Federal Reserve notes (paper money). It is produced and circulated by these agencies (the monetary authority) in the amounts demanded by the nonbank public and commercial banks.

GRAPHICAL PRESENTATION OF BALANCE SHEETS AND MONEY SUPPLY

Figure 13-1 presents in bar-chart form the picture of holdings of financial claims against the banks and the monetary system by the nonbank public, and also the relevant assets and liabilities of the banks, the Federal Reserve System, and the

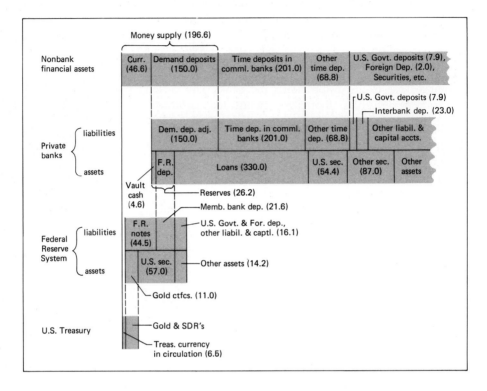

Figure 13-1 Money supply and related items ($ billions as of June 24, 1970).

U.S. Treasury.   Starting from the top, we have the nonbank public holding a money supply of $196.6 billion, composed of $46.6 billion of currency plus $150.0 billion of demand deposits.   This public also holds time deposits in commercial and other banks plus U.S. government and other securities, which are not money claims.   The publicly held currency is a liability of the U.S. Treasury and of the Federal Reserve System; demand and most time deposits are liabilities of the private commercial banks.   On the asset side of their balance sheets, commercial banks hold reserves and earning assets.   Reserves may be held in the form of vault cash ($4.6 billion) and of deposits by the commercial banks with the Federal Reserve banks ($26.2 billion).   These assets earn no interest.   The income-earning assets of the commercial banks are their loans outstanding and their investments in U.S. government and other securities.   As commercial bankers increase their holdings of these earning assets, they increase their deposits or they lose reserves, and thus approach the constraints imposed by the legal requirement of minimum ratios of reserves to deposits.   When they reach the legal limit, the banking system is said to be "loaned up." Normally the private banks hold excess reserves (above the legal minimum) amounting to less than 2 percent of total reserves. It would seem that bankers are good profit maximizers.

The third pair of bars in Fig. 13-1 shows the liabilities and assets of the Federal Reserve System. The assets behind the Federal Reserve note liability ($44.5 billion) and member-bank deposits ($21.6 billion) are largely gold certificates ($11.0 billion) and U.S. government securities ($57.0 billion). Among the other liabilities are U.S. government deposits, for the Federal Reserve System is the bank against which the U.S. government writes its checks. Among the "other assets" are loans to those commercial banks that have borrowed from the Federal Reserve banks to increase their deposits (reserves). Of course, there is a fee, the "Federal Reserve discount rate," charged for such loans.

Finally, the U.S. Treasury, at the bottom of the diagram, has issued claims against itself in the form of gold certificates and Treasury currency (coins and notes). These are backed up by holdings of gold and international reserve assets, such as convertible foreign currencies and International Monetary Fund (IMF) reserve balances and special drawing rights (SDRs).

CREATION OF AND CONSTRAINTS ON MONEY STOCK

Given this picture of the balance sheets of the banks and the monetary system, where one category of liabilities constitutes the money supply, the question arises: *What determines just how large the money supply will be at any specified time?* Commercial banks, of course, are profit-making institutions. Bankers have the opportunity to acquire (extend) loans, which pay them interest, giving in exchange newly created demand deposits, on which the bankers pay no interest. To be sure, they provide some free banking services to their borrowers and other depositors, but on the whole, the opportunity to create, with a stroke of a pen or a computer printer, the demand deposits which are loaned to borrowers at interest would seem to offer opportunities for profit. What would you do if you were a commercial banker?

In the past, business prudence did not always exert sufficient restraint on the volume of bank credit, and hence on the supply of demand deposits or commercial bank notes which were created. Banks failed when business conditions became bad, causing some borrowers to fail to repay loans and some depositors to demand their money in cash. Furthermore, the U.S. Congress has the constitutional power "to coin money, regulate the value thereof, and of foreign coin, and fix the standard of weights and measures." So the federal government has stepped in, notably through the establishment of the Federal Reserve System in 1913, to control the total amount of money created by the banking system while leaving individual banks free to compete within the framework of the overall constraint. Let us examine in more detail how the general controls work and how they are felt by the individual bank.

Money consists of currency and demand deposits held by the nonbank public, including foreign-owned deposits at Federal Reserve banks (FRD_f).

$$M = CU_{NB} + DD_{NB} + FRD_f$$

The money supply changes primarily because of changes in demand deposits at commercial banks (DD_{NB}). And these deposits change when banks alter their holdings of earning assets (USA_C and NBA_C). The limit to money creation is set by the requirement that total reserves R must equal or exceed required reserves RR.

$$R \geq RR = r_D(DD) + r_T(TD)$$

We can check on this requirement by noting that *excess reserves ER must not be negative*, that is,

$$ER = R - RR \geq 0 \tag{13-3}$$

A convenient way to check out the acceptability and legality of a change in the money supply is to show the changes of commercial bank assets and liabilities in a balance-sheet account and then to calculate the implicit change of excess reserves by noting that

$$\Delta ER = \Delta R - \Delta RR \tag{13-4}$$

Example 1

As one example, consider a situation in which the commercial banks hold excess reserves ER_o and are faced by offers from the nonbank public to supply the banks with loan-promises or other securities in exchange for demand deposits. If the bankers decide that the proffered loans or securities provide adequate security, liquidity, and yield, and if they are willing to reduce their excess reserves, they accept the offers.

The following balance-sheet changes take place. (Let Δ specify the absolute magnitude of change in a variable and let a sign be added to indicate the direction of change.)

Commercial Banks		Nonbank Public	
Assets	Liabilities	Assets	Liabilities
$+\Delta NBA_C$	$+\Delta DD_{NB}$	$+\Delta DD_{NB}$	$+\Delta NBA_C$

$$+\Delta M = \Delta DD_{NB}$$

$$-\Delta ER = -r_D(+\Delta DD_{NB}) = -r_D(\Delta M)$$

Money supply and earning assets have increased equally, but excess reserves have declined by a fraction r_D of the increase in money supply.

The change in excess reserves implies a limit to the permissible changes in earning assets and money supply. Since excess reserves cannot become negative, the maximum decline in excess reserves is ER_o. Thus,

$$\max(\Delta NBA_C) = \max(\Delta M) = \frac{ER_o}{r_D}$$

If r_D equals $\frac{1}{6}$, earning assets and money stock can rise by 6 times the initial excess reserves, and no more.

Actually, an individual bank cannot be that liberal. If its borrowers pay out their new demand deposits to persons and businesses that keep their accounts in other banks, the lending bank will lose reserves to those other banks as the checks clear. But those other banks will experience a rise in their excess reserves and can make further increases in their earning assets and demand deposits. In this way, the excess reserves of the initiating bank spread through the system and lead to an overall rise in money supply by ER_o/r_D if the initial excess reserves become fully utilized to support additional demand deposits. The detailed calculations for the diffusion process are presented in the appendix to this chapter. The basic principle is that a single bank with initial excess reserves ER_o can safely increase its earning assets and demand deposits by:

a $\max(\Delta DD_{NB}) = ER_o$ if the new deposits all flow to other banks

b $\max(\Delta DD_{NB}) = (ER_o/r_D)$ if the new deposits all stay in the original bank

c $\max(\Delta DD_{NB}) = \dfrac{ER_o}{1 - f(1 - r_D)}$ where

$f =$ the fraction of the new demand deposits which remain in the original bank

Cases a and b correspond to case c, with $f = 0$ and $f = 1$ respectively.

The commercial bank sector as a whole is like a huge bank which retains all deposits it creates, so that f equals 1 and max(ΔDD_{NB}) equals ER_o/r_D.

Example 2

Consider a situation in which the nonbank public withdraws currency from commercial banks in exchange for demand deposits. Changes in balance sheets, money supply, and excess reserves are as follows.

Commercial Banks		Nonbank Public	
Assets	Liabilities	Assets	Liabilities
$-\Delta CU_C$	$-\Delta DD_{NB}$	$-\Delta DD_{NB}$ $+\Delta CU_{NB}$	

$$\Delta M = +\Delta CU_{NB} - \Delta DD_{NB} = 0.$$
$$-\Delta ER = -\Delta CU_C - r_D(-\Delta DD_{NB})$$
$$= -\Delta CU_C(1 - r_D)$$

Since currency holdings are part of commercial bank reserves, public withdrawal of currency reduces reserves equally. The reduction of required reserves arising from the decline in DD_{NB} is only a partial offset; so excess reserves decline by about five-sixths of the currency withdrawal. Note that the money stock is unchanged. Thus, *the public's preferences for relative proportions of currency and demand deposits in their money stocks affect the reserve position of commercial banks.*

Example 3

As a third example, consider a shift in public preferences toward holding more demand deposits instead of time deposits in commercial banks. The balance-sheet changes are as follows.

Commercial Banks		Nonbank Public	
Assets	Liabilities	Assets	Liabilities
	$-\Delta TD_{NB}$ $+\Delta DD_{NB}$	$-\Delta TD_{NB}$ $+\Delta DD_{NB}$	

$$+\Delta M = +\Delta DD_{NB}$$

$$-\Delta ER = -[(r_D(+\Delta DD_{NB}) + r_T(-\Delta TD_{NB})]$$

$$= -(r_D - r_T)(+\Delta DD_{NB})$$

The money supply has increased, since DD_{NB} are a part of M, while TD_{NB} are not. Excess reserves have declined because the required reserves behind $1 of demand deposits are greater than those behind $1 of time deposits. Once again, *the decisions of the nonbank public can affect the reserve position of commercial banks—and in this case change the money supply also.*

What can the commercial banks do to maintain their reserves in the face of uncontrollable decisions of the nonbank public and to keep their earning assets at as high a level as permissible? Bankers have several degrees of freedom in their choices.

1   Individual bankers compete for deposits, since a gain in deposits from another bank increases the first bank's excess reserves. Of course, this does not help the commercial bank sector as a whole.

2   Commercial banks can maintain some excess reserves to guard against unpredictable fluctuations in requirements, but they keep these reserves at a minimum by organizing efficient trading of reserves from surplus to deficit banks via a "federal-funds" market, even for a day or two.

3   Commercial banks, by services they provide and by interest rates they pay, can influence the nonbank public to shift its liquid asset holdings from currency to demand deposits or from demand deposits to time deposits or other bank securities in order to ease the pressure on the banks' reserves.

4 Commercial banks can increase their reserves by borrowing from the Federal Reserve banks or by selling government securities to the Federal Reserve banks.

The first three choices will help meet temporary or relatively minor problems of adequacy of reserves. But to meet major and long-term changes in money and required reserves, the banks must look to the monetary authority for increases in the supply of reserves, for example, to the fourth choice.

MONETARY POLICY AND THE CONTROL OF RESERVES

To understand how the monetary authority can control the supply of reserves available to commercial banks, we need to examine the balance sheet of the monetary authority, including the Federal Reserve banks and the currency accounts of the U.S. Treasury. The symbol FR will be used to indicate this part of the economy. The principal types of items on this balance sheet were pictured in Fig. 13-1 and are summarized in Table 13-3, where symbols are shown for the various items.

The first two items on the liabilities side provide, respectively, the currency component of the money supply and the reserves for the demand-deposit component. The monetary authority exerts its control over the money supply primarily through transactions which change the deposits of commercial banks with the Federal Reserve banks (FRD_c). On the assumption that commercial bankers are avid profit maximizers, it is a safe bet that increases in reserves will lead to increases in commercial-bank earning assets and demand deposits (money supply) pretty well up to the legal limit set by required reserves. Conversely, decreases in reserves will force contraction of commercial-bank earning assets and money supply nearly in proportion to the reduction in available reserves.

Now let us examine some types of transactions which will change commercial bank reserves and the money supply. In the following examples, we shall assume

TABLE 13-3 Monetary Authority Balance Sheet

Assets	Liabilities and Net Worth
Gold and foreign reserve assets (GO_{FR}) U.S. government securities (USA_{FR}) Loans to commercial banks (BR) Other securities (NBA_{FR}) Property ($PROP_{FR}$)	Federal Reserve notes and Treasury currency (CU) Deposits of commercial banks (FRD_c) Deposits of the U.S. government (FRD_{us}) Deposits of foreigners (FRD_f) FR debt issues owned by nonbank public (FRA_{NB}) Net worth ($FRNW_{us}$)

that (1) *commercial banks always keep excess reserves at some fixed minimum,* so that ΔER equals 0; and (2) *time deposits of commercial banks do not change.*

Example 1

Consider the situation in which commercial banks borrow from the Federal Reserve System (the Fed), giving loan-promises in return for an increase in commercial bank deposits at the Federal Reserve banks. The balance-sheet changes are as follows, always assuming other items to be constant.

Federal Reserve Banks		Commercial Banks		Nonbank Public	
Assets	Liabilities	Assets	Liabilities	Assets	Liabilities
(1) $+\Delta$BR	$+\Delta R$	$+\Delta R$	$+\Delta$BR		
(2)		$+\Delta$NBA$_C$	$+\Delta$DD$_{NB}$	$+\Delta$DD$_{NB}$	$+\Delta$NBA$_C$

The changes in step (1) occur within the Monetary Sector and increase excess reserves of commercial banks by $+\Delta$BR. The changes in step (2) occur in capital-funds markets as commercial banks acquire loans and securities from the public, creating demand deposits as they do so. When the borrowed reserves have all been brought into required reserves, the net changes resulting from the two-step process are:

$$+\Delta M = +\Delta DD_{NB} = +\frac{\Delta BR}{r_D}$$

$$\Delta ER = +\Delta BR - r_D(+\Delta DD_{NB}) = 0$$

Earning assets and money supply have risen by a "reserve multiplier" $(1/r_D)$ *times the increment in borrowed reserves,* and excess reserves are back to their original minimum level. The commercial banks are better off, because the income from their earning assets will more than cover interest on their loan from the Fed (computed at the "discount rate" of interest) plus costs of servicing the new accounts.

Example 2

If commercial banks sell government securities to the monetary authority, the results are similar.

Federal Reserve Banks		Commercial Banks		Nonbank Public	
Assets	Liabilities	Assets	Liabilities	Assets	Liabilities
(1) $+\Delta$USA$_{FR}$	$+\Delta R$	$-\Delta$USA$_C$			
		$+\Delta R$			
(2)		$+\Delta$NBA$_C$	$+\Delta$DD$_{NB}$	$+\Delta$DD$_{NB}$	$+\Delta$NBA$_C$

$$+\Delta M = +\Delta DD_{NB} = +\frac{\Delta USA_{FR}}{r_D}$$

$$\Delta ER = +\Delta USA_{FR} - r_D(\Delta DD_{NB}) = 0$$

Reserves increase by the value of government securities sold to the Federal Reserve banks, and money supply goes up by this amount times the reserve multiplier. The commercial banks gain income from the new earning assets, but they experience both loss of income on the government securities and added costs on the new deposit accounts.

In these two examples, we have assumed that the Federal Reserve banks are willing to take the loans or securities offered by the commercial banks. Even though accommodation at the "loan window" is legally a privilege rather than a right, commercial banks are normally not refused such requests. They may be persuaded not to make them! On the other hand, commercial banks cannot be at all sure that the Federal Reserve banks will buy their government securities, and even if they do, the Fed can thwart attempts to increase reserves by itself selling government securities to bond dealers in the nonbank sector. *Open-market transactions in government securities is a principal tool used by the monetary authority to control commercial bank reserves.* They are not about to let commercial banks preempt the initiative in this area.

Example 3

Let us see how a sale of government securities by the Federal Reserve banks *to the nonbank public* can reduce commercial bank reserves, and hence the money supply.

Federal Reserve Banks		Commercial Banks		Nonbank Public	
Assets	Liabilities	Assets	Liabilities	Assets	Liabilities
(1) $-\Delta USA_{FR}$	$-\Delta R$	$-\Delta R$	$-\Delta DD_{NB,1}$	$+USA_{NB}$ $-\Delta DD_{NB,1}$	
(2)		$-\Delta NBA_C$	$-\Delta DD_{NB,2}$	$-\Delta DD_{NB,2}$	$-\Delta NBA_C$

In step (1), the Federal Reserve banks sell government securities to the nonbank public; the buyers pay for the securities by checks drawn against their demand deposits in commercial banks. The Fed collects by debiting the deposits of commercial banks. The commercial banks lose reserves and reduce the demand deposits of the public by an equal amount. But they lose excess reserves.

$$-\Delta ER_1 = -\Delta R - r_D(-\Delta DD_{NB,1}) = (1 - r_D)(-\Delta USA_{FR})$$

To restore excess reserves to their initial minimum level, either reserves must be rebuilt or required reserves must be reduced by further cutbacks in demand deposits. Since the monetary authority is pursuing a restrictive monetary policy, let us assume that the Fed prevents any recouping of reserves by the commercial banks. Then, in step (2), the further reduction of demand deposits must be enough to lower required reserves by the amount of the decline in excess reserves in step (1). As we would expect, the ultimate decline in money supply equals the reserve multiplier times the value of the initiating sales of securities by the Federal Reserve banks.†

$$-\Delta DD_{NB} = \frac{1}{r_D}(-\Delta R) = -\frac{1}{r_D}\Delta USA_{FR}$$

Thus the monetary authority can control commercial bank reserves closely and with little delay by open-market purchases and sales of government securities. Other tools for controlling reserves and demand deposits include the authority of the Federal Reserve System (1) to alter the ratios for reserves required against demand deposits and time deposits, within limits set by Congress; and (2) to alter the "discount rate," the interest rate charged on loans from the Fed to commercial bank borrowers.

These policy tools influence directly or indirectly the demand or supply of loans and securities in markets for capital funds. Hence, they will affect interest rates and the prices of bonds and other fixed-yield securities. In particular, open-market operations affect prices of government securities, and the impact spreads from there to other security markets by substitution effects among various types of financial claims. *An "easy" money policy aimed at increasing reserves and money supply will tend to raise security prices and lower interest rates; a "tight" money policy aimed at decreasing reserves and money supply will tend to lower security prices and raise interest rates.*

FISCAL POLICY AND THE MONETARY SYSTEM

In the chapter on the Government Sector, we noted that changes in government purchases, taxation, and transfer payments affect GNP by their influence on aggregate demand, directly through G and indirectly by altering sector income flows. We set aside for future discussion the influence of government spending, taxing, and borrowing on capital markets and money supply. The future is now here!

---

† Proof of this statement runs as follows:
$$\Delta ER = -\Delta ER_1 + \Delta ER_2 = 0$$
$$-\Delta ER_1 = (1 - r_D)(-\Delta USA_{FR})$$
$$+\Delta ER_2 = -(-\Delta RR_1) = -r_D(-\Delta DD_{NB,2})$$
From $-\Delta ER_1 + \Delta ER_2 = 0$, we obtain
$$(1 - r_D)(-\Delta USA_{FR}) - r_D(-\Delta DD_{NB,2}) = 0$$
or
$$-\Delta DD_{NB,2} = \frac{(-\Delta USA_{FR})(1 - r_D)}{r_D}$$
Finally,
$$-\Delta DD_{NB} = -\Delta DD_{NB,1} - \Delta DD_{NB,2}$$
$$= -\Delta USA_{FR} - \Delta USA_{FR}\frac{1 - r_D}{r_D}$$
$$= -\frac{1}{r_D}\Delta USA_{FR}$$

State and local governments hold their deposits at commercial banks, and those deposits are in the nation's money supply. Hence, payments between state and local governments and other transactors in the nonbank public sector simply shift demand deposits around within that sector and do not change total money supply or commercial bank reserves. State and local government loans from, or sale of bonds to, commercial banks increase money supply and required reserves, as would similar transactions by persons or businesses, but there is no impact on reserves.

The federal government is different. It keeps its working balances with Federal Reserve banks. Payments into or out of those deposit accounts do alter commercial bank reserves, and perhaps the money supply as well.† Consider a few examples, on the assumption that the monetary authority does not counteract the effects of fiscal policy, that is, assuming a passive monetary policy.

Example 1

If some federal government purchases are financed by running down government deposits at the Reserve banks, then reserves and the money supply will increase.

Federal Government		Federal Reserve Banks	
Assets	Liabilities	Assets	Liabilities
$-\Delta FRD_{US}$			$-\Delta FRD_{US}$
			$+\Delta R$

Commercial Banks		Nonbank Public	
Assets	Liabilities	Assets	Liabilities
$+\Delta R$	$+\Delta DD_{NB}$	$+\Delta DD_{NB}$	

$$+\Delta M = +\Delta DD_{NB} = \Delta FRD_{US}$$

$$+\Delta ER = +\Delta R - r_D(\Delta DD_{NB}) = (1 - r_D)(\Delta FDR_{US})$$

Money supply increases in the amount of the decline in federal government deposits. Excess reserves increase, so that the commercial banks will initiate a further rise in their earning assets and demand deposits until

$$+\Delta M = +\Delta DD_{NB} = \frac{1}{r_D}(\Delta FRD_{US})$$

Note that such current-account spending leads to a decline in the net worth of the federal government and a rise in that of the nonbank public.

---

† The federal government does keep deposits in tax and loan accounts at commercial banks. These are not counted in the money supply, so that payments into them do reduce total money stock. Also, they do enter into calculation of required reserves. However, these accounts are temporary abodes of federal funds and are "called" into the federal government's deposit account at Federal Reserve banks as needed to meet federal expenditures. We shall assume DD_{US} constant.

Example 2

If some federal government receipts from the nonbank public are used to build up government deposits at Federal Reserve banks, reserves and the money supply will decrease. Excess reserves will decline, so that commercial banks will initiate a secondary decrease in earning assets and demand deposits. Net worth of the federal government rises; that of the nonbank public declines. The balance-sheet changes will be as in the preceding example, with signs reversed.

It follows that a balanced federal budget will lead to no changes in reserves or money stock, and to no changes in net worth of the federal government or the nonbank public sectors.

Example 3

Now suppose that the federal government, starting from a balanced-budget position, (1) increases its expenditures, and (2) finances its deficit by selling government securities to the nonbank public. The sector balance sheets show the following changes.

Federal Government		Federal Reserve Banks	
Assets	Liabilities	Assets	Liabilities
(1) $-\Delta FRD_{US}$			$-\Delta FRD_{US}$
			$+\Delta R$
(2) $+\Delta FRD_{US}$	$+\Delta USA_{NB}$		$+\Delta FRD_{US}$
			$-\Delta R$

Commercial Banks		Nonbank Public	
Assets	Liabilities	Assets	Liabilities
(1) $+\Delta R$	$+\Delta DD_{NB}$	$+\Delta DD_{NB}$	
(2) $-\Delta R$	$-\Delta DD_{NB}$	$-\Delta DD_{NB}$	
		$+\Delta USA_{NB}$	

$\Delta M = 0$

$\Delta ER = 0$

The two steps of the process lead to offsetting changes of deposits. The only net effect on balance sheets is a rise in federal government securities outstanding (decrease in government net worth) and a rise in net worth of the nonbank public. So the public is better off[†] if increased federal spending is financed by government borrowing rather than by taxation, though the effect on bank reserves and money supply is nil in both cases.

[†] Some economists argue that the public should not feel better off, because, as taxpayers, they simultaneously acquire a liability for future interest and/or principal payments whose present value equals the value of the government securities. Others point out that the owners of the government bonds are better off and nonowner taxpayers worse off, or that part of the bond costs will be borne by future generations. Here we shall assume that bond financing does not depress consumer spending so much as tax financing would.

Example 4

Finally, suppose that (1) *deficit spending by the federal government is financed by* (2) *borrowing from the Federal Reserve banks.*

Federal Government		Federal Reserve Banks	
Assets	Liabilities	Assets	Liabilities
(1) $-\Delta FRD_{US}$			$-\Delta FRD_{US}$ $+\Delta R$
(2) $+\Delta FRD_{US}$	$+\Delta USA_{FR}$	$+USA_{FR}$	$+FRD_{US}$

Commercial Banks		Nonbank Public	
Assets	Liabilities	Assets	Liabilities
(1) $+\Delta R$	$+\Delta DD_{NB}$	$+\Delta DD_{NB}$	

$$+\Delta M = +\Delta DD_{NB} = +\Delta USA_{FR}$$

$$+\Delta ER = +\Delta R - r_D(\Delta DD_{NB}) = (1 - r_D)(+\Delta USA_{FR})$$

In this case, the initial rise in reserves and money supply has not been reversed by any payments from the nonbank public to the federal government. Since excess reserves have increased, commercial banks will presumably increase earning assets and money supply further until the total rise becomes

$$+\Delta M = +\Delta DD_{NB} = \frac{1}{r_D}(+\Delta USA_{FR})$$

This is the most expansionary method of financing deficit spending, since the stimulus of added government spending is reinforced by a rise in reserves and consequent expansionary monetary effects. It is the modern equivalent of financing government spending by creating paper money on the government printing press.

In the United States, the Federal Reserve banks do not normally buy new federal securities direct from the government. Under our system, the net changes shown in the above accounts would be reached by an additional step. The Treasury would sell new issues to the nonbank public or commercial banks while the Federal Reserve banks would buy an equal value of outstanding issues on the open market. The monetary authority could refuse to permit such an expansionary financing of federal spending by simply not bidding for government securities in the open market. But the Fed is under some pressure to buy, in order to fulfill its responsibility for "maintaining orderly markets" for U.S. government securities.

SUMMARY ON CHANGES IN MONEY SUPPLY

A review of the analyses using the balance-sheet accounts permits some generalizations regarding the effects of intersector transactions on money supply and commercial bank reserves. For the analysis of monetary phenomena, we classify the economic units of the economy into four sectors:

1   Monetary authority—Federal Reserve banks and the monetary accounts of the U.S. Treasury

2   U.S. government—all other parts of the federal government

3   Commercial banks—at present the only private banks that issue demand deposits (checking accounts) which are counted as part of the money supply

4   The nonbank public—persons, nonprofit organizations, nonfinancial businesses, state and local governments, and financial institutions other than commercial banks

Starting at the lowest level, we note that payments between units of the non-bank public just shift money and reserves around without changing total money supply or reserves.   However, when the nonbank public deals with commercial banks, exchanging its debt-promises, securities, or time deposits for demand deposits, the total money supply will change; but the total reserves will be constant, except when the nonbank public changes its holdings of currency.   When the nonbank public makes payment to, or recieves funds from, the United States government's account at the Federal Reserve banks, both the money supply and commercial bank reserves change.   Reserves change because these transactions involve settlement via demand deposits of commercial banks at the FR banks. The open-market transactions between the nonbank public and the monetary authority also involve settlement through reserve accounts and will affect both ordinary money supply and reserves.

Commercial banks do not alter money supply or reserves when they exchange deposits or "federal funds" (reserves) among themselves.   Their transactions with the nonbank public will change money supply but not reserves, except when vault cash is involved.   Commercial bank dealings with the United States government or the monetary authority will change reserves but not the money supply.

Finally, transactions between components of the United States government, or within the monetary authority, or between the U.S. government and the monetary authority will affect neither money supply nor reserves.   Deposits at Federal Reserve banks are involved, but not the deposits (reserves) owned by commercial banks.

The preceding discussion refers just to the primary impacts of the indicated transactions.   If the transactions alter excess reserves of commercial banks, secondary adjustments to money supply and reserves will occur as the commercial banks reestablish their desired level of excess reserves.

This summary of the effects of various transactions on money supply and reserves may be further compressed into the following table, with rows designating paying sectors and columns designating receiving sectors.   The top line in each cell indicates the presence or absence of a primary impact on total money supply; the bottom line refers to reserves.

All entries are " No " along the diagonal from upper left to lower right, since transactions between units in the same sector do not change total money supply

TABLE 13-4 Impact of Transactions on Total Money Supply and Reserves

Paying Sector	Impact on	Receiving Sector			
		Nonbank Public	Commercial Banks	U.S. Government	Monetary Authority
Nonbank public	ΔM?	No	Yes	Yes	Yes
	ΔR?	No	No†	Yes‡	Yes
Commercial banks	ΔM?	Yes	No	No	No
	ΔR?	No†	No	Yes	Yes
United States government	ΔM?	Yes	No	No	No
	ΔR?	Yes‡	Yes	No	No
Monetary authority	ΔM?	Yes	No	No	No
	ΔR?	Yes	Yes	No	No

† Except when currency holdings of nonbank public are changed.
‡ Except when funds from public are held temporarily in U.S. government demand deposits at commercial banks.

or reserves. Entries are the same in cells symmetrically located with respect to that diagonal.

Note that reserves are changed only by transactions in the lower left or upper right quarters of the table. The key to the monetary authority's control over reserves lies in its transactions with the nonbank public or commercial banks. The United States government can alter reserves to a limited extent, and the nonbank public and commercial banks can influence money stock to some degree. However, with short time lags or small fluctuations, the Federal Reserve System can counteract or overwhelm other forces and achieve the money stock which it decides shall be supplied to the economy.

THE SUPPLY CURVE FOR MONEY STOCKS

Given that the monetary authority *can* enforce its decision on the size of the money stock, two questions remain:

What objectives (target variables) and what input signals guide the policy decisions regarding money stock desired by the monetary authority?

How effective are the actions of the monetary authority in achieving its objectives?

This section deals with the first of these questions; the second will be analyzed in later chapters. The objectives of the monetary authority are, in general, the same as those of the federal government—full employment, economic stability along with growth, reasonable price stability, a near-balance of international payments, equitable distribution of incomes, and maximum use of private markets with minimal constraints on private decision making. As was noted previously, some of these objectives are mutually inconsistent at times; for example, full employment and price stability, or economic growth and balance of international payments. In such cases, the simultaneous attainment of multiple objectives may require a mix of fiscal and monetary policy actions in which there is a division of responsibility for achievement of various objectives. In such a division, the monetary authority is usually considered to have special responsibility for (1) *price stability*, (2) *moderation of cyclical swings in aggregate demand*, and (3) *international balance of payments*. Target variables in these areas might be (1) percent rate of change in the consumer price index or the GNP price deflator, (2) size and duration of deviations from real full-employment GNP, and (3) the size and duration of a surplus or deficit in the United States balance of payments.

Since these target variables are not directly under the control of the monetary authority, it must develop some theory as to the linkages between policy actions under its control and the target variables. Since fiscal policy actions and the behavior of the nonbank public also influence achievement of price stability, full employment, and balance of payments, the monetary authority may need to choose some intermediate variables which serve as proximate indicators of its contribution to overall national economic policy. Two indicators frequently suggested as proximate targets are (1) interest rates and other measures of conditions in credit markets, and (2) money stock and other measures of bank credit outstanding.

INTEREST RATE AND CREDIT-MARKET POLICY

According to one point of view, the principal linkages between the monetary authority and its target variables may be traced through the effects of monetary policy actions on the demand or supply of financial claims in capital-funds markets, hence on interest rates. Monetary policy actions will affect aggregate demand via expenditure decisions which are sensitive to the availability and cost of capital funds, for example, residential construction, plant and equipment expenditures, consumer durables, and state and local government capital improvements. Interest-rate changes may also affect personal saving, and hence the consumption function; and they may influence flows of U.S. capital funds abroad or the flow of foreign funds into this nation's capital markets.

Suppose the monetary authority's macroeconomic model indicates that a certain level of interest rates would be optimal for achieving its ultimate target variables, given the outlook for aggregate demand, the inflationary situation, and the balance-of-payments prospects. Note that forecasts are required because the policy actions decided now will operate with a time lag. Suppose a decline in interest rates from i_o to i_α is decided on. The Federal Reserve banks buy U.S. government securities on the open market, driving interest rates down a little and increasing the excess reserves of commercial banks. The commercial bank managers then increase their earning assets, that is, they increase loans outstanding or buy securities from the nonbank public. This drives interest rates down further and increases money supply as banks extend more credit. Such exogenous actions by the monetary authority and endogenous responses by commercial banks continue until interest rates have been driven down to the levels desired by the Federal Reserve Board.

Will continual purchases of securities by the monetary authority be needed to hold interest rates at the new level? Probably not. In the first place, increased aggregate demand stimulated by the exogenous decline in interest rates will lead to increases in production and income flows. These increases will continue until desired withdrawals rise endogenously to equality with the higher level of desired injections resulting from the lower interest rate. In the *new flow equilibrium*, the nonbank public will supply the funds which the Monetary Sector provided initially to start the expansion. In the second place, the nonbank public will reach *a new balance-sheet equilibrium* at the lower interest rate. As we noted in Chap. 12, when the interest rate declines, the nonbank public does desire to shift its portfolio toward fewer securities and more money stocks. But there are limits. After exchanging a certain value of securities for money, the public will reach a new balance-sheet equilibrium at the lower interest rate. The Personal Sector's desired flow of saving will purchase the flow of new securities which the borrowing sectors desire to sell, and no net bank financing will be required. Hence., there will be no further increase in money stocks.

The shift between equilibria is pictured for money stocks in Fig. 13-2. Initial equilibrium is at point *o*, the intersection of the nonbank public's demand for money and a horizontal line at interest rate i_o. This line indicates that the monetary authority has set a desired rate of interest and will supply any amount of

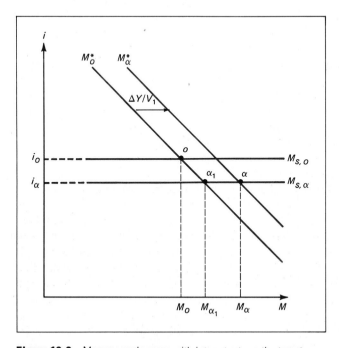

Figure 13-2   Money supply curve with interest rate as the target.

money (will buy or sell securities) as needed to maintain that rate. When the monetary authority decides to lower the interest rate to i_α, it establishes a lower horizontal line at which it will supply any amount of money needed to establish the new interest rate. If the demand curve for money should remain fixed, the new equilibrium point would be at α_1. However, the rise in GNP induced by lower interest rates will increase the demand for transactions balances by $\Delta M_1^* = (1/V_1)\,\Delta Y$, shifting the money demand curve rightward by that amount to M_α^*. So the monetary authorities must provide an additional rise in money stock of $\alpha_1 \alpha = (1/V_1)\,\Delta Y$. The final equilibrium is at point α.

Note that *by setting interest rate as its proximate target, the monetary authority has abdicated control over the money supply.* It must supply whatever amount is demanded at the new interest rate by the nonbank public. *The money supply curve becomes horizontal when interest rates, and not the quantity of money, become the sole target of monetary policy.*

MONEY STOCK POLICY

According to a monetarist point of view, a preferable target for Federal Reserve policy is money stock itself. The policy prescription might well call for a percent growth rate of money stock equal to the growth rate of real GNP less any steady percent per year rise in income velocity. All this would be on a long-term growth basis, not year by year. It is believed that such a policy would help stabilize prices and reduce fluctuations of real GNP and employment around their long-term trend lines. Interest rates would be allowed to fluctuate, and presumably would vary in such a way as to generate countercyclical pressures on aggregate demand.

What supply curve for money stock would correspond to such a policy? In Fig. 13-3a, this policy is represented by two relations between M_S and Y_R.

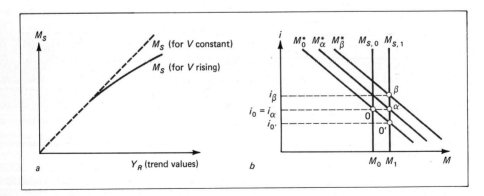

Figure 13-3   Money supply curve with money stock as the target.   (a) Monetarist policy.   (b) Effect on interest rate.

If income velocity stays constant, then the monetary authority would increase money stock by the same percentage each year, which would be equal to the long-term growth rate of real GNP, or about 4 percent per year. Then M_S would be directly proportional to Y_R, and lie on a line through the origin. If income velocity rises by a constant percent per year, then M_S would have a smaller growth rate than Y_R, and the M_S and Y_R relationship would be a curve falling below the preceding line.

Since M_S is fixed independently of interest rate, the money supply curve, when plotted in a graph with interest rate versus money stock, will simply be a straight line parallel to the i axis. (See Fig. 13-3b.) Given the money supply line $M_{S,0}$, the sloping line for money demand M_0^* will yield an intersection determining interest rate i_0 in equilibrium. Suppose that next year the monetary authority increases money stock to $M_{S,1}$ in line with its policy of annual increments. What happens to interest rates depends on the concomitant shift in the money demand line. If it stays at M_0^*, then interest rate declines, as indicated by the equilibrium point 0'. If the demand line shifts rightward by the same amount that money supply increases, the interest rate could stay constant at point α. Or the demand for money might increase faster than money supply and result in a rise in interest rate, as at point β.

There is a countercyclical feedback via money markets at work here. In equilibrium the demand-supply relation for money stocks is $M_S = M_1^* + M_2^*$. Changes in demand for transactions balances (M_1^*) are directly proportional to changes in current-dollar GNP, but changes in money supply (M_S) are directly proportional to real GNP at full employment $(Y_{R,FE})$, according to the policy of the monetary authority which we are assuming here. If current-dollar GNP rises slower than $Y_{R,FE}$ in a recession, then M_1^* will rise slower than M_S, and interest rates will decline (and M_2^* increase). The decline in interest rates will serve to stimulate aggregate demand and raise actual GNP back toward the full-employment level. Conversely, if aggregate demand is strong enough to carry GNP above the value of full-employment output, then M_1^* rises faster than M_S, and interest rates will rise (and M_2^* decline). The rise in interest rates will help to slow the boom in aggregate demand and to bring the economy back toward the full-employment level. So, *a monetary policy which maintains the money supply in a stable relation to real full-employment GNP would act through money markets as a countercyclical stabilizer for aggregate demand*, in a manner analogous to the operation of fiscal policy through automatic stabilizers and full-employment budget balance.†

Other monetary policies are, of course, possible between these two extremes of infinite elasticity and zero elasticity of money supply with respect to interest rates. Their effects on interest rates and aggretate demand will be intermediate between the two policies sketched here.

---

† The reasoning presented here assumes that transactions and income velocities are constant or change at a steady rate, and that prices are stable. However, the general principle would seem sound within a reasonable range of velocity and price changes.

APPENDIX   Diffusion of Excess Reserves and the Rise of Demand Deposits

Suppose that you manage the Last National Bank of Lostville, which has excess reserves of $0.6 million while other commercial banks are "loaned up." It occurs to you that you can increase bank income if you make more loans to local business owners, and you have some leeway to increase loans and deposits without incurring a reserve deficit. How much can you safely lend on the basis of $0.6 million excess reserves?

When you increase loans outstanding by ΔL, you create demand deposits (ΔDD) in your bank to the credit of the borrowers. Let us make the pessimistic assumption that the borrowers will pay all those deposits to persons and companies who deposit the funds in other banks. As the checks clear through the Federal Reserve banks, you lose FR bank deposits (reserves) to other banks ($-\Delta R$), and you also reduce the deposit accounts of your borrowers ($-\Delta DD$). Thus your excess reserves ER will be reduced by the amount of your increased loans ΔL. Consequently, if you started with excess reserves ER_0, you can afford to increase loans only by that amount, unless you wish to meet the drain on your reserves from other sources than your excess reserves.

Table 13-5 shows the balance sheets of the Last National Bank (1) before the loans are made, (2) after loans of $0.6 million have been made and demand deposits increased accordingly, and (3) after the borrowed funds have been paid to depositors in other banks and the checks have been cleared through the Federal Reserve System. As compared to its balance sheet before the loans were made, Last National winds up with $0.6 million more earning assets (loans), $0.6 million less reserves, the same deposit liabilities, and no excess reserves. The demand deposits which it created have been shifted to other banks, along with an equal amount of reserves.

Let us say that the above changes occurred during time period 1, and that the subset of all commercial banks receiving deposits and reserves from Last National in that period is "A-banks." During period 1 the A-banks gained $0.6 million in reserves and in demand deposits. Since required reserves increase by 15 percent of $0.6 million or, $0.09 million, the A-banks experience an increase of $0.51 million in excess reserves during period 1. See the entries for period 1 in Table 13-6.

To maximize their profits, the A-banks increase their loans and demand deposits by $0.51 million in period 2, and those deposits and reserves are shifted to a new subset of banks, called B-banks, by the end of period 2. At that time the B-banks show increases in reserves and demand deposits of $0.51 million and excess reserves equal to 85 percent of $0.51 million, or $0.43 million. In period 3, the B-banks increase loans and demand deposits (money supply) by $0.43 million, and C-banks end the period with $0.43 million more in deposits and an increment of 85 percent of $0.43 million, or $0.37 million, in excess reserves. So the process continues with period-by-period positive increments in loans and demand deposits (money supply) and declining excess reserves. Eventually, when the banking system comes to equilibrium with zero excess reserves, the total increase in loans and money supply for all banks will be

$$\delta L = \delta M = \frac{ER_0}{r_D}$$

where ER_0 is the initial quantity of excess reserves and r_D is the reserve ratio against demand deposits. In our example,

TABLE 13-5 Balance Sheets for Last National Bank ($ millions)

1   Initial balance sheet.   Required reserves are 15 percent against demand deposits and 5 percent
    against time deposits.   Initial excess reserves : $0.6 million.

Assets		Liabilities and Net Worth		(Required Reserves)
Reserves	$1.0	Demand deposits	$2.0	$(0.3)
Vault cash	(0.1)	Time deposits	2.0	(0.1)
Deposit in Federal Reserve bank	(0.9)			
U.S. government securities	2.5			
Other securities	2.0			
Loans	3.0			
Property	0.5	Capital stock	5.0	
	$9.0		$9.0	

2   Last National Bank makes loans of $0.6 million, increasing its demand deposits by $0.6 million.

Assets		Liabilities and Net Worth		(Required Reserves)
Reserves	$1.0	Demand deposits	$2.6	$(0.39)
Vault cash	(0.1)	Time deposits	2.0	(0.1)
Deposit in Federal Reserve bank	(0.9)			
U.S. government securities	2.5			
Other securities	2.0			
Loans	3.6			
Property	0.5	Capital stock	5.0	
	$9.6		$9.6	

3   Customers who borrowed funds issue checks for $0.6 million which are paid to persons and companies who have their deposits in other banks.   Excess reserves of Last National Bank: 0.

Assets		Liabilities and Net Worth		(Required Reserves)
Reserves	0.4	Demand deposits	2.0	$(0.3)
Vault cash	(0.1)	Time deposits	2.0	(0.1)
Deposit in Federal Reserve bank	(0.3)			
U.S. government securities	2.5			
Other securities	2.0			
Loans	3.6			
Property	0.5	Capital stock	5.0	
	$9.0		$9.0	

$$\delta L = \delta M_s = \frac{0.6}{0.15} = \$4 \text{ million}$$

Table 13-6 summarizes the period-by-period process.   The ultimate rise in loans and money supply is just the sum of the increments in each period for each subset of banks.

TABLE 13-6   Increase in Money Supply Arising from Initial Excess Reserves of Last National Bank
($ millions)

Period (t)	Banks	Change in Loans (ΔL)	Change in Money Supply ($\Delta M = \Delta DD$)	Change in Reserves (ΔR)	Change in Excess Reserves (ΔER)	Excess Reserves End-Period (ER)
0	Last National Bank					0.6
1	Last National Bank	+0.6	0	−0.6	−0.6	0
	A-banks	0	+0.6	+0.6	+0.51	0.51
2	A-banks	+0.51	0	−0.01	−0.51	0
	B-banks	0	+0.51	+0.51	+0.43	0.43
3	B-banks	+0.43	0	−0.43	−0.43	0
	C-banks	0	+0.43	+0.43	+0.37	0.37
4	C-banks	+0.37	+	−0.37	−0.37	0
	D-banks	0	+0.37	+0.37	+0.31	0.31

(It should be noted that a given bank may appear in several subsets, that is, it may receive deposits and make loans at several stages of the process—even the Last National Bank.)
In symbolic terms, for period t

$$\Delta M_t = \Delta L_t = ER_0(1 - r_D)^{t-1}$$

The total change from $t = 0$ to the final equilibrium will be the sum of this infinite geometric series of increments, or

$$\delta M = \delta L = \sum_{s=1}^{\infty} ER_0(1 - r_D)^{t-1} = \frac{ER_0}{r_D}$$

PROBLEMS

1 A home buyer pays the contractor $30,000 for a new house, having obtained the funds by taking $5,000 out of his savings account at commercial bank A and by borrowing $25,000 on a mortgage from a savings and loan association which keeps its account in bank B.

Answer the following questions as of the time when all checks have cleared, on the assumption that the contractor keeps his checking account in bank A. Use reserve ratios r_D of 15 percent on demand deposits and r_T of 5 percent on time deposits.

a What change, if any, has occurred in the money supply of the economy?

b What changes in excess reserves occur at bank A and at bank B?

c If banks in the aggregate had no excess reserves before these transactions, what changes might occur afterward in order to bring the banking system back to a position of zero excess reserves.

2 Suppose that the banking system starts with zero excess reserves and that the Federal Reserve System sells government securities to nonbank investors.

   a Use balance-sheet accounts to explain how such open-market operations can reduce the money supply of the economy.

   b Explain what, if any, effects these sales would have on interest rates and on the total income velocity of money ($V = Y/M_S$).

   c After a new equilibrium has been reached, what sector (or sectors) hold less money, and why is it willing to do so?

3 Explain how and why the expansionary effect of an increase in government purchases will differ if the added funds are obtained by:

   a An equal increase in tax revenues

   b Running down the government's bank account at the Federal Reserve System

   c Selling securities, which are bought by the nonbank public

   d Selling securities, which are bought by the Federal Reserve System

In each case, assume that the commercial banking system maintains zero excess reserves, and describe what changes occur in money supply, interest rates, and aggregate demand, as compared with what they were in an equilibrium state that existed before the increase in government purchases.

4 An economy is in an initial equilibrium for income and product flows and for money stock.   Then businessmen decide to increase exogenous investment by δI^*_{EX}, in year 1, and they borrow all the needed funds from commercial banks. Thus money supply rises by δI^*_{EX} in year 1.

   Assume perfect forecasting (or quick output adjustment) by producers, so that GNP rises by $\delta Y_S = k \, \delta I^*_{EX}$ in period 1, with i constant.

   a Explain how the increased money stock comes into existence and who holds it at the end of year 1.

   b By how much does demand for transactions balances increase by year-end?   Assume constant income velocity V_1 for transactions balances.

   c Will demand and supply of money stocks be in equilibrium at year-end if the interest rate remains the same?   Explain the reasons for your answer.

   d Explain in behavioral terms how equilibrium will be reestablished in money markets and what effect, if any, this procedure may have on i and on GNP.

Chapter 14

Flows of Capital Funds and Changes in Money Stocks

At this point we dive into the troubled waters of capital markets and money stocks. So far, we have gently but firmly pushed aside any analysis of the influence of financial markets and monetary policy on aggregate demand and equilibrium GNP. In some models, we have expressed investment demand as a function of interest rates and have indicated the existence of a supply-of-funds schedule, but no equations were included in the model to permit determination of the interest rate. We have noted also that government could influence interest rate and availability of capital funds by its lending programs and by its choice of methods for financing a change in government spending. But the causal linkages were not made explicit in or models.

In this chapter we shall look at the overall picture of the relations between current-account (GNP) flows, capital-account flows, and changes in money stocks. Subsequent chapters present in greater detail the analysis of interactions of capital funds and money markets with the rest of the economy.

SECTOR FLOWS OF FUNDS AND MONEY STOCKS

Back in Chap. 4 we discovered that total saving flows must equal the flows of borrowing by demand sectors to help finance their injections into the stream of aggregate demand. That analysis assumed implicitly that all saving flows went into capital-funds markets to purchase securities issued by borrowers, and also that total borrowed funds were spent to buy current output. Now, however, we must recognize that some of the personal saving flow may be used to increase money stocks held by the Personal Sector, and some of the funds borrowed by other sectors may be used to build up their money stocks. How does this change the flows of funds through capital markets, and how does it relate to demand and supply for money stocks?

Figure 14-1 illustrates the situation. Our four demand sectors are shown as usual, with the GNY and GNP flows indicated by arrows pointing in at the top and out at the bottom of each sector. In addition, each sector has a reservoir for its money stock associated with it. Finally, at the top of the diagram is an oval for capital-funds markets and a rectangle for the Monetary Sector (commercial banks plus monetary authority).

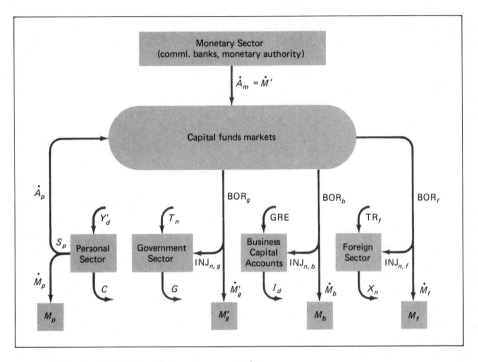

Figure 14-1 Capital-funds markets and money stocks.

A few definitions are needed for the new symbols in the diagram. *The dot above a symbol indicates rate of change of that variable.*

M_j = money stock of sector j. (The M'_g indicates that the U.S. government demand deposits are included here, though they are not included in money stock as usually defined.)

A_j = money value of nonmoney financial assets held by sector j. These are the securities or claims issued by the borrowing sectors to obtain capital funds. $\dot{A}_1$ equals dA_1/dt.

$\dot{M}_j$ = rate of change of money stock of sector j. That is, $\dot{M}_1$ equals dM_1/dt.

BOR_j = flow of money value of securities or claims issued by borrowing sector j, which equals the flow of money it obtains from capital-funds markets.

$\text{INJ}_{n,j}$ = flow of borrowed funds (external financing) used to help finance GNP purchases by sector j.

Government Sector: $\text{INJ}_{n,g} = G - T_n$

Business Capital Accounts: $\text{INJ}_{n,b} = I_d - \text{GRE}$

Foreign Sector: $\text{INJ}_{n,f} = X_n - \text{TR}_f$

Funds may flow in both directions through the channels indicated by the arrows in Fig. 14-1; the tails and heads of the arrows indicate the direction of money flow which is taken as positive. Net flows are normally in the positive direction—implying here that the Government Sector is normally a borrower— but a negative value for BOR_j just implies that sector j is supplying funds to capital markets in the time period involved. The arrow from the Monetary Sector to the capital-funds market indicates that when the Monetary Sector acquires assets (securities and claims issued by borrowers), it increases the money supply M' at the same rate.

If we make the simple equilibrium assumption that all money stocks are constant (or that all $\dot{M}_j$ equal zero), then all personal saving does flow into capital-funds markets and all borrowed funds become net injections into the stream of GNP purchases. The Monetary Sector will not be absorbing any financial assets. So the equality between flows of funds into and out of capital markets becomes:

Flow in $= \dot{A}_p = S_p$

Flow out $= BOR_g + BOR_b + BOR_f = INJ_{n,g} + INJ_{n,b} + INJ_{n,f}$

or

$$BOR = INJ_n$$

Actually, we know from the GNP accounting identities that actual personal saving must *always* equal actual net injections.†

$$S_p = INJ_n \tag{14-1}$$

If we now assume that all money stocks are increasing, we note that the Personal Sector's supply of funds to capital markets ($\dot{A}_p$) will be less than personal saving; also, the borrowing sectors will have to obtain a flow of capital funds in excess of their net injections, in order to increase their money stocks. The Monetary Sector must absorb securities to cover both the excess of borrowings over injections and the reduction in purchases of securities by the Personal Sector. The flows through capital-funds markets become:

Flow in: $\dot{A}_p + \dot{A}_m = (S_p - \dot{M}_p) + \dot{A}_m$

Flow out: $BOR = INJ_n + \dot{M}'_g + \dot{M}_b + \dot{M}_f$

---

† This fact is proved thus:
$$GNY = GNP$$
$$Y'_d + T_n + GRE + TR_f = C + G + I_d + X$$
$$Y'_d - C = (G - T_n) + (I_d - GRE) + (X_n - TR_f)$$
$$S_p = INJ_{n,g} + INJ_{n,b} + INJ_{n,f}$$
$$S_p = INJ_n$$

Will $\dot{A}_m$ equal the rate of rise in total money supply? Equating inflows and outflows above, we find

$$S_p - \dot{M}_p + \dot{A}_m = \text{INJ}_n + \dot{M}'_g + \dot{M}_b + \dot{M}_f$$

or

$$S_p + \dot{A}_m = \text{INJ}_n + (\dot{M}_p + \dot{M}'_g + \dot{M}_b + \dot{M}_f) = \text{INJ}_n + \dot{M}'_s \qquad (14\text{-}2)$$

Since S_p always equals INJ_n, we cancel those terms and obtain

$$\dot{A}_m = \dot{M}'_s$$

The value of the flow of securities absorbed by the Monetary Sector does, indeed, equal the total increase in money stocks held by all four demand sectors.

At first thought, it may be difficult to accept the fact that *the net absorption of securities by the Monetary Sector must always be equal to the net increase in money stocks for the four demand sectors combined.* Refer to Fig. 14-1 and consider the following line of logic.

1 From the GNY equals GNP identity, we proved that

$$S_p = \text{INJ}_n = \text{INJ}_{n,g} + \text{INJ}_{n,b} + \text{INJ}_{n,f}$$

2 Assume money stock is constant for each sector ($\dot{M}_j$ equals zero). Then all personal saving is used to buy securities, and the total flow of securities sold ·by borrowers equals total net injections. So

$$\dot{A}_p = S_p = \text{INJ}_n = \text{BOR}$$

and the Monetary Sector buys no securities on balance ($\dot{A}_m = 0 = \dot{M}'$).

3 Next, assume that money stock is increased by the Personal Sector but is constant for the borrowing sectors. Then Personal Sector net purchases of securities are less than S_p by the flow of funds used to increase personal money stocks. Since $\text{BOR} = \text{INJ}_n = S_p$, Personal Sector purchases of securities will therefore fall short of borrowing by $\dot{M}_p$, and the securities not absorbed by the Personal Sector must be bought by the Monetary Sector. So

$$\dot{A}_m = \text{BOR} - \dot{A}_p = S_p - \dot{A}_p = \dot{M}_p$$

4 Assume that money stocks increase in all sectors. Again, security purchases by the Personal Sector are less than S_p by $\dot{M}_p$. But BOR is now greater than INJ_n by ($\dot{M}'_g + \dot{M}_b + \dot{M}_f$). So the Monetary Sector must absorb securities equal to (*a*) the diversion of personal saving to $\dot{M}_p$ plus (*b*) the diversion of borrowed funds to increases in money stocks of the borrowing sectors. The

general equation for rate of acquisition of financial claims (securities) by the Monetary Sector is:

$$\dot{A}_m = \text{BOR} - \dot{A}_p = (\text{INJ}_n + \dot{M}'_g + \dot{M}_b + \dot{M}_f) - (S_p - \dot{M}_p)$$

$$= (\text{INJ}_n - S_p) + (\dot{M}_p + \dot{M}'_g + \dot{M}_b + \dot{M}_f)$$

$$\dot{A}_m = 0 + \dot{M}'$$

Thus, the flow of financial claims purchased by the Monetary Sector must always equal the rate of change of total money stocks M'.

EQUILIBRIUM IN CAPITAL FUNDS AND MONEY MARKETS

The relations described in the preceding section apply to actual flows of funds and changes in money stocks. They are accounting identities, true in any time period, whether the markets are in equilibrium or not. To investigate equilibrium conditions, we define desired levels for capital-market flows and for money stocks, using the asterisk (*) to indicate desired values of variables. Let us also make the assumption that federal government deposits are kept constant, so that borrowing equals net injections for the U.S. government; then the change of conventional money supply ($\dot{M}$) can be used instead of $\dot{M}'$.

$S_p^* = Y_d - C^*$, or desired personal saving flow

$\text{INJ}_n^* =$ net injections desired by the Government, Business, and Foreign Sectors combined

$\dot{A}_m^* = \dot{M}_s$, that is, the rate of accumulation of financial claims desired by the Monetary Sector from nonbank sectors must equal the rate of increase in money supply desired by the Monetary Sector

$M^* =$ desired change in money holdings for the four demand sectors combined

In equilibrium for the economy, these desired quantities are related to one another by an equation similar to Eq. (14-2), namely,

$$S_p^* + \dot{A}_m^* = \text{INJ}_n^* + \dot{M}^* \tag{14-3}$$

Actually, we can say more than this. *In a static equilibrium situation, the rate of change of money supply desired by the monetary authority ($\dot{A}_m^* = \dot{M}_s$) will be equal to the rate of change of money stocks desired by the four demand sectors ($\dot{M}^*$)—and both of these will be zero.* In other words, in equilibrium the existing stock of money will equal the amount the Monetary Sector desires to supply and also the amount the nonmonetary sectors wish to hold. Hence the desired changes in these amounts will be zero, if we take the concept of equilibrium to imply constancy of stocks as well as flows.

A static equilibrium state for flows of funds through markets for capital funds thus involves two separate balances:

1   Desired personal saving must equal the sum of net injections desired by the borrowing sectors.

$$S_p^* = \text{INJ}_n^* \qquad \text{or} \qquad \dot{A}_p^* = \text{BOR}^*$$

2   Stocks of money which decision makers in the Monetary Sector desire to supply to the economy must equal the demand for money stocks by the nonmonetary sectors of the economy.

$$M_s = M^*$$

Except perhaps for M_s, all these desired values are functions of Y and i, and probably also of wealth, prices, and price expectations for goods and equity claims. An exogenous shift in any one of the functions will lead to changes in market conditions which alter the values of variables desired by other decision makers in these markets.

To simplify the presentation, we assume that, for short-run shifts, Y and i are the most important causal determinants of the desired flows and stocks. Given this simplification, we may write the equilibrium equations:

$$\dot{A}_p^*(i, Y) = \text{BOR}^*(i, Y : \text{INJ}_{\text{EX}}^*) \tag{14-4}$$

and

$$M^*(i, Y) = M_s \tag{14-5}$$

In Eq. (14-4), $\dot{A}_p^*$ varies positively with Y and increases or remains constant as i rises. BOR* varies directly with exogenous injections INJ_{EX}^* but is inversely related to both i and Y, the latter because increased incomes reduce the need for borrowing sectors to obtain external funds to finance a given flow of injections. In Eq. (14-5), M^* is negatively related to i and positively related to Y, as explained in Chap. 12. M_s is exogenous if the monetary authority takes it as a target, but it is determined by demand for money stocks if the monetary authority takes interest rate as a target, as explained in Chap. 13.

Equations (14-4) and (14-5) involve three potential endogenous variables: i, Y, and M_s. If the monetary authority takes M_s as a target (exogenous) variable, then these equations permit simultaneous determination of i and Y. If the monetary authority chooses to maintain a target (exogenous) interest rate, then these equations permit determination of Y and M in equilibrium.

It should be noted that Eq. (14-4) is equivalent, in equilibrium, to

$$S_p^*(i, Y) = \text{INJ}_n^*(i, Y : \text{INJ}_{\text{EX}})$$

Money stocks are constant in equilibrium; so all personal saving is used for purchases of securities and all borrowing is used to finance net injections. This identification permits us to say also that Eq. (14-4) is nothing new. It is just another way of asserting the product-market equilibrium condition that WDL* equals INJ*, or Y_s equals Y^*.

However, Eq. (14-5) is new.   It adds an additional constraint to our set of functions describing the economy.   With it, interest rate i can be made an endogenous variable in the system and solved for in equilibrium along with Y.   Let us consider the behavioral interactions which determine i and analyze graphically the shifts of equilibria which occur (1) when the authorities change the money supply, and (2) when desired injections increase exogenously.

JOINT EQUILIBRIA IN MARKETS FOR CAPITAL-FUNDS FLOWS AND FOR MONEY STOCKS

In Fig. 14-2, the behavioral functions for flows of funds through capital markets are plotted in the left-hand diagram, and the demand and supply curves for money stock are plotted in the right-hand diagram.   The variation of desired quantities with respect to the level of interest rate is shown here.   Each curve is drawn for a constant level of gross national product, say Y_o, and this is indicated by the subscript o in the labels on the curves.

The horizontal distance to the BOR_o^* curve equals the demand for capital funds by the borrowing sectors, hence their supply of securities to capital markets. This borrowing equals the excess of their desired injections over their current income, that is, it equals INJ_n^*.   Desired borrowing increases at lower interest rates because desired injections rise—with sector incomes constant at Y_o.   A rise in GNP would shift the BOR* curve to the left, since higher income flows reduce the need for external financing to pay for any given level of desired injections.

The demand for securities by the Personal Sector ($\dot{A}_{p,o}^*$) is the supply of funds to capital markets.   The line would be vertical if personal saving were not sensitive to interest rates, but it has been pictured here as increasing somewhat at higher rates of interest.   If GNP rises, the $\dot{A}_p^*$ curve shifts to the right, since S_p^* rises as net disposable income goes up along with GNP.

The behavioral curves for money stocks have been described in the two preceding chapters.   If GNP rises, the M^* curve shifts rightward.   Any change in the M_s curve depends on the policy of the monetary authority.

As drawn in Fig. 14-2, the interest rate is such as to be consistent with a demand-supply equilibrium in both capital funds and money markets.   In each of those markets the behavior of market participants is stabilizing in the face of interest-rate fluctuations.   In capital-funds markets, an upward deviation of interest rates would lead to a greater supply of capital funds ($\dot{A}_{p,o}^*$) and a reduction in the demand (BOR_o^*).   Hence, the interest rate would tend to fall back toward the equilibrium level.   Conversely, a downward deviation of interest rates would lead to an excess demand for capital funds and upward pressure on interest rates as borrowers bid against one another for nonbank funds or seek loans from banks.

In the market for money stocks, an upward move of i would reduce desired holdings M_o^* below existing stocks $M_{s,o}$.   Some holders would use unwanted money stocks to buy securities, driving securities prices up and interest rates down.

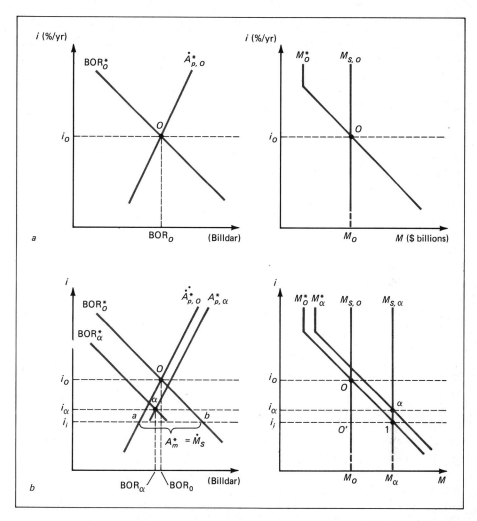

Figure 14-2 Equilibria for flows of capital funds and for money stocks.   (*a*) Initial equilibria.
(*b*) Equilibria after increase in money supply.

Conversely, at interest rates below the equilibrium level, an excess demand for
money stocks develops.   Holders of securities attempt to sell some and to increase their money stocks.   Their attempt drives securities prices down and interest rates up toward the equilibrium level.

But suppose the two markets are not initially in balance at the same interest
rate, and suppose we consider an adjustment process in which Y_s as well as i
changes.   What then?   Will the behavior of market participants cause the economy to adjust toward joint equilibria at the same interest rate for the two markets?

ADJUSTMENT TO A CHANGE IN MONEY SUPPLY

Suppose the monetary authority throws the money market into disequilibrium by increasing the money supply via open-market purchases of government securities. The situation is illustrated in Fig. 14-2b. Starting from equilibria at points 0, as in the preceding diagrams, we imagine the monetary authority desires to increase money supply to $M_{s,\alpha}$ in the right-hand diagram. Open-market purchases by the Fed increase commercial banks' reserves; the commercial banks then increase their earning assets and demand deposits, driving the interest rate down as they do so. If the demand curve for money were unchanged, the new equilibrium interest rate would be i_1, well below the initial equilibrium interest rate i_o. But an interest rate i_1 would lead to a large excess of demand for capital funds over their reduced supply in the left-hand diagram; the excess appears as the horizontal distance ab and arises from an increase in desired injections (hence BOR*) and a reduced supply of funds ($\dot{A}_p^*$) out of personal saving as interest rates decline. The excess demand for funds is met by the securities purchases of the monetary authority and commercial banks. That is, distance ab corresponds to $\dot{A}_m^* = \dot{M}_s$ and accounts for the rightward movement of the M_s line in the right-hand diagram.

At this lower level of interest rates, injections rise and Y increases by a multiplied increment. The higher level of GNP shifts the BOR* curve left and the $\dot{A}_p^*$ curve right, lowering the equilibrium interest rate in capital-funds markets and reducing the excess demand for funds. The higher level of GNP also shifts the M^* curve rightward, raising the equilibrium level of interest rate in the money market. The final outcome, as pictured in Fig. 14-2b, is the emergence of a new equilibrium level of interest rate i_α and a higher income level Y_α, at which both markets are in equilibrium simultaneously. The actual path of adjustment need not involve an excessive reduction of interest rate to i_1 and then a rebound to i_α. More likely, there will be a smoother passage directly from point o to point α in the two diagrams as M_s shifts gradually rightward to its final position.

It seems clear that *normal behavioral responses by market participants to the exogenous changes initiated by the monetary authority will lead to new joint equilibria for flows of capital funds and stocks of money at lower interest rates and higher GNP.* The specific adjustment path will depend on how rapidly the monetary authority increases M_s to its new level and how rapidly GNP increases in response to lower rates of interest.

ADJUSTMENT TO A SHIFT OF THE DESIRED BORROWING CURVE

Now let us investigate the adjustment process when a rise in desired injections shifts the desired borrowing curve to the right. What happens (1) when the monetary authority keeps the money supply constant, and (2) when the monetary authority keeps the interest rate stable?

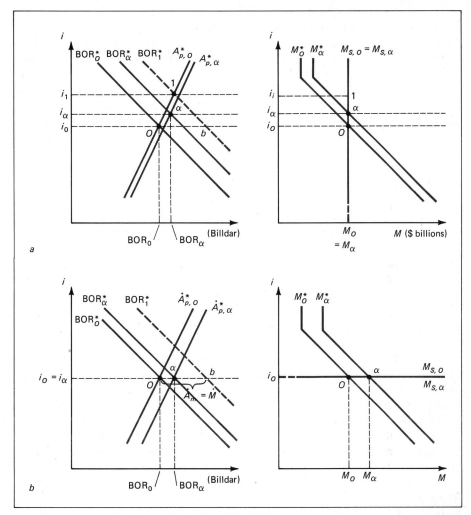

Figure 14-3   Equilibria for capital funds and money after shift in BOR* curve,   (*a*) with M_s held constant, and (*b*) with interest rate held constant.

Example 1   Money Supply Constant

The first case is pictured in Fig. 14-3*a*.   Initial equilibria are at points *o*.   Then an exogenous rise in desired injections, for example, through investment or government purchases, shifts the BOR* curve to BOR_1^*, still consistent with constant GNP equal to Y_o.   The excess demand for capital funds *ob* which would emerge at interest rate i_o cannot be satisfied since the monetary authority will not purchase new securities and increase M_s.   Consequently, borrowers bid the interest rate up to i_1, where BOR_1 is greater than BOR_o by the increment to personal saving induced by the higher interest rate.   Actually, the supply of funds may temporarily be increased more than this because, at higher i, the desired money stocks are less than M_o, and some members of the nonbank public will supply funds to capital markets as they shift their portfolios toward less money and more securities.   But this is a transient effect.

The increased injections will cause Y_s to rise.   This will shift the BOR* curve left from position 1, and the $\dot{A}_p^*$ curve to the right.   The net result is to move the markets toward new equilibria at points α, where the interest rate lies between the initial i_o and the level i_1 indicated for constant Y_o.   The rise in interest rates to i_α reduces the impact of the exogenous rise in injections.   The endogenous decline in INJ* caused by the rise in i makes the equilibrium rise in GNP less than would have been predicted by applying the exogenous multiplier to the initial exogenous rise in INJ*.   In the right-hand portion of Fig. 14-3a, the M^* line shifts right as Y increases and comes to equilibrium with M_s at i_α.

Example 2   Interest Rate Constant

If the monetary authority elects to keep interest rates constant, the economy will experience the full multiplier impact from the exogenous rise in aggregate demand. (Refer to Fig. 14-3b.)   Initial equilibria are at points o again.   The BOR* curve shifts right as before.   But now the monetary authority purchases the added flow of securities (Fig. 14-3b), raising the money supply at a rate $\dot{A}_m^* = \dot{M}_s = ob$ as needed to keep the interest rate at i_o.   Increased injections raise GNP, which shifts the BOR* curve left and the $A_{p,o}^*$ line right continuously until the curves for demand and supply of capital funds intersect at point α.   Creation of money will cease at that point, leaving the total money stock M_α at the level determined by the nonbank public's demand curve for money (M_α^*) at the higher level of Y_α. Injections and GNP rise more than when money supply was kept constant (Fig. 14-3a), because the interest rate in equilibrium is lower than it was in Example 1.

SUMMARY

In this chapter we have considered markets for flows of capital funds and for stocks of money.   We assumed that the demand-and-supply decisions in these markets depend on aggregate incomes Y_s and on the level of interest rates i, as follows:

1   Demand for capital funds (flows) equals BOR*$(i, Y_s: \text{INJ}_{\text{EX}}^*)$, where exogenous injections are a shift parameter, and where both $\Delta(\text{BOR*})/\Delta i$ and $\Delta(\text{BOR*})/\Delta Y_s$ are negative.

2   Supply of capital funds (flow) equals $\dot{A}_p^*(i, Y_s)$, where both $\Delta(\dot{A}_p^*)/\Delta i$ and $\Delta(\dot{A}_p^*)/\Delta Y_s$ are positive.

3   Demand for money (stock) equals $M^*(i, Y_s)$, where $\Delta M^*/\Delta i$ is negative, and $\Delta M^*/\Delta Y_s$ is positive.

4 Supply of money (stock) equals M_s, where M_s is decided by the monetary authority.   Changes in M_s are achieved by open-market transactions and increases in earning assets by the Monetary Sector, that is, $\dot{M}_s$ equals $\dot{A}_m^*$.

From the analysis of such a system we have concluded that:

1   The equation for equilibrium between demand and supply for flows of capital funds is equivalent to the equilibrium condition that aggregate demand equals aggregate supply, or that aggregate desired injections equal aggregate desired withdrawals.

2   However, the equality of aggregate demand and supply for money stock in equilibrium is a new condition imposed on the variables of the system.   It permits us to make one more variable endogenous.   So interest rate i becomes "explained" and can be solved from our system of equations.   Actually, Y and i are jointly determined by interactions of sector decision makers in markets for output, for capital funds, and for money stocks.

3   Markets for flows of capital funds and for money stocks are both stable in the event of erratic fluctuations in either i or Y_s.

4   An increase in M_s, enforced by the monetary authority, will provoke market reactions which lead to new stable equilibria at a lower level of i and a higher level of Y.   Conversely, an exogenous decline in M_s will lead to new equilibria at higher i and lower Y.   The direction of change in BOR at equilibrium depends on the slopes and shifts of the demand-and-supply curves for borrowed funds.

5   An exogenous increase in desired injections, and hence in desired borrowing, will provoke market reactions which result in new equilibria at higher levels of i, Y, and BOR, if the monetary authority keeps the money supply constant. Conversely, an exogenous decline in desired injections and borrowing leads to new equilibria with lower values for i, Y, and BOR, if the money supply is held constant.   For either direction of change in INJ_{EX}^*, the magnitude of change in equilibrium value of Y is less than $k(\delta INJ_{EX}^*)$ because the interest rate changes and induces a change in INJ^* which offsets part of the initial exogenous $\delta(INJ_{EX}^*)$.

6   If the monetary authority follows a policy of altering M_s so as to keep i constant, exogenous shifts $\delta(INJ_{EX}^*)$ will lead to changes in Y in the same direction that amount to the full multiplier impact, that is, δY equals $k(\delta INJ_{EX}^*)$.

7   If the monetary authority responds to an exogenous shift of injections with a policy intermediate between those in items (5) and (6), equilibrium levels of i, Y, BOR, and M will all change in the same direction as desired injections but by amounts that are intermediate between items (5) and (6).

In the next chapter the equilibrium constraint arising from the market for money stocks is combined with the IS (investment-saving) line of Chap. 10 to show in one diagram the simultaneous solution for equilibrium in both product and money markets.

PROBLEMS

1 Assume that the interest rate is always at a level which equalizes the supply and demand for flows of capital funds. A change in interest rate must result from a shift either in the demand curve or in the supply curve for capital funds. What direction of shift for which curve(s) would be consistent with each of the following observations?

   a Flow of funds through capital markets increases even though the interest rate rises.

   b An exogenous rise in investment spending occurs.

   c Interest rates decline and the flow of capital funds rises.

   d The monetary authority increases the money supply.

   e An exogenous rise in personal saving occurs, but interest rates remain stable.

   In each case cite a possible cause for the shifts.

2 Do you agree or disagree with the following statement: "When the money supply is changing, capital-funds markets cannot be in static equilibrium, nor can markets for final output." Explain the reasons for your decision.

3 Interpret the following statement† in terms of the concepts and graphs (or equations) of this chapter: "Short-term interest rates rose sharply in continued reaction to an unexpected surge in credit demand and a toughening in the stance of the Federal Reserve."

4 In Chap. 8, the paradox of thrift was discussed—the possibility that a sudden rise in the public's propensity to save, unexpected by producers, could lead to unplanned inventory investment and a subsequent decline in equilibrium GNP. Trace the effects of such a rise in saving on capital-funds markets and money supply (*a*) during the time before producers change the flow of output (Y_s); and (*b*) during the subsequent adjustment to a new equilibrium in the economy. Indicate what pattern of change you would expect for $\dot{A}_p$, BOR, $\dot{A}_m$, i, and M_s.

5 Historically, some economists have maintained that the interest rate is determined in capital-funds markets as the variable which brings demand and supply of capital funds into equality. That is, the interest rate equals the rate of return on the marginal assets for which funds are borrowed. Other economists have maintained that the rate of interest is determined in markets for money stocks as the variable which brings demand and supply of money stocks into equilibrium. Evaluate these two theories in the light of the analysis presented in this chapter.

---

† *The Wall Street Journal*, March 25, 1971, p. 1.

Chapter 15

General Equilibrium

INTRODUCTION

We have now completed the analysis of sector decision making and equilibrium conditions in all four markets—product markets, markets for labor services, markets for flows of capital funds, and markets for money stocks. And we have discussed the behavior of decision makers when an exogenous change throws some market out of equilibrium. It appears that, for reasonable values of the system constants, adjustments will lead to new stable equilibria, though oscillations may occur if accelerator feedbacks are strong.

Now we shall coordinate the analyses of the individual markets to investigate the overall system response as it moves from one general equilibrium position to another. In particular, we shall try to answer the question: *How can the economy be brought to equilibrium at full employment and maintained at full employment without inflation, even as the economy grows?*

THE COMPLETE SYSTEM OF EQUATIONS

For short-run analysis of the economy, we assume that equilibrium is attainable in one or two quarters. So we estimate the effect of some exogenous stimulus by comparing an initial equilibrium state with the new equilibrium reached when decision makers in all markets have responded fully to the initial disturbance and its induced effects. Implicitly we are assuming no long-continued oscillations from multiplier-accelerator effects in the income and investment feedback paths of the system.

The short-run equilibrium state of the economy will be specified by the values of 11 endogenous variables as follows.

1   Y^* = Aggregate demand, or aggregate purchases of current output desired by the demand sectors (in billdar)

2     $Y_{R,s}$ = Aggregate real output desired by the Production Units (in billdar at 1958 prices or billion base-year units per year)

3   Y_s = Aggregate supply, or value of production expected by Production Units, equal to aggregate incomes distributed from production (in billdar)

4   P_Y = Price for aggregate output (in dollars per base-year unit of output)

5   M^* = Demand for money, or stocks of money desired by all nonbank public sectors (in \$ billions)

6   M_s = Supply of money, or money stock consistent with policy of the monetary authority (in \$ billions)

7   i = Interest rate, presumably a typical long-term rate relevant to investment and portfolio decisions (percent per year)

8   L^* = Demand for labor services, or the flow of labor inputs desired by Production Units (in billion man-hours per year)

9   L_s = Supply of labor services, or the flow of labor output desired by persons in the labor force (in billion man-hours per year)

10   W = Money wage rate, or compensation per man-hour for some representative occupation and grade of worker (in \$ per man-hour)

11   MPL = Marginal productivity of labor (in base-year units per man-hour, or in dollars at 1958 prices per man-hour)

The *concept of general equilibrium* of the economic system is this: *If the behavioral, institutional, and technological relationships of the system are specified and if values of exogenous variables are given, interactions between decision makers in the markets of the economy will result in an equilibrium state described by determinate values of the 11 endogenous variables.* The purpose of this chapter is to present the overall framework for understanding how general equilibrium is established and how it will shift in response to exogenous changes in variables influencing equilibrium. In particular, we want to discover how certain policy variables may be used by the federal government to maintain equilibrium near full employment.

First, consider the set of equations describing the economic system—11 equations to permit determination of equilibrium values of the 11 endogenous variables. The equations are summarized in Table 15-1, where they are grouped according to the market to which they refer.

The first group of four equations deals with markets for final product flows (GNP). The aggregate demand equation indicates that the level of Y^* depends on three endogenous variables (Y_s, i, and P_Y), and on population (N) and the exogenous components of the desired purchases of the demand sectors. Endogenous variables are separated from exogenous items by a colon, and the two right-hand columns in the table classify the causal variables according to whether an increase in that variable would have a positive or a negative influence on aggregate demand. The individual exogenous items will be explained in the next section of this chapter. Usually we display this equation by plotting a line for Y^* versus Y_s; then changes in other causal variables will shift that Y^* line.

The second equation describes the behavior of Production Units in deciding on output levels; it was developed back in Chap. 6. Producers increase output for profit maximization if aggregate demand strengthens or if the marginal-pro-

TABLE 15-1 A Complete Equation Model for Economy in Short-Term Equilibrium†

	Direction of Influence on Dependent Variables	
	Positive	Negative
Product markets (flows)		
1 Aggregate demand		
$Y^* = Y^*\{Y_s, i, P_Y : G_o, X_o, Z_o, N,$	Y_s, P_Y, G_o, X_o, N	i, Z_o
behavioral coefficients$\}$		
2 Aggregate real supply		
$Y_{R,s} = Y_{R,s}\{Y^*, W, \text{MPL}: \beta\}$	Y^*, MPL	W, β
3 Aggregate supply		
$Y_s = P_Y \, Y_{R,s}$	$P_Y, Y_{R,s}$	
4 Equilibrium condition		
$Y^* = Y_s$	—	—
Money markets (stocks)		
5 Demand for money		
$M^* = M^*\{Y_s, i, P_Y: V_1,$	Y_s, P_Y	i, V_1
behavioral coefficients$\}$		
6 Supply of money		
$M_s = M_{s,o}$	$M_{s,o}$	
7 Equilibrium condition		
$M^* = M_s$	—	—
Labor markets (flows)		
8 Demand for labor services		
$L^* = L^*\{W, P_Y, \text{MPL}: \beta\}$	P_Y, MPL	W, β
9 Supply of labor services		
$L_s = L_s\{W, P_Y: N\}$	W, N	P_Y
10 Equilibrium condition		
a. $L^* = L_s$	—	—
b. $W = W_o$	—	—
Production function		
11 $\text{MPL} = \text{MPL}\{L^*: K, \text{TECH})$	K, TECH	L^*

† Endogenous variables are to left of colon; exogenous variables are to the right.

ductivity-of-labor curve is raised; equilibrium output declines if the wage rate W rises or if the marginal coefficient for nonfactor charges (β) increases. The third equation expresses the Production Unit's decisions in terms of expected revenue at the profit-maximizing output, bringing in the price level P_Y at which they expect to sell their output. It is this equation which was plotted in Chap. 6 as the aggregate supply curve. It relates Y_s to $Y_{R,s}$ and has an implied price level at each point on the curve. The curve will shift up if W or β rises, and down if MPL is increased, because those "shift" variables affect marginal costs and hence the price level at which profits are maximized.

Equation (4) imposes the equilibrium condition for product markets. Normally, only three equations are needed to determine equilibrium in a market—one describing the behavior of buyers, one describing the behavior of sellers, and an equilibrium condition. But our equations involve determination of equilibrium values both for real and for money-value flows of output; so an extra equation is needed.

Of our 11 endogenous variables, 7 are involved in the product-market equations (Y^*, Y_s, $Y_{R,s}$, P_Y, i, W, and MPL). The first four of them may be thought of as determined primarily in product markets. Then i would be determined in money markets, W in labor markets, and MPL by the production function. This allocation is not strictly correct, because interactions between markets lead to a simultaneous, joint determination of equilibrium values for all endogenous variables. But it does point to the principal market reactions which determine a consistent set of values for all endogenous variables in general equilibrium.

Equations (5) and (6) add two new endogenous variables M^* and M_s, which embody the behavior of the nonbank public in demanding money stocks and the decisions of the monetary authority in supplying it. Since the equilibrium equation for money markets adds no new variable, this set of three equations may be combined with those for product markets to determine interest rate and money supply and demand, along with the four product-market variables.

Equations (8) and (9) express the behavior of Production Units in choosing the flow of labor inputs consistent with profit maximization and the behavior of the labor force in deciding how large a flow of labor services they would like to supply as a function of wage rate W, price level P_Y, and population N. In Eq. (10), two alternative equilibrium conditions are shown. The first implies that wage rates will adjust to eliminate unemployment, so that the economy will come to equilibrium only at full employment—the classical model. The second alternative indicates that wage rates will be constant in the short run, so that demand for labor services may be less than the supply, leading to equilibrium at less than full employment—the Keynesian model. In either event, the three labor-market equations add only two new endogenous variables (L^* and L_s) and thus permit determination of the money wage rate W as well as L^* and L_s.

Finally, the production function equation expresses marginal productivity of labor MPL as a function of labor input L^* and of shift variables representing capital stock and production technology.

These 11 functional relations permit determination of consistent equilibrium values of the 11 endogenous variables if the behavioral coefficients and the exogenous variables are known. Recall also that equilibrium in product and money markets guarantees equilibrium in markets for flows of capital funds.

Now let us use a graphical approach to analyzing the establishment of economic equilibrium and its shifts when the economy is stimulated by various exogenous changes.

EQUILIBRIUM IN PRODUCT MARKETS

First, consider the four product-market equations. As noted, they contain seven of the endogenous variables. We shall begin by assuming three of them to be determined elsewhere in the model (W, MPL, and i). Then we use the four product-market equations to determine equilibrium values of Y^*, Y_s, $Y_{R,s}$, and P_Y consistent with assumed values for W, MPL, and i.

Aggregate demand is, of course, the sum of purchases desired by the four demand sectors. Let us rewrite the sector demand equations to bring out their dependence on price level explicitly and to allow for some influence of GNP on government, investment, and import expenditures.

In the chapter on the consumption function, it was noted that the intercept of the consumption function would rise as P_Y increased. So we may write a generalized consumption function in the form

$$C^* = a_o + a_1 P_Y + b' Y_s$$

For domestic investment, we noted that general price rises would leave rates of return on projects unchanged but would increase their dollar value. So we may write

$$I_d^* = (h_o - h_1 i) P_Y + h_2 Y_s$$

Some government expenditures may be budgeted in dollar amounts, for example, payrolls in the short run; some may rise as prices rise or as tax revenues increase with GNP. So we write

$$G^* = G_o + g_1 P_Y + g_2 Y_s$$

Net exports may well be separated into exports and imports. For exports depend primarily on economic activity in foreign countries, whereas imports will vary with domestic GNP. Both may depend to some extent on United States price levels. Thus we have

$$X_n^* = X^* - Z^* = (X_o + x_1 P_Y) - (Z_o + z_1 P_Y + z_2 Y_s)$$

If we add these four sector demand equations and combine terms involving the same endogenous variable, we obtain

$$Y^* = (a_o + G_o + X_o - Z_o) + (a_1 + h_o - h_1 i + x_1 - z_1) P_Y$$
$$+ (b' + h_2 + g_2 - z_2) Y_s \qquad (15\text{-}1)$$

Following previous terminology, we designate components of Y^* which do not depend on Y_s as *autonomous* and those which do as *induced*. Then we may write the above expression as

$$Y^* = Y_{\text{AUT}}^* + Y_{\text{IND}}^*$$

where

$$Y_{\text{AUT}}^* = (a_o + G_o + X_o - Z_o) + (a_1 + h_o - h_1 i + x_1 - z_1) P_Y$$

and

$$Y_{\text{IND}}^* = (b' + h_2 + g_2 - z_2) Y_s$$

When we plot Y^* as a function of Y_s, then Y_{AUT}^* becomes the intercept on the vertical axis and Y_{IND}^* is the component of aggregate demand which rises in proportion to Y_s with slope equal to $(b' + h_2 + g_2 - z_2)$. [See panel (a) in Fig. 15-1.] Note that Y_{AUT}^* will increase as prices rise and decrease as interest rate rises. If we assume constant values $P_{Y,\alpha}$ and i_α, then a specific aggregate demand line Y_α^* is determined, and equilibrium for demand sectors will be reached at Y_α, where their aggregate desired expenditures equal their incomes (Y_s).

A question now arises regarding the values of $P_{Y,\alpha}$ and Y_α which are consistent with equilibrium for the demand sectors. Will the price $P_{Y,\alpha}$ motivate Production Units to produce an output with market value Y_α? That is, will $P_{Y,\alpha}$ and Y_α values coexist at some point on the aggregate supply curve? The aggregate supply curve, plotted in panel (b) of Fig. 15-1, is the graphical representation of Eqs. (2) and (3) in the complete equation model of Table 15-1. A specific aggregate

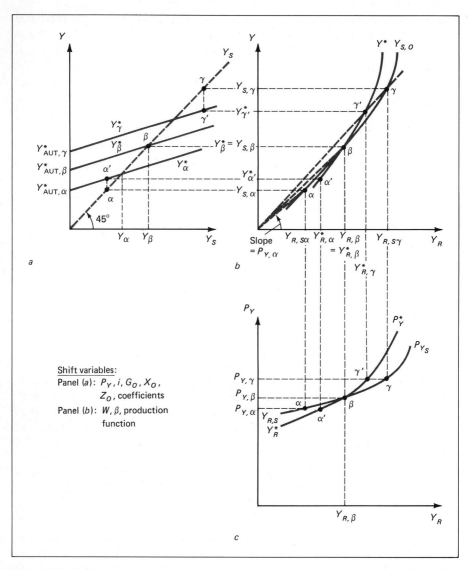

Figure 15-1 Equilibrium in product markets. (a) Aggregate demand. (b) Aggregate supply. (c) Demand and supply price curves.

supply curve relates supplier's expected revenues to real output for fixed values of W, β, and a given aggregate production function. On the aggregate supply curve $Y_{s,o}$ in Fig. 15-1, we can find how much output and incomes producers will generate when prices are at a level $P_{Y,\alpha}$, for example, by drawing a line from the origin at a slope $P_{Y,\alpha}$ to intersect the $Y_{s,o}$ curve at point α. The incomes $Y_{s,\alpha}$ which producers distribute can be projected horizontally to the 45-degree line in panel (a), and the expenditures which the demand sectors desire to make when they receive incomes $Y_{s,\alpha}$ can be determined on the Y_α^* line at point α'. We find that product markets are not in equilibrium at price level $P_{Y,\alpha}$. Aggregate demand exceeds aggregate supply. The excess demand measured in real output can be seen graphically by projecting point α' horizontally across to the constant price line in panel (b). At the intersection point α' in panel (b), we have a point on the aggregate *real* demand curve, since the horizontal distance to α' is $Y_\alpha^*/P_{Y,\alpha} = Y_{R,\alpha}^*$. Real aggregate demand exceeds real aggregate supply $Y_{R,s\alpha}$, and producers experience a decline in inventories and are motivated to increase output.

Now we are faced with the problem of finding a price level at which aggregate real supply and demand are equal. Can we find macroeconomic demand and supply price curves whose intersection determines equilibrium price, output, and expenditures—as we do in microeconomics? The answer is "Yes."

The supply price curve (P_{Y_s} versus $Y_{R,s}$) can be obtained easily, since price and real output are known at each point on the aggregate supply curve $Y_{s,o}$. The supply price curve will show real output rising rapidly as P_Y increases at low levels of output; then real output rises slowly as P_Y increases at high levels of output, near capacity. The curve is much like the supply curve in microeconomics and is shown as the $Y_{R,s}$ curve in panel (c) of Fig. 15-1.

The best method for determining the demand price curve is to assume that Production Units choose successive points along their aggregate supply curve $Y_{s,o}$ in panel (b)—points such as α, β, and γ. At each point the producers' forecasts of demand lead them to produce real output $Y_{R,s}$, offer it for sale at price P_Y, and distribute incomes Y_s from production. Each value of P_Y set by producers will determine a specific aggregate demand line Y^* in panel (a)—for the assumed value of i. The incomes paid out by producers determine the level of incomes (Y_s) received by demand sectors, as located on the 45-degree line in panel (a). This income determines aggregate demand on the Y^* line drawn for the specific value of P_Y. Thus aggregate real demand $Y_R^* = Y^*/P_Y$ is determined—graphically by projecting the value of Y^* horizontally to the P_Y line in panel (b) and reading off Y_R^* on the horizontal axis in panel (b). So we can determine pairs of values of P_Y and Y_R^* (at points α', β, and γ') and can plot the demand price curve P_Y^* shown in panel (c).

The demand and supply price curves plotted in Fig. 15-1(c) imply that at low price and output levels, demand exceeds supply, so that producers are motivated to increase output and prices. At high prices and outputs, there is an excess of supply over demand, so that producers will reduce output and price. In between lies the stable equilibrium point β at which price level $P_{Y,\beta}$ and output $Y_{R,\beta}$ leave both the demand sectors and Production Units satisfied.

In this example, the demand price curve slopes upward, unlike the analogous curve in microeconomics. This upward slope occurs because price rises in macro-economics are coupled with increases in incomes and because aggregate demand is assumed to increase rather strongly when prices rise. If the autonomous component of demand were small, or if it did not increase much when prices rose, it would be possible that real aggregate demand would remain constant or decline as prices and incomes rise along the $Y_{s,o}$ curve; that is, the demand price curve could have a negative slope like that in microeconomics. The likelihood of this would be greater if the initial equilibrium were in the upper portion of the aggregate supply curve, where rising prices cause producers to increase real output very little if at all. In such an inflationary situation, real aggregate demand would decline as prices rise, unless the autonomous component of aggregate demand increases nearly in proportion to price rises.†

---

† These conclusions can be shown by graphical analysis or can be derived algebraically as follows.

Let $f_A = \dfrac{Y^*_{\text{AUT}}}{Y^*}$ at the initial equilibrium point

$f_I = \dfrac{Y^*_{\text{IND}}}{Y^*}$ at the initial equilibrium point, so that

$f_A + f_I = 1$

Let the symbols $\%\Delta Z$ denote the percent change in variable Z and let price elasticities be defined thus:

$E_A = \dfrac{\%\Delta Y^*_{\text{AUT}}}{\%\Delta P_Y} =$ elasticity of autonomous demand, assumed to range from 0 to $+1$

$E^* = \dfrac{\%\Delta Y^*_R}{\%\Delta P_Y} =$ elasticity of real demand

$E_s = \dfrac{\%\Delta Y_{R,s}}{\%\Delta P_Y} =$ elasticity of real supply, assumed to range from high values at the lower end of
the aggregate supply curve to 0 in the upper vertical segment

As producers move along the fixed aggregate supply curve, $\%\Delta Y_s$ equals $\%\Delta Y_{R,s}$ plus $\%\Delta P_Y$, since Y_s equals $P_Y Y_{R,s}$. The consequent percent rise in aggregate demand is a weighted combination of the percent rises in its autonomous and induced components.

$\%\Delta Y^* = f_A(\%\Delta Y^*_{\text{AUT}}) + f_I(\%\Delta Y^*_{\text{IND}})$

$\qquad = f_A E_A(\%\Delta P_Y) + f_I(\%\Delta Y_s)$

$\qquad = f_A E_A(\%\Delta P_Y) + f_I(\%\Delta Y_{R,s} + \%\Delta P_Y)$

$\%\Delta Y^*_R = \%\Delta Y^* - \%\Delta P_Y = (f_A E_A + f_I - 1)(\%\Delta P_Y) + f_I(\%\Delta Y_{R,s})$

Dividing through by $\%\Delta P_Y$ to obtain elasticities, we find

$E^* = (f_A E_A + f_I - 1) + f_I E_s$

If E_s equals zero (vertical segment of aggregate supply curve), then

1 E^* is less than 0 if E_A is less than 1; that is, the aggregate real demand curve is negatively inclined if autonomous demand rises less than in proportion to price increases.

2 E^* equals 0 if E_A equals 1; that is, the aggregate real demand curve is vertical, on top of the aggregate real supply curve.

If E_s is larger than zero, the slope of the aggregate real demand curve (proportional to E^*) will be negative if

$f_A E_A + f_I(1 + E_s) < 1$

and positive if

$f_A E_A + f_I(1 + E_s) > 1$

The tendency toward positive slope is stronger if autonomous spending and real output respond strongly to prices and if induced demand is a large proportion of the total (that is, if f_I is large).

As may be seen from this discussion, the nature of the demand price curve changes depending on where the equilibrium point lies along the aggregate supply curve. And it will change if the aggregate supply curve shifts position because of a change in wage rates. As a consequence of this interaction between demand and supply price curves, they are not used so much in macroeconomic analysis as are the aggregate (current-dollar) demand and supply curves plotted in panels (a) and (b) of Fig. 15-1. However, the preceding discussion has established that *normal behavior of the demand sectors and Production Units will lead to determinate, stable equilibrium values for four endogenous variables in product markets*—Y^*, Y_s, Y_R, and P_Y. These equilibrium values are conditional on the assumed values of other shift variables affecting the positions of the aggregate demand and supply curves— namely i, G_o, X_o, Z_o, system coefficients, W, β, and the production function. We proceed next to work those variables into our graphical general equilibrium analysis, starting with money markets and the interest rate.

EQUILIBRIUM FOR PRODUCT AND MONEY MARKETS

The actions of decision makers in product markets and in money markets are interconnected. The level of GNP determined in product markets affects demand for money stocks, and hence the equilibrium rate of interest. The level of interest rate determined in money markets affects aggregate demand, and hence the equilibrium GNP in product markets. Interaction between the two markets determines levels of GNP and interest rate, which are consistent with simultaneous equilibria in both markets.

Consider demand and supply in money markets—Eqs. (5), (6), and (7) in the equation model of Table 15-1. In Chap. 12, demand for money was presented as a sum of transactions demand M_1^* and asset demand M_2^*. If we now make explicit the influence of price level on asset demand, the demand equation may be written

$$M^* = M_1^* + M_2^* = \frac{1}{V_1} Y_s + (\mu_o - \mu_2' i)P_Y$$

Transactions demand is directly proportional to GNP. Asset demand is directly proportional to P_Y (to maintain the real stock of liquid assets), but it declines at higher interest rates as liquid assets are shifted into securities. At an interest rate i of μ_o/μ_2', asset demand would become zero.

Money supply is determined by the monetary authority. It may be that an equation could be developed to describe how the monetary authority determines M_s in the light of interest rates, unemployment rate, price level and change, and balance of payments. But we shall content ourselves in this model with taking these policy decisions as exogenous and write

$$M_s = M_{s,o}$$

Use of the equilibrium condition

$$M^* = M_{s,o}$$

then permits us to write the solution equation. For $i < \mu_o/\mu_2'$,

$$\frac{1}{V_1}Y_s + (\mu_o - \mu_2' i)P_Y = M_{s,o}$$

Solved for Y_s, we may write the relation in the form

$$Y_s = V_1(M_{s,o} - M_2^*) = V_1[M_{s,o} - (\mu_o - \mu_2' i)P_Y] \qquad (15\text{-}2)$$

Initially we assume that P_Y is constant and that $M_{s,o}$ and the behavioral coefficients are fixed. Then this equation determines what pairs of values for Y_s and i will make transactions plus asset balances total to the available money supply $M_{s,o}$. This equation determines what has traditionally been called the LM line— L for liquidity (demand for money) and M for money supply. A specific LM line is drawn for fixed values of $M_{s,o}$, P_Y, and behavioral coefficients. Under those constraints, an increase of Y_s in product markets will increase M_1^* and drive interest rates up in money markets until M_2^* has been reduced by as much as M_1^* increased. So the graph of the LM line will show i increasing as Y_s increases, as plotted in Fig. 15-2.

The vertical segment of the line at high interest rates indicates that with a fixed money supply $M_{s,o}$, there is a ceiling level of Y, achieved when all money stock is in transactions balances utilized with a fixed velocity V_1—that is, $Y_{\max}$ equals $V_1 M_{s,o}$. For a value of Y below the maximum level, say Y_o, the demand for transactions balances is less. Assume the unneeded money stocks are used to buy securities or to repay loans. This leads to a decline in interest rates and an increase in M_2^* until $M_{1,o}^*$ plus $M_{2,o}^*$ equals $M_{s,o}$ at point o in the diagram.

Note that horizontal distances on this graph are measured in billdar but are proportional to money stocks, the proportionality factor being transactions velocity V_1. So we may interpret the diagram as representing demand and supply for money as a function of interest rate. At point o we can identify the horizontal segment ao as measuring desired transactions balances, the horizontal segment ob as measuring desired asset balances, and the total distance ab as measuring money supply.

Now imagine that an exogenous increase in aggregate demand raises GNP to a level Y_α. Demand for transactions balances would increase to $M_{1,\alpha}^* = (1/V_1)Y_\alpha = (1/V_1)(ac)$ in Fig. 15-2. With interest rate still at i_o, $M_{2,o}^* = (1/V_1)$ (ob) remains unchanged. So the demand for money stocks exceeds the supply; that is, graphically,

$$ac + ob > ab \qquad \text{or} \qquad M_{1,\alpha}^* + M_{2,o}^* > M_{s,o}$$

So nonbank borrowers bid up interest rates in an attempt to secure added money stocks via borrowing or sales of securities in capital markets. (The Monetary

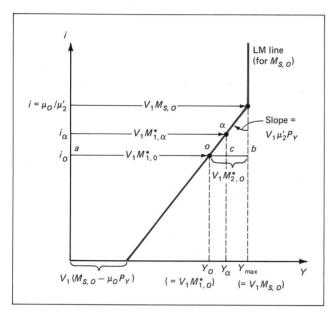

Figure 15-2   Money market equilibrium and the LM line.

Sector refuses to buy securities, since to do so would increase M_s.)   As i rises, M_2^* declines until a new equilibrium is reached at point α in the diagram, where aggregate demand for money again equals the fixed supply.   The rise in interest rates (δi) will be $(1/V_1 \mu_2' P_Y)\delta Y$, since the slope of the LM line is $V_1 \mu_2' P_Y$ with respect to the i-axis.

An increase in money supply will shift the whole LM line rightward by a distance $\delta Y = V_1 \delta M_s$.   The value of $Y_{\max}$ will increase by this amount, shifting the vertical segment of the LM line to the right by $V_1 \delta M_s$.   Along the sloping portion of the LM line, M_2^* will stay the same at each level of interest rate; so all the increased money stock must be absorbed into transactions balances, implying an increase in equilibrium GNP by $V_1 \delta M_s$ also.   This illustrates the potential benefits from monetary policy.   An increase in money supply can hold interest rates down in the event of a rise in GNP, or it can lower interest rates (if GNP stays constant initially) and thus stimulate a rise in aggregate demand.

To make explicit the interaction between money markets and product markets, let us recall how GNP varies with interest rate for equilibrium in product markets. If the aggregate demand equation [Eq. (15-1)] is solved for Y under the equilibrium condition of Y^* equal to Y_s, the result is

$$Y = k Y_{\mathrm{AUT}}^* = k[(a_o + G_o + X_o - Z_o) + (a_1 + g_1 + h_o - h_1 i + x_1 - z_1)P_Y] \qquad (15\text{-}3)$$

where the autonomous expenditure multiplier is

$$k = \frac{1}{1 - b' - h_2 - g_2 + z_2}$$

On the assumption that P_Y, exogenous components of demand, and the behavioral coefficients are fixed, this equation relates equilibrium GNP to interest rate i. If i increases, Y will decrease because investment spending will be reduced. This decrease reduces Y^*_{AUT} (the intercept on the vertical axis for the Y^* line in Fig. 15-1) and hence lowers equilibrium GNP.

The graph of Eq. (15-3), with P_Y constant, is called the IS line—indicating that points on the line correspond to equality between desired investment and desired saving, that is, to equilibrium for the demand sectors. As shown in Fig. 15-3, the IS line slants downward. Hence, it will quite surely intersect the upward-sloping LM line. For specific initial values of money supply, price level, exogenous expenditures, and behavioral coefficients, the IS_o and LM_o lines in Fig.15-3 intersect at point o. The values of interest rate (i_o) and GNP (Y_o) determined at the intersection are consistent with equilibrium for product demand sectors and for money markets.

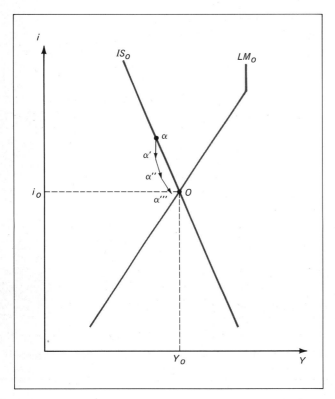

Figure 15-3   General equilibrium in product and money markets.

IS equation: $Y = k[(a_o + G_o + X_o - Z_o) + (a_1 + g_1 + h_o - h_1 i + x_1 - z1)P_Y]$

LM equation: $Y = V_1 M_{s,o}$ for $i \geq \mu_o/\mu'_2$
$Y = V_1[M_{s,o} - (\mu_o - \mu_2 i)P_Y]$
for $i < \mu_o/\mu'_2$

Further analysis indicates that this equilibrium point is stable. Suppose that the economy is operating at point α—perhaps because it was in equilibrium there with a lower money supply before an increase to $M_{s,o}$ shifted the LM line rightward to the position shown. At α, demand sectors are in equilibrium. But $M_{s,o}$ is greater than M^*, and the excess supply of money drives interest rates down. If Y holds constant temporarily, the economy moves to a point such as α'. There both demand sectors and money markets are in disequilibrium. At α', aggregate output is below the equilibrium level and Production Units will raise Y_s. At the same time, the continued excess supply of money drives interest rates down further. The economy moves to α", say. Again market behavior in this new disequilibrium state lowers interest rate and increases output further, perhaps shifting the economy to α'''. Continued shifts move the economy along a spiral path inward toward equilibrium at o—assuming that the constants underlying the IS and LM lines do not change.

In general, the movement from any disequilibrium point will be horizontally toward the IS line and vertically toward the LM line. Such shifts will lead to a convergent inward spiraling toward the equilibrium point at o. If the adjustment in one market is very rapid, the economy will shift directly to the equilibrium line for that market and then slide along that line toward point o as the slower adjustments toward equilibrium proceed in the other market.

As noted, the equilibrium point at the intersection of the IS and LM lines guarantees that the money market is in equilibrium and that aggregate demand equals aggregate incomes for the demand sectors. But it does not follow that the price level and GNP at that intersection point will be consistent with profit maximization for Production Units. To determine an equilibrium of demand and supply for product markets, we need to relate the IS-LM diagram to the aggregate supply curve, as was done for the aggregate demand and supply curves in the preceding section.

In Fig. 15-4, the IS-LM diagram has been rotated 90 degrees counterclockwise in panel (a) and aligned alongside the aggregate supply curve in panel (b), so that GNP can be projected horizontally between the two panels. Initial equilibrium in panel (a) is at point o, where interest rate is i_o, aggregate demand is Y_o^*, and an implicit price level $P_{Y,o}$ underlies the IS and LM lines. In panel (b), it is assumed that the price level $P_{Y,o}$ would lead Production Units to produce an output Y_{R,s_o}, which is well below the real aggregate demand $Y_{R,o}^*$ determined at point o on the constant-price line with slope $P_{Y,o}$ in panel (b). So producers increase output, prices, and incomes distributed, moving up along the $Y_{s,o}$ curve. As product prices rise, two reactions show up in panel (a). First, the rise in P_Y increases some components of autonomous demand and thus increases equilibrium GNP at each value of i; that is, the IS line is shifted upward. Second, the rise in P_Y increases asset demand for money at each value of i since

$$M_2^* = (\mu_o - \mu_2' i)P_Y$$

This rotates the LM line downward, with the pivot at the kink of the LM curve.

The new lines IS_α and LM_α in panel (a) intersect at point α, where i_α and Y_α

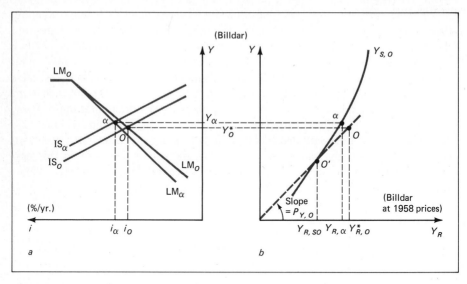

Figure 15-4 General equilibrium for product demand and money markets. (a) IS and LM lines. (b) Aggregate supply curve.

are both higher than at the initial equilibrium at point o. It is assumed that the price level $P_{Y,\alpha}$, corresponding to point α in panel (a), will be the price level at the point α in panel (b) located by projecting Y_α across to the aggregate supply curve. Points α will then display values of Y^*, Y_s, Y_R, P_Y, and i which are consistent with equilibria of demand and supply in both product and money markets. We may be more confident that an equilibrium will be reached in this case than in the previous section, where the interest rate was constant; for here an increase in incomes and prices will raise interest rates and slow the rise in aggregate demand by reducing investment.

EQUILIBRIUM FOR PRODUCT, MONEY, AND LABOR MARKETS

We may now conclude the graphical representation of the complete macroeconomic system by adding diagrams for the aggregate production function and for labor markets, corresponding to Eqs. (8) through (11) of the model given in Table 15-1. Actually, this was done in the diagram at the end of Chap. 6 and is repeated in Fig. 15-5 with the graphs rotated to permit easy projection from one panel to another.

Panel (a) contains the IS and LM lines described in the preceding section—rotated 90 degrees counterclockwise so that the value of Y can be projected across horizontally to panel (b). The equilibrium values of i_o and Y_o in panel (a) are consistent with equilibrium in money and product markets simultaneously. When the Y_o value is projected across to the aggregate supply curve in panel (b), it determines values of $Y_{R,o}$ and $P_{Y,o}$ at which producers are in equilibrium. Thus,

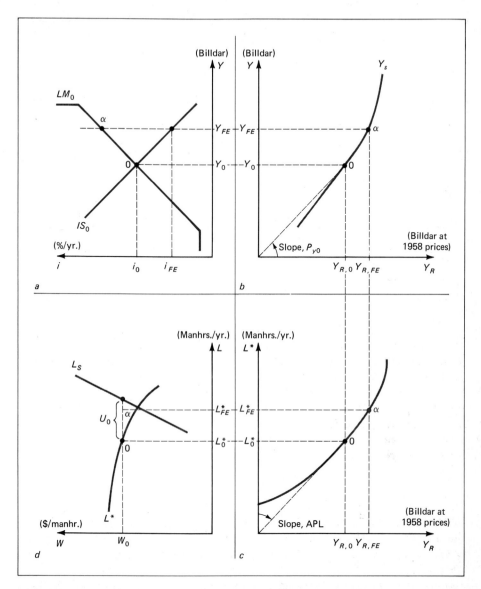

Figure 15-5   General equilibrium diagrams.   (*a*) IS and LM lines.   (*b*) Aggregate supply curve.
(*c*) Production function.   (*d*) Labor demand and supply.

money GNP is split into price and real GNP components when producers' decisions are taken into account.

The equilibrium value $Y_{R,o}$ determined in panel (b) is projected down to the aggregate production function in panel (c) to determine the equilibrium demand for labor consistent with $Y_{R,o}$.[†]

The labor demand determined in panel (c) is projected across to panel (d)—also rotated counterclockwise 90 degrees to align axes properly. In panel (d) the equilibrium demand for labor (L_o^*) corresponds to the point at which the wage-rate line W_o cuts the labor demand curve L^*. Consistency of the equilibrium points in panels (b), (c), and (d) is guaranteed, because producers take wage rates and the production function (MPL) into account as they decide on output and labor input to employ for profit maximization.

The labor supply curve L_s shows the flow of labor services which the labor force desires to provide at each wage rate, assuming price level constant, so that W is proportional to real wage rate. At wage rate W_o, an excess of supply over demand for labor services is indicated in Fig. 15-5d. The excess equals unemployment U_o. If we designate a specific unemployment rate U_{FE} as consistent with "full employment," we can locate a point α which corresponds to full employment at wage rate W_o. If the economy were to reach a full-employment state at W_o, the labor demand curve would have to shift leftward until employment at wage rate W_o rose to L_{FE}^*. The implications of this shift include (1) higher real output $Y_{R,FE}$ in panels (c) and (b); (2) higher prices in panel (b); and (3) higher money GNP, indicated by Y_{FE} in panels (a) and (b). As becomes clear in panel (a), the economy can reach equilibrium at full employment only if either the IS or the LM line, or both, shifts upward so that their intersection falls on the horizontal dashed line at the Y_{FE} level on the GNP axis. It is the objective of fiscal and monetary policies to achieve such shifts.

Reaching full-employment equilibrium in the real world involves more than shifts in the IS and LM lines. The other schedules in panels (b), (c), and (d) will probably be shifting at the same time. For example, technological advances will shift the production function and the aggregate supply curve down, and will raise the L^* curve, since it is the marginal-net-revenue curve for labor inputs. A rise in wage rates will shift the aggregate supply curve up, and tend to raise prices. An increase in prices will shift both the L^* curve and the L_s curve leftward. A price rise will probably shift the IS line upward also, and rotate the LM line downward.

To bring out the shift variables underlying each of the curves in Fig. 15-5 and to show how exogenous changes alter equilibrium positions, the following section analyzes various ways of reaching equilibrium at full employment.

[†] The production function is drawn here with the L-axis vertical and the Y_R-axis horizontal—because it makes projections between panels easier. In this form it shows labor requirements as a function of output, and labor productivities appear as slopes with respect to the vertical axis. The intercept on the vertical axis indicates the existence of overhead labor, which would be employed even at zero output in the short run.

HOW TO ACHIEVE EQUILIBRIUM AT FULL EMPLOYMENT

The variables determining equilibrium in labor markets, the production function, and the aggregate supply curve cannot be influenced much by government policy makers in the short run. These variables include marginal productivity of labor (dependent on capital stock and technology), price level, wage rates, and marginal nonfactor charges. Short of direct controls over prices and wages, the government cannot change these rapidly.

So government full-employment policy focuses principally on influencing aggregate demand. In our model, aggregate demand is determined at the intersection of the IS and LM lines. Shifting one or both of these curves can move the equilibrium point toward full-employment aggregate demand. *Government actions which shift the IS line are usually referred to as fiscal policy; government actions which shift the LM curve are referred to as monetary policy.*

FISCAL POLICY

Recall that the equation of the IS line is

$$Y = k Y_{\text{AUT}}^* = k[(a_0 + G_o + X_o - Z_o) + (a_1 + g_1 + h_0 - h_1 i + x_1 - z_1)P_Y] \quad (15\text{-}3)$$

where

$$k = \frac{1}{1 - b' - h_2 - g_2 + z_2}$$

An increase in government spending G_o would raise aggregate demand by $k\,\delta G_o$ at all levels of i, and hence would shift the IS line upward by that amount in Fig. 15-6. A decrease in taxes or an increase in transfers and subsidies could raise the autonomous components of consumer demand $(a_0 + a_1 P_Y)$ or of investment spending $(h_0 P_Y)$, or they could stimulate exports X^*. Finally, a decrease in tax *rates* could raise the multiplier (k). All these policy actions would move the IS line out to higher values of Y at each level of i. Reductions of G_o, increases of taxes, or reductions of transfers and subsidies would move the IS line downward.

The effect of a stimulative fiscal policy is pictured in Fig. 15-6, panel (a). A tax cut, increase in transfers, or increase in government spending has shifted the IS line upward from an original position IS_0 until it intersects the fixed LM line at point α on the full-employment line. Let the needed increase in Y to reach full employment be

$$\delta Y_\alpha = Y_{\text{FE}} - Y_o = oa$$

According to simple multiplier theory, a rise in government spending by $\delta G_\alpha^* = (1/k)\,\delta Y_\alpha$ would be sufficient to raise equilibrium GNP by δY_α, if the interest rate, and hence investment spending, is constant. If the monetary authorities did increase money supply so as to keep interest rate at i_o in Fig. 15-6, then indeed an upward shift of the IS line until it passes through point a would lead to equilibrium at Y_{FE}. But this would involve monetary policy actions to increase M_s and shift the LM line upward to go through point a. We are assuming a pure fiscal policy here, with M_s and the LM line fixed.

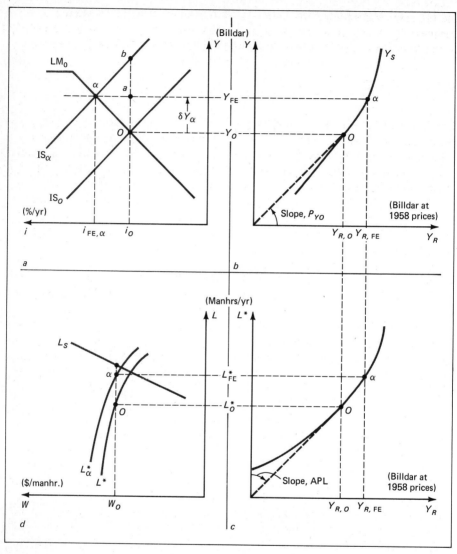

Figure 15-6 Full employment by fiscal policy. (*a*) IS and LM lines. (*b*) Aggregate supply curve. (*c*) Production function. (*d*) Labor demand and supply.

With M_s fixed, an increase in G^* and the resulting expansion of GNP will raise transactions demand for money and force interest rates up until asset balances are reduced by as much as transactions balances rise. The rise in interest rates would induce a decline in investment spending and hence a reduction of equilibrium GNP *along* the new IS_α line. To bring the equilibrium GNP up to the full-employment level, G^* must increase enough to offset the endogenous decline in I_d^* as well as to expand aggregate demand in toto. From the equation of the IS line, we calculate the requisite increase of G^* in Fig. 15-6(a) to be

$$\delta G_\alpha^* = \frac{1}{k}\,\delta Y_\alpha + h_1\,\delta i_\alpha = \frac{1}{k}\,(oa + ab)$$

The magnitude of the decline in I_d^* is $h_1\,\delta i_\alpha$. For fiscal policy to be effective, this decline should be small. Graphically, this will occur if the IS line becomes nearly perpendicular to the Y-axis, implying low sensitivity (h_1) of investment to interest rate; it will also occur if the LM line swings toward parallelism with the Y-axis, so that a given rise in Y leads to a small change in interest rate.

Conversely, if investment is very sensitive to interest rate (the IS line nearly parallel to the Y-axis) and if asset balances do not change much with interest rate (the LM line nearly perpendicular to the Y-axis), a fiscal stimulus is largely offset by a decline in investment demand, and fiscal policy is ineffective.

Before going on to a consideration of monetary policy, let us assume that fiscal policy does increase equilibrium GNP to the full-employment level at point α in Fig. 15-6a. Will the Production Units Sector and the labor markets come into equilibrium at full employment?

If the production function, wage rates, and marginal nonfactor costs remain fixed, then the aggregate supply curve in panel (b) is stable. Projecting Y_{FE} in panel (a) across to panel (b), we locate the profit-maximizing point α on the aggregate supply curve. Both real output and GNP price index increase from their previous values at point o.

The increase in real output to $Y_{R,FE}$ requires larger labor inputs, as may be determined by projecting $Y_{R,FE}$ down to panel (c) and determining the required labor inputs L_{FE}^* corresponding to point α on the production function.

In panel (d) the demand curve for labor shifts up, because the price increase in product markets from $P_{Y,o}$ to $P_{Y,\alpha}$ raises the marginal-net-revenue productivity of labor. The new labor demand curve L_α^* will intersect the wage-rate line W_o at L_{FE}^*, the same labor inputs as were determined at point α on the production function in panel (c). This must be true because producers decide on real output and required labor inputs as coordinate decisions, taking into account the constraint imposed by the production function. The point α in panel (c) involves the minimal unemployment consistent with the definition of full employment, perhaps 4 percent.

In the move from underemployment to full-employment equilibrium, we assumed wage rates constant at W_o and the production function fixed. The implications of removing these assumptions will be taken up after we have discussed achievement of full employment by monetary policy. Also, the rise in price level during the expansion to full employment would cause some shifts in the IS and LM lines, but previous discussion indicated the adjustments would lead to a new stable equilibrium, and they might be small, in any event.

MONETARY POLICY

At a given level of Y, an increase in money supply will depress interest rates. This occurs as the Monetary Sector buys securities to increase M_s and drives interest rates down. The decline in i induces an increased demand for asset balances. However, the decline in i will also induce an increase in investment spending and in any other components of demand which are sensitive to interest rates. The consequent increase in GNP will increase the demand for transactions balances. When equilibrium has been reached at the new, higher money supply, part of the increased money stock will have been absorbed into M_1^* and part into M_2^*, as indicated by the relation:

$$\delta M_s = \delta M_1^* + \delta M_2^* = \frac{1}{V_1}\,\delta Y + \mu_2(-\delta i)$$

In Fig. 15-7, an increase in money supply has been assumed, sufficiently large to shift the LM line upward until it intersects the fixed IS line at point β, corresponding to full-employment demand Y_{FE}. The upward shift of the LM line is $V_1\,\delta M_s$. When the new equilibrium is reached, the increment of money supply (δM_s) is divided between transactions balances and asset balances in proportion to line segments oc and cd respectively, since $oc = V_1\,\delta M_1^* = \delta Y_\beta$, and $cd = V_1\,\delta M_2^*$.

As may be seen graphically, the increase in δM_s required to bring about a given increase in equilibrium GNP depends on V_1 and on the slopes of the IS and LM lines. V_1 determines the needed increment to transactions balances, that is, δM_1^* equals $(1/V_1)\,\delta Y$. The slopes of the IS and LM lines determine the increment to M_s needed to meet added demand for asset balances.

Monetary policy will influence equilibrium Y most strongly when the IS line is nearly parallel and the LM line nearly perpendicular to the Y-axis. Under those conditions, a neeeded rise in GNP to the full-employment level can be achieved with a relatively small increase in money supply. Monetary policy is very effective. As we noted in the preceding section, these conditions are fulfilled in the upper part of the LM line, where the line is perpendicular to the Y-axis.

On the other hand, *monetary policy will be ineffective if aggregate demand is not sensitive to interest rate*, that is, if the IS line is perpendicular to the Y-axis. Then increases of money supply will indeed lower interest rates, but this will not stimulate an increase in aggregate demand and equilibrium GNP. All the increment in M_s will be absorbed by increased asset holdings at the lower interest rates.

Also, monetary policy will be ineffective if the LM line is nearly parallel to the Y-axis. Whether the IS line is vertical or not, an increase in M_s, shifting the LM line upward, would then leave interest rate and equilibrium GNP unchanged. All the increment in M_s will be absorbed into asset holdings of money at the constant interest rate.

If we assume that expansion of the money supply has brought equilibrium

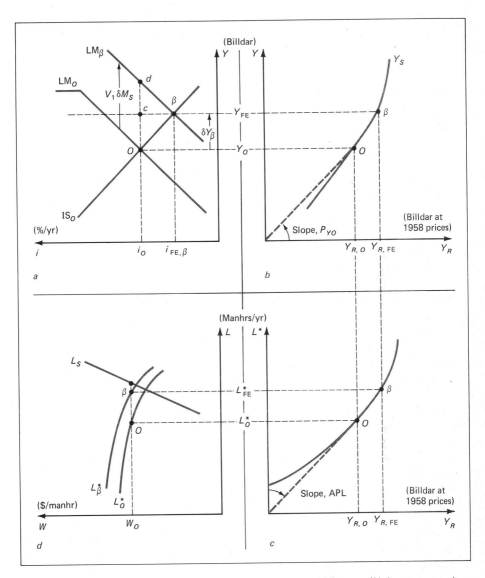

Figure 15-7 Full employment by monetary policy.   (*a*) IS and LM lines.   (*b*) Aggregate supply curve.   (*c*) Production function.   (*d*) Labor demand and supply.

GNP up to Y_{FE} in panel (a) of Fig. 15-7, concomitant changes must have occurred in other panels, as discussed previously in the fiscal policy section. The equilibrium point on the aggregate supply curve in panel (b) will move to higher price level and real output ($Y_{R,FE}$). The increase in price level shifts the labor demand curve upward and moves the equilibrium output to the full-employment point β at the constant wage rate W_o in panel (d). Full-employment output and labor demand converge at the point β on the production function in panel (c). And the price rise, in moving to full employment, will cause some further movements in the IS and LM lines as the final equilibrium is reached.

THE EFFECT OF WAGE-RATE CHANGES

In the preceding discussions of fiscal and monetary actions to stimulate the economy to full employment, it was assumed that wage rates stayed constant at W_o. Now let us assume that wage rates rise as the economy rises toward full-employment levels. If the production function, and hence the average and marginal productivity of labor curves, remains unchanged, then a rise in wage rates will shift the aggregate supply function upward. If income shares to labor, to capital, and to nonfactor costs remain the same at given output, prices must rise by the same percentage as wage rates, for labor share is $WL/P_Y Y_R = W/P_Y(APL)$. Let us assume that the aggregate supply function does shift up by the same percent that wage rates rise.

Let the economy start from an equilibrium (point o) at less than full employment, and assume that fiscal and/or monetary policy shifts aggregate demand up to the level $Y_{FE,\alpha}$ which would call for the full-employment output on the original aggregate supply curve $Y_{S,o}$ at point α. See panels (a) and (b) in Fig. 15-8. Prices would be higher at α than at o, which would lift the labor demand curve from L_o^* to L_α^* in panel (d) and yield full employment of labor at point α if wage rates stay at W_o.

Now, starting from α, assume that wage rates and prices both go up by the same percent and that both the wage-rate line and the L_α^* curve in panel (d) shift leftward proportionally. It might seem that the new equilibrium would be at α', where demand for labor remains at the full-employment level. But consider the state of affairs back in panel (b). The aggregate supply curve shifts up by the same percent as wage rates, to a position $Y_{s,\gamma}$. If the IS and LM lines in panel (a) do not shift, the equilibrium value of money GNP remains the same. But this rate of expenditures, projected across to the new aggregate supply curve, leads Production Units to turn out a profit-maximizing output of $Y_{R,\gamma}$, which is somewhat below the full-employment level.

Since the price level at point γ is below that at α', the labor demand curve in panel (d) will shift up by a smaller percent than needed to go through point α'. Rather, the L_γ^* curve will shift up just enough to intersect the new wage rate at point γ. The labor demand L_γ^* at this point is less than full employment and, when

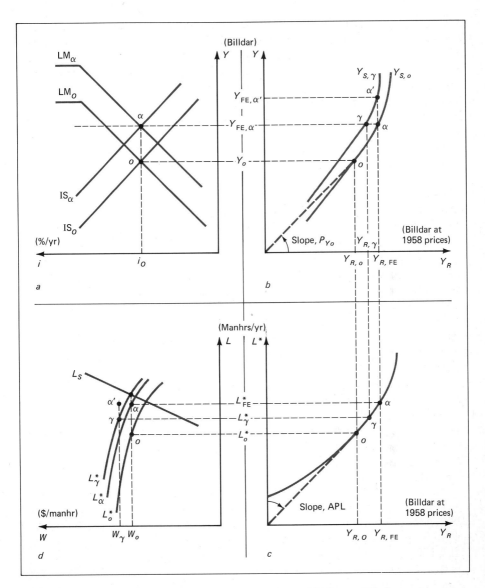

Figure 15-8   Expansion with rise in wage rates.   (*a*) IS and LM lines.   (*b*) Aggregate supply curve.   (*c*) Production function.   (*d*) Labor demand and supply.

projected across to the production function, determines the same real output as was found from point γ on the aggregate supply curve.

In summary, *fiscal and monetary policies which would raise aggregate expenditures to the full-employment level on the original aggregate supply curve will fail to generate full employment if rises in wage rates shift the aggregate supply curve up.* The price rise stemming from increased wage rates will increase the money value of full-employment output and make the original target for aggregate demand ($Y_{\text{FE},\alpha}$) inadequate to generate full employment.

There may be some further changes in IS and LM curves resulting from the rise in price in the movement from α to γ, but they tend to offset each other in their influence on Y^* and could hardly restore the economy to full employment.

DYNAMIC EQUILIBRIUM IN A GROWING ECONOMY

In a steadily growing economy, some variables will change relatively slowly and regularly. So decision makers and markets anticipate the changes and adapt to them in a stable manner. Such a smooth, premeditated adaptation to long-term trends is different from the economic situation when unexpected deviations produce short-term disequilibrium. *The state of an economy which is adjusting smoothly to expected long-term changes may be described as one of dynamic equilibrium.*

Two important continuing changes in a growing economy are (1) rise in labor productivity stemming from technological advance, increase in capital stock, and qualitative improvements in the labor force; and (2) increases in the labor force which accompany population growth. Let us see how market adjustments in our general equilibrium model can accommodate these exogenous changes in the production function and in the labor supply curve.

Qualitatively, it is clear that increases in labor productivity and in total labor force must increase the full-employment output of the economy ($Y_{R,\text{FE}}$). Unless prices decline proportionately, aggregate demand must increase, requiring upward shifts of the IS and/or LM lines. Price change will depend on how the changes in marginal productivity of labor and in wage rates shift the aggregate supply curve and how far output expands along the new supply curve.

The analysis of these changes can be clarified by use of the general equilibrium diagrams of Fig. 15-9. Let us make two reasonable assumptions to simplify the discussion: (1) The production function shifts in such a way that MPL is increased by the same percentage at each level of labor input. (2) Wage rates rise by the same percentage that MPL rises. The economy starts in equilibrium at points α in the four panels. Then the production function shifts outward to a position $Y_{R,\beta}$. This shifts the labor demand curve leftward by the same percentage as MPL rose, since ordinates of the labor demand curve are MNRL $= (1 - \beta)P_Y(\text{MPL})$. Temporarily, we assume β and P_Y unchanged in this equation. The labor demand curve, $L_b^*(P_{Y,\alpha})$, would lead to an increase in labor demanded

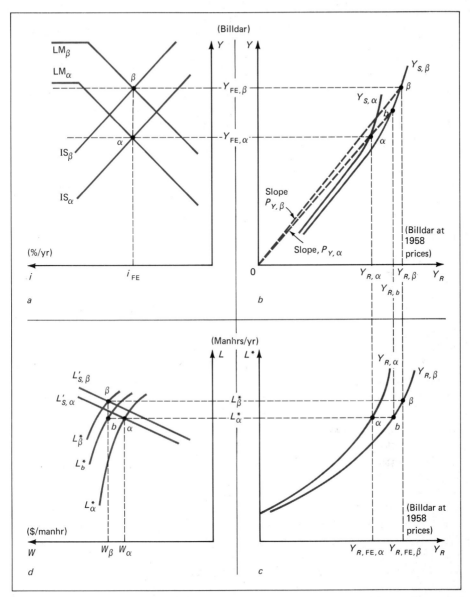

Figure 15-9   Equilibria in a growing economy.   (*a*) IS and LM lines.   (*b*) Aggregate supply curve.
(*c*) Production function.   (*d*) Labor demand and supply.

at the initial wage rate W_α, but a rise in wage rates to W_β will yield equilibrium at point b and leave the quantity of labor demanded unchanged at L_α^*. (Remember that the percent rises in MPL and W are equal.) Projection from point b across to the new production function determines real output, $Y_{R,b}$. Projection upward to panel (b) locates a point b at the intersection of the vertical line $Y_{R,b}$ and the constant-price line 0α extended. This point b will be on the new aggregate supply curve $Y_{s,\beta}$. One may assume different price levels and their corresponding L^* curves in panel (d) and thus determine additional points on the new aggregate supply curve. Clearly $Y_{s,\beta}$ lies under the $Y_{s,\alpha}$ curve.

Where would the full-employment point be on the new aggregate supply curve, and how can the economy be brought to it? In panel (d) of Fig. 15-9, two curves for full-employment labor supply are shown. They are designated with "primes" because they are labor supply adjusted for the unemployment rate taken as consistent with full employment, that is, L_S' equals $(1 - U_{FE})L_S$. The initial adjusted labor supply line $L_{s,\alpha}'$ cuts the initial labor demand curve L_α^* on the wage-rate line W_α. The new adjusted labor supply line $L_{s,\beta}'$ is shifted upward to indicate growth in labor force from the initial to the final equilibrium. At point b the labor demand is at the initial full-employment level. But this falls short of full employment in the new equilibrium because population growth and a higher real wage have both increased the flow of labor supply. Achievement of full employment at point β on the new labor supply curve $L^*(P_{Y,\beta})$ requires an upward shift of the labor demand curve stemming from a rise in output prices. How much of a price rise is needed depends on how rapidly the L^* curve declines, that is, on how rapidly MPL drops along the production function as L^* increases. Little price rise will be required if the L^* curve is nearly vertical in the region between b and β, that is, if the production function is nearly straight in that region.

If we project from point β in panel (d) across to the production function and up to the aggregate supply curve, we determine level of output, the price level, and money GNP at the full-employment point β in panel (b). The aggregate demand consistent with full employment is higher than at point α, both because of the increased real output from a larger labor force working at higher productivity and because of some increase in price level. Can such an increase in aggregate demand be expected to occur "naturally," or will the economy need to be prodded continuously to keep demand in step with rising full-employment output?

In panel (a) of Fig. 15-9, an increase of equilibrium aggregate demand to the new full-employment level $Y_{FE,\beta}$ is pictured as having been achieved at constant interest rate by simultaneous upward shifts of the IS and LM lines. It would not seem possible to increase aggregate demand in the long run by continuing declines in interest rates. Instead, the historical record seems to show that long-term increases in money supply and income velocity have shifted the LM line upward about enough to match GNP growth, keeping interest rates approximately constant. Monetary policy and international markets for capital funds seem to have worked in this direction.

If interest rates have been reasonably constant in the long run, then what forces might operate to shift the IS line upward to yield a rising level of aggregate demand at stable interest rates? Recall the equation of the IS line:

$$Y = k Y^*_{AUT}$$

$$Y = k[(a_o + G_o + X_o - Z_o) + (a_1 + g_1 + h_o - h_1 i + x_1 - z_1)P_Y] \qquad (15\text{-}3)$$

where

$$k = \frac{1}{1 - b' - h_2 - g_2 + z_2}$$

We cannot count on long-term growth in the multiplier k. But growth in population and price levels may increase the autonomous components of consumer spending. Population growth also stimulates investment spending; for example, it increases the demand for houses, the demand for plant and equipment to produce and distribute the flow of consumer goods which is rising and also changing as new products are introduced. Population growth stimulates government spending also, for example, for schools, streets and highways, urban services, utilities, crime prevention, hospitals, and now environmental improvement. Technological advances and rising wage rates also provide exogenous stimuli for investment spending to produce new products or increase labor productivity for existing products. *So there are strong long-term forces which shift the aggregate demand curve (or the IS line) upward. Though not automatic and by no means steady, growth in aggregate demand does seem to catch up eventually with growth in potential full-employment output of the economy.* Lapses below full employment have lasted up to 10 years; they are economically wasteful, and involve a great deal of human suffering. Surges above full employment have lasted for several years during wartime and left a legacy of inflation and economic inequalities and maladjustments which interfere with attainment of national economic goals. The following chapters discuss the nature and effectiveness of policies to maintain near-full employment with price stability in a growing economy involved in international economic relations.

PROBLEMS

1 a Aggregate demand determines the money value of the flow of spending for current output. How are the price and real components of aggregate demand determined?

b How is it possible to have an upward-sloping real aggregate demand curve?

c It is sometimes maintained that (1) if GNP and interest rate are both rising, fiscal policy is stimulating the economy, whereas, (2) if GNP rises while interest rates are falling, monetary policy is stimulating the economy. Do you agree or disagree? Explain.

d It is sometimes maintained that equal percentage increases in wage rate and price level would leave the economy in equilibrium at the same real output, assuming the production function fixed. Do you agree or disagree? Explain.

2 A five-sector economy is described initially by the following equations.
Sector income equations (billdar)

$$Y_d = 21 + 0.63\,Y_S$$

$$T_n = -19 + 0.25\,Y_S$$

$$\text{GRE} = -7 + 0.12\,Y_S$$

$$\text{TR}_f = 5$$

Product demand functions (billdar)

$$C^* = 0.95\,Y_d = 20 + 0.60\,Y_S$$

$$G^* = 200$$

$$I_d^* = 265 - 25i \qquad \text{where } i \text{ is in percent per year}$$

$$X_n^* = 5$$

Money market equations ($ billions)

$$M_1^* = \frac{1}{6}\,Y_S$$

$$M_2^* = \begin{cases} 0 & \text{for } i \geq 12 \\ 100 - 8.33i & \text{for } i < 12 \end{cases}$$

$$M_S = 200$$

a Find the equations for the IS and LM lines under the initial conditions, and solve for Y_o and i_o.

b Tabulate sector incomes, GNP purchases, and saving or borrowing for the four final demand sectors (to the nearest billdar).

c Assume that full-employment GNP is 930 billdar. Calculate what rise in G^* would bring the economy to full employment with X_n^* and M_S held constant and with the consumer demand function unchanged. What would happen to the interest rate?

d Tabulate sector incomes, GNP purchases, and saving or borrowing for the four final demand sectors (to the nearest billdar).

e If G^*, X_n^*, and the income and demand functions were constant, what increase in M_S would bring the economy to full-employment GNP of 930 billdar? What would happen to the interest rate?

f Tabulate sector incomes, GNP purchases, and saving or borrowing for the four final demand sectors.

g In view of the above results, point out the differences in impact of government spending and monetary policy, and discuss the circumstances under which one policy would be preferred over the other, or under which a mixture would be better.

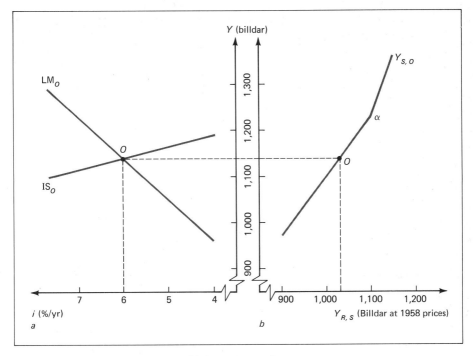

Figure 15-10 (*a*) IS and LM lines. (*b*) Aggregate supply curve.

3 An economy is at initial equilibrium o, with Y_o equal to 1,140 billdar and $Y_{R,o}$ equal to 1,030 billdar at 1958 prices. Hence, $P_{Y,o}$ equals 1.107 (1958 equaling 1.00). See Fig. 15-10. Initial equations for the model are:

Aggregate demand: $Y^* = (50 + 0.6 Y_s) + (250 - 10i) + 216$

Aggregate supply:

Below full employment: $Y_s = -200 + 1.3 Y_{R,s}$

Above full employment: $Y_s = -1,630 + 2.6 Y_{R,s}$

IS line: $Y = 1,290 - 25i$

LM line: $Y = 600 + 90i$

Unfortunately, the initial equilibrium is below the full-employment output,

$Y_{FE} = 1,230$ billdar   or $Y_{R,FE} = 1,100$ billdar at 1958 prices

Present *graphical* or *algebraic* or *verbal logic* to answer the following parts of the problem.

a By how much would the government need to increase government purchases G^* to bring the economy to equilibrium at full employment (point α), assuming other injections constant (hence interest rate constant) and the consumption function unchanged? What would be the rises in GNP and in its price index from o to α?

b How would the answers in part (*a*) be changed if we keep the same δG^* but remove the assumption that other injections are constant and assume, instead, that money supply is kept constant? Give reasons for the changes, sketch the IS_β and LM_β lines on the chart, and label the equilibrium points β.

c Along with the changes above, suppose that wage rates increase faster than average productivity of labor, so that the aggregate supply curve is shifted up by 5 percent. If the IS and LM lines remain as in part (*b*), describe the new equilibrium values Y_γ, and $P_{Y,\gamma}$. Will the economy be at full employment? Illustrate situation (*c*) on the graphs, labeling the new equilibrium points γ.

4 An economy in initial equilibrium below full employment experiences an exogenous rise in Y^*_{AUT}, that is, an upward shift of the aggregate demand curve. Assume that money supply, wage rate, and behavioral coefficients stay constant and that the production function remains fixed.

The adjustment to new equilibrium values of the endogenous variables is assumed to take place by steps as follows:

1. i and P_Y remain constant.
2. i varies, but P_Y stays constant.
3. Both i and P_Y vary.

In each of the four panels of the general equilibrium diagrams for the economy, locate the point at which the economy would be operating at the end of each of the three steps, for example, by labeling points 1, 2, and 3 in each panel.

For each of the 11 endogenous variables, tell whether its value increases, decreases, or remains constant from the initial to the final equilibrium.

5 An economy in initial equilibrium at less than full employment experiences a rise in wage rates, perhaps as a lagged response to a previous price rise. Assume that money supply, autonomous demand, and behavioral coefficients stay constant and that the production function remains fixed.

The adjustment to new equilibrium values of the endogenous variables is assumed to take place by steps as follows:

1. i and P_Y remain constant.
2. P_Y varies, but i stays constant.
3. Both P_Y and i vary.

In each of the four panels of the general equilibrium diagrams, locate the point at which the economy would be operating at the end of each of the three steps, for example, by labeling points 1, 2, and 3 in each panel.

For each of the 11 endogenous variables, tell whether its value increases, decreases, or remains constant from the initial to the final equilibrium.

Part Four

Economic Policy: Targets and Instruments

Chapter 16

Policy Targets and Instruments

NATIONAL ECONOMIC GOALS

The economic goals of most nations include such items as near-full employment of persons desiring and able to work; sustainable long-term growth in real GNP per capita; near-stability of the general price level; equitable distribution of incomes; and reasonable balance of international payments. In the United States we might add such objectives as a reasonable balance between private and government purchases of current output; maintenance of vigorous competition in most markets; and conservation of natural resources against exploitation and pollution.

These goals are not derived from economic analysis or a macroeconomic model. They are sociopolitical value judgments which develop from the cultural emphases and political processes of society. In the United States, they develop from public priorities, pressure-group advocacy, and legislative enactments, as in the Employment Act of 1946. In totalitarian countries, the goals might be formulated by the ruling group and undergirded by public support developed through propaganda or punishment.

POLICY INSTRUMENTS

The central government is the sector most directly involved with policy actions to achieve the national goals. It has the power to influence macroeconomic activity directly by exercising control over its own economic behavior and indirectly by influencing the economic decisions of the private sectors. Its policy instruments may be classified under broad headings as *fiscal policy, monetary policy, incomes policy*, and *economic regulation*.

Fiscal policy instruments include *control over government spending, tax levels and structure, transfer payments and subsidies*, plus *government borrowing and lending programs*. By and large, these actions change equilibrium GNP by shifting aggregate demand curves, that is, by shifting the IS line. Changes in G^* alter one component of aggregate demand directly; changes in taxes, transfers, and subsidies influence private demands by altering income streams and marginal costs or revenues; borrowing and lending programs affect private spending via their impact on the cost and availability of credit to borrowers.

Monetary policy instruments include *changes in reserves, reserve requirements, and central-bank discount rates.* They affect the cost and availability of credit to private and government borrowers and disturb balance-sheet equilibrium in the economy, evidenced by shifts of the LM line and changes of interest rates. Monetarist economists maintain that changes in money supply also exert sizable wealth or portfolio effects on private purchases of current output. This would correspond to a shift of the IS line in our equilibrium diagrams.

Incomes policy refers to *government attempts to influence directly private decisions on prices, wages, rents, dividends, and perhaps interest charges.* Usually undertaken in an effort to slow and reverse a continuing inflation in its "cost-push" phase, the measures taken may range from government persuasion and mobilization of public opinion, through wage-price guidelines administered by publicity and confrontation, to legislated control of wages and prices for parts or all of the economy. In terms of our general equilibrium diagrams, these instruments are used to slow the upward shifts of the aggregate supply curve.

Other economic regulation of business covers policies aimed at preserving competition (antitrust laws), protecting the interests of customers against "natural monopolies" (public utility, telephone, some transportation agencies), protecting the public from unsafe products, pollution, and high credit charges, and setting minimum-wage rates, quotas on imports, and other restrictions. These instruments are aimed primarily at the long-term growth, productivity, and quality of markets rather than at short-range adjustments of aggregate demand to full-employment levels.

THE ROLE OF MACROECONOMIC MODELS

The choice of policy instruments required to achieve predetermined economic targets obviously involves a *quantitative* understanding of the macroeconomic system. The targets need to be clearly specified, and data must be available to measure progress toward achieving them. *A quantitative model is needed by the policy makers to calculate the types and magnitudes of instruments which will lead to a desired goal.* Perhaps two or more instruments will be used at the same time—both tax changes and alterations of money supply. Then policy makers need to estimate the joint effects of the two instruments working together— or perhaps against each other.

A macroeconomic model is needed also to calculate the side effects of a given policy instrument and to estimate the trade-offs among several targets. From our analysis of the macroeconomic system, it should be abundantly clear by now that a change in any of the exogenous variables of the system will probably send shock waves throughout the economy, affecting all the endogenous variables in varying degrees. Hence, *a change in some policy instrument will probably affect other target variables in addition to the one at which the instrument was primarily aimed.* For example, government deficit spending to produce full employment may lead to price rises and may interfere with investment by raising interest rates. Or an increase in money supply to lower interest rates and stimulate investment may worsen the balance of payments as capital funds flow abroad, where rates become relatively higher.

These considerations point to two important principles regarding policy instruments and the achievement of multiple economic targets.

1 *In general, the government must use as many policy instruments as the number of targets it wishes to achieve simultaneously.* Suppose two targets are to be achieved—full employment of labor and a level of investment spending that will lead to desired growth of capital stock. Tax cuts might be used to raise aggregate demand to the full-employment level (causing an upward shift of the IS line). But this might raise interest rates and reduce investment below the desired level. So money supply might be increased to hold interest rates down (seen in the upward shift of the LM line). The use of the second policy instrument is required to achieve the second target, or to maintain it against the unfavorable side effects of the first policy, aimed at a different target.

2 *In cases where available instruments are inadequate to achieve multiple economic targets simultaneously, policy makers need to evaluate the trade-offs between targets and choose instruments which achieve the best available partial realization of various goals.* In the preceding example, if monetary policy could not be altered (say because of international repercussions), policy makers might need to evaluate the rate of substitution between employment and investment as taxes were changed, then to choose tax cuts which would give the best available weighted combination of employment and investment. "Best" depends on the weights, or value judgments, assigned by the policy makers to marginal changes in the target variables. In this particular instance, the policy makers might also try to tailor the tax cuts to support investment by cutting taxes on profits or giving investment tax credits to stimulate investment spending more than other components of aggregate demand, thus offsetting the negative impact of interest rate rises.

TARGETS, INSTRUMENTS, AND THE GENERAL EQUILIBRIUM MODEL

In the preceding chapter, the general equilibrium model was presented in graphs and also in the form of 11 equations involving 11 endogenous variables (Table 15-1). Table 16-1 shows the endogenous variables and the equations determining them. A simultaneous solution of this set of equations would yield equations expressing each endogenous variable as a function of the exogenous variables and the behavioral constants of the system. (Exogenous variables and constants follow the colons in the functional equations in the table. Many system constants have not been shown explicitly.)

Note that from the point of view of policy makers when they have specified their targets and are selecting the instruments to achieve them, the nature of some of the variables shifts. *The target variables are usually endogenous variables in*

TABLE 16-1 Equations for General Equilibrium

Endogenous Variable		Equation
1	Aggregate demand	$Y^* = Y^*\{Y_s, i, P_Y : G_o, X_o, Z_o, N, \text{behavioral coefficients}\}$
2	Real output	$Y_{R,\,s} = Y_{R,\,s}\{Y^*, W, \text{MPL}: \beta\}$
3	Aggregate supply	$Y_s = P_Y Y_{R,\,s}$
4	GNP price level	$Y_s = Y^*$
5	Demand for money	$M^* = M^*\{Y_s, i, P_Y : V_1, \text{behavioral coefficients}\}$
6	Supply of money	$M_s = M_{s,\,o}$
7	Interest rate	$M^* = M_{s,\,o}$
8	Demand for labor	$L^* = L^*\{W, P_Y, \text{MPL}: \beta\}$
9	Supply of labor	$L_s = L_s\{W, P_Y : N\}$
10	Wage rate	*a.* $L^* = L_s$ *b.* $W = W_o$
11	Marginal productivity of labor	$\text{MPL} = \text{MPL}\{L^* : K, \text{TECH}\}$

the causal model of the system, for example, L^*, I_d^*, or P_Y ; but when the policy makers specify fixed target values for them, those variables become exogenous in the policy model. On the other hand, the instrumental variables, say G^*, M_s, or tax rates, become endogenous analytically, because the policy makers solve the set of equations to find out what values of the instruments are consistent with the specified values of the target variables. Thus, suppose that L_o^* and $I_{d,\,o}^*$ are the desired values of two target variables and that G_o and $M_{s,\,o}$ are the instrumental variables to be used to achieve the targets. Then the causal model embodied in the equations in Table 16-1 would be transformed into the policy model by making L_o^* and $I_{d,\,o}^*$ (hence i_o) into exogenous variables and by shifting G_o and $M_{s,\,o}$ into the endogenous category. Solution of this "policy model" set of equations leads to determination of G_o, $M_{s,\,o}$ and the other nine endogenous variables in terms of L_o^*, i_o, and the other exogenous variables and behavioral constants.

Consideration of this policy model provides the mathematical support for the assertion made earlier that attainment of multiple targets requires adjustment of as many instrumental variables as the number of targets to be reached. If policy makers take j endogenous variables as targets, making them exogenous in the policy model, the set of 11 equilibrium equations would contain only $(11 - j)$ endogenous variables. Then it would not, in general, be possible to find unique values of the $(11 - j)$ endogenous variables which satisfied all 11 equations simultaneously. The situation would be like that for a set of three equations for equilibrium in the market for a given product.

Demand: $Q_D = f\{P: Y\}$

Supply: $Q_S = g\{P: \text{ULC}\}$

Equilibrium: $Q_D = Q_S$

Given the behavioral coefficients in the demand and supply equations, we could solve the three equations for the three endogenous variables Q_D, Q_S, and P in terms of the exogenous variables Y (income), ULC (unit labor cost), and the coefficients of the system. However, if the government legislated a target price, making it exogenous, then Q_D and Q_S can indeed be determined, but there will be no assurance that their values will satisfy the equilibrium relation $Q_D = Q_S$. In general, there will not be two values for the two endogenous variables which satisfy three equations simultaneously. The upshot of the analysis is that *if policy makers wish to take* j *endogenous variables as targets, they must convert* j *of the previously exogenous variables to instrumental variables if all the set of equations are to be satisfied simultaneously.*

As a corollary of this proposition, it follows that *if policy makers control* j *instrumental variables, they can achieve predetermined values for only* j *endogenous (target) variables.* They will have to accept the equilibrium solution values for other endogenous variables, some of which they might desire to take as targets also.

ILLUSTRATIONS OF USE OF POLICY VARIABLES TO ACHIEVE TARGETS

Example 1

Target: Full employment of labor, $L^* = L^*_{FE}$.

Analysis: In the short run, with the production function fixed, the demand for labor depends on output decisions of producers.

$$Y_{R,s} = Y_{R,s}\{Y^*, W, \text{MPL}: \beta\}$$

Of the variables entering into production decisions, the government has most control over Y^*.

$$Y^* = Y^*\{Y_s, i, P_Y : G_o, X_o, Z_o, N, \text{behavioral coefficients}\}$$

Policy: For use of fiscal policy to achieve full employment, G_o might be taken as the instrumental variable. With $L^* = L^*_{FE}$ taken as exogenous and G_o as endogenous in a policy model of the general equilibrium equations, we could solve for the value of G_o needed to bring the economy to full employment.

Implications: To solve for G_o, we must forecast the values of all other exogenous variables and coefficients in the model—perhaps we assume them unchanged from recent past periods. We must accept the equilibrium solution values for the other 10 endogenous variables. Implicitly, we also accept the equilibrium values of other variables not shown explicitly in this reduced set of equations, for example, domestic investment I_d^*, government surplus or deficit S_g, balance of payments, and sectoral income distribution.

Example 2

Targets: Domestic investment expenditures shall be at a level to generate optimal growth in capacity and labor productivity, $I_d^* = I_{d,\alpha}^*$. Full employment is to be maintained also.

Analysis: Assuming the interest-rate model for investment, the achievement of a target investment depends on establishing the corresponding interest rate.

$$I_{d,\alpha} = g - hi_\alpha$$

Of the exogenous variables determining i, the government exerts most short-term influence over money supply. (Y_s is being controlled to maintain full employment.) From the LM-line equation [Eq. (7) in Table 16-1] we have

$$i = i\{Y_s, P_Y, M_{s,o} : V_1, \text{behavioral coefficients}\}$$

Policy: With L_{FE}^* and i_α taken as exogenous in the equations of the policy model, we can solve for values of G_o and $M_{s,o}$ which would lead to achievement of the two targets simultaneously. In terms of the graphs of the general equilibrium equations, we note that L_{FE}^* determines $Y_{R,FE}$ on the production functions and this value projected up to the aggregate supply curve determines Y_{FE}. In the IS and LM diagram, once Y_{FE} has been determined, we can think of G_o positioning the IS line and $M_{s,o}$ positioning the LM line so that both of them intersect the Y_{FE} line at interest rate i_α.

Implications: Again, we must forecast or assume constant all exogenous variables other than G_o and $M_{s,o}$ and all model coefficients. And we must accept the equilibrium solution values for the other nine endogenous variables shown in the model, and the many others not appearing explicitly. Actually, the policy makers should be well content with $Y_{R,s}$ and MPL because they are closely associated with the two targets which are being achieved.

Example 3

Targets: Other targets may be added to the preceding two, perhaps the balance of the government budget at full employment and the balance of international payments.

Analysis: Additional instrumental variables would be needed, equal in number to the added targets. Tax policy would be a prime candidate and might provide multiple instruments. The revenue requirements to balance budgets at full employment may be met by many different structures of taxes, transfers, and subsidy payments. So a structure may be chosen which will affect international payments or domestic consumption in such ways as to help promote other goals while meeting full-employment revenue requirements.

Policy: Suppose G_o and $M_{s,o}$ have been adjusted to yield desired levels of L_{FE}^* and $I_{d,\alpha}^*$. Now personal taxes are raised to achieve a balanced budget. Higher taxes lower the C^* versus Y_s line and shift the IS line downward. Hence G_o may need to be increased to offset the decline in consumer spending and keep aggregate demand at the full-employment level. But the change in G_o alters the budget balance. So it goes.

Implications: Note that changing a given policy variable affects the equilibrium solution for several endogenous variables—in varying degrees depending on the channels of influence and the magnitudes of coefficients in the macroeconomic system. *There is not a one-to-one causal connection between targets and instruments.*

When we fix the values of target variables in the policy model and solve for the required values of policy variables, the solution is a general equilibrium solution. *The combined set of values of instrumental variables yields the combined set of targets.* To be sure, some instruments are more closely connected to certain targets than others, but the general principle is valid.

Example 4

Targets: Suppose that the policy makers desire to hold prices at a given level, along with maintaining the targets previously set, that is, they add $P_Y = P_{Y,\,\alpha}$.

Analysis: From the general equilibrium set of equations summarized early in this section, the equations relevant to price determination are:

$$P_Y = \frac{Y_s}{Y_{R,\,s}}$$

where

$$Y_s = Y^* = Y^*\{Y_s, i, P_Y : G_o, X_o, Z_o, N, \text{behavioral coefficients}\}$$

$$Y_R = Y_R\{Y^*, W, \text{MPL} : \beta\}$$

As was noted above, for a given production function, Y_R is determined when the target of full employment of labor is specified. According to the real aggregate supply equation (last equation, above), the full-employment level of real output must be consistent with values of Y^*, W, MPL, and β. Heretofore we have adjusted Y^* by fiscal and monetary policies to call forth the required output and labor demand from producers. But then P_Y was determined endogenously as that price level which exists at the full-employment output on the equilibrium aggregate supply curve, that is, for the equilibrium values of W, MPL, and β.

Unless we break through the restrictive assumptions involved above, we are in the position of the customer for a Ford car back in the 1920s who was told he could order any color he wished—so long as he specified black. We can have any price level we desire—so long as it is the full-employment level.

Policy: If we really want a lower price level, it appears that we must accept an output lower than full employment; we "trade off" lower prices for higher unemployment.

Our alternative, of course, is to shift the aggregate supply curve or to bring in some forces influencing price level which have not been included in the model to this point. A decline in wage rates, a rise in productivity of labor, or a decline in the marginal share of nonfactor costs (β) would shift the aggregate supply curve downward, that is, would lead profit-maximizing producers to achieve full-employment output at lower prices. According to Eq. (10) in the general equilibrium model, wage rates are either exogenous or determined by P_Y, MPL, β, and the labor supply function. This points to government intervention in labor market settlements or in the supply of labor function.

A shift of the production function which raises MPL will probably raise APL of labor also, and hence the level of $Y_{R,\,FE}$. The downward shift of the aggregate supply curve depends on relative increases of wage rates and APL, since ULC equals W/APL. If wage rates rise more than prices, the full-employment labor *supply* will increase, causing a further increase in $Y_{R,\,FE}$. The ultimate effect on P_Y is not clear, but depends on the relative increases in wage rates and average productivity of labor and on coefficients of the functions involved.

Prices might be helped to fall by decreases in β, for example, by a decline in sales and excise tax rates, production subsidies from the government, or a decline in prices of imported materials used in production. But whether the authorities can affect β enough to influence price level appreciably is questionable.

Implications: The upshot of this analysis is that *the price level would seem to be different from the targets previously considered. It is so closely linked to the full-employment target that separate control of price level and of full employment is difficult.* The trade-off between them is determined by production decisions and labor-market reactions in closely knit causal interactions, so that the blunt tools of fiscal and monetary policies cannot alter the trade-off in a clear-cut, predictable manner. These problems will be discussed at greater length in the chapter on inflation. This introduction may indicate why governments, in trying to achieve targets of full employment and price stability concurrently, may find it necessary to move into more direct channels for influencing prices and wages (incomes policy) and into attempts to influence the operation of labor markets as well as product markets.

PROBLEMS IN ACHIEVING MULTIPLE TARGETS

This discussion has pointed to a few of the practical problems the government faces as it attempts to steer the nation in the direction of its economic goals. First, *use of a policy variable to attain one objective may detract from achievement of another target.* For example, progress toward full employment may be accompanied by increased inflation, or faster economic growth may lead to a balance-of-payments deficit. Second, *use of several instruments simultaneously to achieve multiple targets concurrently poses difficult problems of coordination of policy actions in timing and magnitude.* The Federal Reserve Board may not be willing to adapt monetary policy to fiscal actions by Congress, or vice versa. Or one arm of the federal government may try to slow an inflation at the same time that the Agriculture Department is supporting farm prices.

A third practical difficulty lies in *inadequacy of the policy instruments.* The policy makers may not know the system coefficients well enough to predict the effects of their policy instruments with required accuracy.   Or it may be impractical to vary the instrument over the range needed to achieve desired targets.   For example, government spending in wartime cannot be cut appreciably, and even in peacetime a large share of federal spending is uncontrollable.   Or private sectors may be relatively insensitive or even perverse in their response to instrumental variables.   This could be the reaction to attempts to use monetary policy when investment spending does not respond to changes in interest rate, or to attempts to use fiscal policy when the LM line is nearly perpendicular to the Y-axis.

Fourth, there are *problems of time lags in the operation of the policy instruments*, especially when the delays are of uncertain length.   Time lags are sometimes classified as internal and external.   The *internal lags* are delays within the policy-making agency—delays in data gathering and in identifying an emerging problem, then the time needed to analyze the situation, to formulate policy prescriptions, to select instruments, and to implement the actions selected.   The *external lags* are delays in response of the economic system to policy actions taken—time for private decision makers to perceive the change in the policy variables and to alter their decisions and behavior, and for markets to adjust to new equilibria consistent with the new policy variables.   The situation is analogous to that for an automobile driver approaching a road hazard.   The internal lags are the time intervals required for the driver to perceive the hazard, to analyze the current and prospective situations, to decide on some desired combination of braking and steering policy actions, and to implement those decisions by his or her own movements.   The external lags would be mechanical response lags in the braking and steering mechanisms and the time required for deceleration and changed wheel direction to achieve the new desired values of speed and direction for the car.

For national economic problems, our procedures for gathering and processing data have been facilitated by sampling and computers, so that lags in recognizing problems have been reduced to a few months.   For *fiscal policy* the additional internal lags may be very long, particularly if congressional action is required. In the late 1960s, Vietnam war expenditures brought the economy to full employment by early 1966, and President Johnson proposed a tax increase in mid-1967 to counter inflationary pressures.   But congressional analysis of the problem and legislative procedures delayed the imposition of a 10 percent surtax on incomes until the third quarter of 1968, by which time inflation was well established. In the case of fiscal policy via changes in government purchases, there will also be legislative lags and time required to set up administrative apparatus, prepare plans, take bids, let contracts, and complete other arrangements before the impact on private behavior becomes potent.   External lags may not be long for fiscal policy.   Demand sectors may respond quickly to—or even anticipate—income changes induced by shifts in taxes or transfer payments.   Production sectors may respond quickly to a spate of government contracts, not immediately by delivery of goods but initially by building up their inventories of materials and parts required to meet the contracts.   Yet even here, time is required for production to reach full flood and for the multiplier and accelerator feedbacks on spending in other sectors to take full effect.

For *monetary policy*, internal lags are shorter because the Federal Reserve Board is an independent agency which can take action without waiting for congressional approval. And once its analysis and choice among policy instruments have been made, the Board can implement open-market actions quickly or enforce a change in discount rate or in reserve requirements on short notice. But the external lags seem to be large for monetary instruments, perhaps 6 to 18 months. The initial impact is on government bond dealers' portfolios and on reserves of commercial banks. When these institutions perceive the changes to be a lasting policy change, they readjust their balance sheets and change interest rates over widening circles of financial securities. In particular, banks extend credit more freely and more cheaply to persons, businesses, and state and local governments, thus stimulating their spending on goods and services. Multiplier and accelerator feedbacks extend the stimulus throughout the economy. But these adjustments to changed conditions in markets for credit and capital funds seem to proceed more slowly than do reactions to the income changes and government spending involved in fiscal policy actions.

For either fiscal or monetary policy, the internal and external lags are different in different historical situations. If this variability is great enough—and some economists believe it to be so—successful use of economic policy to maintain full employment and achieve other economic targets becomes very difficult in a dynamic economy subject to endogenous business cycles. In such an economic system, policy measures adopted to stimulate an economy in recession may come to fruition at a time when they reinforce an inflationary boom in the private sectors. Conversely, steps taken to rein in a boom may act to restrict aggregate demand after the economy has moved downward into its recession phase.

The existence and variability of time lags between emergence of a problem and the impact of policy instruments adopted to solve the problem raise the question of *economic forecasting*. Earlier in this chapter we noted the role of a macroeconomic model in helping policy makers to choose policy instruments and evaluate their impact. The policy makers also need the model to forecast the prospective movements of the economy without use of any instruments of control. Only then can they judge whether their actions will be helpful in achieving desired targets *at the time when the policy variables take effect*. Certainly the art of economic forecasting has not been perfected yet, perhaps especially in the timing of turning points. But notable developments have been made in the postwar years. Surveys now provide lead-decision data on spending plans for consumer durables and for plant and equipment. And computerized macroeconomic models, some involving more than 100 equations, have been developed and are being used to prepare forecasts on a regular schedule. In addition to generating forecasts, these complex models are used to make simulation runs to estimate the impact of a wide range of individual policy variables, or combinations of them.

The models, of course, still have their margins of error. Unexpected exogenous events occur, and human learning and expectations lead to changes in the behavioral responses within the economic system. As one economist wrote ruefully in the late 1960s: "After the economic instability we have experienced in the past five years, the parameters of the system cannot be located with precision and may well be in flux." Nevertheless, *with the forecasting tools available and given the inertia in the economic system, it seems that future developments can be foreseen well enough to permit effective use of policy instruments in spite of their time lags.*

Some shifts in system coefficients operate to moderate movements away from an equilibrium position. These shifts are known as *automatic stabilizers.* Thus, an income shift toward profits during a business expansion will tend to slow the rise in personal incomes and consumption expenditures, and the sharp decline in profits during a recession helps sustain personal income. A progressive tax system and transfer payments, such a unemployment and welfare benefits, also help stabilize consumer incomes against cyclical swings of GNP. Then the tendency of consumers to maintain their spending at the expense of saving during cyclical swings of income, that is, the countercyclical movements of MPC, helps further to stabilize aggregate demand. The operation of these automatic stabilizers can to a large extent be written into our econometric models.

It is the *nonautomatic destabilizers* which interfere with achievement of economic targets through the use of policy instruments. Prominent among these are *expectations of consumers and businessmen.* Consumer expectations of future inflation, of their own employment and incomes, of war, tax changes, economic controls—these seem to cause unpredictable shifts of the consumption function. For example, a 10 percent surtax on incomes took effect in the last half of 1968, in a delayed effort to curb the inflationary surge which accompanied intensification of the Vietnam war. Yet the Council of Economic Advisers records in its 1969 report:

> Consumer spending . . . was particularly stimulative in the third quarter, jumping $13 billion (annual rate) at a time when the surcharge was first affecting paychecks. . . . In fact, saving declined far more than had been anticipated in the third quarter. . . . During the second half of the year the saving rate averaged $\frac{3}{4}$ of a percentage point lower than in the first half—nearly twice the decline that would have been consistent with both an unchanging basic strength of consumer demand and the anticipated lagged response to the tax increase.

Despite this short-term perverse response of consumers, there was no doubt that a restrictive fiscal policy was needed; and a slowing of real growth rates did occur in 1968 and 1969, followed by the recession in 1970.

Business firms' expectations can also be destabilizing. An expected expansion of aggregate demand and prices may lead to an increase in investment expenditures to build up stocks of inventories and fixed capital to new desired levels.

This accelerator effect intensifies the increase in aggregate demand and may worsen inflationary pressures. Expectations of an inflationary environment by both business managers and labor leaders may also promote price rises via the labor-management bargaining process. Labor leaders will push for larger wage increases to offset anticipated increases in consumer prices, and managers will be more willing to grant large wage increases because of their expectations that the market will permit increases in their product prices. Such self-realizing expectations can be destabilizing and can work against the thrust of policy variables; they tend to amplify movements which have been started by other forces. But their existence does not change judgments regarding the *direction* of economic policy, only the magnitude of policy instruments needed to achieve desired targets.

Unexpected exogenous shocks are another source of instability and uncertainty in achieving national economic targets—shocks such as technological developments, widespread crop failures, international economic changes, war or internal civil strife, crises in money and credit markets, or surges of growth in population and labor force. For some countries, changes in terms of trade, foreign competition, or international capital movements may upset domestic equilibrium. For the United States, perhaps only a war will cause enough of an exogenous shock to dominate economic developments and require major policy changes—on the assumption that major breakdowns of money and credit mechanisms have been ruled out by changes in the economy since 1929.

Despite all these problems in achieving multiple targets for the national economy, it seems that we have done better in the post-World War II period than previously. A better understanding of the economic system, improvements in timely availability of economic data, use of computerized econometric models, strengthening of automatic stabilizers, and a clearer commitment to economic goals by the federal government are some elements that seem to have contributed to our improved performance. The outstanding nexus of problems requiring solution was defined in these terms in the *Economic Report of the President* in January 1972: "In the summer of 1971 the American economy was beset by a conflict among four objectives—faster growth, higher employment, greater price stability, and a more balanced external position. The danger was that steps to speed up growth and boost employment by expanding demand would worsen both the inflation and the balance-of-payments deficit." The simultaneous attainment of these four objectives on a sustainable basis has, to date, been beyond the capacities of policy makers in the United States and other industrialized nations.

PROBLEMS

1 Define what is meant by policy targets and instruments, and give three examples of each. Describe the role of a macroeconomic model in choosing and implementing economic policy. Where are policy targets and instruments found in the model, or how are they related to it?

2 Describe the differences between the 11 general equilibrium equations considered as a policy model and as a forecasting model. If employment, government deficit, and interest rate are taken as target variables to be achieved with G^*, tax rates, and M_s as policy instruments, what will be the 11 endogenous variables in the set of policy equations?

3 What is meant by automatic stabilizers and by nonautomatic destabilizers? Define and give three examples of each. Can you think of any nonautomatic stabilizers? Explain.

4 List some of the principal problems which the government would face in using fiscal and monetary policies to achieve simultaneously full employment, a balanced government budget, and price stability.

5 "The necessity to accept a trade-off between targets of high employment and low price rise suggests that the government is not using enough policy instruments." Do you think this is the case, or is there some other reason why high employment and rate of inflation cannot be controlled separately?

Chapter 17

Maintaining Full Employment: Keynes versus the Monetarists

THE CLASSICAL MODEL

Having considered possible policy instruments in relation to the achievement of economic goals, and having noted some of the problems in trying to achieve multiple targets simultaneously, let us attempt to compare and evaluate alternative policy models which are currently advocated for achieving full employment with growth but without inflation. In particular, we shall focus on the claims of the Keynesian versus the monetarist policy. To appreciate the differences more clearly, we shall outline briefly the theoretical development from classical to Keynesian and then to modern monetarist models. Although no one economist laid out a clear and complete macroeconomic model based on classical principles, a synthesis of the theories of economists classed in this tradition during the nineteenth and early twentieth centuries can be developed. It is here expressed in terms of IS and LM lines, aggregate supply curve, and so on, to permit comparison with later models, though the classical economists did not use this terminology or analytical apparatus.

The special assumptions implicit in the theories of classicists are as follows:

1 Aggregate *real demand* was assumed kept equal to *real output* by means of market price adjustments. It was assumed that suppliers produce goods only to use them personally or to exchange them for goods produced by others. (This is Say's law of markets: "Supply creates its own demand.")

2 Both real consumption and real investment were assumed to increase if interest rate declined. So the IS line would show increases in equilibrium Y at lower interest rates—the relation used in the general equilibrium model of Chap. 15.

3 Demand for money was assumed proportional to GNP, that is,

$$M^* = \frac{1}{V_1} Y = \frac{1}{V_1} P_Y Y_R$$

In some theories this was justified by reasoning which implied that money demand was for transactions balances only and that the transactions velocity was nearly constant in equilibrium states. In other theories, it was hypothesized that the nonbank public desired to hold real cash balances in a constant proportion to real incomes.   Here,

$$M_R^* = \frac{M^*}{P_Y} = k\,Y_R$$

Both yielded the same money demand function.

With money supply exogenous, both theories led to the conclusion that changes in M_s would lead to proportionate changes in Y—either in P_Y or in Y_R, or a combination of the two.

4   Labor markets were assumed to be competitive; wage rates adjusted flexibly to keep demand and supply for labor services in equilibrium.   Thus, *full employment of labor was assured by wage-rate flexibility.*   Both demand and supply of labor are stable functions of real wage rate (W/P_Y).

5   The production function was assumed fixed.   Wage flexibility led Production Units always to supply a full-employment level of output.

Given such a model, let us analyze its response to exogenous changes, including fiscal and monetary policy instruments.

Example 1

How do equilibrium values of endogenous variables change if some exogenous component of aggregate demand shifts?

Suppose the economy starts in a position of full employment at points o in Fig. 17-1.   In panel (a), the LM line is perpendicular to the Y-axis because money demand was assumed to be determined by Y and not to vary with interest rate. Along the IS line, the equilibrium value of Y increases when i decreases because both investment and consumption respond inversely to interest rate.   The IS line intersects the LM line at a full-employment GNP and interest rate i_o, which the classical economists called the *natural rate.*

The money GNP determined at the intersection of the IS and LM lines projects across to the full-employment point o on the aggregate supply curve of panel (b).   This point o projects down to the production function and across to the full-employment point o in the labor-market diagram of panel (d).   The labor supply curve $L'_{s,\,o}$ graphs the flow of labor services derived from 96 percent employment of the labor force, and it intersects the initial labor demand curve at point o.

Now suppose that one of the exogenous components of aggregate demand declines, shifting the IS line downward to position IS_α.   Since money supply is assumed constant, money GNP must be unchanged in the new equilibrium at point α.   Constant Y_o will determine an unchanged full-employment equilibrium point on the aggregate supply curve.   There will be no change in P_Y, nor in W, nor in the L^* and L'_s curves in panel (d).

How can equilibrium remain at full-employment output after the decline in an exogenous component of aggregate demand?   The answer is given in panel (a). As the IS line shifts downward, the interest rate declines, and the equilibrium intersection moves rightward along the fixed LM_o line.   At the new equilibrium

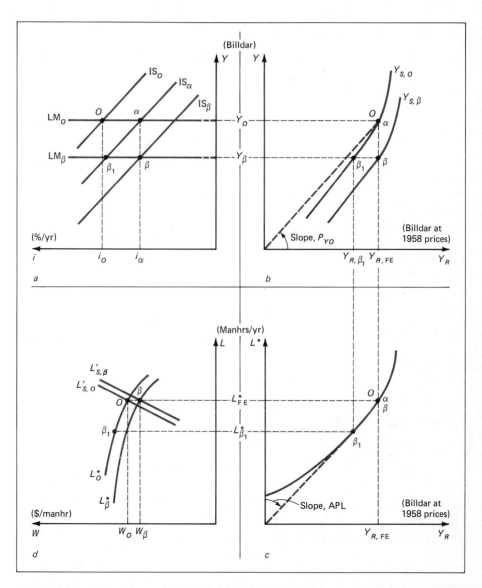

Figure 17-1 Classical model with declines in IS line or in money supply.   (*a*) IS and LM lines.
(*b*) Aggregate supply curve.   (*c*) Production function.   (*d*) Labor demand and supply.

point α, interest rate has declined just enough to *induce* increases in investment and consumption that together offset the initial *exogenous* decline in aggregate demand. *So long as the LM_o line remains fixed, the money value of aggregate demand is fixed, and full-employment output is maintained at a constant-price level. The exogenous demand shift and the change of interest rate would cause offsetting changes of various demand components within the fixed total.*

If government purchases were increased to shift the IS line back to the IS_o position, then a rise in interest rate would cut back interest-sensitive expenditures by an amount equal to the rise in G^*. Aggregate demand and price level would be unchanged in the new equilibrium. So *fiscal policy is ineffective; it can alter the division of GNP between private and government purchases but cannot change aggregate output and employment.*

Classical economists sometimes granted the possibility that a sudden decline in aggregate demand accompanied by a slow adjustment of interest rate could lead to a temporary slump. So long as i remained above its new "natural" level of i_α, aggregate demand would fall below the original level, and producers would choose real outputs below the full-employment level. This would bring several equilibrating mechanisms into play. Unemployment would emerge and wage rates decline. The decline in demand and downshift of the aggregate supply curve (because of lower W) would lead to both lower product prices and some recovery of *real* output (for given Y). Also, with money GNP below the full-employment level, demand for money stocks would fall below the fixed supply, and interest rate would decline. Ultimately the decline in i would restore aggregate demand to the original level. If P_Y had dropped below the original level, then money aggregate demand Y_o would lead to demand for real output in excess of $Y_{R, FE}$, and wage rates, prices, and the aggregate supply curve would move back up to their original equilibrium levels at full employment.

If one assumes wage flexibility, a fixed money supply, and a demand for money proportional to Y, *shifts of aggregate demand might produce "lapses from full employment" in this classical model, but the economy would return to equilibrium at full employment.*

Example 2

How do equilibrium values of the endogenous variables change if the money supply is altered?

Suppose that the economy starts originally in full-employment equilibrium at point α in Fig. 17-1. Then money supply is reduced to M_β, which shifts the LM line downward to a position LM_β, corresponding to aggregate demand Y_β, The reduction in M_s drives interest rates up and induces a decline in interest-sensitive components of aggregate demand. The IS-LM equilibrium shifts leftward along the IS line to point β_1. The reduced level of money aggregate demand (Y_β) would lead producers to reduce prices and output from α to β_1. As unemployment emerges, wage rates fall, and the aggregate supply curve in panel (*b*) shifts down. Note that *the price decline also leads to a downward shift of the IS line*. The classical model assumed that a given value of i led to a definite value for *real* aggregate demand, so that a change in P_Y would change *money* aggregate demand in proportion to the price change at each level of i. The downward movements of wages, prices, aggregate supply, and the IS line continue until a new equilibrium is reached at points β. In the new equilibrium, Y, P_Y, and W will all have declined by the same percentage that money supply was reduced. The IS line and the aggregate supply curve will shift downward; the labor demand and supply curves will have shifted rightward—all by the same percentage. The equilibrium values of i_∞ and $Y_{R,\,FE}$ will remain unchanged at their original full-employment equilibrium values.

In summary, *a change in money supply in the classical model leads to equal percentage changes in all money values in the economy, but real values and interest rate are unchanged, including the real money stocks.* Again, wage flexibility and a horizontal LM line hold the economy to its full-employment equilibrium.

This was a rather reassuring model. Attainment of a full-employment equilibrium was shown to be plausible, and once reached, it seemed to be remarkably stable. Exogenous shifts in aggregate demand, even changes in fiscal policy instruments, would be fully countered by compensatory changes in interest rate. Changes in money supply would cause shifts in aggregate demand in money measure, but normal demand-supply reactions in product and labor markets would eventually render the lowered money expenditure consistent with full-employment output at lower price and wage levels.

The theory did not trace the path of adjustment from one equilibrium position to another. In particular, the ways in which changes in money supply generated changes in aggregate demand were unclear. Also, the time lags involved in adjustments of interest rates, prices, and wage rates were not spelled out.

In addition, some of the implied connections between equilibrium values of variables seemed odd. An exogenous decline in aggregate demand (downshift of the IS line) leads to a decline in interest rate but no change in product prices. Conversely, an exogenous decline in money supply (downward shift of the LM line) leads to a decline in product prices but no change in the interest rate. One might expect demand changes to affect prices and money supply changes to affect interest rates. The explanation for these unexpected relations lies in three characteristics of the model:

1 Flexibility of wage rates guarantees that the economy is in equilibrium only at full employment. Initial market reactions might lead to the expected changes in prices and interest rates, but the initial movements are reversed as the economy returns to equilibrium.

2 The assumption that demand for money stock is rigidly linked to aggregate demand flow Y, coupled with the model's requirement that equilibrium output always be at the full-employment level, requires that price level be proportional to money stock in equilibrium, that is,

$$Y = P_Y Y_{R,\,FE} = M_s V_1$$

3   The assumption that *real* aggregate demand is a function solely of aggregate real income and interest rate, coupled with the model's requirement that equilibrium income must always be at the full-employment level, requires that interest rate is determined by the exogenous components and coefficients in the *real* aggregate demand function so as to make

$$Y_R^* = C_R^* + I_{d, R}^* + G_R^* = Y_{R, \text{FE}}$$

THE KEYNESIAN MODEL

The classical synthesis was not upset because of economists' dissatisfaction with its theoretical shortcomings.   Rather, the major and prolonged depression in Western industrialized nations during the 1930s provided an example of "another good theory spoiled by the facts."   In the United States, the unemployment rate rose from 3.2 percent in 1929 to 8.7 percent in 1930 and then exceeded 14 percent in each of the next 10 years, with a peak rate of 24.9 percent in 1933.   Needless to say, to unemployed workers and political leaders, this seemed like more than a temporary "lapse" from full employment.   Policy makers sought policy prescriptions from economists, and the British economist John Maynard Keynes developed a new macroeconomic model which purported to explain the deficiencies of the classical model and to offer possibilities for stimulating an economy seemingly mired in a state of less-than-full employment.

The special assumptions underlying Keynes's model are as follows:

1   Aggregate *real demand* was considered to be a function of *real income* (for consumption) and of long-term rate of interest (for investment).   Thus, equilibrium *real* GNP would be a function of i, and equilibrium *money* GNP would be price times the real GNP, that is, in equilibrium,

$$Y = P_Y Y_R(i)$$

The only change here was to downgrade the dependence of consumption (and personal saving) on interest rate.

2   Demand for money stock was analyzed into component demands for transactions, precautionary, and speculative purposes—the first two components proportional to money GNP and the last dependent on interest rate.   So

$$M^* = M^*(Y, i)$$

or, in the additive form used in this book,

$$M^* = \frac{1}{V_1} Y + M_2^*(i)$$

This marked an important departure from the classical model.   *Aggregate demand was no longer constrained to change in strict proportion to money supply.*   Changes in M_s might be absorbed into speculative balances M_2^*, which have been called asset balances in this book.   Since M_2^* is inversely related to i, this change in the model leads to an LM line which has a positive slope in the normal range of interest rates instead of being perpendicular to the Y-axis. Keynes also suggested that at some very low interest rate, members of the nonbank public might be willing to increase their asset holdings of money almost indefinitely, although with no change in i. Such behavior would lead to a segment of the LM line which is parallel to the Y-axis at a low interest rate—the region we have called the "liquidity trap" in our previous discussions and graphs of the LM line.

3   In labor markets *Keynes assumed that money wage rates would adjust downward very slowly when unemployment rose above "frictional" levels.* This "stickiness" of wage rates removed the flexibility which guaranteed a full-employment equilibrium in the classical model. He assumed, as the classical economists had done, that the supply of labor is a stable function of the real wage rate.

4   The production function was assumed fixed in the short run, as in the classical model.

Now let us see how the Keynesian model would respond to exogenous changes in demand and to fiscal and monetary policy instruments.

Example 1

How do equilibrium values of endogenous variables change if some exogenous component of aggregate demand shifts?

Suppose that the economy starts in a state of full employment at points 0 in Fig. 17-2. Then an exogenous decline in aggregate demand shifts the IS line from IS_o to IS_{α_1}, while money supply and the sloping LM line stay fixed. If the interest rate declines along with the downward movement of the IS line, the equilibrium point slides down the LM line in panel (*a*) to point α_1. The decline in aggregate demand to Y_{α_1} leads producers to cut back output and lower prices, as seen by the movement from point 0 to point α_1 in panel (*b*). The price decline shifts the IS line down further because of the assumption that a given interest rate determines *real* aggregate demand. This interaction between demand and supply is shown reaching a new equilibrium at point α on the two diagrams. Through these changes in product and money markets, the wage rate is assumed to hold constant at W_o. However, the decline in product prices shifts the labor demand curve from L_0^* to L_α^* and causes a decline in labor inputs from the full-employment level at point 0 to a reduced level at point α. Labor demand L_α^* of course projects via the production function to the point α on the aggregate supply curve in panel (*b*). An unemployment gap $\alpha\alpha'$ has opened up in panel (*d*), both because of the decline in demand for labor and because lower prices increase real wage rates and shift the supply-of-labor curve to the right.

Now the economy is stuck in an equilibrium state below full employment. The situation differs from the classical model because the downward shift of the IS line does not lower interest rates enough to induce a rise in aggregate demand sufficient to offset all the exogenous decline. Also, wage rates are inflexible here. Keynes believed that unions and individual workers would resist and postpone wage reductions, so that his model offered a plausible explanation for the persistence

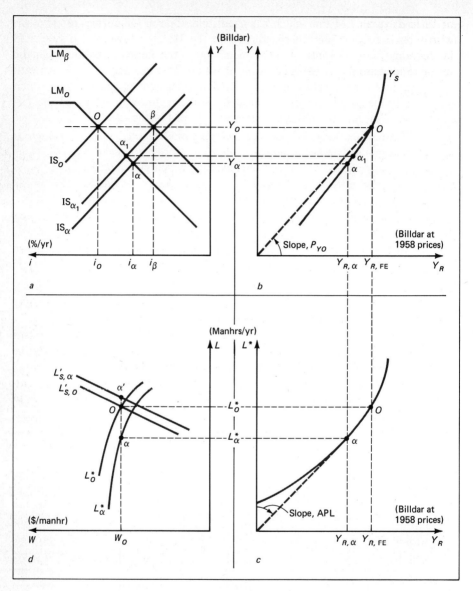

Figure 17-2 Keynesian model with exogenous decline in aggregate demand. (*a*) IS and LM lines. (*b*) Aggregate supply curve. (*c*) Production function. (*d*) Labor demand and supply.

of unemployment. He went further and argued that workers were probably wise to oppose reductions in money wages. Pay cuts might well lead to reductions in incomes, if real aggregate demand were constant, and this might lead to pessimistic expectations and further downward shifts of the IS line.

Example 2

Would fiscal policy help pull the economy out of its equilibrium below full employment?

Keynes believed that tax cuts or increases in government purchases of goods and services could restore the economy to full employment. As we have noted in the discussion of general equilibrium in Chap. 15, such fiscal policy actions would cause exogenous increases in C^* and G^*. These increases would shift the IS line upward. If the dosage of fiscal policy were large enough to offset the original exogenous decline in aggregate demand, the economy in Fig. 17-2 would shift back from under-full employment at point α to the full-employment state at point 0. Prices would rise as output expanded, lifting the labor demand curve to L_o^* and restoring labor markets to full-employment equilibrium at point 0.

Example 3

Could monetary policy lift the economy out of the doldrums and restore full employment?

Keynes advocated use of monetary policy under favorable circumstances. Thus in Fig. 17-2, starting in a recession at points α, an increase in the money supply would shift the LM line upward and lower the interest rate . The decline in i would induce an increase in aggregate demand along the IS line, and the consequent rise in price level would shift the IS line upward from IS_α to IS_{α_1}. Final full-employment equilibrium would be reached at point β. By then the curves in the labor-market diagram in panel (d) would be back at their original positions, intersecting at the full-employment point 0.

However, Keynes had some reservations about the effectiveness of monetary policy in a serious protracted depression when pessimistic expectations had become entrenched in the minds of business managers and consumers. Note in Fig. 17-2 that the decline into depression reduced the interest rate from i_o to i_α. The attempt to extricate the economy by monetary policy would drive interest rates down further to i_β. Keynes believed that the nonbank public might react to very low interest rates (high bond prices) by simply holding large stocks of speculative (asset) balances, absorbing whatever increased money supply the authorities provided without appreciable change in the interest rate. If this occurred, then interest-sensitive segments of aggregate demand would not be stimulated much by increases in money supply, and monetary policy would fail to generate the desired economic recovery. This situation is pictured in Fig. 17-3a.

Example 4

From an initial full-employment equilibrium at point 0, economic activity recedes to a depressed state at point α because of a decline in aggregate demand (downward

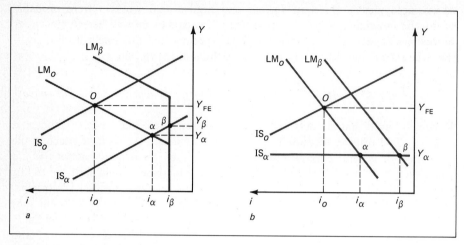

Figure 17-3 Ineffective monetary policy. (a) Keynesian "liquidity trap." (b) $I*$ insensitive to interest rate.

shift of the IS line with money supply, and hence the LM line, fixed). Now the authorities increase money supply to stimulate the economy, moving the LM line upward to LM_β. For a while, the interest rate declines and induces an increase in aggregate demand (downward movement along the IS_α line). But then the nonbank public becomes willing to hold additional asset balances without further decline in the interest rate. The interest rate holds constant at i_β; no further rises in induced investment occur; and the economy is stuck at point β in an equilibrium with Y_β less than the full-employment level.

Keynes envisioned another explanation for possible failure of monetary policy to stimulate the economy. In a depression, business owners would be operating well below capacity level of output from their plant and equipment, and they might become very pessimistic in their forecasts of future sales. Under these circumstances, the expected rates of return on many investment projects would drop to very low levels, or become negative. Fixed investment might well decline to or below replacement levels, and inventory investment might become negative for a time and then hold near zero. In such a state of affairs, reductions of interest rate would probably not make many potential projects into profitable prospects for investment spending. In terms of our analytical apparatus, this implies that investment demand would become insensitive to interest rate, that is, coefficient $h_1 \cong 0$ in the function

$$I_d^* = (h_0 - h_1 i)P_Y$$

This implies an investment demand line and an IS line which are nearly perpendicular to the Y-axis in the IS-LM diagram.

Example 5

As may be seen in Fig. 17-3b, a horizontal IS line would make it impossible for an increase in M_s to raise the level of GNP. Interest rate would decline from i_α to i_β when the larger money supply shifts the LM line from LM_0 to LM_β, but the decline in interest rates has no influence on the level of investment or equilibrium GNP in the assumed state of affairs. Indeed, it might be that even with an interest rate of zero, investment would not come up to the level required to generate a full-employment level of aggregate demand.

Keynes's policy prescription for situations in which monetary policy proved ineffective was, of course, " Use fiscal policy." Increases in government purchases, tax cuts, and increases in transfer payments to persons would increase aggregate demand independently of the sensitivity of interest rates to money supply or of investment to interest rates. As may be appreciated from the graphs of Fig. 17-3, an upward shift of the IS lines from the IS_α position would raise the equilibrium value of Y. If money supply were increased at the same time to keep interest rates down, a given vertical shift in the IS line would increase Y by more than if M_s were kept constant. But fiscal policy could theoretically be large enough to move the economy back to full employment at point 0 consistent with the original money supply.

In view of Keynes's reservations regarding the efficacy of monetary policy during a depression, the Keynesian model came to be identified with a reliance on the use of fiscal policy instruments to achieve and maintain full employment.

THE MODERN MONETARIST POLICY MODEL

Since World War II, the emphasis on the importance of monetary variables in explaining macroeconomic phenomena and as policy instruments has been reasserted on the basis of both theoretical and empirical considerations. Largely developed and advocated by Milton Friedman and the " Chicago School" of economists, the modern monetarist model emphasizes the stability of *demand for real money balances*, the importance of *adjustments in stocks of real and financial assets* following a monetary disturbance, and *the impact of wealth and inflation on asset rates of return and on consumer expenditures*. Progress has been made in explaining the *transmission mechanism*, that is, the causal linkages through which monetary policy leads to changes in aggregate demand and output. And the hypotheses involved have been subjected to some statistical testing—individually and as behavioral equations in econometric models of the total economy. Finally, the models have led to recommendations for monetary policy targets and instrumental variables.

According to Milton Friedman:†

---

† *International Encyclopedia of the Social Sciences*, vol. 10, p. 434, 1968.

The quantity theory... is the empirical generalization that changes in desired real balances (in the demand for money) tend to proceed slowly and gradually or to be the result of events set in train by prior changes in supply, whereas, in contrast, substantial changes in the supply of nominal balances can and frequently do occur independently of any changes in demand. The conclusion is that substantial changes in prices or nominal income are almost invariably the result of changes in the nominal supply of money.

Using an asterisk to denote desired values of variables and subscript R to denote real (constant-dollar) magnitudes, Friedman's statements may be expressed symbolically as

$$M^* = P_Y M_R^*$$

where M_R^* is a stable function of such slowly changing variables as population, real permanent income, total nonhuman wealth, rates of return on money, bonds, and equities, and the expected rate of inflation. Since the monetary authorities can change money supply M_s rather quickly, they can cause a discrepancy to emerge between actual and desired money stocks. Call the resulting gap an *excess supply of money*, that is,

$$\mathrm{EM}_s = M_s - M^* = M_s - P_Y M_R^*$$

As we have noted in previous chapters, such an excess money supply leads to changes in flows and prices in markets for capital funds, in such directions as to move the economy toward a new equilibrium state. In the Keynesian model, the initial impact fell on interest rates, as security (bond) purchases or sales led to a change in security prices. Later, the response of aggregate demand to the changes in interest rates led to GNP changes. And the combined effects of interest rate and GNP changes altered M^* until it was brought into balance with the new M_s.

In the monetarist model, the excess money supply leads to readjustments of holdings of many more assets and to a more direct impact on aggregate demand than via the interest-rate link to investment spending. To understand this broader model, we need to develop concepts of the *nation's intersectoral balance sheet*, *of rates of return on various assets, and of an overall balance-sheet equilibrium* in addition to flow equilibria in product, labor, and capital-funds markets.

THE NATION'S INTERSECTORAL BALANCE SHEET

Table 17-1 presents the major balance-sheet items for sectors of the economy arranged in a matrix form, so that the sector issuing a given claim appears at the top of its column and the sector owning that claim appears at the left end of its row. *Items within the vertical lines are liabilities of the column sector and assets of the row sector; they would cancel out in a consolidation of balance sheets for all sectors.* Real assets plus gold and foreign exchange assets are shown in the

TABLE 17-1   The Nation's Balance Sheet—an Intersectoral Matrix

(Summation across a row yields a sector's total assets. Summation down a column yields a sector's total liabilities and net worth.)

Sector Owning Assets	Financial Claims by Issuing Sector					Real Assets‡‡
	Monetary Authority†	Commercial Banks‡	Federal Government§	Borrowing Sectors¶	Personal Sector††	
Monetary authority *(m)*		BR	USA$_{FR}$			GO$_m$ PROP$_m$
Commercial banks *(c)*	R		USA$_c$	BA$_c$	PA$_c$	PROP$_c$
Federal government (US)	FRD$_{US}$	DD$_{US}$		BA$_{US}$		PROP$_{US}$
Borrowing sectors *(B)*	CU$_B$	TD$_B$ DD$_B$	USA$_B$		PA$_B$	PROP$_B$
Personal sector *(p)*	CU$_p$	TD$_p$ DD$_p$	USA$_p$	BA$_p$		PROP$_p$
Net worth		CNW$_p$	USNW	BNW$_p$	PNW	

† Under Monetary Authority, *R* designates reserves owned by commercial banks; FRD$_{US}$, Federal Reserve deposits owned by the federal government; and CU, currency, with the subscript denoting the owning sector.

‡ Under Commercial Banks, BR represents borrowed reserves; DD, demand deposits and TD, time deposits, with their subscripts denoting the owning sectors; and CNW$_p$, commercial bank net worth, assumed owned by the Personal Sector as equity securities.

§ Under Federal Government, USA designates federal government securities, with the subscript denoting the owning sector; and USNW, the net worth of the federal government.

¶ The Borrowing Sectors column includes the Business Sector, state and local governments, the Foreign Sector, and financial intermediaries.   BA represents the sectors' securities issued, with the subscript denoting the owning sector; and BNW$_p$, the net worth of borrowing sectors, assumed owned by the Personal Sector as securities.

†† Under Personal Sector, PA represents this sector's securities issued to the sectors denoted by the subscripts; and PNW, the net worth of the Personal Sector.

‡‡ Under Real Assets, GO$_m$ represents the stock of gold and foreign exchange owned by monetary authorities; and PROP, the real property owned by sectors denoted by the subscripts.

right-hand column; they, along with other net claims against foreigners, constitute national wealth.   The balance-sheet equation for any sector may be constructed by setting the sum of items across its row (assets) equal to the sum of items down its column (liabilities and net worth).   In the case of the Personal Sector, the equity securities representing net worth of commercial banks (CNW$_p$) and of the borrowing sectors (BNW$_p$) need to be added to the other assets of that sector.   Thus,

> Currency and demand deposits *plus* time deposits *plus* loan securities *plus* equity securities *plus* real property *equals* debts plus net worth,

or

$$(CU_p + DD_p) + TD_p + (USA_p + BA_p) + (CNW_p + BNW_p) + PROP_p$$
$$= (PA_c + PA_B) + PNW$$

The symbols used in the two-way table are defined in its footnote.   In general, *A* is used for loan claims, NW for equity claims (net worth), *D* for deposits, CU for currency, and PROP for real property.   Prefixes denote sectors issuing claims: US for the federal government, FR for Federal Reserve System, *C* for commercial banks, *B* for borrowing sectors, *P* for the Personal Sector.   The same symbols are used as subscripts to denote the sector owning a given financial claim or real asset.

The borrowing sectors' balance-sheet equation is like that for the Personal Sector but with subscripts *B* instead of *p*.

For the federal government we have:

Federal government deposits at Federal Reserve banks and commercial banks *plus* business loans held by the U.S. government *plus* federal property *equals* U.S. government securities outstanding *plus* net worth of the federal government,

or

$$(\text{FRD}_{\text{US}} + \text{DD}_{\text{US}}) + \text{BA}_{\text{US}} + \text{PROP}_{\text{US}} = (\text{USA}_{\text{FR}} + \text{USA}_c + \text{USA}_B) + \text{USA}_p + \text{USNW}$$

For commercial banks:

Total reserves *plus* loans outstanding *plus* commercial bank property *equals* borrowed reserves *plus* federal government demand deposits *plus* demand and time deposits held by the nonbank public *plus* net worth (including equity claims held by the Personal Sector),

or

$$R + (\text{USA}_c + \text{BA}_c + \text{PA}_c) + \text{PROP}_c = \text{BR} + \text{DD}_{\text{US}} + (\text{DD}_B + \text{DD}_p) + (\text{TD}_B + \text{TD}_p) + \text{CNW}_p$$

Finally, for the monetary authorities (the Federal Reserve System and the U.S. Treasury):

Borrowed reserves *plus* Federal Reserve holdings of federal government securities *plus* gold and foreign exchange holdings *plus* property *equals* total reserves *plus* federal government deposits *plus* currency outside banks,

or

$$\text{BR} + \text{USA}_{\text{FR}} + \text{GO}_M + \text{PROP}_M = R + \text{FRD}_{\text{US}} + (\text{CU}_B + \text{CU}_p)$$

The decision makers in each of the sectors can alter their assets and liabilities by *capital-account transactions* in markets for securities, deposits, and real property. Presumably, they will do so whenever they expect that such portfolio adjustments will increase the present value of their expected income streams and capital gains.

In addition, decision makers may alter their assets and liabilities by *saving* out of current income or by *borrowing* to consume more than current income. Such transactions relate income and expenditure flows to changes in the stock items on balance sheets. They will change net worth of the sector—upward if saving is positive, downward if the sector engages in net borrowing to finance consumption expenditures. (Borrowing to finance purchases of financial claims or real property is a capital-account transaction and does not change net worth of the sector.)

What factors enter into decisions to alter balance sheets, and what variables in the system will be affected by the transactions involved in portfolio adjustments? The answers to these questions involve concepts of rates of return on various types of wealth items.

RATES OF RETURN ON VARIOUS ASSETS

Rational decision makers would allocate their total wealth among asset items in such a manner that they could not be made better off by selling a dollar's worth of one asset and buying a dollar's worth of another asset. This implies that *in equilibrium, the marginal money rate of return must be the same for all assets.* We need a definition of this marginal money rate of return which is broad enough to permit comparison of marginal rates of return on a wide range of assets—durable goods, bonds, common stock, even money.

Any wealth item yields a net flow of real services which is desired by the owner. Assume for expository purposes that the net flow of services (net of any operating costs of using the item) remains constant indefinitely far into the future. (Streams varying with time in future years can be expressed as an equivalent steady, infinite stream, where "equivalent" indicates equal present value.) Let F_{sw} stand for the net flow of services from a wealth item, where the unit used for measuring F_{sw} is the real flow of services obtainable by rental payment of $1 per year at prices of a base year b. The flow of services is provided by a wealth item whose real measure is Q_W, the unit of measure being the amount of that item purchasable for $1 in base-year prices. For example, a house worth $20,000 at base-year prices might yield a net stream of housing services measured as worth a rent of $800 per year at rental rates of the base year. Then Q_W equals $20,000 at the base-year price level, and F_{sw} equals $800 per year at prices of the base year.

A small increment to the stock of a given wealth item (ΔQ_W) will yield an increment to the constant, infinite flow of services (ΔF_{sw}). The ratio $\Delta F_{sw}/\Delta Q_W$ is, of course, the marginal productivity of the given wealth item, measured in units per year of added services produced per unit of added stock.

$$\text{MPW} = \Delta F_{sw}/\Delta Q_W$$

To obtain the *marginal-productivity rate of return in current dollars*, we simply multiply the service flow by the price for that service and multiply the real stock of wealth by the price of the wealth item. Thus, we may write

$$r_W = \frac{P_{sw} \, \Delta F_{sw}}{P_W \, \Delta Q_W} = \frac{P_{sw}}{P_W} \, \text{MPW} \tag{17-1}$$

where r_W is the *marginal-productivity rate of return* for that wealth item.

Thus, if house-rental rates should rise 15 percent from the base year and house prices 10 percent, the current marginal-productivity rate of return for the house just described would be

$$r_W = \frac{1.15(800)}{1.10(20,000)} = \frac{920}{22,000} = 4.16 \text{ percent per year}$$

In the base year, the rate of return and marginal productivity are equal, since F_{sw} and Q_W are measured in base-year units, that is, in units of physical quantities equal to amounts purchasable for \$1 at base-year prices. Thus,

$$r_{W,o} = \frac{800}{20,000} = 4 \text{ percent per year}$$

In addition to the marginal-productivity rate of return defined by Eq. (17-1), the owner of a wealth item will experience a capital gain if the price of the item rises—and a loss if price declines. The *overall marginal rate of return on the wealth item* W then becomes

$$r'_W = \frac{P_{sw}}{P_W} \text{MPW} + \frac{1}{P_W} \frac{\Delta P_W}{\Delta t}$$

or

$$r'_W = r_W + \frac{\dot{P}_W}{P_W} \tag{17-2}$$

Thus r'_W *measures the rate of increase in the money value of the holder's wealth resulting from ownership of the given asset.*

The capital gains component of return from owning an asset may be expressed in terms of factors determining the price of the asset. Instead of saying that r_W is the ratio of the value of services flow to the market value of an asset, we can say that the price of an asset is simply the present value of the infinite stream of returns per unit of the asset.

$$P_W = \frac{1}{r} \frac{P_{sw} \Delta F_{sw}}{\Delta Q_W} = \frac{P_{sw}(\text{MPW})}{r}$$

where r is the appropriate rate of interest to use in discounting future streams of returns to present values, a kind of general or average rate of return on assets.

The rate of change in price of the asset will equal the sum of the rates of change for P_{sw} and MPW minus the rate of change in r.

$$\frac{\dot{P}_W}{P_W} = \frac{\dot{P}_{sw}}{P_{sw}} + \frac{\dot{\text{MPW}}}{\text{MPW}} - \frac{\dot{r}}{r}$$

Inserting this expression into Eq. (17-2), we find

$$r'_W = \frac{P_{sw}}{P_W} \text{MPW} + \frac{\dot{P}_{sw}}{P_{sw}} + \frac{\dot{\text{MPW}}}{\text{MPW}} - \frac{\dot{r}}{r} \tag{17-3}$$

In a static equilibrium state, only the marginal-productivity rate of return exists.

TABLE 17-2 Total Rates of Return for Various Assets

$$r'_w = r_w + \frac{\dot{P}_w}{P_w}$$

Asset	r_w	$\dfrac{\dot{P}_w}{P_w}$
Durable goods	$\dfrac{P_{SD}}{P_D}$ (MPD)	$\dfrac{\dot{P}_{SD}}{P_{SD}} + \dfrac{M\dot{P}D}{MPD} - \dfrac{\dot{r}}{r}$
Bonds and loans	$\dfrac{\Delta INT}{\Delta B}$	$-\dfrac{\dot{r}}{r}$
Equity claims	$\dfrac{\Delta DIV}{\Delta E}$	$\dfrac{\dot{P}_c}{P_c} + \dfrac{M\dot{P}E}{MPE} - \dfrac{\dot{r}}{r}$
Real money stock	$\dfrac{P_{SM}}{P_Y}$ (MPM)	$\dfrac{\dot{P}_Y}{P_Y}$
Nominal money stock	$\dfrac{P_{SM}}{P_Y}$ (MPM)	

Table 17-2 summarizes the marginal-productivity rate of return r_w and the capital gains rate $\dot{P}_w/P_w$ for various assets. In the case of financial assets, the flow of services yielded is the flow of real goods which can be purchased by the additional income derived from the asset, that is $\Delta INT/P_c$ or $\Delta DIV/P_c$, where P_c is the consumer price index. The value of the incremental asset is $P_B \Delta Q_B = \Delta B$ for bonds or $P_E \Delta Q_E = \Delta E$ for the equity claims, that is, the expenditure made for the increment to the portfolio of real assets. So the marginal-productivity rates of return are, respectively,

$$\frac{P_c(\Delta INT/P_c)}{P_B \Delta Q_B} = \frac{\Delta INT}{\Delta B}$$

and

$$\frac{P_c(\Delta DIV/P_c)}{P_E \Delta Q_E} = \frac{\Delta DIV}{\Delta E}$$

A word of explanation may be added with regard to the wealth item we call *money*. Does it have a meaningful marginal productivity and rate of return? The real stock of this wealth item is $M_R = M/P_Y$, measured in units at base-year prices. Since P_Y equals 1.00 in the base year, $M_{R,o}$ equals M_o. For a year in which P_Y equals 1.25, we would calculate that M_R equals $0.80M$ in units of purchasing power at base-year prices.

The flow of services provided by the real money stock is its reduction of transactions costs by providing a reservoir of generally acceptable legal tender which is easily transferable by coin or check, and its provision of protection against costs of meeting uncertain fluctuations of streams of receipts or outlays. Such flows of services from a money stock might be measured by the maximum constant-dollar service charges a bank customer would pay for equivalent services of cost reduction, liquidity, and protection, perhaps obtained by an open line of credit. If the flows of services F_{SM} are measured in terms of dollars per year at base-year prices, then we may define marginal productivity of real money stock as

$$MPM = \frac{\Delta F_{SM}}{\Delta M_R}$$

The marginal-productivity rate of return is obtained by multiplying the numerator by a price index for the flow of services of money, say P_{SM}, and the denominator by the price of a real unit of money. Since one real unit of money is $1 with purchasing power equal to $1 in the base year, the price of this unit is just equal to the general price index P_Y. Thus, the marginal-productivity rate of return on real money stock becomes

$$r_{M_R} = P_{SM} \frac{\Delta F_{SM}}{P_Y(\Delta M_R)} = \frac{P_{SM}}{P_Y} MPM$$

What about capital gains or losses on money stock? If we fix our attention on a real unit of money, the amount of money capable of purchasing goods and services worth $1 in base-period prices, then the price index for this asset equals the price index for a general market basket of goods and services, that is, P_{M_R} equals P_Y. Consequently, the asset experiences a rate of capital gain given by $\dot{P}_{M_R}/P_{M_R} = \dot{P}_Y/P_Y$. The overall marginal rate of return for real money stock becomes

$$r'_{M_R} = r_{M_R} + \frac{\dot{P}_Y}{P_Y}$$

If, however, the stock of money is measured in terms of nominal (current-dollar) units, a price rise will not lead to any capital gains because the number of real units of money declines at the same rate that their price rises. *The overall rate of return on a nominal money stock* M *will be limited to the marginal-productivity rate of return.*

$$r'_M = r_{M_R}$$

The preceding paragraphs have defined the concept of rate of return for the principal types of *assets in a nation's balance sheet. Each of them has a marginal-productivity rate of return defined as the ratio of the money value of an infinite stream of returns to the money value of the asset. In addition, assets may have a capital gains rate of return, defined as the rate of increase of the price of the asset.* Durable goods, equity claims, and real money stock experience capital gains during inflation in product markets; bonds and nominal money stock do not. Durable goods, bonds, and equity claims decline in price if the general rate of return r rises. The emergence of differences in rates of return on various assets will lead owners of wealth items to buy and sell assets, altering market prices of the assets in such a way as to change the rates of return back toward equality. In static equilibrium, the capital gains rate of return vanishes.

BALANCE-SHEET EQUILIBRIUM

Previous chapters have emphasized flow equilibria for the economy—in product markets, in labor markets, and in markets for flows of capital funds. In the chapters dealing with demand and supply for money, and with the related LM line, we did deal with balance-sheet adjustments to a limited extent. Excess supply or demand for money led to purchases or sales of securities and resulted in changes of interest rates. The modern monetarist economists point out that *there is a wide range of assets on the balance sheet of the economy and that an excess demand or supply in the market for one of them should logically lead to readjustments in holdings of all of them in order to restore an equality of rates of return on all wealth items.* For example, an exogenous increase in money supply, with its demand unchanged, will lower the marginal productivity of money and hence lower the rate of return to money. Assuming no change in the supply-demand situation for other assets, the rates of return for bonds, equity claims, and durable goods will all exceed the rate of return on money. Hence, holders of money will exchange it for these other assets, driving their prices up and marginal productivities down until equality is restored between the rates of return on all assets.

The concept of a balance-sheet equilibrium may be pushed even further to encompass a balance between asset holdings and streams of current expenditures. Consider a consumer's choice of obtaining a flow of services either by direct expenditure of rent or by purchase of the durable good yielding the services. This is similar to a producer's lease-or-buy decision regarding plant and equipment items. In one case the services are obtained by a flow of rental payments, in the other case by a flow of interest payments on the funds required to purchase the asset source of the services. Rental payments amount to $P_{sw} \Delta F_{sw}$ dollars per year to secure a small incremental flow of real services ΔF_{sw}, measured in dollars per year at base-year prices. The same flow of expenditures paid out as interest would permit purchase of a quantity of the asset source with market value $P_W \Delta Q_W$, subject to the equality

$$i(P_W \Delta Q_W) = P_{sw} \Delta F_{sw}$$

where i is the interest rate on borrowed funds.

Ownership of the increment ΔQ_W of the asset would yield an incremental flow of services with value $r_W(P_W \Delta Q_W)$ dollars per year. The ratio of ownership flow to rental flow of services obtained by the same annual rate of expenditures becomes:

$$\frac{\text{Ownership flow}}{\text{Rental flow}} = \frac{r_W(P_W \Delta Q_W)}{P_{sw} \Delta F_{sw}}$$

$$= \frac{r_W(P_W \Delta Q_W)}{i(P_W \Delta Q_W)} = \frac{r_W}{i}$$

If r_W is larger than i, the incremental flow of services, and hence the marginal utility per dollar per year of expenditures, is greater from ownership than from rental.

From this example we can induce a general rule: *When the rate of return on an asset exceeds the interest rate at which funds may be borrowed, a rational utility maximizer will secure the desired flow of services by buying the asset source rather than by purchasing the services (renting the asset).* This is the rule that we derived for investment decisions back in Chap. 9, and we see now that it applies to consumer durable goods and to financial assets as well. When stocks of an asset are accumulated as buyers act according to this rule, several changes occur to bring r_W and i into equality. Recall that, apart from capital gains, r_W equals (P_{SW}/P_W)(MPW). As purchases of the asset rise, P_W may well rise if costs of production of the asset increase. Also, as the stock of the asset increases and the flow of its services rises, P_{SW} may well decline, and MPW will also fall if we assume diminishing marginal productivity. All these changes tend to lower r_W. On the other hand, increased borrowing of funds may well drive i up, also tending to bring i and r_W together. *When i equals r_W, the flow of services obtained by market purchases of services (rental) will equal the flow of services obtainable by equal interest expenditures to own the asset source.* So an equilibrium is reached between flow of expenditure to buy services and the existing stock of the asset providing those services.

THE TRANSMISSION MECHANISM

Now we are in a position to answer a question which the classical monetarists did not treat adequately: *By what causal connections will an exogenous change in money supply alter the flows of aggregate demand, production, and employment?* Whose behavioral responses and what market reactions are involved in the transmission mechanism?

Assume an economy in an initial equilibrium in markets for goods and services, labor, flows of capital funds, and in sectoral balance sheets. Now let the monetary authority increase money supply by open-market purchases of government securities. The bond dealers who sell their securities to the Federal Reserve System now have a more liquid portfolio; presumably they desire this because of the favorable price (lower rate of return) on the securities they sold. When the bond dealers deposit the Federal Reserve's checks received from the sale of their bonds, the excess reserves of commercial banks increase. *The commercial banks are thrown into disequilibrium*, because they then hold excess reserves which would permit an increase in profits. The banks increase loans and purchases of securities in order to raise earning assets to an optimal level in relation to reserve assets. In so doing, they lower loan-interest rates and drive up prices of securities. The initial recipients of deposits, obtained by borrowing and sale of securities, may well search out other assets whose prices have not yet risen (rates of return declined) and buy them for their portfolios with those deposits. So the initial stimulus spreads throughout financial markets, raising prices and lowering rates of return on all assets which are held in portfolios along with money. *The monetary authority, by increasing* M_s, *lowers the rate of return on money, and optimizing behavior of wealth holders leads to lower rates of return (higher prices) for other financial assets.*

The adjustments to portfolios of financial assets will extend to durable goods also. The rate of return on durable goods comes to exceed the lowered interest rate on borrowed funds and the rate of return on financial assets. So it becomes rational to borrow funds to buy durable goods instead of buying (renting) their services, and it becomes profitable to sell some bonds and stocks to acquire durables. Thus, *some of the funds provided to the nonbank public when commercial banks increase their loans and investments will provide purchasing power to buyers of consumers' and producers' durable goods.* This is a stimulus to aggregate demand for current output, and it increases GNP and employment.

The balance-sheet changes originating from a change in money supply may affect demand for current output in still another way. Holders of bonds and equity claims issued before the rise in money supply experience a capital gain from the increase in prices of those assets as market rates of return decline. Assuming little change in product prices during the initial portfolio adjustment, the increase appears to the asset holders as an increase in real wealth, and they may well decide to save less and spend more out of current income. However, this gain to asset holders is largely matched by the loss to the issuers of the old securities. The decline in r raises the present value of the stream of future payments which issuers are committed to paying. The increased value of their liabilities may lead debtors to save more and reduce purchases as a proportion of current income. The net effect on spending out of current income would seem to be small and indeterminate, except for the nature of some debtors. The spending of the federal government is not likely to be reduced because of such changes in present value of their debt service obligations. And one wonders whether business managers will cut back, since the change does not appear in their balance sheets or operating statements. *The net effect of an increase in money supply on consumption and saving will probably be to shift the consumption function up and the saving function down.*

Another impact of interest rates on consumption comes from substitution effects between current and future purchases. The current cost of a future purchase is $PV = (P_Q Q)/(1 + i)^n$ if the purchase is to be made n years in the future. When interest rate i declines, the present cost of the future purchases increases, with P_Q assumed constant through time, because more funds must be set aside now to amount to $P_Q Q$ after n years. Given the consumers' time preference for present rather than future consumption, such a rise in the relative cost of future purchases will raise present consumption out of current income. As we have noted, durable goods buying will be stimulated particularly because they provide the opportunity to purchase future services now. But other forms of saving will be discouraged by the higher prices (or lower rates of return) for financial assets.

In summary, changes in money supply influence markets and aggregate demand in three principal ways:

1   Beginning with Federal Reserve purchases of government securities and with commercial bank purchases of securities and increase in loans, wealth holders readjust their portfolios of assets in response to the excess of money supply over demand. Prices of bonds and equity claims rise; rates of return and interest rates decline.

2   As a part of the balance-sheet adjustment process, demands for consumers' and producers' durable goods increase, because their rates of return exceed the lowered interest rates.

3   The lower rates of return on assets lead to an increase in the proportion of income consumed, which may be interpreted either as a wealth effect or as a rise in relative prices of future goods as compared to presently purchased goods.

These three effects are consistent with constant prices for goods and services. If product prices rise as a consequence of the increases in aggregate demand, the capital gains components of asset prices will be affected, and the differential effects on different assets will lead to further portfolio adjustments and changes in asset prices. These inflationary aspects of the process will not be traced out here. The important conclusions for policy considerations are that *changes in money supply will alter rates of return, demand for durable goods, and the consumption function.* They can lead to movement along an IS line as the interest rate changes and also to a shift of the IS line as the consumption and investment demand curves shift. The interest rate which prevails in the new equilibrium depends on the magnitude of the shift of the IS line.

An initial decline in interest rate caused by a rise in M_s may be followed by stability or even a rise as the IS line shifts upward. Some monetarists identify three phases in the movements of interest rates in response to an increase in money supply. First comes the *liquidity effect*, an initial decline in rates caused by increased M_s before incomes, prices, or money demand have changed appreciably. Second comes the *income effect*, a rise in interest rates accompanying rising incomes and demand for money. And finally comes the *price expectations effect*, a further rise in rates occurring when anticipation of inflation causes lenders to demand and borrowers willingly to pay an inflationary premium on borrowed funds.

The product price level remains indeterminate. *The increase in* M_s *leads to a rise of current-dollar aggregate demand, but the division of the rise between price change and increase in real output depends on aggregate supply conditions.*

It should be noted that *measured velocity will depart from the desired long-term level during the transition to a new equilibrium.* Measured velocity is the ratio of GNP to money supply for the given period, that is, V equals Y/M_s, or M_s equals Y/V. Recalling that in equilibrium the desired velocity relates demand for money to aggregate *permanent* income, that is, M^* equals $\overline{Y}/V^*$, we derive the relation

$$M_s - M^* = \frac{Y}{V} - \frac{\overline{Y}}{V^*}$$

If M_s is increased and M^* rises slowly, the right side of the equation initially becomes large and positive and then declines to zero as equilibrium is reestablished. The early impact may be a sharp decline in V, since Y rises with a lag. Later Y rises and V could move upward as the excess money supply is reduced. Near the end of the adjustment, $\overline{Y}$ will also rise and help close the gap.

SUPPORT FOR THE MONETARIST VIEW

Adherents of the monetarist position support their views by empirical evidence regarding the importance of monetary influences on economic activity, by criticisms of the usefulness of fiscal policy, and by prescriptions for monetary policies.

EMPIRICAL EVIDENCE

As for empirical evidence, the historical records of many countries exhibit many instances in which increases in money supply were associated with strong aggregate demand and inflations. Such conditions existed, for example, in sixteenth-century Europe following the importation of gold from the New World, in Australia and the United States following gold discoveries in the 1840s, in Germany, Austria, and Russia after World War I, in South American countries before and after World War II, and in the United States following the doubling of money supply during World War II. Conversely, there are many examples in which restrictions on the growth of money supply accompanied the reversal of inflationary trends or movement into recession or depression, for example, the contraction of money supply in the United States from 1929 to 1933 and again in 1938, and the monetary reform in Germany after World War II. The clear evidence of changes in the rate of rise of money supply in advance of peaks and troughs of United States business cycles has led to the designation of this series as a leading indicator by the National Bureau of Economic Research. Friedman summarizes:[†] *"Every severe contraction has been accompanied by an absolute decline in the stock of money, and the severity of the contraction has been in roughly the same order as the size of the decline in the stock of money."* (Italics added.)

Of course, the usefulness of a variable as a policy instrument depends on its controllability by the authorities, its causal impact on target variables, and its freedom from unwanted side effects. Money supply is technically controllable by the Federal Reserve System. As to its causal impact, some critics of the monetarist viewpoint assert that borrowing from banks increases when aggregate demand rises, and decreases when demand weakens—so that the observed correspondence between movements of money stock and national product arises because demand for capital funds by the nonbank public increases the money stock rather than vice versa. Total private borrowing is used by the National Bureau of

---

[†] Milton Friedman, "Quantity Theory," *International Encyclopedia of the Social Sciences*, vol. 10, pp. 443–444, 1968.

Economic Research as a leading indicator in its classification of indicators also! In rebuttal, Friedman indicates that, on the basis of his extensive studies:† *"Although changes in the rate of growth of the stock of money have to some extent reflected the contemporaneous course of business, on many occasions they have quite clearly been the result of independent forces, such as the deliberate decisions of monetary authorities.* The clearest examples are probably the wartime increases and the decreases from 1920 to 1921, 1929 to 1933, and 1937 to 1938." (Italics added.)

It should be noted that *the correspondence between money supply and income found in the historical record is a relation between money stock and current-dollar GNP.* The correspondence is best over long periods of time, since the measured velocity ($V = Y/M_s$) is quite variable in the course of a business cycle. For long-term movements, it seems generally accepted that real demand factors determine the growth of real output Y_R, so that the growth in money stock correlates well with the aggregate price level P_Y, allowance being made for the secular trend in velocity in the relation

$$M_S V = P_Y Y_R$$

In short-term fluctuations there may well be more impact on Y_R, but the variability of time lags for the influence of M_s on Y (reflected in the variability of V) makes it difficult to establish reliable relations between M_s and other economic variables in the short run. As Friedman and Schwartz describe the situation:‡

> The case for a monetary explanation is not nearly so strong for the minor U.S. economic fluctuations that we have classified as mild depression cycles as the case is for the major economic fluctuations. Indeed, if the evidence we had were solely for the minor movements, it seems to us most unlikely that we could rule out—or even assign a probability much lower than 50 percent to—the possibility that the close relation between money and business reflected primarily the influence of business on money.

Nevertheless, they believe that since "money plays an independent role in major movements," it is not "likely to be almost passive in minor movements." To hold otherwise would imply that the policy actions of the Federal Reserve Board could be made an endogenous variable in an economic model, or that changes in money supply would not affect aggregate demand in any way.

POLICY IMPLICATIONS

The uncertainty regarding the timing and magnitude of the effects of monetary policy in the short run leads some monetarists to downgrade the importance of monetary policy for short-run stabilization of the economy. Indeed, given the

---

† Ibid.
‡ Milton Friedman and Anna J. Schwartz, "Money and Business Cycles," *Review of Economics and Statistics*, Supplement, February 1963.

inadequacies of economic forecasting, they foresee grave danger that monetary policies to stabilize the economy at one phase of the business cycle might well come to fruition in a later phase of the cycle when their impact would be destabilizing. As a consequence, the dominant policy recommendation emerging from monetarist theory looks to stabilization of long-term growth of the economy. It emphasizes controlled growth of money stock as its intermediate target, rather than interest rates, investment, or government spending. *Money stock should grow steadily at close to the rate of advance of full-employment real GNP, after allowance for the trend rate of change in desired velocity.* If the wage-price determining mechanism of the economy were such as to introduce a secular trend to product prices, some allowance for this might be made in determining optimal growth in money supply. Thus, if full-employment real GNP rises at $4\frac{1}{2}$ percent per year, if desired velocity rises 2 percent per year, and if a secular price rise of $2\frac{1}{2}$ percent per year were acceptable to the government, a steady growth in money supply at about 5 percent per year would be indicated. Such a policy would provide automatic stabilization because increases in aggregate demand faster than increases in $M_s V^*$ would lead to restrictive credit conditions and rising interest rates, whereas a lagging economy would provoke "easy money" conditions.

Monetarists believe that stable growth of the economy can be achieved more readily by controlled growth of M_s *than by fiscal policy measures aimed at keeping aggregate demand growing in step with rising full-employment GNP.* This belief is based on at least three propositions: (1) The policy parameter V^* on which monetary policy is hinged is more stable and predictable than is the multiplier parameter k which serves as the fulcrum for fiscal policy; (2) the impact of fiscal policy measures becomes unclear and uncertain if proper allowance is made for the impact on private spending of the government's mode of financing added expenditures; and (3) the internal time lags and controllability of monetary policy are much more favorable.

1   The first proposition is an empirical assertion which should be testable if adequately specified. One problem arises in measuring V^*, since it is a desired, long-run coefficient relating permanent income to desired money stock, that is, V^* equal $\overline{Y}/M^*$. Current-year values of Y and M_s are not appropriate values for numerator or denominator. Problems arise in measuring the multiplier k also. By definition, k equals $\Delta Y/\Delta Y_{\text{AUT}}^*$. Time lags between desired and actual expenditures and between changes in autonomous and induced demand make observed values of ΔY and ΔY_{AUT}^* differ from the values required to test the theory.

There are also the difficult problems of deciding what expenditure components to include in autonomous expenditures and what assets to include in the definition of money. Also, for both V^* and k, monetary and Keynesian economists do not assert constancy of the parameters over time, but maintain that they are stable functions of a few variables, for example, interest rates and inflation rate in the case of V^* and tax rates in the case of k.

2   The second proposition points out that many fiscal policy models neglect to take account of side effects of changes in government spending, taxing, and

transfer payments—side effects which depend on the concomitant government actions in capital-funds markets. For example, if an increase in government expenditures leads to deficits financed by borrowing from the central bank, the upward shift of the IS line should be accompanied by an upward shift of the LM line as money supply increases. Depending of the relative shifts of the two lines, interest rates may rise or fall, with repercussions on private spending for durables (movement along the new IS line).

To isolate the effects of fiscal policy from those of monetary changes, one needs to assume that a change in government spending is financed by an equal offsetting change either in taxes or in government securities outstanding—so that money supply stays constant. If increased government purchases are financed by an equal increase in taxes, then the IS line will shift upward not by the full amount of G^* but by a lesser amount dependent on the decline in private purchases occasioned by the reduction of after-tax income. In the absence of effects from tax-induced income redistribution, it is usually hypothesized that, if ΔG^* equals ΔT_n, after-tax incomes and expenditures of the private sectors will stay constant. Then ΔY equals ΔG^* initially. With money supply constant, however, this rise in GNP would raise interest rates and cause a decline in the private expenditures which are sensitive to interest rates and to the decline in value of private securities caused by the rise in i. The first of these effects is caught in the usual IS-LM representation; the second is not and would amount to a downward shift of the IS line.

The case of *financing increased government purchases by sales of government securities to the nonbank public* is implicitly assumed in the usual treatment of fiscal policy in the IS-LM framework. The IS line shifts up by ΔG^*; the LM line is fixed. Interest rate rises, leading to a reduction of interest-sensitive expenditures by private sectors (a leftward movement along the new IS line). Some monetarists assume an LM line nearly perpendicular to the Y-axis, so that the rise of interest rates reduces private expenditures enough to offset the rise in government purchases—a full "crowding out" of private by government spending. Others bring in wealth effects on private spending, so that the full array of influences includes:

1   Reduction of spending on consumer and producer durables caused by higher interest rates
2   Reduction of all consumer spending because higher interest rates raise saving as a proportion of disposable income
3   Rise in consumer spending caused by an increase in holdings of newly issued government bonds
4   Decline in consumer spending because increased interest rates lower the market value of holdings of "old" government bonds and other securities
5   Decline in consumer spending because consumers anticipate added tax liabilities to pay interest on the new government bonds[†]

[†] Aside from income redistribution effects, it would seem that this should be offset by the fact that the bond holders receive the interest payments back from the government.

In addition, some suggest that the increased government spending may upset "confidence" in the private sectors and reduce their investment spending or increase their demand for liquid assets (money). Many point out that this "crowding out" of private expenditures will probably be more complete when the economy is operating near full employment. Then inflation sets in, interest rates rise more to include an inflationary premium, and the government is able to maintain its desired real purchases better than is the private sector. On the whole, monetarists conclude that fiscal policy is a rather weak reed to depend on in attempts to control economic activity.

3   The third proposition regarding the inadequacies of fiscal policy points to the perverse, uninformed, and tardy reactions of popularly elected legislators. They are slow to recognize the need for policy actions; they do not comprehend the relation of policy instruments to performance of the economy; and they are frequently guided more by selfish interests or pressure groups than by the informed counsel of impartial experts. For example, consider the nearly two years' delay in increasing taxes (from 1966 to 1968) when inflationary momentum was building up during the Vietnam war. By contrast, the monetary authority is largely independent of political pressures and has an expert staff to detect and analyze economic problems quickly. Action by the Federal Reserve Board of Governors and the Open-Market Committee can follow quickly. The monetary base (currency and commercial bank reserves) can be altered in short order, and the money supply soon thereafter. To be sure, the external lags are long and uncertain, but the ultimate outcome of monetary policy is assured—at least in terms of current-dollar GNP. If the economy can be kept on track close to full-employment output, most of the rise in GNP, it is hoped, will be in real output rather than in price rises.

Friedman summarizes the policy implications of modern monetarist theory thus:† *"Acceptance of the quantity theory clearly means that the stock of money is a key variable in policies directed at the control of the level of prices or of money income.* Inflation can be prevented if and only if the stock of money per unit of output can be kept from increasing appreciably. Deflation can be prevented if and only if the stock of money per unit of output can be kept from decreasing appreciably." (Italics added.)

CRITIQUE OF THE MONETARIST MODEL

The renewed interest in monetary phenomena has contributed to economic theory, to empirical understanding, and to policy decisions, as discussed in the preceding section. Nevertheless, as might be expected, nonmonetarist economists have not been reticent in expressing their reservations regarding the renewed emphasis on monetary economics. Granting that major changes in money supply have been an important and exogenous influence on business activity at some times, many

---

† Milton Friedman, "Quantity Theory," *International Encyclopedia of the Social Sciences*, vol. 10, p. 445, 1968.

economists object to the overwhelming and ubiquitous influence of monetary causation which some monetarists assert. Surely, surges of government and private demands in real terms have caused business expansions which propel the economy beyond full-employment output and into inflation, for example, in wartime or when major innovations stimulate investment spending. To say that monetary measures can restrain or abet the expansion and inflation is quite different from saying that these measures have caused them. Also, there are maladjustments between desired and actual stocks of inventories or fixed capital goods which have their impacts on rates of return and on flows of goods to adjust those stocks (accelerator investment). It is hard to believe that these maladjustments arise solely from monetary causes or that monetary policy can prevent the real consequences. And it is doubtful whether monetary factors can be blamed for the cost-push inflationary pressures our economy has experienced during and since the Vietnam war—or that practicable monetary measures can control them.

In a similar vein, monetarists are charged with claiming too much (or too little?) when they assert the impotence of fiscal policy unless it is accompanied by increase in money supply. Theoretical considerations and historical experience suggest that tax cuts can stimulate aggregate demand and that increases in government spending financed by borrowing from the public are expansive forces. Again, reverse causation—the influence of aggregate demand on money supply—is probably a larger factor than monetarists often admit, though the interactions are certainly difficult to disentangle. Apart from theoretical possibilities, two economists at the Federal Reserve Bank of St. Louis,† using regression analysis, found that changes in Federal Reserve holdings of government securities, and hence changes in the monetary base, were influenced, in descending order of importance, by (1) market rates of interest, (2) changes in the amount of United States government debt outstanding, and (3) economic stabilization objectives. Thus money supply responds to credit conditions, to federal government deficit financing, to unemployment, to inflationary pressures, and to balance-of-payments problems. It is not entirely exogenous.

Though other economists do not refute the monetarist theory dealing with the impact of changes in M_s on rates of return for various assets and on the consequent balance-sheet adjustments, they do question the practical importance of these effects. In the first place, *the purchases and sales of assets involved in the portfolio adjustment are transient flows, existing only during the disequilibrium state and vanishing in the new equilibrium.* The durable goods purchases stimulated by this adjustment process are quite analogous to accelerator-investment flows— perhaps important but not permanent. And, like accelerator-investment expenditures, they are contingent on expectations that the change in rates of return will not be reversed during the life of the asset. The permanent equilibrium flows of aggregate demand arising from monetary stimuli stem from the shifts of the

---

† Michael Keran and Christopher Babb, "An Explanation of Federal Reserve Actions," *Federal Reserve Bank of St. Louis Review*, July 1969, pp. 7–20.

consumption function induced by the wealth effect and from changes in interest-sensitive durable goods expenditures. *Nonmonetarist economists remain agnostic regarding the practical importance of wealth effects on propensities to save and to spend, at least within the range of wealth changes which can be brought about by a reasonable change of money supply.* The interest-rate effects on investment were, of course, part of the Keynesian model.

Economists interested in short-term forecasting find the monetarist model of limited use. As Friedman says, the long and variable lags involved in transmission of monetary changes to demand, output, prices, and employment vitiate the use of money supply as an exogenous variable in forecasting. Attempts have been made to develop distributed-lag models with money supply as an independent variable, but their usefulness is an open question as yet. Another deficiency is that monetary variables, except for interest rates, have been related only to aggregate GNP and price level. Many forecasters are interested in forecasting components of GNP, and *monetary factors have not yet been brought into forecasts of sector demands—again with the exception of interest rates used in investment and housing demand equations*, sometimes also for state and local government expenditures.

Finally, in the area of policy considerations, we have already noted Friedman's reservations regarding the usefulness of monetary measures for short-term stabilization of business fluctuations. The irregular response of the economy to monetary policy actions makes it difficult to quantify that response adequately for policy determination. On a quarterly basis, the velocity factor in $MV = Y$ changes more (in percentage) and accounts for more of the variation in Y than does money supply. Also, some of the side effects of monetary policy may be undesirable at times, for example, the impact on plant and equipment expenditures, residential construction, and short-term international flows of funds. The prospects for short-term stabilization painted by the monetarists seem overly pessimistic. Both monetary and fiscal policy measures have been found wanting; yet our postwar experience suggests that the economy was rather stable in the wake of some major shocks.

In summary, we might conclude that the monetarists have made a plausible case for modification of the money demand function and perhaps of the consumption and investment demand functions in our macroeconomic model, though the forms, coefficients, and time lags for the monetarist functions need to be tested empirically. Perhaps the Keynesian and monetarist theories are passing into a phase of consolidation and empirical testing to distinguish and validate hypotheses.

PROBLEMS

1 Describe the different assumptions underlying the classical and Keynesian theories regarding demand for money stocks, and point out the consequences of those assumptions with respect to the efficacy of monetary policy and fiscal policy in controlling aggregate demand.

2 Describe the different assumptions with regard to determination of the wage rate in the classical and Keynesian models. Discuss the consequences of these assumptions with respect to the nature of the aggregate supply curve and the determination of equilibrium GNP.

3 Describe the nature of the "transmission mechanism" by which the influence of a change in money supply is transmitted to aggregate demand in the modern monetarist model.

4 Consider a growing economy in which labor force and labor productivity increase steadily through time. For three years, at successive stages of growth, sketch the labor demand and supply curves, the production functions, and the aggregate supply curves—all on the assumption that full employment of labor is maintained at a constant price level. Label equilibrium points 0, α, and β in each panel. Can money wage rates change without causing a price rise? Explain.

5 Assume that the government aims at a target of full employment without inflation in a growing economy in which labor force and labor productivity rise continuously, that is, full-employment real GNP increases steadily.

a State and explain briefly a policy rule for controlling the money supply so as to help achieve this target.

b What would be a comparable rule for fiscal policy?

c In what circumstances might stimulation of aggregate demand by monetary policy be inadvisable—because ineffective or harmful?

d In what circumstances might stimulation of aggregate demand by fiscal policy be ineffective or harmful?

Chapter 18

Prosperity, Inflation, and Controls

INTRODUCTION

Recent Gallup polls in several free-world nations asked leading government officials, educators, scientists, corporate executives, economists, bankers, physicians, attorneys this question: "Which of these problems (from a list of 15) do you regard as the five most urgent problems facing your nation?" Based on frequency of mention, the responses indicated that the problem considered most serious was inflation, followed by the problem of unemployment. From such facts emerge the cruelest conflict facing economic policy makers in recent years. *A prosperous economy with low unemployment usually becomes an inflationary economy.*

The dilemma may be illustrated by the Phillips Curve, named for the British economist who discovered evidence in English records that low unemployment clearly is associated with strong increases of wage rates. A comparable relation between unemployment rate and inflation in the United States is shown in Fig. 18-1. Except in 1952 and 1953 when price controls were influential, and except in the years between 1970 and 1974, which we shall discuss later, it appears that *reduction of unemployment below perhaps $4\frac{1}{2}$ percent does result in price rises in excess of 2 percent per year, whereas unemployment rates greater than $4\frac{1}{2}$ percent are accompanied by price rises of less than 2 percent per year.*

In Chap. 17 we noted that the aggregate supply function implicitly relates price levels to real output. So, *for a given production function and given wage rate, the equilibrium of aggregate demand and aggregate supply (in current dollars) jointly determines real output and the price level. Fiscal and monetary policy instruments which bring the economy to full-employment output must accept the price level which exists at the full-employment point on the aggregate supply curve.* Those instruments can influence aggregate demand, but not real GNP and price separately.

This policy dilemma raises such questions as the following: Is it important to control inflation? What are the causes of inflation? How can it be stopped? Are there instruments other than monetary and fiscal policies which will act on inflation separately?

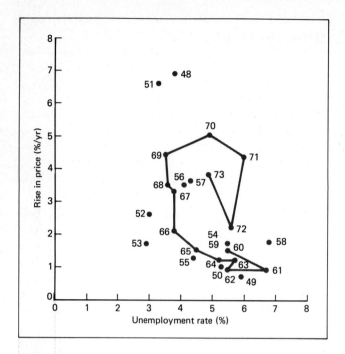

Figure 18-1   Rise in price versus percent unemployment for private
nonfarm GNP

DEFINITION AND MEASUREMENT

Inflation may be defined as *a rising average level of prices for goods and services
currently produced in the economy.* It is a dynamic phenomenon. A gauge of
inflation is *the rate of rise of some general product-price index, measured in percent
per year.* Conversely, deflation is measured by the rate of decline of a general
price index for goods and services.

Note that the absolute level of prices is not an indication of inflation; the
rate of rise of prices is. Note that rising prices for *current output* are involved;
rising prices for land, existing durable goods, bonds, and equity claims do not of
themselves constitute inflation—though they may well accompany it. Also, the
price rises for currently produced goods and services must be widespread, though
not universal—widespread enough to dominate the movements of comprehensive
price indexes for goods and services. Changes in relative prices of various products
may accompany an inflation but are not necessary. Thus inflation may not lead
to substitution of one good for another, as does the change of relative prices
analyzed in microeconomics.

Several general price indexes are employed in measuring inflation. The
consumer price index (CPI) is frequently used. It is a composite index of prices
for a market basket of goods, calculated from indexes for some 400 items weighted

by the relative importance of each category in the consumption of wage and salary earners in urban areas during a recent base year (currently 1967). The market basket comprises both commodities and services, and includes the effects of sales and property taxes but not income taxes. It has been criticized as a measure of inflation: (1) because the failure of its producer (the Bureau of Labor Statistics) to make adequate allowance for quality improvements over time gives the index an inflationary bias of perhaps 1 to $1\frac{1}{2}$ percent per year; and (2) because the index is restricted to finished consumer goods and omits price changes for business capital goods, for government purchases, and for intermediate goods and services.

The *wholesale price index* (WPI) of the Bureau of Labor Statistics is another frequently used gauge of inflation. It is based on producers' quotes for some 2,300 items priced at the level of the first significant commercial transaction. It combines prices of raw materials and intermediate goods as well as final goods, using weights proportional to the value of "shipments for sale" of each item as recorded in a recent base-year industrial census. The wholesale price index also has shortcomings: (1) It relies largely on producers' list prices and does not reflect accurately market discounts during periods of price weakness or market premiums and changes of product mix in periods of tight supply; and (2) it does not cover prices of services, either those provided as intermediate goods or those flowing to final demand sectors.

The *implicit GNP price index* (P_Y) is the most comprehensive in coverage. It is the ratio of current-dollar to constant-dollar GNP. In effect, it is a weighted composite of prices for GNP components, with the weights reflecting the up-to-date, current-year expenditures for the various components. This price index shows two deficiencies: (1) It could change even if there were no price changes of the components, because of shifts of current-year expenditure weights for various components; and (2) it includes incorrect price indexes for government output and, to a lesser extent, for the construction components of GNP, because inadequacies of output (and hence productivity) measures in these sectors make their price indexes reflect prices of inputs rather than outputs.

The *private GNP price index* covers all GNP components except compensation of government employees. Hence it avoids the problem of trying to assign a price index to government output, where market prices are nonexistent and productivity measurements are hard to make. However, it too is a current-weighted composite index which reflects changes in the mix of outputs as well as true price changes of the components, and it does include components, such as services and construction, where changes in quality and productivity are hard to measure. Sometimes the private *nonfarm* GNP price index is used, since it reduces the influence of erratic fluctuations in farm prices and permits more meaningful analyses of the effects of wage rates, man-hours, and productivity on inflation.

Finally, it should be noted that *the rates of rise of price indexes are not always adequate measures of inflationary pressures in the economy*. Aggregate demand may exceed full-employment supply during wartime or in other strong expansions, but price rises may be prevented or reduced by price controls, rationing, allocations of materials, or forced loans to the government.   During such periods of *suppressed inflation*, the government attempts to restrain the operation of private decision making in free markets and to impose some other mechanism than the price system for determining output flows and allocating the output among demand sectors. The costs of a suppressed inflation are very high—they include gray and black markets with their strains on personal morality and enforcement efforts, inefficient allocation of materials and of final goods, the development of a backlog of demand which may generate a virulent inflation when controls are removed, and a divisive struggle over shares of output which may lower productivity and threaten social stability.   Most economists would prefer an *open inflation* to the suppressed variety, and as little as possible in either case.

CAUSES OF INFLATION

Many causes of inflation have been suggested by different analysts.

1   Excess of aggregate demand (in current dollars) over the value of full-employment output at present prices
2   Expansive monetary policy—"too much money chasing too few goods" or too rapid an increase of the money supply relative to the rise of real output of goods and services
3   Government deficit spending, or expansionary fiscal policies
4   Faster increases in labor compensation (wage rates plus fringe benefits) than in labor productivity
5   Administered pricing practices in concentrated industries
6   Expectations of inflation—with the consequent speculative buying and struggle over shares of the national income
7   Reduction of supply—in wartime, strikes, or natural disasters.

Let us examine these theories in relation to what has happened in the United States in the postwar years, especially since 1963.

EXCESS DEMAND INFLATION

Figure 18-2 shows how the full-employment GNP of the economy, on a real (constant-price) basis, has risen steadily during the postwar years with the rise in labor force and in output per man-hour.   The solid line shows how actual constant-price GNP, which may be taken as real aggregate demand, has fluctuated from well below to just slightly above the full-employment output of the economy.   The four periods when actual output rose to full-employment levels correspond to four periods of inflation in the United States—1948 (the end of the World War II inflation), from 1951 to 1953 (the Korean war boom), 1956 and 1957 (the surge of consumer and producer durable goods buying), and between 1966 and 1969 (the Vietnam war boom).

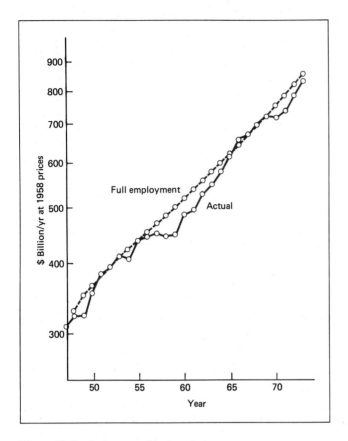

Figure 18-2 Real gross national product.

Figure 18-3 shows the correspondence between the four periods of postwar inflation and the four periods when GNP was at or near its full-employment (potential) values. The top panel plots the percent rise from the previous year for the GNP price index. The four periods of inflation show up clearly. The solid line in the bottom panel plots the percent ratio of actual to potential (full-employment) GNP. As indicated by the left-hand scale, the economy dropped to 92 percent of potential in the recessions of 1949, 1958, and 1961; it rose to, or slightly exceeded, potential in the prosperous years 1951 through 1953, in 1955, and in 1966 through 1968. The dotted line in the bottom panel presents the level of unemployment as a percentage of the civilian labor force, and is read against the inverted scale at the right. In general, it traces out the same periods of prosperity and recession but indicates that the labor market was tighter (unemployment less) than the "GNP as a percent of potential" curve would suggest in 1948, 1951 through 1953, 1956 and 1957, and 1969 and 1970. Except for 1952 and 1953, when price and wage

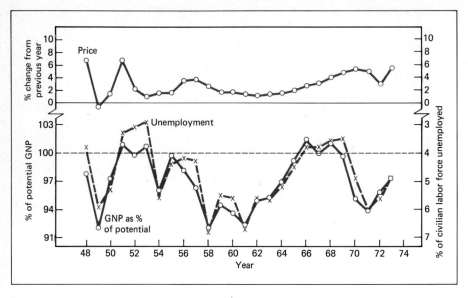

Figure 18-3   GNP price change and related factors.

controls were in effect, price rises seem to correlate more closely with unemployment than with the "GNP as a percent of potential" curve.   In 1970 and 1971 and in 1973 and 1974, inflation continued at a higher rate than would have been expected from either of the indicators in the bottom panel.   But usually there does seem to be a significant correspondence between high levels of demand (or employment) and high rates of inflation.   *Inflationary periods (with the GNP price rising faster than 2½ percent per year) have occurred in all postwar expansions except the brief recovery from 1958 to 1960, and inflation subsided in all business-cycle recessions except that in 1970, following four years of inflation.*

MONEY SUPPLY AND INFLATION
Figure 18-4 relates the rise in money supply (top) to inflationary episodes in the economy (middle panel).   Both series are measured as the percentage change from the previous year.   There is some rough correspondence between movements of the two curves, but it is not close.   Rise in money supply was slow during the inflations of 1948 and 1956–1957 and was moderate from 1964 to 1967, while the most recent inflationary movement was building up.   The pattern of movement of the two series was inverted between 1968 and 1972.   Also, the increased rate of expansion of the money supply in years following recessions does not seem to give rise to inflation; see 1950, 1955, 1959, and 1962.   It may be that the effects of rapid rise of the money supply appear in aggregate demand (or real GNP as a percent of potential GNP), or that they appear with variable time lags, or are

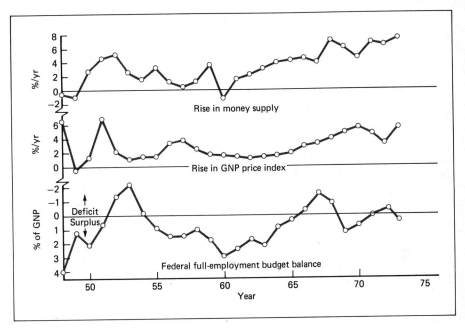

Figure 18-4   Monetary and fiscal policy related to inflation.

partially masked by the influence of more powerful forces making for inflation, for example, unemployment.   At any rate, the relation of change in money supply to inflation does not stand out so clearly as does that of the two factors graphed previously, though it might prove significant in a multivariable analysis.

FISCAL POLICY AND INFLATION

The inflationary impact of fiscal policies is often attributed to government spending in excess of its revenues, that is, to deficit spending, whether it arises from increased spending or reduced taxes, and however the deficit may be financed.   It is clear, however, that deficits sometimes arise from decreased revenues during a recession and that they should be considered counterdeflationary rather than inflationary at such times.   A better measure of the economic thrust of fiscal actions is the full-employment budget imbalance, that is, the surplus or deficit which would emerge with current government purchases held constant but with net revenues adjusted to the level that would occur at full-employment GNP.

In the bottom panel of Fig. 18-4 is a line showing the full-employment position of the federal budget during the postwar years, with surplus or deficit expressed as a percentage of GNP.   The scale is inverted because a surplus is considered deflationary and a deficit inflationary.   There does not seem to be a close, dependable relation between the full-employment budget situation and the onset or persistence of inflation.   The budget was moving from surplus toward deficit in 1948 and in

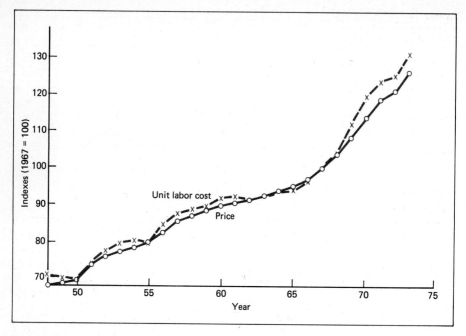

Figure 18-5   Private nonfarm GNP: price and unit labor cost.

1951, and hence may have contributed to inflationary pressures, but it moved further toward a deficit position in 1949 and 1952, when inflation subsided.   The fiscal posture was one of restraint (surplus) during the inflation of 1956 and 1957. The swing from a surplus in 1963 to near-balance in 1965 and then to a deficit for 1966 through 1968 may well have contributed to the inflationary build-up in the latter years.   However, the surplus of 1969 and 1970 did not reverse the price rise, and the slower inflation of 1971 and 1972 was accompanied by a neutral fiscal policy.   It seems clear, then, that *the state of the federal full-employment budget is not a dominant force in determining the existence of inflation.*   As with monetary policy, the influence of fiscal policy may need to be examined in a multivariate analysis along with other causal variables, and it may work through its effects on aggregate demand.

INCREASES IN UNIT LABOR COSTS

From 1970 through 1974, inflation continued even though aggregate demand was well below the value of full-employment output and unemployment was high. Under such conditions, price rises would seem to be motivated by cost pressures, notably labor costs.   The next three charts present measures of labor-cost pressures and analyze their sources and effects.   Figure 18-5 shows that for private nonfarm GNP, where labor productivity can be meaningfully measured, indexes of

TABLE 18-1

REVENUE = LABOR COSTS + COST OF MATERIALS AND SERVICES
        + DEPRECIATION + INTEREST + TAXES + PROFITS

$$R = P \times Q = LC + CMS + D + INT + T + PR$$

$$\frac{R}{Q} = P = \frac{LC}{Q} + \frac{CMS}{Q} + \frac{D}{Q} + \frac{INT}{Q} + \frac{T}{Q} + \frac{PR}{Q}$$

$$\text{UNIT LABOR COST (ULC)} = \frac{LC}{Q} = \frac{W \times MH}{Q} = W \div \left(\frac{Q}{MH}\right)$$

$$= W \div \text{OUTPUT PER MAN-HOUR}$$

prices and of unit labor costs have moved closely together during the postwar period. They must, because labor costs are the major element of costs of national output—about 75 percent. It is noteworthy, however, that in the early stages of postwar inflationary intervals, the price index has risen above the unit labor-cost index (in 1950 and 1951, 1955, and in 1963 through 1965); then unit labor costs have caught up and overshot. The 1967–1970 period of catch-up and overshoot is the most noteworthy of such postwar periods.

What determines unit labor cost, and why does it influence prices so strongly? Table 18-1 expresses in equation form the allocation of a firm's revenue to various costs, to taxes, and to profits after taxes. Since revenue may be expressed as the product of price times physical quantity of output (Q), it is possible to divide all terms by Q and obtain an equation expressing price P as the sum of the cost, tax, and profit items *per unit of output*. The dominant term is the first one: unit labor cost (ULC). Unit labor cost may be expressed as a ratio of hourly rates of compensation (W) divided by output per man-hour (Q/MH). Thus, *if hourly rates of compensation rise only in the same proportion as output per man-hour (labor productivity), their ratio (ULC) will stay constant*. If ULC is constant, on the right-hand side of the price equation, then labor costs do not exert much upward pressure on prices. However, if W rises faster in percentage than output per man-hour does, the ULC term on the right increases, and either some other term on the right side must decrease (notably, profits per unit of output) or prices must rise. This is the situation economists refer to as one of *labor-cost pressure* on prices.

Figure 18-6 shows that year-to-year percentage changes in prices have fluctuated quite closely in step with unit labor costs for private nonfarm GNP. ULC is more volatile, but over the whole of the postwar period the series have moved closely in step on a cumulative basis. The lower half of the chart shows the annual percentage changes of the two factors entering into the ULC ratio. To a close approximation, the vertical gap from the output per man-hour curve to the compensation per man-hour curve equals the height of the ULC curve in the top panel. Note how, from 1966 to 1970, both the high growth in compensation per man-hour and the low gains in productivity contributed to the high rates of rise in ULC, and hence to cost-push inflationary pressures. In 1971 a higher gain in productivity narrowed the gap and led to a slower rise in ULC. Indeed, ULC rose

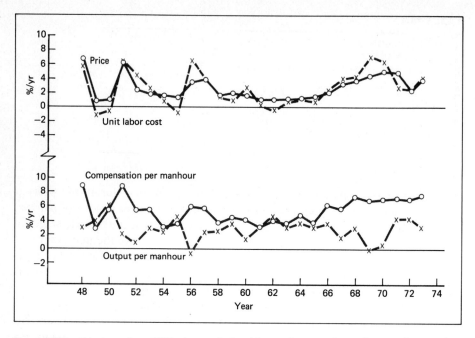

Figure 18-6   Private nonfarm GNP prices and related factors (percent change from previous year).

less than prices; so there was some easing of inflationary pressures.   It was hoped that continued high unemployment plus Pay Board restraints would lower the rate of rise in compensation per man-hour in 1972 and that a strong rise in output would lead to further gains in labor productivity.   In the event, ULC did rise more slowly in 1972.   This, coupled with restraints by the Price Commission, helped lead to a considerably slower price rise.

ADMINISTERED PRICING PRACTICES
As may be noted from the preceding charts, there has been no full year of the post-war period when comprehensive price indexes declined, with the exception of 1949 for some indexes.   Even allowing for some upward bias of the price measures, this suggests that producers may have the power to keep prices up even during periods of weak demand and recession.   Especially in industries dominated by a few large producers, some economists have hypothesized, company managers have learned to minimize price competition and "administer" prices upward by tacit coordination of policies or by "follow-the-leader" pricing practices.   It is some-times maintained that strong, industrywide unions have abetted this development by negotiating similar contract settlements and placing all firms in an industry under the same cost pressures.   Managers are willing to grant settlements that raise unit labor costs because they feel sure that price rises will permit them to pass the added

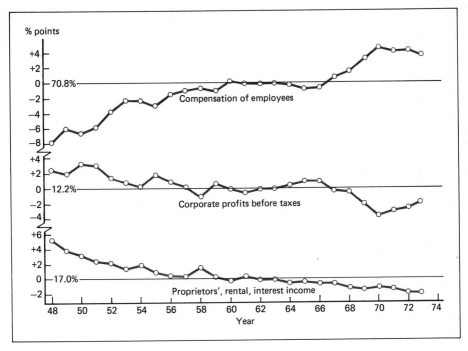

Figure 18-7   Percent shares of national income (before tax).

costs on to customers.   Thus, large firms and large unions are linked in an economic power play against the small, less-organized customers—and prices are ratcheted upward.   It is difficult to develop a clear-cut statistical test which disentangles the working of an administered price mechanism from other forces operating on prices in the postwar era, but some studies indicate that price movements do reflect average costs and "normal markups" as well as competitive marginal-cost causation.

This theory as to the causation of inflation does point out the underlying struggle over shares of national income that is involved.   Table 18-1 showed the cost and profits allocations of revenue of a firm.   If this equation is aggregated over all production units, the term for intermediate goods and services is netted out, and we obtain an equation for the allocation of gross national income on a "before income tax" basis:

GNY *equals* [compensation of employees *plus* proprietors', rental, and interest incomes *plus* corporate profits] *plus* depreciation charges *plus* indirect taxes

The terms in brackets constitute *national income*, or payments for services of labor, capital, and natural resources used in production.   The percentage breakdown of national income during the postwar years is shown in Fig. 18-7.   The year 1963 has been taken as a base year, and the percentage for each component in 1963 is shown on a base line at the left of the chart.   Annual deviations for each component from its 1963 percentage have been plotted for the years 1948 through 1973.†

Typically, we observe a shift to profits at the expense of labor compensation just before, or early in, an inflationary period—a typical response of the system to rise in demand and output flows with wages stable or rising at "normal" rates. Later in the inflationary interval, rises in compensation rates outstrip productivity rises plus the rise in prices, and the labor share increases at the expense of profits. Such a shift marks off the interval of cost-push inflationary pressures.   From 1967 through 1969 we experienced both excess demand and a rapid rise in wages which drove labor costs per unit of output upward.   This was a period of both demand and cost pressures on prices.   For 1970 through 1974, fiscal and monetary policies had reduced aggregate demand below full-employment levels, but the inflation continued, largely because of cost pressures.   It seems clear that labor has gained ground in the tug-of-war over shares of national income—at least on a pretext basis.

Figure 18-8 shows what happened to income shares of the four demand sectors on an after-tax basis.   The net income flow to each sector equals income receipts from production minus income tax payments plus transfers (gifts) from other sectors.   Business gross retained earnings equals the flow of capital-consumption allowances plus retained (undistributed) corporate profits, that is, the flow of "internal funds" from current operations.   In 1948, net disposable personal income was very high following postwar tax cuts.   After the Korean war started in mid-1950, income tax rates were raised very quickly, as is indicated by the shift of income to the Government Sector.   Presumably this shift, together with price and wage controls, limited the inflation to a single year, though demand pressures persisted in 1952 and 1953.   No marked shifts in after-tax incomes were observed during the inflationary interval of 1956 and 1957.   In the early stages of the Vietnam war, income tax rates were not raised; large government deficits emerged in 1967 and the first half of 1968.   Beginning in late 1965 aggregate demand exceeded full-employment output valued at stable prices, and our longest postwar inflation was under way.   The 10 percent income surtax in the second half of 1968 and in the full year 1969, along with restrictive monetary policy, brought aggregate demand below potential output in 1970 through 1972.   So tax rates were cut back to stimulate the economy, and the percentage of GNP flowing to the Personal Sector as disposable income rose sharply.   Government spending was reduced moderately, but large federal deficits have emerged.

---

† "Compensation of employees" covers wage and salary payments plus fringe benefits and social security tax payments.   "Corporate profits before taxes" is the aggregation of reported profits plus a correction by the national income accountants to remove capital gains or losses on inventories. "Proprietors', rental, and interest income" is the combined income stream for farm owners, proprietors of unincorporated businesses (doctors, lawyers, dentists, other self-employed individuals), and rental and interest incomes for property owners and financial investors.   This series has a downward trend because of the declining importance of agriculture and the shift toward the corporate form of business enterprise.   Back in 1929 the percentage split was: compensation of employees, 58.9 percent; corporate profits, 12.0 percent; proprietors', rental, and interest incomes, 29.1 percent.

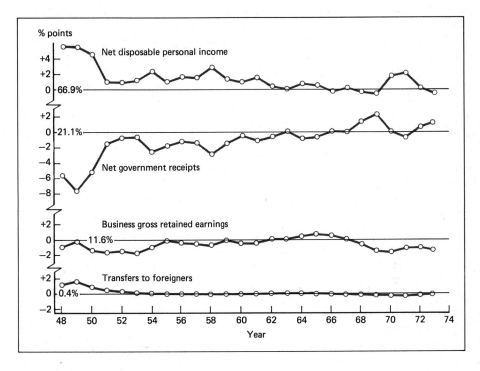

Figure 18-8   Sector after-tax incomes as percent of GNP.

EXPECTATIONS OF INFLATION

Between 1970 and 1974, prices and wages continued to rise at a rapid rate, although aggregate demand was well below the full-employment level and unemployment was in the 5 to 6 percent range.   Wage and price behavior in those years was clearly abnormal, as may be noted in the Phillips Curve diagram of Fig. 18-1 or the time-series chart of Fig. 18-3.

Many economists believe that the persistence of price rises in those years was explainable in large part because the preceding inflation had continued for four years, leading decision makers to believe that it would continue.   *Expectations of continuing inflation provoke behavior which tends to perpetuate the price rises and fulfill the expectations.*

1   Consumers may react by spending an unusually large proportion of disposable income to stock up on goods before they become more expensive. (Actually, this seems not to have occurred on a large scale; personal saving rates were above the postwar average for most of 1970 through 1974.)

2   Workers, unionized or not, will demand wage raises adequate to cover normal productivity increments *plus* expected increases in the cost of living.

3 Employers will be more willing to grant large increases in wages and fringe benefits because they expect to charge higher product prices to cover those costs and because they know their competitors will be faced with higher costs also.

4 Purchasers of all types—consumers, governments, businessmen buying intermediate goods or capital goods—accept price rises with less resistance because they know that costs of production are rising and expect even higher prices later.

Overall, resignation to forces uncontrollable by individual decision makers, plus a struggle to protect one's own (family, firm, union), leads to behavior tending to weaken price competition, to strengthen cost and profit pressures, and to promote realization of the inflationary expectations. Such a situation may develop into a progressively more rapid spiral of prices or may lead to imposition of some type of control of prices, wages, and property incomes in an effort to break the feedback between actual and expected inflation.

REDUCTION OF SUPPLY

On rare occasions, excess demand may arise because of a reduction in the flow of output, for example, because of war damage to productive equipment, a prolonged nationwide strike, widespread crop failures, an oil embargo by producing states, or a major epidemic. Of course, reduction of production cuts off income flows and purchasing power, but transfer payments or use of savings can sustain demand in the face of reduced supply, at least in the short run, and lead to price rises as buyers compete for the available output. Such rare occurrences pose no problem of economic analysis. If aggregate demand (in current dollars) declines less than physical output, higher prices are required to raise the money value of output flow to equality with aggregate demand.

GRAPHICAL ANALYSIS OF INFLATIONARY PROCESSES

These descriptions of postwar inflationary episodes suggest the following typical steps in the causal process.

1 Aggregate demand (in current dollars) increases until it exceeds the value of full-employment output at existing prices. The surge of demand may arise from an expansion of money supply, from a stimulative fiscal policy, from an autonomous rise in investment demand or perhaps consumer durables, or even from a surge of demand for exports.

2 Assuming that wage rates and the production function stay constant temporarily, producers will increase output and prices, moving along the original supply curve beyond the full-employment point. The division of an increment in spending between output and price rises shifts toward the latter as output approaches and then exceeds the full-employment level. This initial phase of the process is called *demand-pull inflation.*

3   As the demand for labor expands with output, employers bid wage rates up in a tightening labor market, and workers demand wage increases because of the prior rise in prices.   Increased wage rates, of course, shift the aggregate supply curve up—assuming a fixed production function and thus no productivity increase to offset the effect of wage rises on unit labor costs.

4   Faced with a constant aggregate demand, producers adapt to their higher costs by raising prices and reducing output.   This secondary rise in prices stemming from the upward shift of the supply curve is called *cost-push inflation.*

5   The way is now open for labor to demand additional wage hikes based on the secondary price rise.   Thus steps 3 and 4 can repeat in unending interaction, though successive increments would become progressively smaller.   The process could converge to a new stable price level at full employment—if aggregate demand stays constant in current dollars.

6   If the above process continues long enough so that inflationary expectations take root, then the aggregate demand curve may shift up again and initiate a new surge of demand inflation.   Or, if some of the demand sectors stubbornly maintain their *real* purchases, increasing their expenditures in step with price rises, then the aggregate demand curve will continually shift upward.   (This is possible in macroeconomics because incomes rise in step with price increases, for given physical output; the total value of production is always distributed as income to demand sectors.)   *The inflation will die out only if the initial real demand (in excess of full-employment output) is brought down to the full-employment level as price rises, income redistribution, changes in interest rates, or automatic stabilizers squeeze real demands in some sectors.*

The inflationary process just described may be visualized by use of the general equilibrium diagrams developed in Chap. 15.   In Fig. 18-9, the economy is assumed to start in equilibrium at full employment at points designated α in the four panels.   Then the inflationary process develops as described.

1   Both fiscal and monetary policies are assumed to become expansionary; the IS and LM lines both shift up, keeping interest rate constant.   The new intersection at β determines an aggregate demand Y_β^* which is greater than the initial values of full-employment supply $Y_{FE,\alpha}$.   The IS and LM lines are assumed to remain fixed thereafter.

2   Aggregate demand Y_β^*, when projected across to panel (*b*), determines a point β' on the original supply curve.   Output and prices at β' are both higher than at α.   Projection of point β' downward to the production function and across to panel (*d*) determines an increased demand for labor at point β', where the constant wage-rate line W_α is cut by the labor demand curve L_β^*.   [The labor demand curve has shifted left from L_α^* in proportion to the price rise accompanying the move from α to β' in panel (*b*).]

3   The strong demand for labor drives unemployment down and wage rates up to $W_{\beta''}$.   (Wages are assumed to rise in the same proportion as prices did in step 2; so the new wage line $W_{\beta''}$ cuts the L_β^* line at point α' directly to the left of α.)

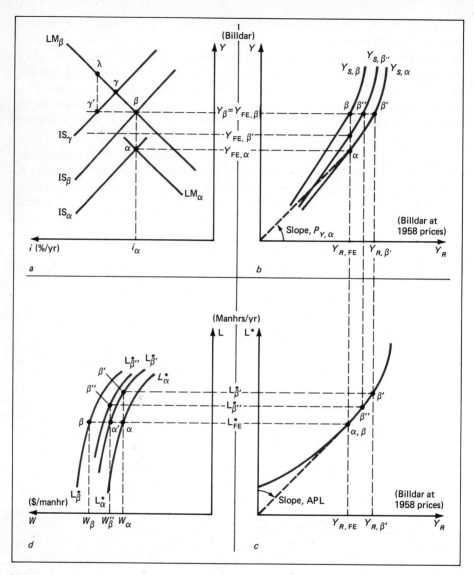

Figure 18-9 The inflationary process. (*a*) IS and LM lines. (*b*) Aggregate supply curve.
(*c*) Production function. (*d*) Labor demand and supply.

This rise in wage rates shifts the supply curve in panel (b) upward by the percentage rise in wage rates—to $Y_{S,\beta''}$.

4   Producers readjust real output downward and prices upward to maximize profits at point β'' where the new supply curve crosses the unchanged level of aggregate demand determined in panel (a).   The new price rise (from β' to β'') shifts the labor demand curve up proportionally in panel (d) and determines a new labor-market equilibrium at β''.

5   Since demand for labor at β'' exceeds the full-employment level, another round of wage increases ensues, and the supply curve shifts up again in panel (b). Steps 3 and 4 are thus repeated with diminishing increments until an equilibrium is finally reached at points β on all diagrams.   Here wage rates and prices have increased in equal proportions, so that the demand for labor and the flow of real output are again at the full-employment levels.   *The percentage increase in prices and wages is equal to the initial percentage rise in aggregate demand*, from α to β in panel (a).

Note that *as prices rise, the current-dollar value of full-employment real output increases in proportion*:

$$Y_{FE} = P_Y \, Y_{R,FE}$$

In the new equilibrium, the excess demand has been removed because price increases shift the horizontal Y_{FE} line in panel (a) upward until it goes through point β.   All the rise in current-dollar aggregate demand from α to β has been absorbed in higher prices; the real output and employment of labor are the same at points β as at α initially.   Also, wage rates, prices, total incomes, aggregate demand, and the value of output have all risen by the same percentage.   The distribution of incomes and of GNP purchases by sectors may be altered, but we are neglecting the impact of these changes on the aggregates.

The diagrams of Fig. 18-9 may be used to elucidate an inflationary process under many other assumed conditions.   For example, starting from full-employment equilibrium at points β, assume that aggregate demands increase (the IS line shifts upward) but that money supply is kept constant (the LM line remains fixed). Then the equilibrium point for product and money markets shifts upward along the LM line.   Interest rates rise, and prices and wages chase each other upward again. The horizontal line for value of full-employment output shifts upward in panel (a) as prices rise, until the three curves intersect again at a new equilibrium.   Two assumptions regarding aggregate demand will illustrate the range of outcomes.

1   Assume that the current-dollar aggregate demand curve shifts a given amount and then remains constant at the higher level.   Then the new IS_γ line determines the new equilibrium point at its intersection with the fixed LM line, say point γ in Fig. 18-9.   The horizontal line for the value of output at full employment drifts upward, as prices rise, until it goes through point γ.

2   Assume that the aggregate demand line shifts up a given amount and then drifts upward further in step with price rises as the final demand sectors attempt to maintain their real purchases in spite of inflation. Graphically, the IS line shifts initially to IS_γ, intersecting the full-employment line at γ'. Then the IS line and the horizontal full-employment line Y_{FE} shift upward together as prices rise. Equilibrium is attained when the intersection point γ' moves upward to intersect the fixed LM line at point λ.

In both these cases the constant money supply led to higher interest rates as inflation progressed. This reduced some components of aggregate demand and brought the total back to the full-employment level—at which price stability is restored. Wage rates, of course, are assumed to rise with some lag during the inflation, but to catch up by the time the new equilibrium is reached.

How could we picture the case of *unrestrained inflation* in this diagram? Two limiting cases may illustrate the range of possibilities.

1   Suppose an initial rise in aggregate demand shifted the IS line to IS_γ, intersecting the LM line at γ. Now assume that aggregate demand and money supply both rise in step with prices. Then in Fig. 18-9, all three lines forming the triangle $\beta\gamma\gamma'$ will drift upward in proportion to price rises. Excess demand will never be eliminated, and inflation will continue indefinitely.

2   Suppose that aggregate demand is not affected by high interest rates in an inflationary situation, so that the IS line is horizontal. Then an initial shift of the horizontal IS line from β to γ, if followed by an upward drift in step with price rises, will keep the IS-LM line intersection moving upward to higher interest rates on the fixed LM line but *always* above the full-employment line Y_{FE}, which is drifting upward at the same rate. Aggregate demand, prices, wages, and interest rates rise continuously, and excess demand persists.

The diagrams could also be used to show *the operation of fiscal policy to control inflation*. Increases in tax rates or reductions in government spending would restrain the upward movement of the IS line—or even shift it down—and offset inflationary impulses stemming from shifts in private demand functions or from monetary policy. If the nonbank public during an inflationary period desires to hold practically no asset balances of money, so that the LM line becomes well-nigh horizontal, the effectiveness of fiscal policy restraints might be eroded by large declines in interest rates when the IS line shifts down. But low interest rates are not likely in a prolonged inflationary period, and so long as money supply is held constant, the inflation will be limited by monetary restraints in any case.

These examples have all dealt with inflationary movements started by an excursion of aggregate demand to an excess over the value of full-employment output. The inflation ended only if some restraints led to a new equilibrium at full-employment output with a higher price level. But it may be that wage negotiations in an inflationary setting, along with below-average productivity gains, result in wage rates and unit labor costs such that producers maximize profits at outputs below full employment. *If, despite aggregate demand at less than full-employment levels, both labor and management expect inflation to continue and attempt to maintain their shares of national income, an inflationary spiral may persist along with high unemployment.*   It is this situation which confronted policy makers in the United States, Canada, and other countries from 1970 through 1974.   Theoretically, a long-continued period of weak demand should slow rises in prices and wage rates, but the length and depth of recession needed to accomplish the goal seem to be unacceptably long, given current price- and wage-setting mechanisms.   The policy options that remain open to political leaders in such a situation will be discussed after a brief review of the effects of inflation.

EFFECTS OF INFLATION

Why worry about inflation?   Unemployment is clearly undesirable.   It involves loss of potential output which can never be recouped, and it involves economic deprivation of families, loss of human dignity and respect, perhaps loss of work skills and motivation.   If acceptance of inflation is necessary to prevent unemployment, why not pay the price, and gladly?   Indeed, inflation so often accompanies a golden era of prosperity that it may become tinged with an aura of "gilt by association."   Certainly, persons and groups whose incomes plus capital gains rise faster than prices are not so likely to object vigorously to the source of their relative gains.

Nevertheless, there are several harmful effects of inflation which need to be kept in mind.

1   *Income effects.*   Any price rise alters the real income of the seller favorably and of the buyer unfavorably.   So persons and groups whose money incomes are relatively fixed lose income shares in an inflation as compared to those whose output of labor or goods rises in price more rapidly.   Persons receiving incomes from pensions, bond interest, social security benefits, insurance policies, and similar sources are disadvantaged.   State and local governments will be also, until they can raise tax rates.   On the other hand, the federal government will benefit because its tax revenues are largely based on incomes and sales, and inflation raises more incomes into higher tax brackets.

2   *Asset effects.*   As was brought out in the preceding chapter, the prices and rates of return of various real and financial assets respond differently to inflation.   Consumer and producer durables whose physical productivity remains unchanged rise in price because the market value of their output goes up.   Prices of shares of stock which represent an equity in real productive capital goods should rise also.   But holders of financial claims denominated in money units suffer, because the real purchasing power of their principal declines.   Thus, holders of savings accounts, bonds, mortgages, loans, insurance policies, and other fixed-dollar items experience a decline in the purchasing power of their savings between the time they are set aside and the time the principal funds are paid back.   Business managers experience a similar effect when they find that depreciation allowances

accumulated on an original-cost basis are inadequate to purchase a replacement for a capital good used in production. Of course, the creditors' loss is the borrowers' gain, and some maintain that this wealth redistribution is desirable because borrower-debtors are the active, productive, growth-producing units in our economy. It should be noted, however, that the Government Sector is one of the largest debtor groups in the economy, but not necessarily the most productive.

3   *International effects.*   To the extent that the rate of inflation in the United States exceeds that in other countries, importers are better off and United States exporters are worse off, assuming that U.S. and foreign products are substitutable in domestic and foreign markets. In the aggregate, an inflation tends to worsen a country's balance of trade. There may be an offset if high interest rates accompany the inflation and lead to a net inflow on capital account. But such an inflow may be largely short-term speculative funds which may quickly be withdrawn and cannot be relied on as a source of funds for long-term capital investment.

4   *Inflationary expectations.*   If an inflation lasts so long that business analysts and workers come to expect its continuation, demands for wage increases will escalate, employer resistance may weaken, and labor-management negotiations may become a farce played out at the expense of customers, who foot the bill via higher product prices. In addition, if the inflation continues while labor demand is short of full employment, equal percentage rises in wages and prices will leave labor demand unchanged—assuming labor productivity unchanged. Any justification for inflation would then seem to be removed. The situation may degenerate into a never-ending struggle over shares of the national income accompanied by worsening effects of income redistribution, by speculative markets for assets, and by a deteriorating international balance of payments. At the end of this route lie massive labor unrest, disruption of capital markets, and devaluation of the currency.

All in all, acceptance of an inflation is not costless, even if it could cure unemployment—which seems doubtful. Can't we find a cure for inflation without causing high unemployment?

STOPPING INFLATION

In microeconomics, excess demand for a given product is cured by its price rise, because customer incomes are assumed constant and because substitutes are assumed available at near-constant prices. In macroeconomics, however, general price increases can raise income flows proportionately, and prices of substitute products may well be rising also. So *real* aggregate demand may not be cut back by the mechanisms which operate so effectively in individual product markets. General price rises may make some contribution to eliminating excess demand if they redistribute incomes in favor of groups or sectors with a below-average marginal propensity to spend, if they lead to a shift toward saving, or if they raise interest rates. But the generation of inflationary expectations and of capital gains may outweigh those restraining forces.

The obvious way to stop inflation is to reduce excess demand by fiscal and monetary policies, that is, by actions which shift the IS and LM lines downward. Increased taxes, reduced government expenditures, restraint on credit availability, and reduction or slow growth of money supply are standard counterinflationary measures which have been used effectively on many occasions and, along with automatic stabilizers, will undoubtedly continue to constitute our first line of defense.

In addition, increases in supply of goods may help—if the increased supply does not generate equal increments to domestic incomes and expenditures. Increased imports and reduced exports provide an effective and semiautomatic offset of this type. An increased flow of imports absorbs domestic expenditures without generating domestic incomes from production; whereas reduced exports is a direct decline in aggregate demand. Increased labor productivity is the other helpful element on the supply side. By softening the impact of wage rises on unit costs, productivity gains may moderate cost-push pressures and lessen the struggle over income shares; they also increase full-employment output.

From 1968 to 1970 we used monetary and fiscal policy to eliminate demand in excess of full-employment output. Then in 1970 through 1974 the economy operated at 94 to 97 percent of full employment but with prices rising 3 to 9 percent per year. Levels of unemployment, in the 5 to 6 percent range, were unacceptably high. Apparently there is need for some policy instruments that go beyond the fiscal and monetary impacts on aggregate demand.

The objective of the new policies would be to reverse inflationary expectations and to modify the wage- and price-setting mechanisms that generate inflation when output is below full-employment levels. The new instruments are referred to generally as *incomes policy*. One variant is the announcement by the federal government of "guidelines" for wage and price decisions. Such guidelines were set forth in the 1962 *Economic Report of the President* as follows:†

> The general guide for noninflationary wage behavior is that the rate of increase in wage rates (including fringe benefits) in each industry be equal to the trend rate of over-all productivity increase.... The general guide for noninflationary price behavior calls for price reduction if the industry's rate of productivity increase exceeds the over-all rate—for this would mean declining unit labor costs; it calls for an appropriate increase in price if the opposite relationship prevails; and it calls for stable prices if the two rates of productivity are equal.

It was recognized that modifications were needed for particular industries to reconcile the guidelines with objectives of equity and efficiency, because of their different positions in an industry growth cycle, or because of peculiar initial conditions. But the major thrust was clear. Increases in rates of compensation were to average out to the trend rate of increase in national output per man-hour (later set at 3.2 percent per year) so that unit labor costs would remain constant

† P. 189.

overall.   Different rates of increase of productivity for various industries were to result in differential price rises or declines, rather than in different rates of increase in labor compensation.   Attempts were made to influence private decision making to conform with the guidelines by

1   Government persuasion and requests, backed by publicity
2   Government intervention in markets—by government contracts and sales from government stockpiles of raw materials
3   Government mediation and conciliation services during labor disputes
4   Government threats of antitrust investigations or other harassment of noncooperators

Some statistical analyses indicated that the guidelines contributed to the period of price stability in the early 1960s, but voluntary restraints proved unenforceable when excess demand pressures built up in 1965 as the United States military role in Vietnam expanded.

In 1971, after fiscal and monetary policy measures had brought demand below full-employment output again, legislated controls on wages, interest, dividends, rents, and prices were instituted to curb the persistent inflation.   The controls, plus weak demand and rapid gains in productivity, combined in unknown proportions to slow the inflation in 1972, but inflation at over 5 percent per year continued into 1973 and 1974.   The federal administration requested a year's extension of controls to quench inflationary expectations more completely and to moderate wage rises in a year with an unusually large number of major contract negotiations, but controls were removed progressively in early 1974 and expired on April 30 of that year.

Other policies aimed at reducing unemployment without stimulating price rises include government programs to improve the operation of labor markets. Government employment services are developing computerized data banks to match job openings with characteristics of unemployed workers.   Government-funded programs for retraining and relocating workers are being expanded.   Education and on-the-job training for chronically unemployed workers are being developed.   In addition, industrial courts or mandatory arbitration may be utilized to bring third-party intervention into wage settlements when management-labor conflicts threaten the public with serious interruptions of vital output or with unacceptable price rises.

FUTURE PROSPECTS

Many economists feel that we are in a period of economic history when the inflationary bias is stronger than it used to be, so that we shall experience recurrent periods of inflation.   Their case is rather persuasive.   The Full Employment Act of 1946 committed the federal government to a policy of maintaining a high level of production, incomes, and employment, and it is well-nigh essential that any federal administration achieve this, at least by the November election every second or fourth year. This imperative, coupled with economists' analyses of how the federal government can stimulate the economy, does raise the probabilities that aggregate demand will periodically be pushed up to or above the full-employment level.

On the supply side of the market for output, it may be maintained that cost-push forces are stronger now than they were a few decades ago. Labor legislation of the 1930s and the growth of labor unions (though their membership has stabilized at about one-fourth the labor force) has strengthened employees in wage-setting negotiations with employers. So the struggle over income shares may well lead to a quicker, stronger response of pay rates to any initiating price or profits rise.

Also, it is argued by many, the power of large companies in many industries is greater than it used to be. A few large competitors in a market will recognize the dangers of strong price competition, and they will experience the same cost pressures from industrywide patterns of labor settlements. So we get fewer price cuts and more industrywide advances in prices.

In this view, then, large companies and large unions can negotiate contracts in which rates of pay increase more than productivity, with the consequent rise in unit labor cost being passed on to customers in higher prices. This unholy alliance of big labor and big employers engenders inflation with its harm to customers, creditors, persons with relatively stable incomes, exporters, persons with savings in fixed-money claims, and others.

The diagnosis of the malady is somewhat persuasive, though it may well be that a period of excess demand and the emergence of expectations of further inflation are needed to trigger the epidemic in its most virulent forms.

It is to be hoped that business managers, workers, and government officials will have learned some lessons from our current bout of inflation—the most severe and prolonged since World War II. Perhaps business executives will learn that general price rises provoke a counterattack on the pay front that may lower their profits for years to come. Perhaps employees will learn that wage increases in excess of productivity rises, plus cost-of-living adjustments, lead to further price rises that erode their pay gains and to the loss of some of their jobs either to other domestic or to foreign workers. Perhaps government officials will learn the vital necessity of balancing their budgets when the economy is operating at full employment and of preventing or quickly reversing a rise in aggregate demand beyond full-employment output at current prices.

However, learning is a slow, difficult process, especially in complex situations such as macroeconomic interactions where it is so difficult to establish simple, clear-cut, readily understandable cause-and-effect relations. It is so much easier and compelling to understand one's own self-interest, that many may have taken from this inflationary episode only the knowledge of how better to protect themselves and take advantage of the situation the next time around.

The problems of achieving full employment and price stability simultaneously have not been solved. Further outbreaks of inflation in the future seem likely. The challenge is to find policy instruments which can help stabilize prices without worsening national performance on the full-employment objective—and to make these policies prevail against special group interests.   Considerable social, political, and economic ingenuity will be required to modify the private wage- and price-setting mechanisms to permit achievement of both goals without a degree of government intervention that is unacceptable in a democracy and inimical to economic efficiency.

Some economists recommend a policy of adapting to inflation rather than trying to control it.   They maintain that the harmful effects of price rises on income distribution and allocation of resources can be neutralized to a large extent by building price-escalator provisions into all long-term contracts.   Cost-of-living adjustments have already been legislated for social security benefit payments, and escalator clauses are being inserted more frequently in union contracts and in long-term industrial purchasing agreements.   Similar provisions could be devised to escalate the value of bank accounts, mortgages, bonds, insurance policies, retirement annuity contracts, and similar items.   If all income flows and the money value of items of financial wealth were subject to adjustment in proportion to the general price level, and assuming that money supply was increased rapidly enough to prevent interest rates from rising, what would be the harm in permitting inflation to continue?   Real output and standard of living can be maintained, and we would avoid the periodic bouts of high unemployment, shifts in income distribution, and government interference in economic decision making which accompany current efforts to control inflation.

Against such proposals it may be argued that the balance of payments would be hard to maintain if the rate of inflation in the United States differed appreciably from that experienced by its trading partners.   A floating exchange rate for the dollar would help maintain balance, but it may be argued that a reserve currency such as the U.S. dollar should not be subject to frequent exchange-rate adjustment.

The difficulties with proposals for a general "indexing" of incomes and financial claims would seem to be practical ones.   What price index should be used to calculate adjustment factors?   Would there not be a prolonged transitional period of piecemeal indexing with accompanying uncertainty, injustices, and conflict?   Would not business costs and uncertainty increase because of, for example, costs of making price changes frequently, costs of getting information and taking action to adapt to fluctuating rates of inflation, and uncertainty over accumulation of reserves to meet pension obligations?   Under a system of general indexing, would the rate of inflation be steady or fluctuating, and would expectations of continuing inflation lead to a rising rate and ultimately runaway inflation and a breakdown of the monetary system?   Some argue that indexing has worked in Brazil with a declining rate of inflation, but others contend that this experience suggests that a rigid, centralized control of the economy is required for application of general indexing.

Inflation and unemployment are problems of economic dynamics.   Theoretical models are still unsatisfactory in these areas, and empirical analysis is difficult because human expectations and learning are involved.   For the present, this area of study must be left open and the policy problems unresolved.

PROBLEMS

1 Define the demand-pull and cost-push varieties of inflation. Tell how you would distinguish between them (*a*) on the basis of the theoretical model and diagrams of aggregate demand and supply curves, and (*b*) on the basis of historical, empirical data for the United States economy.

Are there any other varieties of inflation? Explain.

2 The wage-price guidelines enunciated in the early 1960s specified that, except under special circumstances, wage rates (including fringe benefits) should rise in all industries at a rate equal to the long-run average national rate of gain in output per man-hour. It was asserted that such increases would not be inflationary.

  a Discuss the logic underlying this wage guideline and point out any inadequacies you see in it.

  b Since rates of rise in output per man-hour differ widely among industries, what would be the implications of this policy for changes in unit labor costs and in prices for various industries?

3 It is sometimes asserted that (*a*) inflations are caused simply because money supply is increased too fast, and (*b*) an inflation can be stopped by holding money supply constant.

Explain the logic underlying these assertions—the economic behavior or causal chains involved, and the definition of "too fast." Point out any limitations or inadequacies, theoretical or practical, that you see in these linkages between monetary policy and inflation.

4 It is sometimes asserted that (*a*) inflations are caused simply because the federal government runs a deficit, and (*b*) an inflation can be stopped by deflationary fiscal policies.

Explain the logic underlying these assertions—the economic behavior or causal chains involved, and what is meant by "deflationary policies." Point out any limitations or inadequacies that you see in these linkages between fiscal policy and inflation.

5 Some persons argue that an inflation arising from excess demand can be cured if producers will increase supply to eliminate the excess-demand gap. Others maintain that when producers increase output, they increase incomes also, so that demand will rise with output and the excess-demand gap will remain. Resolve this argument or misunderstanding, if you can.

6 List a few of the principal costs and benefits of inflation, indicating which persons or groups are made better off and which worse off by inflation.

Chapter 19

Foreign Sector and the Balance of Payments

INTRODUCTION

Transactions between a nation's residents and foreign persons and organizations fall into the same categories as transactions within the national economy except that there will be no cancellation of intermediate goods and services. Major categories are:

1. Purchases and sales of currently produced goods and services, including payments or receipts for flows of factor services between countries

2. Transfer payments to and from foreigners, for example, private gifts to foreigners for humanitarian purposes, remittances by immigrants to their families abroad, and government grants to foreigners for relief, technical assistance, military aid, and other purposes

3. Capital transactions, such as private purchases and sales of financial claims and real assets, and government loan and securities transactions or exchanges of real assets

4. Transactions which "settle" the net balance of those claims for payment between nations that arise from the three preceding types of transactions

The first two categories of transactions are current-account items and appear in the gross national product and income accounts. Type 1 transactions are recorded as exports and imports, and their net value is the nation's balance of trade, or net exports X_n. Type 2 transactions appear as net transfers to foreigners (TR_f) and provide foreigners with purchasing power to buy exports from the donor nation. The current-account balance, net exports less net transfers to foreigners equals the net increase (or decrease) in the nation's wealth as a result of foreign transactions. Hence, it is called *net foreign investment*.

$$I_f = X_n - TR_f$$

(As usual, human beings are not counted as items af wealth; so net immigration or emigration is not counted as a change in national wealth.)

Transactions of types 3 and 4 do not change a nation's wealth; they do alter its portfolio of assets and hence may change its international liquidity position.

Type 3 transactions involve acquisition by residents of title to financial claims against foreigners and to real assets located abroad, and the reverse acquisition by foreigners of title to financial claims against residents and to real assets in the home country.   These transactions are presumably made for investment motives.

The balance of international claims for payment which result from the preceding three types of transactions is called a nation's *balance of payments*.   "Settlement" of the preceding transactions is accomplished by a net flow between nations of internationally acceptable means of payment, which may be liquid reserve assets —gold, convertible currencies, International Monetary Fund (IMF) gold *tranche*, and special drawing rights (SDR's)—or may be short-term liquid claims.   The net change in a nation's portfolio arising from type 4 transactions is a change in liquidity position, though not in total wealth.

Let us begin the analysis of international transactions and liquidity by considering the determinants of trade flows and the rate of exchange.

INTERNATIONAL TRADE AND THE EXCHANGE RATE

Presumably persons, businesses, and governments buy goods and services from foreigners because it is cheaper or more profitable to do so, or because the items are not available domestically.   For example, U.S. tourists must buy foreign services in order to view the Alps or the cathedrals of Europe, and U.S. companies must import coffee beans from abroad.   But for goods producible either in the United States or abroad, what determines whether they will be bought from domestic or foreign suppliers?   This question brings us to a consideration of *comparative advantages* and of *exchange rates* between countries.

Imagine two countries which have differences in natural resources (climate, mineral deposits, soil fertility, waterways, forests, wildlife), in population (age distribution, education, technological and managerial knowledge and skills), and in stock of capital goods acquired during their economic development.   Aside from products unique in each country, the two economies will probably produce many of the same items to meet the needs of their people.   Of these common items, some will be transportable at a cost that will make them candidates for international trading.   *What determines which products will flow which way between countries?*

Because of the different endowments of factors of production, the ratios between prices of various products in the two countries would probably differ. (Absolute price comparisons are not possible because the countries have different currency units and, we assume, no established exchange rate initially.)   It seems reasonable that the products for which a country has relatively low prices will become its exports and those which are relatively high priced will be imported. Let us see how we can verify and quantify this judgment.

Imagine that we obtain prices of tradable goods 1, 2, 3, ..., n in currencies of the two countries before any trading takes place between them.   We rank these commodities according to the magnitude of the ratio of price in the foreign country

(p_f) to that in the United States (p_d).   (Subscript f stands for "foreign" or for "franc"; subscript d stands for "domestic" or "dollar.")

$$\frac{p_{1f}}{p_{1d}} > \frac{p_{2f}}{p_{2d}} > \frac{p_{3f}}{p_{3d}} > \cdots > \frac{p_{nf}}{p_{nd}}$$

All these ratios have units of "francs per dollar," and are simply the number of francs obtainable abroad from the sale of an amount of commodity j purchasable for \$1 in the United States.   *The ratios are essentially currency exchange rates obtainable via purchase of commodity* j *in one country and its sale in the other.*

Now assume that there are exchange markets in the currencies of the two countries, with an (unknown) *exchange rate R* in francs per dollar; this is the price of the dollar in foreign exchange markets, in terms of francs.   Consider a trader who buys commodity 1 in the United States, sells it abroad, and converts the francs received into dollars in international currency markets.   The trader's transactions will be as follows:

> Buy q_1 in the United States for $q_1 p_{1d}$
>
> Sell q_1 abroad for $q_1 p_{1f}$
>
> Exchange foreign revenue for $(q_1 p_{1f}/R)$ dollars

The ratio of the trader's final receipts to initial outlay will be:

$$\frac{q_1 p_{1f}}{R} \div q_1 p_{1d} = \frac{p_{1f}/p_{1d}}{R}$$

The trader could realize a gain only if the price ratio p_{1f}/p_{1d} is greater than R. This seems reasonable.   *The commodity rate of exchange must exceed the currency rate of exchange if a profit is to be made*—and then only if the margin is great enough to cover transportation, insurance, handling, and marketing costs.

If for some commodity n, the price ratio is less than the exchange rate, that is, if p_{nf}/p_{nd} is smaller than R, it becomes profitable to *import* such a commodity into the United States.   In this case, the ratio of trader's revenue to initial outlay is $R/(p_{nf}/p_{nd})$.

This suggests the possibility of partitioning the commodities produced in the two countries into potential exports or imports.   In Fig. 19-1, the pretrade price ratios, or commodity exchange rates, have been plotted for commodities ranked by magnitude of that ratio.   When the currency exchange rate is R_α (francs per dollar), commodities 1 through 5 are potential exports from the United States, and commodities 7 through 10 are potential U.S. imports.   Commodity 6 would not normally be traded.   A rise in the exchange rate to R_β would cut commodity 5 out of the list of potential exports and add commodity 6 to the import list.   The United States has a *comparative advantage* in production of commodities j with high price ratios p_{jf}/p_{jd}.   Foreigners have the comparative advantage for products with low price ratios.

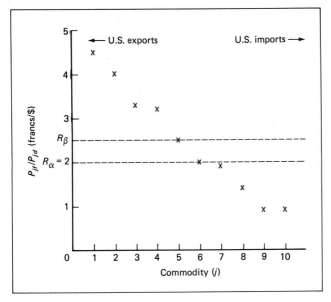

Figure 19-1   Price ratios for tradable commodities, and exchange rates.

When commodities flow in international trade, the potential gains from trade as estimated from pretrade price ratios will be reduced by the costs of moving and marketing them.   Also, *if the amount of the commodity traded is large relative to the total production or consumption in either country, the price ratio will shift down for U.S. exports and up for imports as trade progresses.*   For U.S. exports, the increased output to meet foreign demand will raise marginal costs and prices in this nation, and the added supply abroad may lower their prices in the foreign market—with both effects acting to shift the price ratio downward.   For U.S. imports, the added flow may raise foreign marginal costs and prices while lowering U.S. prices—with both effects acting to raise the price ratio.

The quantity of a given commodity that will be traded internationally depends on the pretrade price ratios, costs of transportation and marketing, elasticities of demand and supply in receiving and producing countries, plus any subsidies, tariffs, or quantitative restrictions applying to that commodity.   Let f_j be the physical volume flow of commodity j in foreign trade between the United States (d) and the rest of the world (f).   If the export commodities are numbered from $j = 1$ to m_α and if import commodities are numbered from $j = m_\alpha + 1$ to n, when the exchange rate is R_α, U.S. receipts for exports and payments for imports will be:

$$X = \sum_{j=1}^{m_\alpha} p_{jd} f_j$$

and

$$Z = \sum_{j=m_\alpha+1}^{n} p_{jd}f_j$$

The U.S. balance of trade will be $X_n = X - Z$.

Note that the *trade balance depends on the exchange rate*. A rise in the exchange rate will tend to increase the foreign price of U.S. exports and lower the domestic price of our imports. If elasticities of demand in domestic and foreign markets are greater than 1, this would normally cut the dollar value of U.S. exports and raise the dollar value of our imports. The balance of trade would shift toward a deficit. The converse would occur if the exchange rate were lowered, again assuming elastic demand for U.S. exports and imports. It seems reasonable that there should be some level of the exchange rate which will lead to equality between dollar values of export and import flows.

One other effect of changes in the exchange rate should be noted—its influence on the gains from trade. When the exchange rate (price) of the dollar rises, one dollar's worth of U.S. goods exchanges for a larger physical volume of foreign goods. The price of the U.S. imports, in terms of physical exports needed to purchase them, is lowered. Thus, *a rise in the exchange rate for the dollar constitutes an effective rise in the productivity of U.S. factor inputs, hence, a rise in the standard of living.* The converse is true when the exchange rate declines

The *effects of different rates of rise of prices at home and abroad* can be analyzed in a similar fashion. Suppose U.S. prices rise faster than foreign prices. Then all price ratios for commodities in Fig. 19-1 will decline. Assuming that the exchange rate remains constant, foreigners will receive less U.S. goods per franc while U.S. buyers will obtain more nearly the same amount of foreign goods per dollar. The physical terms of trade have become more favorable to the United States. But the volume of exports may drop off and that of imports may rise, so that the balance of trade will shift toward deficit. The effects are similar to the effects of a rise in exchange rate. As may be seen in Fig. 19-1, a decline in all price ratios with R_α constant would lead to the same relative shifts as a rise in exchange rate to R_β with price ratios constant. The effects would be reversed if prices in the United States should rise more slowly than prices abroad—for the internationally traded commodities.

Finally, *the impact of trade flows on international currency markets, and hence on the balance of payments*, should be brought out. Foreign buyers of U.S. exports must buy dollars to pay U.S. producers. So payment for U.S. exports involves a demand for dollars and a supply of francs in transactions with international currency dealers. Conversely, U.S. importers supply dollars and demand francs in their dealings with international currency dealers. If U.S. export and import flows are equal in dollar value, the demand and supply of dollars balance. *If the United States has an export surplus, the demand for dollars exceeds the supply arising from current trade transactions.* If the nation has an import surplus, the supply of dollars to international currency markets exceeds the demand insofar as current trade transactions are concerned. This excess demand for, or supply of, dollars on trading account is one component in the overall balance, or imbalance, of payments.

UNILATERAL TRANSFERS

Gifts from persons and from private and government organizations in one country to entities in another country do not leave the donor any claim for future payments. The recipient country experiences a gain in wealth and the donor a loss. If we assume that the transaction is in money, a gift from the United States to a foreign recipient involves a supply of dollars to international currency dealers to buy francs for the foreign recipient; a gift from a foreign source to a recipient in the United States involves a demand for dollars and a supply of francs to the foreign exchange market. On balance, *the net outflow of transfers from the United States will result in a net supply of dollars to foreign exchange dealers,* or a net demand for francs. This is one more component in our overall balance of payments.

It may be noted in passing that most of our U.S. government grants to foreigners are conditional upon purchase of goods or services from the United States. So a deficit item for unilateral transfers is to a large extent offset by return flows of payments for U.S. exports. The net effect on the overall balance of payments is small.

CAPITAL TRANSACTIONS

Capital transactions include private and government purchases and sales of existing real assets, which are held abroad rather than imported to the purchaser's country. They also include transactions in financial claims—long-term and short-term, liquid and nonliquid—by persons, by financial and nonfinancial businesses, and by governments and central banks.

Capital transactions do not lead to net changes in wealth for the countries of the participants, but they may lead to changes in their international liquidity positions. Thus a U.S. buyer of a foreign bond or automobile plant supplies dollars to foreign exchange markets to obtain the foreign currency needed for his purchase. If this transaction is not offset by other transactions involving reverse payments, some foreign individual, business, commercial bank, government, or central bank will wind up with a liquid claim against the United States, a claim for payment in some internationally acceptable medium of exchange. A net increase of foreign-held claims against U.S. foreign exchange reserves constitutes a decline in this country's international liquidity position.

The impact of capital transactions on the U.S. international liquidity position may be traced by noting how those transactions affect the demand for and supply of dollars in foreign exchange markets. U.S. purchases of capital assets from foreigners involve a supply of dollars to exchange dealers, just as did U.S. imports or unilateral transfers. Sales of U.S. real assets or financial claims to foreigners involve foreign demand for dollars from foreign exchange dealers, just as did foreign purchases of U.S. exports or their unilateral transfers to U.S. recipients. The net effect of all these international transactions is shown in the balance-of-payments accounts, discussed in the next section.

However, before leaving the discussion of capital transactions, we should note that changes in currency exchange rates will affect them also. *A rise in the exchange rate of the U.S. dollar means that $1 will buy more francs, hence more real and financial assets in foreign markets.* The rise might well stimulate the flow of U.S. funds abroad—increasing the supply of dollars to international currency markets and worsening the U.S. balance of payments and international liquidity position. A decline in the exchange rate (price of the dollar) would have opposite effects—always assuming that the demand for foreign assets is elastic.

We should note also the influence of interest rates, or asset rates of return generally, on international capital flows. Obviously, rates of return that are higher in the United States than abroad (for assets of comparable liquidity and risk) will lead foreigners to invest capital funds here. Higher rates of return imply lower prices in the United States for assets yielding a given income stream, and investors will buy assets (real and financial) where their prices are low, just as they will buy lower-priced commodities. Thus, differential movements of interest rates in various countries imply differences in asset price changes and lead to shifts in international asset flows—just as differential movements of commodity prices lead to shifts in export and import flows for commodities. *Relatively high interest rates in the United States will normally increase the demand for dollars to buy U.S. assets and will thus shift the balance of payments toward a surplus.* Relatively low interest rates in this country will lead to an outflow of capital funds, and the supply of dollars to exchange markets will tend to move the balance of payments toward a deficit.

It is interesting to note that *domestic economic developments during a business cycle have opposite effects on a country's trade balance and capital-funds flows.* Commodity price rises during the full-employment phase of the cycle will shift the trade balance toward a deficit, but the rises in interest rates in that phase will shift the balance on capital account toward a surplus. These effects will both be reversed during the depression phase of the cycle.

DEMAND AND SUPPLY FOR DOLLARS

International transactions by U.S. nationals may be classified according to whether they involve a demand for dollars in foreign exchange markets or a supply of dollars to foreign exchange dealers. *In balance-of-payments accounting, items involving a demand for dollars are considered positive (+) and those involving a supply of dollars are taken as negative (−).* Major items in each class are as follows:

Demand for dollars (+)	Supply of dollars (−)
U.S. exports of commodities	U.S. imports of commodities
U.S. exports of services	U.S. imports of services
U.S. investment income from abroad	Foreign investment income from the United States
Foreign transfers to the United States	U.S. transfers to foreigners
Foreign direct investments in the United States	U.S. direct investments abroad
Foreign loans to the United States and net purchases of U.S. securities	U.S. loans to foreigners and net purchases of foreign securities

If the total dollar flows for these two classes of transactions do not balance for a given time period, the net difference must be financed (1) *by changes in U.S. reserve assets or* (2) *by changes in U.S. liquid liabilities owed to foreigners.* Changes in these assets and liabilities absorb any excess supply of dollars in foreign exchange markets or provide dollars to meet an excess demand in foreign exchange markets.

In Fig. 19-2, curves are drawn for the aggregate demand for and supply of dollars in international transactions of types 1, 2, and 3, listed at the beginning of this chapter. As we have noted previously, these flows will be affected by the exchange rate for the dollar. *As the exchange rate (price of the dollar) rises, the*

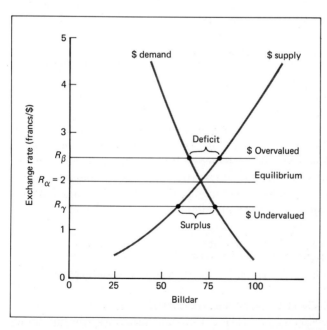

Figure 19-2   Demand and supply for dollars versus the exchange
rate.

demand for dollars declines and the supply of dollars (*demand for foreign currencies*) *rises*—assuming demand elasticities greater than 1. At an exchange rate R_α, the demand for and supply of dollars are shown as equal, and the balance of payments is zero. At a higher exchange rate R_β, the supply of dollars exceeds the demand, and some U.S. entities must buy the excess dollars from the market in "settlement" transactions by providing foreigners with U.S. reserve assets or with some acceptable liquid claims (presumably interest-bearing) which they are willing to hold. At the lower exchange rate R_γ, the demand for dollars exceeds the supply, the United States has a surplus in its balance of payments, and settlement of the surplus will involve an increase in U.S. reserve assets or a reduction in foreigners' liquid claims against the United States. At exchange rate R_β, the dollar is said to be overvalued; at R_γ, it is undervalued.

Of course, *a fixed exchange rate* R_α *will not lead to a zero balance of payments continuously over a long period of years.* The curves for the demand for and supply of dollars will shift as the U.S. economy or foreign countries move through their respective business cycles; or as prices of internationally traded commodities change at different rates in the United States and in foreign countries; or as rates of return (prices of assets) vary differently; or as tariffs, export subsidies, or quantitative controls distort free flows of commodities between countries; or as the ebb and flow of confidence in the dollar or in foreign currencies motivate speculative flows of capital funds. The instabilities in the demand for and supply of dollars (and other currencies) in international money markets lead to many problems regarding international transactions, exchange rates, and an international medium of exchange. For example:

1   How should a nation's balance of payments be measured, and how should a "basic," long-term imbalance be distinguished from a transient disturbance?

2   How should persistent imbalances be corrected—by internal economic adjustments, by changes in tariffs and subsidies, by restrictions on imports and capital flows, or by changes in exchange rates? Should surplus countries as well as deficit countries act to restore balance?

3   Under what conditions and by what procedures should changes in exchange rates occur? For example, should they be effected by unconstrained market changes of floating rates, or only after defense of existing parities until payments disequilibrium exceeds objective criteria?

4   What shall be accepted as an international medium of exchange, and how shall its supply be determined?

THE U.S. BALANCE-OF-PAYMENTS ACCOUNT

The balance-of-payments accounts provide the statistical basis for analyzing these problems, much as the national income and product accounts provide the framework for analyzing domestic economic activity. Table 19-1 presents figures

TABLE 19-1 United States Balance of Payments, 1946–1973 (in billdar)

[*Plus (+) items involve demand for dollars; minus (−) items supply dollars to foreign exchange markets*]

		1946–1949 Average	1950–1954 Average	1955–1959 Average	1960–1964 Average	1965–1969 Average	1970	1971	1972	1973
Current account										
1	Exports	+13.3	+12.6	+16.9	+21.7	+31.3	+41.9	+42.8	+48.8	+70.3
2	Imports	−6.4	−10.5	−13.2	−16.2	−28.5	−39.8	−45.5	−55.8	−69.8
3	Balance of trade	+7.0	+2.2	+3.7	+5.4	+2.8	+2.2	−2.7	−7.0	+0.5
4	Military transactions, net	−0.6	−1.8	−2.8	−2.4	−3.0	−3.4	−2.9	−3.6	−2.2
5	Investment income, net	+1.1	+1.7	+2.5	+3.0	+4.0	+3.8	+5.0	+4.5	+5.3
6	Other services, net	+0.6	−0.0	−0.4	−0.2	+0.2	+0.4	+0.4	+0.1	+0.8
7	Balance on goods and services	+8.0	+2.1	+3.0	+5.8	+4.0	+2.9	−0.2	−6.0	+4.5
8	Transfer payments, net	−3.9	−3.0	−2.4	−2.6	−2.9	−3.3	−3.6	−3.8	−3.9
9	Current account balance	+4.1	−0.9	+0.6	+3.3	+1.1	−0.3	−3.8	−9.8	+0.7
Capital account										
10	Private long-term, net	−0.8	−0.7	−2.0	−3.0	−1.8	−1.4	−4.4	−0.1	+0.1
11	Government loans, net	−2.2	−0.2	−0.7	−1.0	−1.9	−2.0	−2.4	−1.3	−1.5
12	(Balance on current account & long-term capital)	+1.1	−1.8	−2.0	−0.7	−2.6	−3.8	−10.6	−11.2	−0.7
	Private short-term									
13	Nonliquid assets, net	−0.1	−0.1	−0.2	−1.2	−0.2	−0.5	−2.3	−1.5	−4.3
14	Liquid assets, net	n.a.	n.a.	n.a.	+0.7	+3.4	−6.0	−7.8	+3.5	+2.5
Miscellaneous items										
15	Allocation of SDR'S	0.0	0.0	0.0	0.0	0.0	+0.9	+0.7	+0.7	0.0
16	Errors and omissions	+0.7	+0.2	+0.5	−1.0	−0.5	−0.5	−9.8	−1.8	−2.8
17	Official settlements balance	n.a.	n.a.	n.a.	−2.2	0.0	−9.8	−29.8	−10.4	−5.3
Settlement transactions										
18	Decrease in U.S. reserves	−1.5	+0.1	+0.3	+1.0	0.0	+2.5	+2.3	0.0	+0.2
19	Increase in U.S. liabilities to foreign official agencies	n.a.	n.a.	n.a.	+1.2	+0.1	+7.4	+27.4	+10.3	+5.1

Sources: Economic Report of the President, for historical data; Economic Indicators or Survey of Current Business, for current data.

for flows of U.S. international payments during the postwar years. Component flows are grouped under the main categories of transactions noted in the first section of this chapter—current-account, capital-account, and settlement transactions. In addition, some miscellaneous items and significant measures of imbalance are tabulated. All items except exports and imports are net flows. Flows are given a *plus* sign if the transactions involve a demand for dollars in foreign exchange markets (like U.S. exports) and a *minus* sign if they involve a supply of dollars (like U.S. imports).

The balance-of-payments record is illumined by recollection of major events in the United States and other economies during the postwar era.

1   Between 1946 and 1949 the United States dominated world markets and helped rebuild war-damaged economies in Europe and the Far East while restoring its own domestic capital stock also.

2   From 1950 to 1954 the United States was involved in the Korean war and in cold-war competition with communist nations in atomic energy developments, defense alliances, and economic aid to uncommitted nations.

3   The period between 1955 and 1959 witnessed the emergence of the European Economic Community and Euratom, along with the beginning of United States–Soviet competition in space technology. The U.S. economy slowed after a post-Korean durable goods boom.

4   From 1960 to 1964 the U.S. economy experienced near-stability of prices but underfull employment. U.S. investment abroad was high, notably in Common Market countries. Confrontations with communist nations continued in Cuba and at the Berlin Wall; Russia led in orbital space flights.

5   The years between 1965 and 1969 were marked by growing United States involvement in the Vietnam war, accompanied by social unrest and a prolonged inflation. European economic growth and integration continued, and Japan emerged as a major economic power. The U.S. government imposed restrictions on capital outflows. Space exploration shifted to lunar landings, with the United States landing the first astronauts.

6   In 1970 the United States experienced an inflationary recession and major outflows of capital funds.

7   Persistent inflation, worsening of the U.S. trade balance, and massive capital outflows led to a new economic policy on August 15, 1971. A wage-price freeze was imposed on the U.S. economy, followed by Phase II controls in November. Gold payments to foreign governments were stopped, and the dollar was floated to permit determination of new exchange rates against other currencies in international money markets. In December a devaluation of the dollar by about 8 percent changed the official gold price from $35 to $38 per troy ounce.

8   In 1972 the U.S. economy experienced a strong cyclical recovery, and inflation slowed somewhat. Arms limitations talks with the Soviet Union led to some initial agreements, and trade pacts with the Soviets followed Nixon's visit to

Moscow. His visit to China started the rebuilding of cultural and economic contacts with the People's Republic of China.

9   In early 1973, economic controls were relaxed and U.S. troops were withdrawn from Vietnam. Inflation accelerated and renewed speculation against the dollar led to another devaluation (about 10 percent) and floating of the exchange rates for the dollar and other currencies. Economic controls were tightened again in midyear, but strong worldwide demand for food, feed grains, and raw materials, coupled with some crop shortages, led to renewed acceleration of inflation. The oil embargo from November 1973 to March 1974 caused further supply problems and upward pressures on prices into 1974.

The specific events and the tides of change in international affairs noted here are reflected in the balance-of-payments accounts in Table 19-1.

Lines 1 through 3 show our large export surplus during the early postwar years when European and Japanese economies were weak, followed by smaller surpluses during the Korean war (1950–1953) and by renewed strength to the mid-1960s. From 1965 through 1972, there was a progressive decline in the U.S. surplus on merchandise trade and actual deficits in 1971 and 1972—the first annual trade deficits since 1893. There are many contributing causes:

1   Expanded production capacity and economies of scale in the more rapidly developing economies
2   Spread of technological knowledge, so that rising productivity with lower labor costs abroad has cut into the U.S. competitive advantage for low-technology manufactures and automobiles
3   Aggressive export promotion by foreign nations
4   Discriminatory trade arrangements against the United States by individual countries or trading blocs
5   A bias in exchange rate adjustments toward devaluation against the dollar
6   Faster rise in export prices for the United States than for foreign countries in the 1960s
7   Rapid rise in U.S. demands for foreign raw materials and fuels .

For 1973 the devaluations of the dollar, strong demand for our agricultural exports, and widespread foreign prosperity and inflation resulted in an amazing 44 percent rise in U.S. exports and a positive balance of trade. However, higher petroleum import prices and weakened foreign demand for our exports seem likely to swing the balance negative again in 1974.

Line 4 in Table 19-1 reflects the U.S. commitment to military defense of the free world, following a relaxation in the early postwar years. It consists of Department of Defense purchases of goods and services from foreign suppliers, net of foreign payments for military goods and services supplied by the United States. Indochina expenditures have swelled the total in recent years.

The net investment income flow (line 5) is a bright spot in the records. As U.S. residents have increased their direct (real) investments abroad and their holdings of financial claims, the return flow of interest, dividends, and profits has provided a rising income stream which helps finance U.S. purchases from foreigners. The rate of rise may slow, however, if foreign investments in the United States continue their rapid rise of recent years. Included as an outflow in this item are U.S. government payments of interest on official holdings of dollar securities.

Line 6 is a small balance resulting from a net U.S. deficit for tourist and travel expenditures combined with a net U.S. surplus for fees, royalties, and payments for other services.

The balance on goods and services is shown on line 7. When private and government transfers to foreigners are subtracted (line 8), one obtains the net excess demand for dollars (+) or supply of dollars (−) to foreign exchange markets as a result of current-account transactions (line 9). The pattern is one of surplus in the early postwar years, a small deficit during the Korean war period, renewed build-up to high levels in the early 1960s, a decline during the Vietnam war becoming precipitous between 1970 and 1972, and the sharp improvement in 1973.

The next items in the balance-of-payments table (lines 10 through 16) cover capital transactions, involving both real and financial assets and liabilities, short- and long-term claims, and the Private and Government Sectors. Private long-term capital flows (line 10) cover direct investments in real assets plus purchases of long-term financial claims. Since U.S. investments abroad exceed the reverse flow of foreign purchases of U.S. assets and claims, the entry is negative, indicating a net supply of dollars to foreign currency markets for these transactions. Similarly, there is a net outflow of dollars for U.S. government investments and loans to foreigners (line 11).

For a country to be basically solvent in its international transactions, its current-account plus long-term capital flows should average near zero over a period of years. Short-term capital flows may become positive or negative because of cycles of business activity, interest-rate differentials, or speculation on currency markets, but the *basic balance* of current-account and long-term capital flows determines the viability of a country's exchange rates and international liquidity position over the long run. From this perspective, line 12 reveals the serious and fundamental deterioration of the U.S. international payments position in postwar years. From 1946 through 1949, our large surplus on goods and services exceeded the deficits from transfer payments and long-term capital outflows. During the 1950s and early 1960s, our reduced surpluses on goods and services could not cover transfers and long-term capital movements; so the basic balance became moderately negative. From 1965 through 1969, the situation worsened but was masked by restraints on the flow of U.S. capital abroad. From 1970 to 1972, the basic deficit worsened rapidly as the trade balance swung far into deficit, and the rise in net income from investments did not offset negative shifts in the remaining items. The improvement in 1973 reflected the swing in current-account balance discussed above, and hence may be reversed in 1974.

Movements of private short-term funds are probably best gauged by adding lines 13, 14, and 16, since the "errors and omissions" item probably reflects untraceable "hot money" and other speculative flows.   These flows were apparently largely offsetting and cyclical up through 1969.   Then, in 1970 and 1971, there was a massive outflow of short-term funds, which resulted in recurrent dollar crises, a devaluation of the dollar in late 1971, and the breakdown of the postwar international monetary system.   There was some respite in 1972 as far as short-term capital movements were concerned, but the huge basic deficit led to another dollar crisis and devaluation in early 1973.

If all items down through line 16 are combined (including the small credits for allocations of Special Drawing Rights by the IMF in 1970–1972), we obtain the net figure for all transactions involving current expenditures, gifts, and investments among the Personal, Business, and Government Sectors in the United States and abroad.   *A positive balance indicates that those transactions resulted in an excess of demand over supply of dollars in international currency markets; a negative balance indicates an excess supply of dollars.*   This net figure is called the *Official Settlements Balance.*   If it is positive, other nations must pay the United States that amount in some form of international "money" to obtain the excess of dollars they demand for the preceding transactions.   If the Official Settlements Balance is negative, the United States must pay foreign countries that amount to "buy back" the excess supply of dollars which flowed into international currency markets in the preceding transactions.   How are these settlement transactions made?

SETTLEMENT TRANSACTIONS AND INTERNATIONAL LIQUIDITY

Basically, there are two means for settling balances of payments between nations. First, the deficit nation may transfer to its creditor some internationally acceptable means of payment—gold, convertible currencies, rights to draw currencies from the International Monetary Fund (the IMF *tranche*), or special drawing rights (SDR's) on the IMF.   Second, the deficit nation may induce the creditor nation to hold liquid claims against it, the inducement being adequate interest payments. These official I.O.U.'s may take the form of interest-bearing bank deposits or federal government securities.

In Table 19-1, these two types of settlement transactions are labeled "Decrease in U.S. Reserves" (line 18) and "Increase in U.S. Liabilities to Foreign Official Agencies" (line 19).   *Either a decrease in U.S. reserves or an increase in U.S. liabilities to foreign governments and central banks constitutes a demand for dollars in currency markets*; hence they are plus items in balance-of-payments accounting and can offset a negative Official Settlements Balance resulting from other types of transactions.

As may be seen in line 18, U.S. reserve assets increased in the early postwar years at a rate of $1.5 billion per year.   This was the period of strong demand for U.S. output to provision and rebuild war-shattered economies in Europe and Japan;

and it was a period of "dollar shortage" as those economies struggled to earn or borrow the funds to pay the United States for its goods and services. The situation was reversed in the 1950s and early 1960s as U.S. reserves decreased (line 18 positive) and as dollar claims against the nation expanded rapidly. [Although reported figures do not separate the changes in private liquid assets and official claims against the United States prior to 1960, we can calculate the combined increase in U.S. liabilities (line 14 and line 19) as follows: 1946 through 1949, −0.2 billdar; 1950 through 1954, +1.6 billdar; 1955 through 1959, +1.4 billdar.] It is clear that our payments deficits have been "settled" primarily by foreigners accepting claims against the United States rather than demanding U.S. reserve assets; that is, line 19 exceeds line 18. The flows of claims to foreign official agencies became huge in 1970 and 1971, both because of weakness in the basic balance (line 12) and because private owners of liquid dollar claims exchanged those holdings for claims denominated in foreign currencies (lines 14 and 16). Foreign governments and central banks were obliged to acquire these dollar liabilities in order to maintain official exchange rates for their currencies against the dollar (line 19), but they did cash some of them for U.S. reserve assets (line 18), and the dollar was devalued in late 1971. In 1972, private flows of liquid dollar claims were nearly in balance (lines 14 and 16). But the huge U.S. basic deficit ($11.2 billdar) continued, and foreign official agencies absorbed another $10.3 billion of U.S. liquid liabilities. In early 1973, private parties sold off holdings of liquid dollar claims at a high rate again; the dollar was devalued another 10 percent; and its exchange rate against other currencies was permitted to float in order to establish some level consistent with equilibrium in international currency markets.

Currently the United States faces both a flow problem and a stock problem in its international transactions. The flow problem is the deficit in the basic balance on current-account and long-term capital transactions. The stock problem is the nation's relatively small holdings of reserve assets and the much larger stock of outstanding liquid liabilities to foreigners. Table 19-2 traces the decline in U.S. international reserve assets from their high at the end of 1949, and the accompanying rise in U.S. liquid liabilities to foreigners. From the end of 1949 to the end of 1969, U.S. reserve assets fell about 35 percent and then by 20 percent more in the next three years—this while the dollar value of U.S. foreign trade was expanding more than fivefold.

Even so, most of U.S. payments deficits were financed by an accumulation of our I.O.U.'s abroad, as shown in the right-hand columns of Table 19-2. By the end of 1960, total liquid liabilities exceeded U.S. reserve assets, and by 1970 U.S. liquid liabilities to official agencies exceeded reserve assets. The position worsened drastically in 1970 and 1971 as foreign private creditors unloaded their portfolio of short-term claims against the United States, leading to a tripling of official holdings in two years. The dollar was devalued in December 1971 by 7 percent to 14 percent against our major trading partners. Short-term flows of funds stabilized in 1972, but renewed deficits in the basic balance and speculation against dollar claims led to a further devaluation of the dollar by 10 percent in early

TABLE 19-2 United States International Liquidity Position ($ billions at end-period, with gold valued at $35 per oz through 1971, $38 per oz in 1972, and $42.22 per oz beginning in March 1973)

| | U.S. Reserve Assets | | | U.S. Liquid Liabilities | | |
Year	Gold	Other	Total	To Central Banks and Governments	To Private Holders†	Total
1949	24.6	1.5	26.0	3.4	3.5	6.9
1954	21.8	1.2	23.0	7.5	4.9	12.4
1959	19.5	2.0	21.5	10.1	9.3	19.4
1964	15.5	1.2	16.7	15.8	13.6	29.4
1969	11.9	5.1	17.0	16.0	29.9	45.9
1970	11.1	3.4	14.5	23.8	23.2	47.0
1971	11.1	2.1	13.2	50.6	17.2	67.8
1972	10.5	2.7	13.2	61.5	21.4	82.9
1973	11.6	2.7	14.4	66.8	25.8	92.6

Sources: International Monetary Fund, *International Financial Statistics*, Washington, D.C., Sept. 1974 and earlier issues.
† Including International Agencies.

1973 and to floating exchange rates. Late in 1973 and early in 1974 the basic balance became positive, and short-term capital outflows reversed as foreigners became more willing to hold the outstanding stock of liquid claims against the United States.

In the early postwar years, liquid dollar claims were welcomed by foreigners. World production of gold (about $1 billion per year) and its movement into official stocks did not increase international monetary reserves nearly as fast as the value of world trade. As shown in Table 19-3, gold stocks held by governments and central banks rose from $33 billion at the end of 1949 to a peak at about $42 billion in 1965, and then declined to $39 billion at the end of 1972. World exports rose nearly sevenfold over this span of time.

TABLE 19-3 International Monetary Reserves (U.S. $ billions at year-end, with gold valued at $35 per oz through 1971, $38 per oz in 1972, and $42.22 per oz beginning in March 1973)

Year	Gold	Reserve Position in IMF	SDR's	Foreign Exchange	Total
1949	33.3	1.7	0	11.2	46.1
1954	34.9	1.8	0	16.7	53.5
1959	37.9	3.2	0	16.4	57.6
1964	40.8	4.2	0	24.1	69.0
1969	39.1	6.7	0	32.4	78.3
1970	37.2	7.7	3.1	44.5	92.6
1971	39.2	6.9	6.4	78.2	130.6
1972	38.8	6.9	9.4	103.6	158.7
1973	43.1	7.4	10.6	122.0	183.2

Sources: International Monetary Fund, *International Financial Statistics* Washington, D.C., Sept. 1974 and earlier issues.

The Bretton Woods Agreement of 1945 set up the International Monetary Fund to supervise the expansion of international money. Under its sponsorship, countries were given rights to draw needed reserves from the capital subscribed by member nations—one-fourth in gold or dollars and the rest in the member's own currency. Drawings are limited in relation to a country's quota and its efforts to correct its payments deficit. More recently, the IMF has issued unrestricted special drawing rights (SDR's) to nations in proportion to their subscriptions to the Fund's capital—the so-called "paper gold." As shown in Table 19-3, these types of "IMF money" were more than 40 percent as large as gold reserves by 1972. Even with this supplement to gold, international monetary reserves would have risen much more slowly than the value of international trade since World War II were it not for the third major form of international money—convertible foreign exchange.

Some currencies are so widely used in international trading that they are normally convertible into other currencies on demand and hence become an acceptable medium of exchange for settling international transactions. As is shown in Table 19-3, such holdings of foreign exchange increased more than tenfold from 1949 to 1973, or from one-third to nearly three times the value of gold reserves. Reserves held as convertible currencies have provided the major growth in international money in the postwar years, and U.S. dollar holdings provided by far the greatest part of those holdings. Note that the U.S. dollars held as reserves arise as I.O.U.'s by which this country has settled a major part of its balance-of-payments deficits since 1949. U.S. deficits provided the needed expansion of international money.

The special role of the U.S. dollar was based not only on its widespread use in world trade but on the monetary system set up by the Articles of Agreement for the IMF. The monetary experts wished to maintain some links between international money and gold, as had been the case under the prewar gold exchange standard. In the postwar era, only the United States guaranteed to exchange its currency for gold, or vice versa, at a fixed ratio—$35 per ounce. (At the end of 1949 the United States held $24.6 billion of gold out of the world's total gold reserves of $33.3 billion.) So the dollar was tied to gold, and other currencies were linked to the U.S. dollar at official exchange rates, which were to be altered only after consultation with the IMF. Foreign countries maintained actual exchange rates in international currency markets within 1 percent (plus or minus) of the official parity by buying or selling dollars in exchange for their own currencies.

The international monetary system under the IMF did provide an orderly, multilateral payments mechanism, which facilitated a great expansion of world trade. But there were potential weaknesses. The special position of the U.S. dollar proved untenable in the long run. The expansion of international reserves was dependent on continued deficits in this country's balance of payments; yet *foreign countries could demand settlement of U.S. deficits in gold, thus reducing the reserves behind the dollars held abroad and undermining confidence that the dollar is "as good as gold."* The United States had the special advantage that it could pay for some imports or foreign investments simply by increasing its I.O.U.'s outstanding, without having to earn the reserves by foreign trade.   Most of these I.O.U.'s (dollars) held by private or official agencies abroad did earn interest, but some foreigners resented our ability to borrow at short-term rates to make long-term investments in their nations.   The fact that the United States could run a continual deficit also implied that the dollar was overvalued in international exchange markets, so that this nation could buy more goods and services abroad for $1 than would have been the case if it had been forced to adopt an equilibrium rate of exchange.

On the other hand, *the IMF formula did deprive the United States of the power to alter the exchange rate of the dollar as an economic policy instrument.*   Even if we were to obtain consent to increase the dollar price of gold, the exchange rate of the dollar vis-à-vis other currencies would not be altered—since they were tied to the dollar rather than to gold.   Other nations could and did change their official dollar parities, and the bias in such changes was a net devaluation against the dollar.   This stimulated the foreign economies by promoting a surplus in their balance of payments—and of course their surpluses were simply the reverse measure of the U.S. deficit.

Continued U.S. deficits over two decades reduced the nation's gold and other reserves and raised liquid liabilities until reserves were less than 25 percent of official liabilities by the end of 1973.   This, coupled with a stubborn inflation in the United States and emergence of the first trade deficits of the century, led to a breakdown of confidence in the dollar as a stable international money.   Once that faith was gone, prudent men—speculators, bankers, corporate treasurers, oil-rich sheiks, and assorted capitalists—hurried to sell dollar-denominated claims which they felt might depreciate.   Their expectations were self-realizing.   Foreign central banks accumulated dollar claims from the prudent men, as they were obliged to do to maintain official parities, but the flow reached rates of several billion dollars per day on occasion.   In August 1971 and February 1973, the situation became untenable, and foreign exchange markets were closed.   When they reopened, the dollar had been devalued and was allowed to float for a period to permit the exchange markets to guide central bankers in selection of new official parities between other currencies and the dollar.   In 1971 the United States terminated its promise to buy and sell gold at $35 per ounce.   In December 1971, the price of gold was changed to $38 per ounce and in February 1973 to $42.22 per ounce. (In private markets gold prices rose to more than $160 per ounce in early 1974.)

Before ending this discussion of the U.S. balance of payments, a few words of caution may be helpful.   First, the emphasis on U.S. payments deficits and on the deterioration in its liquidity position should not be taken to indicate that the United States is an international pauper.   The fact that, until 1971 and 1972, this nation had a surplus on current-account transactions indicates that it was acquiring foreign real or financial assets; that is, U.S net foreign investment was positive.   The total value of U.S. investments abroad (private and public, long-term and short-term) amounted to about $30 billion in 1950, $86 billion in 1960, and $167 billion in 1970. This nation's liabilities to foreigners amounted to about $20 billion in 1950, $41 billion in 1960, and $98 billion in 1970. Consequently, the international net worth position of the United States rose from perhaps $10 billion in 1950 to $45 billion in 1960, and to $69 billion in 1970. It is the nation's short-term liquidity position which has deteriorated so badly—something like the situation of a growth company whose total assets are increasing rapidly but which experiences an excess of short-term liabilities over liquid assets.

A second caution is that the causes of the United States situation are multiple and complex, and the positive and negative components of the overall balance of payments are frequently interrelated. So no simple, single-item explanations or solutions are to be expected. For example, it might appear obvious that balance-of-payments deficits could be ended simply by restrictions on imports, while keeping exports constant. But our import reductions would reduce the export earnings and trade surpluses of foreign nations, which might well react by restricting their imports from the United States. Sometimes cuts in government grants (transfers) to foreign countries are suggested, but it should be realized that most of such funds are used for purchases from the donor nation, so that reductions in our exports would offset most of the gain from cutbacks in grants. Some proposals for restrictions on U.S. investments abroad fail to recognize that, in the long run, these investments provide the return flow of interest and dividends which has been a large and rising plus item in our current-account flows. In addition, U.S. real investments abroad often lead to related exports of materials and machinery from this country. Finally, restraints on the flow of U.S. capital abroad may simply lead foreign borrowers to absorb funds from their own capital markets which might otherwise have been invested in the United States. All this is not to say that the cited proposals will have no net effect on the U.S. payments deficit, but that the effects may well be smaller and more complex than a "one-line" considera-tion of the balance-of-payments table would suggest.

TOWARD A NEW SYSTEM FOR INTERNATIONAL TRADE, PAYMENTS, AND MONEY

At the present time, most world currencies are floating, implying that governments and central banks are not intervening in foreign exchange markets in an attempt to maintain fixed exchange rates. The "two-tier" valuation of gold was officially ended in November 1973, but central banks have not yet revalued their gold stocks from the $42.22 per ounce level nor sold gold into private markets where the price is over $150 per ounce. The huge increases in world petroleum prices have led to large balance-of-payments surpluses for oil-exporting countries and deficits for importers. So the international trade and payments system in mid-1974 was in what might be called a disorderly transition period, otherwise known as a mess.

The restructuring of international economic relations which is now under way involves at least three major problem areas:

1   Rules in markets for current-account and capital transactions
2   Policies for correcting disequilibria in a nation's balance of payments
3   Creation and regulation of international money

RULES IN MARKETS FOR CURRENT-ACCOUNT AND CAPITAL TRANSACTIONS

Economic analysis indicates that maximum gains from comparative advantages in production will be realized when goods and services, including capital services, flow between countries without constraints of tariffs, subsidies, quotas, or other barriers than normal costs of transportation, handling, insurance, and similar expenditures.   Exceptions are recognized for production related to national defense and for infant industries which may eventually have a comparative advantage though they are not competitive initially.   The General Agreement on Tariffs and Trade (GATT) was organized in the postwar period to provide a set of rules to maintain relatively unconstrained flows of goods and services among nations and to facilitate negotiations to resolve conflicts and reduce barriers to trade.   GATT has achieved major successes in freeing international trade.   However, many nations set up barriers against U.S. exports and subsidized exports to this country in the early postwar years when dollars were scarce, and the European Common Market has preferential tariffs, specially to protect agricultural production.   Of course, the United States has some quotas, restrictive government-procurement policies, agricultural subsidies, restrictive design and performance standards for imports, and other barriers in addition to tariffs on its own part.   Also, companies and unions which have been hurt by import competition are increasingly vocal in their demands for protection.   But, on the whole, postwar administrations in the United States have attempted to reduce restraints on international flows of goods and services, believing that this relaxation would help ease our balance of payments difficulties.   And in recent negotiations under GATT, the U.S. bargaining position has emphasized multilateral reductions in tariff and nontariff impediments to the flow of world trade and capital funds.

POLICIES FOR CORRECTING DISEQUILIBRIA IN A NATION'S BALANCE OF PAYMENTS

Short-term fluctuations in a country's balance of payments, for example, those arising from business cycles, can be met by changes in stocks of reserve assets or of liquid liabilities to foreigners.   That is what money stocks and credit lines are for—to bridge temporary inequalities between inflows and outflows of funds. But prolonged, basic deficits must be rectified by more lasting, fundamental economic adjustments.   Long-term disequilibria arise because economic developments at home and abroad lead to differential changes among countries in prices of internationally traded goods, in productivity and costs for tradable goods, in demands for imports, in long-term investment flows, in transfer payments, in new-product development, in war-related expenditures, and other conditions.

Historically, three types of policies have been identified for dealing with persistent official settlements deficits.   (1) The nation may deflate its economy to reduce demand for imports and lower prices of its exports.   (2) The nation may raise import tariffs, export subsidies, and quantitative and qualitative barriers to imports and place restrictions on capital flows in order to bring demand and supply for its currency into equilibrium in foreign exchange markets.   (3) The nation may devalue its currency to a level which will bring the international demand for and supply of its currency into balance.

Needless to say, recommendation 1 is not favored in an era when central governments have accepted responsibility for maintenance of high levels of employment and incomes—and when the political longevity of the party in power depends on reasonable success in achieving those goals.   A recession is bitter medicine, and in some circumstances it may have little effect, for example, when demand for a country's exports is inelastic with respect to price or when a domestic recession leads to a large increase in outflow of capital funds.   Nevertheless, consideration of this policy does point up the fact that stabilizing, noninflationary domestic policies will contribute to achievement of balance-of-payments objectives, and vice versa.

Protectionist policies of type 2 are more often chosen if a nation's leaders are unwilling to permit a devaluation of its currency.   Higher tariffs act like a sales tax and will reduce payments to foreigners because the higher domestic price reduces physical flow of imports demanded and/or because foreign suppliers reduce prices in their currencies in order to offset part of the increased tariff.   Export subsidies will increase foreign demand for dollars if the foreign demand for exports is elastic, but the cost is the domestic transfer of incomes from taxpayers to recipients of the subsidy.   Direct controls include quotas on imports, requirements that governments purchase from domestic suppliers, discriminatory specifications on imports so that domestic producers have an advantage, "tieing" of grants and loans to purchases from the nation supplying the funds, restrictions on tourist expenditures or capital investments abroad, and bilateral barter transactions. In addition, governments may support a nation's exporters by obtaining marketing information in foreign countries, conducting trade fairs, supplying financing on favorable terms to purchasers of exports, providing commercial insurance and guarantees against expropriation of foreign facilities, and similar activities.   The list of nonprice influences on international transactions is as long as the ingenuity of businessmen and government officials—hence, it grows over time.

All these interferences with normal commercial operations in international markets for goods and services and for capital funds tend to reduce the possible gains from trade and involve administrative costs.

Despite the economists' objections to interference with free markets, such policies are widely supported by domestic producers and labor unions who would be protected from import competition, by exporters who would be subsidized or otherwise aided by the government, and by government officials with a strong fear of unemployment or devaluation. Somehow the interests of consumers and taxpayers are often inadequately weighed by policy makers!

Nevertheless, there is still a widespread understanding of the advantages of free trade among officials at top levels of government, and an appreciation that self-serving policies by one nation may well lead to retaliatory action by its trading partners. Also, many channels are available now for multilateral consultation and decision on mutually agreeable procedures for correcting serious disequilibria—the IMF, GATT, the Organization for Economic Cooperation and Development (OECD), the Organization of American States, the World Bank, and regional trade groups. Prospects for removing restraints on trade may well improve also because the stigma associated with adjustments of the exchange rate of a nation's currency is disappearing.

The third policy to correct payments deficits is devaluation, that is, a lowering of the exchange rate or price of a nation's currency unit in terms of foreign currencies. Given elastic demands for goods and securities, the devaluation would lead to an increase in the demand flow and a decrease in the supply flow of dollars in foreign exchange markets.

In the real world, there may be lags of a year or more before major effects of devaluation show up—presumably because of long-term supply contracts, inventories, and time lags in recognition of new opportunities and in planning changes in production and marketing activities.

Exchange-rate adjustments constitute a recognition that the relative price levels of two countries have become inconsistent with equilibrium in international market payments, and they provide a means for adjusting relative prices internationally without major impacts on domestic economic policy. It may well be desirable to reinforce the exchange-rate adjustment with changes in domestic policies, for example, through counterinflationary measures, or temporary assistance to companies and workers harmed by the effects of shifts in the exchange rate.

As to the timing, magnitude, and procedures for changes in exchange rates, a great deal of discussion and experimentation has gone on in recent years. The Bretton Woods and IMF agreements specified fixed parities between currencies which were to be maintained within narrow limits over long periods of time and were to be changed by relatively large percentages only after consultation with the IMF. On the other hand, free-market economists maintain that the relative prices of national currencies should be determined day by day in international currency markets without any consultation or price-fixing agreements between central bankers and government officials. In the reconstruction of world trade and money arrangements now under way, it seems clear that *exchange rates will be more flexible*—perhaps by allowing wider fluctuations around official parities, perhaps by permitting more frequent and smaller changes in official parities (the "sliding peg"), perhaps by permitting periodic, temporary floating to allow a currency in disequilibrium to find a new parity, perhaps by continuous free floating of exchange rates. Quite surely *the role of the U.S. dollar as a reserve currency will be reduced* and the United States will be given some power to devalue or revalue the dollar.

Most of the proposals discussed here imply that the country experiencing a balance-of-payments *deficit* should change its economy, foreign trade policies, or exchange rate to restore equilibrium.   U.S. negotiators have proposed that countries in *surplus* should also act to reduce their surpluses.   Such symmetry in the adjustment process is urged on the basis of equity, efficiency, and speed of adjustment in maintaining international payments near balance.   The proposal recommends that the need for a country to take action be gauged by an objective measure, such as the deviation of its stock of international reserves from an agreed " base " level.

A great deal of discussion, negotiation. and legislative action remains before the international monetary and settlements system can be rebuilt in a new, flexible framework which can maintain acceptable and stable balances of payments for all major trading nations within a dynamic pattern of reasonably free transactions among nations.

CREATION AND REGULATION OF INTERNATIONAL MONEY

The third problem area in developing the new international economic institutions involves the international medium of exchange.   Historically, gold was the money used for international settlements, as it was the base for domestic money.   However, the twentieth century has witnessed a decline in the role of gold in international monetary systems.   Why has this occurred?   What can replace gold?

International reserves are helpful in providing confidence to traders that the foreign currencies they use in their transactions will be convertible without loss. From the viewpoint of national governments, international reserves allow time to make market adjustments if their payments fall into deficit temporarily—thus postponing the impact of international developments on domestic policies and helping keep trade free of crippling restrictions that might otherwise need to be imposed.   For both purposes, a nation's international reserves need to grow in some relation to the expected magnitude of settlement requirements, which in turn may be related to a nation's volume of trade.

In recent decades, the stock of gold available for monetary reserves has risen too slowly and proved too variable to serve as the international reserve asset.

Gold is a commodity (1) which must be discovered, mined and refined, and (2) which has uses in industry, in the arts, and as an asset desired by hoarders and speculators.   Given a fixed price for gold in monetary uses, (1) its production will drop off when its production costs rise in inflationary periods, and (2) more of the newly mined gold—and perhaps some of the existing money stocks—will flow to persons wishing to purchase it for nonmonetary uses at prices higher than the fixed monetary rate.   It would seem undesirable to rely on an international money whose supply depends on variations in costs of production and fluctuations in private demands for industrial, artistic, and speculative purposes.   Actually, private demand has outstripped new mine production in the Western world since 1968.

What can take the place of gold? In domestic monetary systems, gold has been replaced in coinage by fractional-value base-metal coins, by paper money, and by demand deposits (which are simply liabilities of commercial banks).

In short, *we have shifted domestically from money which had a commodity market value to a fiduciary money which costs much less to create, which depends on confidence in the solvency of the issuer, and whose supply must be regulated by the central government.*

The parallel development for international money is striking. Following World War I some countries agreed that currencies of nations with extensive trade and adequate gold reserves would be acceptable in settlement of payments deficits. The pound sterling and the U.S. dollar were the principal reserve assets, and these were simply deposit liabilities or short-term securities of the reserve countries. The creation of such money cost much less than the production of gold; the essential ingredient which determined its value was confidence—and loss of confidence would destroy it.

By the end of 1949, total world reserves were nearly one-fourth in convertible currencies, nearly three-fourths in gold, and a small fraction in a new form of international money—a country's right to draw currencies of other nations from the IMF up to a maximum amount of twice its quota. This IMF *tranche* was fiduciary money supplied by the IMF in return for interest and the borrowing nation's promise to repay. This line of credit at the IMF was reinforced for some nations by so-called swap arrangements whereby central banks agreed to exchange their own currencies up to specified amounts.

The move toward fiduciary money was furthered in 1969 after market indications of lack of confidence in the dollar threatened the continued usefulness of that major component of the world's money supply. The members of the IMF voted to create special drawing rights (SDR's) and to distribute these among member nations in proportion to their quotas. SDR's are guaranteed in terms of gold, and they yield interest. They are backed by the promises of member nations to accept them in settlement of international debts up to an amount equal to three times the recipient nation's allocation.

Some $9\frac{1}{2}$ billion of SDR's have been created, in three installments on January 1 of 1970, 1971, and 1972. Note that *this "paper gold" constitutes a permanent addition to the international money stock, whereas the regular drawing rights are simply intermediate-term credits which must be repaid.*

By the end of 1973, less than 24 percent of world reserves were held in gold, 10 percent were claims against the IMF, over 66 percent were convertible foreign exchange (claims against other governments or banks). Some analysts feel that the creation of international money has been made too cheap. Reserves do not have to be "earned" either by producing gold or by supplying goods and services in international trade. How will the total amount of international reserves be determined and controlled? Theoretically, the IMF could calculate the "base" reserves needed by each country and add these to obtain a world total. Then it might be given power as an international central bank to control the outstanding total of some or all components of world money supply to maintain the desired total. But will member nations agree to give this power to an international monetary institution? And can the total be controlled when three components exist—gold, convertible currencies, and IMF credit? Won't there be a tendency for nations experiencing balance-of-payments deficits to pay with I.O.U's to other nations or to the IMF, to avoid the necessity of domestic restraint, devaluation of their currencies, or imposition of controls? Perhaps surplus nations or the IMF membership as a whole will be able to resist such pressures, but the real possibility emerges that the supply of reserves will come to be influenced by international power politics to an increasing extent.

An immediate problem in this area is what to do with the stock of liquid claims against the United States which form a major part of world reserves but which have lost the confidence which is required for an acceptable money asset. Perhaps two devaluations of the dollar and a continued floating exchange rate will bring the U.S. balance of payments back into equilibrium or surplus, so that the stock of dollars outstanding can be stabilized or reduced. Then a restoration of confidence might lead to shifts of some dollar claims from official to private ownership and a willingness of central banks to hold the remaining stock as part of their reserves.

This story must be left unfinished. It will probably take years for international monetary arrangements to be agreed upon. Problems of national sovereignty and freedom of domestic economic policies are involved. And interrelated problems of trade restrictions, procedures for changing exchange rates, and other mechanisms for redressing balance-of-payments disequilibria must be worked out at the same time.

Chapter 20

Economic Growth

THE NATURE AND IMPORTANCE OF ECONOMIC GROWTH

In the lecture he delivered in Stockholm when he received the Nobel Prize in Economic Science (December 1971), Simon Kuznets stated:† "A country's economic growth may be defined as a long-term rise in capacity to supply increasingly diverse economic goods to its population, this growing capacity based on advancing technology and the institutional and ideological adjustments that it demands." He went on to say that the modern epoch of economic growth dates back to the late eighteenth century and was limited largely to some 15 to 18 countries—most of Europe, the overseas offshoots of Western Europe (Australia, Canada, New Zealand, and the United States), and Japan—barely one-quarter of the world's population. Four of the principal characteristics of this modern epoch were listed and discussed by Kuznets.

1 Modern rates of growth of per capita product and of population in the developed countries are much higher than in previous centuries.

2 The rate of rise of productivity, that is, of output per unit of all inputs, is much higher than past rates.

3 The rate of structural transformation of the economy has been rapid in growing economies; for example, the shift from agricultural to non-agricultural occupations and more recently from industry to services, increasing importance of large-scale production units, related shifts from personal enterprise to corporate organization and from self-employment to employee status, and change in relative shares of domestic and foreign suppliers.

4 The structures of society and its ideology have changed rapidly in growing economies; for example, increased urbanization, secularization and depersonalization of human relations, breakdown of family stability with increased geographical and occupational mobility, emphasis on education in science, technology, and business management, development of money

---

† Simon Kuznets, "Modern Economic Growth: Findings and Reflections," *American Economic Review*, June 1973, p. 247.

and credit markets, and higher valuation of economic motivations and status.

Rough comparisons of growth rates for noncommunist developed countries are estimated as follows:

	Middle Ages to Mid-Nineteenth Century	Period of Modern Growth
	(Approximate growth rate, % per year)	
Population	0.2 to 0.25	1
Output	0.4 to 0.45	3
Output per capita	0.2	2

The modern growth rates imply that in one century population would be multiplied roughly 3 times, output nearly 20 times, and output per capita over 7 times.

Kuznets suggests† that "...some new major growth source, some new epochal innovation, must have generated these radically different patterns. And one may argue that this source is the emergence of modern science as the basis of advancing technology—a breakthrough in the evolution of science that produced a potential for technology far greater than existed previously."

Growth is usually one of the most important economic objectives of a nation. An expanding flow of output is considered desirable because it permits an increase in real consumption by the nation's inhabitants and in the resources available for providing national security and power in international relations, for example, by giving assistance to underdeveloped nations. If the rate of economic growth exceeds that of population, an increase in real income per capita occurs, making possible a rising standard of living for the nation—more or better food, housing, health care, education, recreation, governmental services, and leisure.

Despite its obvious benefits, the desirability of high growth rates has been increasingly questioned in recent years. Even back in 1960, the President's Commission on National Goals implied that increasing costs should set limits on the rate of growth:‡ "The economy should grow at the maximum rate consistent with primary dependence upon free enterprise and the avoidance of marked inflation." Since that time, other costs of economic growth have been stressed, including pollution of air and water, deterioration of the life span and quality of life in congested urban areas, depletion of natural resources (land, minerals, fuels), growing inequality of income between advanced industrial nations and underdeveloped economies, and the social and political disruptions stimulated by rapid technological and economic changes.

Some ecologists and economists maintain that a continued compound-interest growth of world population and output is impossible. In a system with finite stocks of some essential economic ingredients (land, fresh air and water supplies,

† Ibid., p. 249.
‡ President's Commission on National Goals, *Goals for Americans*, p. 10, Prentice-Hall, Inc., Englewood Cliffs, N.J., 1960.

minerals, fuels), exponential growth of population and output becomes impossible; even a stable system must recycle the essential materials; and competition for scarce resources will intensify, leading to elimination of some competitors. Others argue that technological advances can permit continued growth and that government tax policy and controls can stop the growth of pollution and waste by imposing their costs on the private parties responsible.

Given the emergence of such problems, an understanding of economic growth becomes a crucial tool for intelligent analysis and action on the questions: Is economic growth desirable? How fast? In what directions?

In addition, there are the related problems:

1   What fraction of our lives do we want to devote to economic pursuits, compared with the fraction spent as dependent children, getting a formal education, home-making, or in retirement?

2   How do we want to allocate our time during the years we are in the labor force, for example, how many hours per year for work, bodily rest and maintenance, cultural and recreation activities?

3   How many children do we want, or how fast should the population grow?

MEASURING ECONOMIC GROWTH

Economic growth has been defined as the increase of potential output of the economy, and it is often measured by the *trend rate of rise of real GNP at full employment*. Several inadequacies of this definition and measurement should be noted.

1   It does not correct for population increase. For some purposes it is more relevant to define growth as the rate of increase in potential output per capita, which will equal the growth rate defined above minus the rate of rise of population.

2   It does not count increases in voluntary leisure. Leisure is a desired good, along with the products which are included in GNP. So we might measure growth as the rate of rise in potential output, holding constant the percentage of population in the labor force and the average hours worked per year. Essentially, the resulting growth rate becomes the rate of rise in output per man-hour, that is, rise in the average productivity of labor.

3   Inadequacies of GNP statistics will be reflected in the measurement of growth to the extent that the desired output measure changes its ratio to GNP. (*a*) Errors in price indexes may cause true aggregate output to rise somewhat more rapidly than constant-dollar GNP. (*b*) Net national product is more relevant for measuring economic growth but capital consumption allowances are quite surely improperly measured. (*c*) Some government services provided to businesses should probably be counted as intermediate products and deducted from GNP to obtain a truer measure of final output, for example, services of roads and aviation facilities, police and fire protection, provision of economic statistics and research used by businesses. (*d*) A shift of production from home to commercial establishments

may raise GNP; for example, laundry, once done at home, may be sent to a commercial laundry. Or it may lower GNP, as when a family does its own housework, and thus does not pay wages to a domestic servant. (*e*) Finally, the deterioration of the environment caused by productive processes is an unrequited input to production whose value should properly be deducted from GNP to arrive at net product of the economy. That is, net social product equals measured net national product less unmeasured "depreciation" of environmental capital.

4 It should be noted that the growth in real GNP is not necessarily a measure of rise in well-being or standard of living, even if calculated on a per capita basis. Divergences may occur if the distribution of incomes and expenditures differ from the allocations which would exist in competitive markets. For example, taxes to finance military expenditures, rationing and controls, government regulation of business, or inflation may distort optimal flows. Also, the population movement to the cities, accompanying economic growth, involves increases in personal expenditures for transportation, medical care, sanitation, police and fire protection, and other urban costs which should be deducted from increased output to arrive at net gains in well-being.

With all these reservations and qualifications in mind, real GNP still provides the most comprehensive and useful single indicator of economic growth. Its growth rates may be estimated back one to two centuries for some developed nations, and United Nations efforts during the postwar years now permit some cross-section comparisons for many countries in the noncommunist world.

Table 20-1 summarizes data for 14 developed nations, from Simon Kuznets's book *Economic Growth of Nations*. Growth rates in real GNP, population, and product per capita are given for each country during its own years of modern economic growth. Countries are arranged from top to bottom roughly in the historical sequence of entry into their eras of growth. The arithmetic average of the growth rates for total product is 2.8 percent per year, but it is noteworthy that in the first seven countries only one exceeds the average rate, and in the last seven countries none falls below the average. Apparently, countries which began their growth later were able to grow faster, perhaps because they could import advanced technological knowledge and equipment and also capital funds from the older developed nations.

Population growth averages a little over 1 percent per year for the whole group of countries. Above-average rates occur primarily in the overseas offshoots of European nations—the United States, Canada, and Australia. High rates of natural increase and immigration both contributed to the high growth rates.

As for product per capita, average growth was 1.7 percent per year. This growth rate is at or below the average for the seven older industrialized countries, and at or above the average for the younger countries except Australia. Within these 14 developed countries, rapid population growth does not seem to depress the rate of increase in output per capita; rather, it accelerates growth of output.

The right-hand column gives the average GNP per capita for each country

TABLE 20-1   Growth Rates for Developed Countries in Their Periods of Modern Economic Growth

Country	Periods Covered	Growth Ratio (% per year)			1965 GNP per Capita (U.S. dollars per yr)
		Total Product	Population	Product per Capita	
Great Britain–United Kingdom	1765–1785 to 1963–1967	2.2	1.0	1.2	1,870
France	1831–1840 to 1963–1966	2.0	0.3	1.7	2,047
United States	1834–1843 to 1963–1967	3.6	1.9	1.6	3,580
Germany	1850–1859 to 1963–1967	2.7	1.0	1.7	1,939
Belgium	1900–1904 to 1963–1967	1.9	0.5	1.3	1,835
Netherlands	1860–1870 to 1963–1967	2.5	1.3	1.2	1,609
Switzerland	1910 to 1963–1967	2.4	0.8	1.5	2,354
Sweden	1861–1869 to 1963–1967	3.2	0.6	2.6	2,713
Denmark	1865–1869 to 1963–1967	2.9	1.0	1.9	2,238
Norway	1865–1869 to 1963–1967	2.8	0.8	2.0	1,912
Canada	1870–1874 to 1963–1967	3.5	1.8	1.7	2,507
Japan	1874–1879 to 1963–1967	4.0	1.2	2.8	876
Italy	1895–1899 to 1963–1967	2.8	0.7	2.1	1,100
Australia	1900–1904 to 1963–1967	3.0	1.7	1.2	2,023
(Averages)		(2.82)	(1.04)	(1.75)	(2,040)

Source: Simon Kuznets, *Economic Growth of Nations*, tables 1 and 2, Harvard University Press, Belknap Press, Cambridge, Mass., 1971.

in the year 1965, measured in U.S. dollars per person. The arithmetic average is $2,040 per year per person. The seven older developed countries had a little higher average than the younger seven, $2,175 compared with $1,910, but the higher annual growth rate for output per capita in the younger countries, 2.04 percent compared with 1.46 percent, would close the gap in less than 25 years.

Table 20-2 summarizes data for seven less-developed countries for which usable data are available for more than 50 years. Countries have been listed in order of their income per capita in 1965. As shown in the right-hand column, the values are much lower for these countries than for the developed countries in the preceding table. However, for all the countries except India and Egypt,

TABLE 20-2 Growth Rates for Less-Developed Countries

Country	Periods Covered	Growth Rates (% per year)			1965 GNP per capita (U.S. dollars per yr.)
		Total Product	Population	Product per Capita	
India	1861–1869 to 1881–1889	1.8	0.5	1.2	86
	1881–1889 to 1901–1909	0.1	0.5	−0.3	
	1901–1909 to 1952–1958	1.6	0.9	0.6	
	1952–1958 to 1963–1967	3.5	2.3	1.2	
Egypt	1895–1899 to 1945–1949	1.1	1.4	−0.2	185
	1945–1949 to 1963–1966	5.2	2.5	2.6	
Philippines	1902 to 1938	3.0	2.1	0.9	255
	1938 to 1950–1954	1.8	2.2	−0.4	
	1950–1954 to 1963–1967	5.9	3.2	2.7	
Ghana	1891 to 1911	2.8	1.0	1.8	312
	1911 to 1950–1954	3.8	2.5	1.3	
	1950–1954 to 1963–1967	4.1	2.7	1.3	
Mexico	1895–1899 to 1925–1929	1.8	0.6	1.3	461
	1925–1929 to 1963–1967	4.7 to 6.1	2.7	2.0 to 3.3	
Jamaica	1832 to 1930	1.0	1.0	0	504
	1929–1931 to 1950–1952	2.5	1.6	0.9	
	1950–1952 to 1963–1966	7.7	1.7	5.9	
Argentina	1900–1904 to 1925–1929	4.6	3.4	1.1	811
	1925–1929 to 1963–1967	3.5	2.5	0.9	
(Averages for most recent periods)		(5.0)	(2.5)	(2.5)	(373)

Source: Simon Kuznets, *Economic Growth of Nations*, table 3, Harvard University Press, Belknap Press, Cambridge, Mass., 1971.

the levels were higher than Kuznets's estimates of $200 to $250 per capita (at 1965 prices) which seem to have been the minimum levels for the presently developed countries at the times when they entered on their periods of rapid growth. Indeed, the data in Table 20-2 indicate that, except for Argentina, total product in these less-developed countries has accelerated greatly in the most recent periods covered, the average being about 5.0 percent per year. Population has risen more rapidly in recent periods also, averaging 2.5 percent per year. Nevertheless, the average rate of growth in output per capita has been about 2.5 percent per year in recent periods, above the average growth rate of the younger developed countries in Table 20-1—2.0 percent per year for per capita output.

If these data are typical of economic progress in other less-developed countries, many of them may be moving into an era of modern economic development. It should be noted, however, that higher percentage growth rates for less-developed countries are consistent with a widening gap in levels of output per capita. Thus, a 2.5 percent per year rise at a level of $400 per person annually amounts to an increase of $10 per person each year, whereas a rise of 1.7 percent per year on a

TABLE 20-3 Economic Growth of the United States Economy, 1800 to 1972

Time Period	Rates of Growth (% per year)		
	Real GNP	Population	Real GNP per Capita
1. 1800 to 1840	4.3	3.0	1.3
2. 1834–1843 to 1859	4.8	3.1	1.7
3. 1859 to 1879–1888	4.0	2.3	1.6
4. 1880–1889 to 1910–1914	3.4	1.9	1.5
5. 1910–1914 to 1963–1967	3.1	1.3	1.7
6. 1929 to 1954	3.0	1.2	1.8
7. 1954 to 1972	3.8	1.4	2.4

Sources: Lines 1 through 5: Simon Kuznets, *Economic Growth of Nations*, table 1, p. 13, Harvard University Press, Belknap Press, Cambridge, Mass., 1971. Lines 6 and 7: Council of Economic Advisors for Potential GNP, in *Long Term Economic Growth, 1860–1970*, p. 182, U.S. Department of Commerce, and in *Business Conditions Digest*, July 1974, p. 95; and *Economic Report of the President*, Feb. 1974, for population including armed forces overseas.

base of $2,000 per person annually amounts to an increase of $34 per person each year.   It would require a growth rate 5 times as great in the less-developed country whose income per capita is one-fifth that in a developed country before the dollar level of income would rise by the same absolute increment as in the developed country.   In more ways than one, "the curse of the poor is their poverty."

Since we shall be discussing economic growth with primary reference to the United States, we may well examine its record in a little more detail.   Table 20-3 shows the record for the United States since 1800, when this country was just entering its period of modern economic growth.   (Business-cycle influences have been largely eliminated by averaging several years for the beginning and ending points or by using potential GNP instead of actual values.)   The growth rate of real GNP accelerated until 1859 and slowed thereafter, except from 1954 to 1972, a period which may have been unusual.   Population growth slowed continuously after 1859 because of declining birth rates and a slowing of immigration after World War I.   Real GNP per capita accelerated in the early 1800s and then maintained a near-constant rate of 1.5 to 1.8 percent per year for more than a century.   The rise in the latest period may be associated with the Southeast Asia war or with a rise in the ratio of labor force to population as the youth born in the first decade after World War II moved into the labor force.   It seems unlikely that such a high growth rate of GNP with a low rate of rise of population can persist.

SOURCES OF GROWTH IN POTENTIAL OUTPUT

Ultimately, economic growth springs from many persons' desires to increase their knowledge and improve their own living standards, or those of their children or other persons.   It consists of an increase in economic product per man-hour worked, that is, an increase in labor productivity.

What factors determine the growth of output per man-hour? The *personal characteristics* of the working population are vital—their health, education, motivation and discipline, initiative, attitudes toward innovation and risk, and confidence in, and expectations of, future improvement. The *characteristics of the economic, legal, and social systems* are important—laws regarding property rights and contracts, forms of business organization, rewards to innovators and entrepreneurs, size and competitiveness of markets, ability and willingness to adapt social institutions to realize the economic potentials of science and technology, well-developed monetary system, government-business relations which are stable and supportive, reasonable balance in income distribution and power of labor and management, an educational system that supports science and technology and development of needed occupational skills.

Natural resources and their development, of course, affect output per man-hour also—climate, arable land, forests and minerals, fossil fuels, rainfall, and water resources for use in agriculture, energy generation, human consumption, transport, fishing, and manufacturing.   Finally, *capital goods*, the man-produced goods used in production, are a vital determinant of labor productivity—improvements to farmland plus agricultural machinery, improved varieties of seeds, fertilizers, and pesticides; plant and equipment for mining and refining ores and fuels and for generating mechanical and electric energy; homes, commercial, school, and public buildings; streets and roads, air and water navigation systems and terminals; manufacturing plants and equipment; transportation and communication equipment.   To the extent that capital goods augment the output of workers using them by more than the inputs required to produce the capital goods, there is a net gain in productivity from using "round-about," capital-intensive production processes.

Changes through time in these determinants of economic growth affect our macroeconomic model by shifting the production function upward to higher outputs at any given labor input.   Since economic growth is measured by comparing *full-employment* or potential outputs at two time periods, we would be calculating the rate of increase of real GNP between points such as A and F in Fig. 20-1. Part of the increase in full-employment output stems from the rise in full-employment man-hours from $L_{FE,1}$ to $L_{FE,2}$, presumably because of a rise in labor force. But this increase would have amounted only to $A'B$ if the original production function had remained unchanged.   Distance BF represents the increment to full-employment output which is attributable to changes in variables other than man-hour inputs, for example, increased education of the labor force, use of more energy or capital per worker, shifts of workers from lower to higher productivity industries, or improvements in the organization or technology of production processes.   Such gains are represented graphically by a counterclockwise rotation of ray $0A$ to $0F$, since the slope of such lines from the origin are proportional to the average productivity of labor APL.   Some economists have attempted

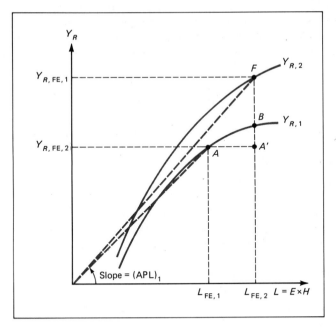

Figure 20-1 Shift of the production function during economic growth.

to assess statistically the contribution of various factors to the rise in APL, but "growth accounting" becomes very tricky. In addition to the inadequacy of measures of real GNP, statistical estimation of contributions made by separate factors is made difficult because some causal factors have no clear-cut historical data and many are closely interrelated.   For example, measures of health and education of the labor force may be lacking, and major innovations are usually accompanied by changes in quality and quantity of capital goods and by occupational shifts in the labor force.   Note too that some inputs to production, for example, clean air and rainfall, government services, and managerial innovations, are not purchased in markets in a manner that permits determination of their marginal productivities.

In algebraic form, we may write the short-term production function at a given period in time as

$$Y_R = f(L) = \text{APL} \times L$$

where

Y_R = real GNP (billdar at 1958 prices)
L = flow of labor input (man-hours worked per year)
APL = average productivity of labor ($ at 1958 prices per man-hour worked)

The measurements of economic growth compare full-employment points on a succession of shifting short-run production functions.

$$Y_{R,\text{FE}} = (\text{APL})_{\text{FE}} \times L_{\text{FE}}$$

If we designate g as growth rate (percent per year), for relatively small rates of rise (say less than 10 percent per year), we may write approximately

$$g_{Y_R} = g_{\text{APL}} + g_L$$

where *full-employment conditions are assumed*. In recent years, the Council of Economic Advisers estimates that the full-employment GNP for the United States is growing about 4.0 percent per year (g_{Y_R}), and that this is accounted for by a 2.5 percent per year rise in output per man-hour (g_{APL}) and a rise in labor inputs of 1.5 percent per year (g_L). The growth in labor supply arises from a 1.75 per cent per year increase in labor force or employment coupled with a 0.25 percent per year decline in average hours worked per year. Clearly, changes in labor productivity account for more of our economic growth than does increase in labor inputs; these changes are responsible for practically all the growth in real GNP per capita, since the ratio of labor force to population has stayed nearly constant.

Presumably, additional elements contributing to growth might be isolated. Sumner Slichter, in his *Economic Growth in the United States*, summarizes his survey of technological trends in the American economy as follows:[†] "The industrial revolution breaks up into three phases—a first phase when principal reliance was placed upon the increase in capital (1800 to 1900); a second phase when principal reliance was placed upon the use of energy (1900 to 1929) and a third phase characterized by revolutionary changes in the art of management and a sensational growth of technological research (1929 to present)." He suggests that a fourth phase may be beginning, "characterized by machines which, instead of supplementing man's muscles, will help to think and will perform thinking operations for him"[‡]—a phase emphasizing computers and cybernetics. It might be possible to utilize series for energy utilization, managerial education, research and development efforts, and computer capabilities in the economy and to estimate their contributions to the growth rate of output. Other factors which have been suggested include attitudes of business executives toward risk taking and innovation, work attitudes and incentives, and government capital, but measurement problems are currently very great.

THE MATHEMATICS OF ECONOMIC GROWTH AND INCOME DISTRIBUTION

To relate economic growth to income distribution, capital stock K is sometimes entered separately in a *production function* which specifies that output depends on

---

† Sumner H. Slichter, *Economic Growth in the United States*, p. 78, Collier Books, The Macmillan Company, New York, 1961.
‡ Ibid., p. 84.

inputs of labor L and of capital K and is augmented over time by a factor representing productivity gains (A). Thus

$$Y_R = Af(L, K) \tag{20-1}$$

where Y_R and A are measured in billdar at constant prices, L is in man-hours per year (sometimes adjusted for quality changes), and K is \$ billions at constant prices.

The formula for a small change in Y_R from one year to another may be written in terms of changes in the quantities of factor inputs and in their marginal productivities.

$$\Delta Y_R = (\text{MPL}) \, \Delta L + (\text{MPK}) \, \Delta K + \frac{Y_R}{A} \Delta A \tag{20-2}$$

The last term comes from the equality of percentage changes in Y_R and A when L and K are constant, that is, from $\Delta Y_R / Y_R = \Delta A / A$. For percentage changes, we divide all terms in Eq. (20-2) by Y_R and alter the terms on the right to obtain percentage changes of the input.

$$\frac{\Delta Y_R}{Y_R} = \text{MPL} \, \frac{L}{Y_R} \frac{\Delta L}{L} + \text{MPK} \, \frac{K}{Y_R} \frac{\Delta K}{K} + \frac{\Delta A}{A}$$

If the percentage changes are all measured on a per year basis, they become annual growth rates, and we have

$$g_{Y_R} = \text{MPL} \, \frac{L}{Y_R} g_L + \text{MPK} \, \frac{K}{Y_R} g_K + g_A \tag{20-3}$$

Thus, the production function tells us that *the growth rate of total output will be a weighted sum of the growth rates for labor input, for capital stock, and for output attributable to productivity gains.* The weights on the growth rates for labor and capital are the elasticities of output with respect to each input in turn, with other growth determinants constant.

Now consider the *distribution of total income.* Income payments are solely to suppliers of labor and capital.

$$Y_R = wL + \rho K \tag{20-4}$$

where

$w =$ constant-dollar wage rate, that is, dollars per man-hour in the base year
$\rho =$ constant-dollar rate of return to capital (including land), or dollars per year per dollar of capital stock, all at base-year prices

Initially, let us combine the information from the production function and from the income distribution equation on the basis of two simplifying assumptions:

1   That there is no increase in productivity, so that ΔA and g_A are zero

2 That the income payments per unit of labor and capital equal their marginal productivities, that is, $w = \text{MPL}$ and $\rho = \text{MPK}$

Under these conditions, Eq. (20-3) becomes

$$g_{Y_R} = \frac{wL}{Y_R}g_L + \frac{\rho K}{Y_R}g_K$$

or

$$g_{Y_R} = f_L g_L + f_K g_K$$

where f_L and f_K are the total income shares of labor and capital. As is obvious from Eq. (20-4),

$$f_L + f_K = 1$$

Thus, under the stated assumptions, we find that the growth rate for total output (g_{Y_R}) is a weighted average of the growth rates for labor and capital, the weights being the income shares of the respective factors of production. If growth rates g_L and g_K are equal, then g_{Y_R} will also be the same; in other words, our assumptions imply constant returns to scale. Under these assumptions, one can also conclude that factor payments based on marginal productivities will distribute all of an increment of output as incomes to labor and capital.

$$\Delta Y_R = (\text{MPL})\,\Delta L + (\text{MPK})\,\Delta K = w\,\Delta L + \rho\,\Delta K \qquad (20\text{-}5)$$

When we remove assumption 1 above, we must remove assumption 2 also. When A changes, the incremental production function relation, Eq. (20-2), taken with the incremental income distribution equation from Eq. (20-4), leads to the equality

$$\Delta Y_R = (\text{MPL})\,\Delta L + (\text{MPK})\,\Delta K + \frac{Y_R}{A}\Delta A = \Delta(wL) + \Delta(\rho K) \qquad (20\text{-}6)$$

The increase in output exceeds the sum of marginal productivity times incremental inputs of labor and capital, the excess being the increased output stemming from productivity gains. This increment of output will be allocated as income to labor or capital, or both, so that either w or ρ increases, or both do.

These growth, productivity, and income distribution relations are illustrated in Fig. 20-2. In the initial period, the economy is on the production function curve BC at point B in panel (a), with full-employment labor input L_1 and output $Y_{R,1}$. Capital stock is K_1 and the productivity level A_1. In the final period the economy is on the production function curve FG at point G, with full-employment labor input L_2 and output $Y_{R,2}$. Capital stock has risen to K_2 and the productivity level to A_2. To show the components of increase in Y_R, we may insert an intermediate production function curve DE, assumed to correspond to capital stock K_2 but to productivity level A_1. (To illustrate the concept of constant returns to scale, the ratio K_2/K_1 is assumed to be the same as L_2/L_1. It follows

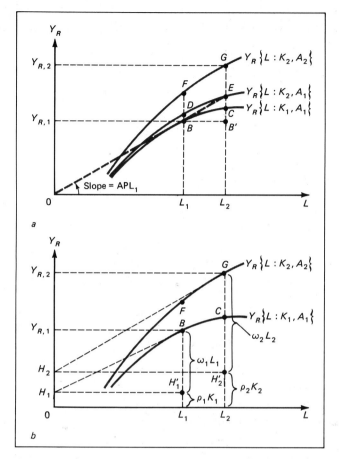

Figure 20-2   Growth, production functions, and income distribution.
(*a*) Production functions.   (*b*) Income distribution.

that point E on this production function at L_2 is on the ray $0B$ from the origin, so that the output ratio $L_2 E/L_1 B = 0L_2/0L_1 = K_2/K_1$.) We may then identify the components of rise in output as follows:

$$\Delta Y_R = (\text{MPL}) \, \Delta L + (\text{MPK}) \, \Delta K + \frac{Y_R}{A} \Delta A$$

or

$$B'G = B'C + \text{CE} + \text{EG}$$

$B'C$ measures the rise in output caused by increased labor input, with capital stock and technology constant.   CE measures the increment to output from the rise in capital stock with labor input and technology constant.   EG measures the increase in output attributable to productivity improvements, with labor and capital inputs constant at L_2 and K_2.

To visualize the allocation of income, refer to Fig. 20-2(b). Draw a tangent to the initial production function curve BC at point B and extend it to intersect the vertical axis at H_1. Since the slope of this tangent is the marginal productivity of labor, the distance $H_1 Y_{R,1}$ (or $H_1' B$) will equal $(\text{MPL}_1) L_1$. If unit payments for factor inputs equal marginal productivities, the graphical interpretation of the income distribution equation

$$Y_R = w Y_R + \rho K$$

becomes

$$0 Y_{R,1} = H_1 Y_{R,1} + 0 H_1$$

where $H_1 Y_{R,1}$ is income to labor and $0 H_1$ is income to capital (including land or natural resources).

Similarly, a tangent to the final production function curve (FG) at G will cut the vertical axis at H_2, and the allocation of total output (incomes) to labor and capital will be proportional to line segments $H_2 Y_{R,2}$ and $0 H_2$ respectively. If we assume unit-factor incomes w and ρ to be equal to their marginal productivities both in initial and final states (B and G), then any change in income shares depends on how the increases in labor, capital, and technology affect the marginal productivity of labor. The increase in L will tend to reduce MPL, the rise in K will tend to increase MPL, and improved technology would be expected to raise MPL unless the innovation were strongly laborsaving.

To keep income shares constant in the move from B to G, the real wage rate w (assumed equal to marginal productivity of labor) would have to increase in the same proportion as the average productivity of labor. This follows from

$$\text{Labor share} = \frac{wL}{Y_R} = \frac{w}{\text{APL}} = \frac{\text{MPL}}{\text{APL}}$$

During this century it appears that the share of labor in national income has risen, especially after 1929. Apparently a rise in capital stock per man-hour or the labor-saving nature of technological innovations has raised the marginal productivity faster than average productivity of labor, or increased bargaining power of labor has raised rates of pay above MPL, or some combination of these influences. At any rate, it does appear that labor has managed to acquire its share of the productivity gains from technical advance.

The mathematics of the production function gives some basis for decomposing the growth rate of full-employment real GNP into the contributions arising from growth in labor supply, capital stock, and total factor productivity (A). In Eq. (20-3) the growth rate of Y_R was expressed as

$$g_{Y_R} = \text{MPL} \frac{L}{Y_R} g_L + \text{MPK} \frac{K}{Y_R} g_K + g_A \tag{20-3}$$

If we assume that MPL equals w and MPK equals ρ, then $(MPL)(L/Y_R)$ equals f_L and $(MPK)(K/Y_R)$ equals f_K, where labor income share is approximately 0.75 and property income share is approximately 0.25 for the postwar United States. Full-employment man-hours have risen about 0.8 percent per year (g_L) in this period and gross capital stock about 3.6 percent per year (g_K).   From the known growth rate of 4.0 percent per year (g_{Y_R}), we can then calculate productivity rise (g_A) as a residual.

$$g_{Y_R} = f_L g_L + f_K g_K + g_A \tag{20-7}$$

or

$$4.0 = (0.75)(0.8) + (0.25)(3.6) + g_A$$
$$4.0 = 0.6 + 0.9 = g_A$$
$$g_A = 4.0 - 1.5 = 2.5\% \text{ per year}$$

Rough though the calculations may be, it is clear that *productivity gains contribute more to economic growth than do increases in labor and capital combined.*

Some of the factors underlying productivity gains can be identified, but evaluation of their independent contributions and their interactions is extremely difficult.   Some of the influences may be regarded as quality changes in labor services and capital stock, for example, rising educational attainments of the labor force or more efficient, higher-speed machinery.   Attempts have been made to "embody" such changes into quality-adjusted measures of labor or capital inputs, thus explaining more of the rate of growth in terms of factor inputs and reducing the share attributed to the unexplained productivity factor.

However, some determinants of the growth rate seem to be inescapably "disembodied," for example, shifts of proportions of output among industries, economies of scale, changes in terms of international trade, rationalization of production, and government support or hindrance of efficient utilization of resources.   Also, there are complex interactions and feedback among the factors affecting the growth of output and productivity.   Educational advances affect the rate of technological and managerial innovation, and technological advances increase the demand for educated workers.   New products and processes stimulate demand for consumer and investment goods; growth of aggregate demand provides markets for output of new capital goods, saving to finance investment, and profits to stimulate introduction of further innovations.

One can only conclude that the mathematics of economic growth displays some interesting and meaningful relationships but that the growth of an economy is a far more complex phenomenon than economists can analyze with available data.   The causal factors extend strictly economic factors into the demographic, technological, cultural, political, ideological, institutional, and natural environments and structures of a nation.

AGGREGATE DEMAND AND ECONOMIC GROWTH

It is hardly conceivable that *potential output* of a country could increase over a long period of time at a rate appreciably different from the growth rate of *aggregate real demand*.   If demand should tend to rise faster than output, then shortages of goods and inflation, plus fiscal and monetary policies, would restrain the growth of real aggregate demand, and the profitability of expansion of output would stimulate the growth of potential output.   Ultimately the two rates would approximate each other.   If demand should tend to rise slower than potential output, then emerging unemployment might slow the rise in labor inputs (in average hours or in labor force); or reduced investment (in view of idle plant and equipment) might slow the rise in capital stock; or introduction of innovations might be reduced—all tending to slow the growth in potential output.   Meanwhile, price declines and government policy might stimulate real aggregate demand.   These changes would bring the growth rates of full-employment output and aggregate demand into balance.

The forces underlying growth of aggregate real demand may be brought to mind by thinking of the four final demand sectors.   Real consumption will, of course, grow with population, but the earlier chapter on consumer behavior suggests that, over the long run, consumption rises more closely in step with real disposable income.   The urge to consume also supports economic growth by motivating participation in the labor force.

Investment is particularly related to economic growth.   For a first reason, the technological progress which underlies economic development opens potentially profitable business opportunities—new products, cost-reducing improvements in current methods of production, or new methods of production.   And realization of these opportunities usually requires investment in new or improved plant and equipment.   For a second reason, increases in aggregate demand stimulate net investment to keep stocks of capital goods rising to expand production capacity—the accelerator effect on investment spending.

As for net exports, over the long term, exchange rates should adjust to keep export and import flows nearly equal in money measure, unless there is a sustained net flow of capital funds or a continuing balance-of-payments deficit.   In real terms, a country may experience a stimulus to its growth if foreign demand for its exports rises more strongly than domestic demand for imports, but the effect would not be large for a country such as the United States.

Government Sector purchases of goods and services have risen faster than total output in the last-half century, from about 10 percent of GNP in 1929 to 20 percent recently.   The expansion of population, plus urbanization and increased emphasis on education and national security, has led to a faster rise in government than in private spending.   The postwar commitment of the federal government to maintain high levels of employment and incomes should assure strength of government demand when needed to keep aggregate demand rising in step with full-employment output, and it may accelerate economic growth by preventing prolonged depressions that slow the accumulation of capital stocks.

What growth rate for real aggregate demand will be required to keep pace with the growth of full-employment output? The growth in real full-employment GNP has been expressed previously [Eq. (20-6)] as a weighted combination of the growth rates for labor services, capital stock, and total factor productivity.

$$g_{Y_R} = f_L g_L + f_K g_K + g_A \tag{20-7}$$

In this relation, the growth rates of labor supply and of total productivity may be considered largely autonomous, not dependent on the growth rate of real aggregate demand. However, the growth rate of capital stock is influenced by the growth rate of real aggregate demand (g^*). In a simple model, we may assume that capital stock grows at the same rate as real demand, that is, g_K equals g^*. Using this assumption and imposing the equilibrium condition that real demand grows at the same rate as full-employment output ($g^* = g_{Y_R}$), we obtain:

$$g^* = f_L g_L + f_K g^* + g_A$$

Solving for g^*, we obtain a kind of multiplier relation relating growth rate of real demand to the growth rates of the autonomous elements underlying output growth.

$$g^* = \frac{1}{1 - f_K}(f_L g_L + g_A) = \frac{1}{f_L}(f_L g_L + g_A) \tag{20-8}$$

or

$$g^* = g_L + \frac{g_A}{f_L}$$

Figures for the U.S. economy in recent years are approximately f_L of 0.75, g_L of 1.0 percent per year, and g_A of 2.5 percent per year. From Eq. (20-8) we calculate

$$g^* = g_L + \frac{g_A}{f_L} = 1.0 + \frac{2.5}{0.75} = 4.3\% \text{ per year}$$

This growth rate is in excess of the growth rate of population or labor force because of rising productivity and because capital stock is increasing faster than labor inputs.

Can aggregate demand be kept growing at this rate? Presumably so. Investment spending includes replacement components which increase with capital stock, innovation-dependent components related to technological advances, and components motivated by capacity needs—all of which should grow with the economy. Consumer purchases may be taken as responding to rising incomes. Both exports and imports reflect growth abroad and at home. Government purchases are subject to upward pressures from population and urbanization, and in recent years they have come to be considered a balance wheel to be controlled deliberately to keep overall aggregate demand growing at a rate equal to the growth rate of full-employment output in the long run.

In the *Economic Report of the President* for 1971 (chap. 3), the Council of Economic Advisers estimated full-employment output of the economy for five years ahead. (Its $g_{Y_{FE}}$ was 4.3 percent per year.) Then the Council projected prospective claims on GNP, that is, aggregate demand by Private and Government Sectors of the economy in accordance with " present programs and tendencies." The sectors' projected demands fell short of potential output in 1975 and 1976 by only 1 to 2 percent, indicating a very narrow margin of resources available for allocation. Three possible policies were indicated to raise aggregate demand to full-employment levels.

1 Monetary policy might be used to lower saving (raise consumption) and increase investment spending. (It is worthy of note that, with stable rates of growth, the net saving ratio of the economy is calculable. See the appendix to this chapter.)

2 Taxes might be reduced to increase private consumption and perhaps investment.

3 An increase in government expenditures might absorb the unallocated output.

On the other hand, if projected demands should exceed potential output, then inflation or restrictive monetary and fiscal policies could slow the rise in aggregate demand to keep it in step with full-employment GNP.

Thus by long-term demand management, the Government Sector can hope to keep aggregate purchases expanding in line with growth of full-employment output.

RETROSPECT AND PROSPECT

Early in the history of economic theory, dating from Adam Smith's *Wealth of Nations* in 1776, much thought was directed to national economic growth in wealth, population, and production. In the late 1800s and early 1900s, the attention of economists was focused predominantly on microeconomics—marginal analysis of consumption, production, and markets. Beginning perhaps with the world-wide depression of the 1930s, the development of national income accounting, and advances in analysis of business cycles and macroeconomic theory, attention has swung back in recent years to problems of economic growth.

Nineteenth-century political economists believed that beyond the " progressive state " lay a " stationary state," enforced by limitations of land and food supply, by a bound to capital accumulation, or perhaps by voluntary restriction of population growth in order to maintain a high and cultured standard of living. Now twentieth-century political economists are speculating anew about the problems of economic growth. Are there limits to the rate of growth? What price and and what kind of growth are desirable? What would a slow-growth economy be like?

As for the limits to economic growth, we moderns have a greater understanding of the creative possibilities of scientific knowledge and its practical application. A nation can feed itself and export agricultural products with less than 4 percent of its labor force employed in agriculture. The stock of capital goods used in production is approaching $20,000 per worker in the United States. Automation and computerization have helped raise man's production capabilities to heights unimagined by nineteenth-century economists. Developments of synthetic materials and atomic energy promise to push back the limitations threatened by scarce natural resources. We have actually placed men on the moon. Are there limits to output per man-hour?

Some analysts believe the boundaries have been pushed back but not removed. Supplies of arable land, high-grade ore deposits, fresh air, pure water, and marine life may well be limiting factors. Remember that only one-quarter of the world's population were involved in the rapid pace of modern economic growth by the middle of the twentieth century, and note that world population currently doubles every 30 to 35 years. An exponentially growing quantity reaches a fixed limit very quickly late in the process. Consider the story of the water lily.† "Suppose you own a pond on which a water lily is growing. The lily plant doubles in size each day. If the lily were allowed to grow unchecked, it would cover the pond in 30 days, choking off the other forms of life in the water. For a long time the lily plant seems small, and so you decide not to worry about cutting it back until it covers half the pond. On what day will that be? On the twenty-ninth day, of course. You have one day to save your pond." Are we in something like this situation with respect to use of arable land, pollution of air, water, and land, and resources of natural gas and petroleum?

Such considerations lead into the question as to the pace and kind of economic growth that are desirable. Increasingly, we become aware of the costs of economic progress. There are *human costs* of urban living, absorption of human energy and attention in getting an education and "getting ahead," problems of occupational health and safety, the psychological and physical strains arising from rapid change, accentuation of "generation gaps" and family breakdowns, and loss of privacy. Then there are *environmental costs* of air and water pollution, noise, destruction of "green" areas and wildlife, depletion of earth and water resources. There are *social problems* such as the decay of inner cities, a "hereditary" poverty class and large welfare costs, urban transport and crime problems, the increased potential for disruptions of supply which accompanies increased specialization and interdependence, the huge burden of spending for national defense, increased government controls over economic decision making, and the political power of large business and labor institutions. Some economists have tried to correct measured GNP for some of these costs to obtain a rough "Measure of Economic Welfare" (MEW). One such calculation by Professors Tobin and Nordhaus at

† Cited in Donella Meadow, et al., *The Limits to Growth: a Report for the Club of Rome's Project on the Predicament of Mankind*, Universe Books, New York, 1972.

Yale University finds that per capita MEW rose by 1.1 percent per year from 1929 to 1965, in contrast to a 1.7 percent per year increase in per capita net national product.† If the marginal costs of rapid economic growth rise in the future and marginal benefits decline some growth critics maintain that a reduced growth of output will become preferable. Similarly, the marginal productivity of capital may decline until it no longer justifies the sacrifice of present consumption involved in additional net saving. And a slowing of the rate of capital accumulation would reduce the growth rate of GNP.

Along with a slower pace of growth, some analysts recommend a changed allocation of resources. Perhaps a diversion of funds from armaments to foreign economic assistance would provide greater protection against the explosive disparities in living standards and competition for food and scarce natural resources. Perhaps tax and subsidy policies could help direct economic demand away from energy and goods with large "pollution coefficients" and toward services and cultural pursuits with smaller impacts on the environment.

Certainly we need more analysis of the nature of a slow-growth economy for an advanced nation. What would be the consequences of accepting the present standards of affluence insofar as our use of energy and manufactured goods is concerned? With little or no growth in population and labor force, technological progress might be devoted to producing more leisure instead of more goods, to improving the quality of the environment (or counteracting its deterioration), and to raising output in education, scientific research, medicine, art, music, recreation, municipal services, and other beneficial activities. Measured productivity might slow or cease, but the quality and security of life might still improve in a slow-growth or no-growth society.

The case for quality of living and a stationary economy was expressed well by John Stuart Mill in his *Principles of Political Economy*,‡ published in 1875.

> There is room in the world, no doubt, and even in old countries, for a great increase of population, supposing the arts of life to go on improving, and capital to increase. But even if innocuous, I confess I see very little reason for desiring it. The density of population necessary to enable mankind to obtain, in the greatest degree, all the advantages both of co-operation and of social intercourse, has, in all the most populous countries, been attained. A population may be too crowded, though all be amply supplied with food and raiment.... A world from which solitude is extirpated is a very poor ideal.... Nor is there much satisfaction in contemplating the world with nothing left to the spontaneous activity of nature;... every flowery waste or natural pasture ploughed up, all quadrupeds or birds which are not domesticated for man's use exterminated... every hedgerow or superfluous tree rooted out.... If the earth must lose that great portion of its pleasantness... for the mere

† William Nordhaus and James Tobin, "Is Growth Obsolete?"*Economic Growth*, Fiftieth Anniversary Colloquium, vol. 5, National Bureau of Economic Research, New York, 1972.
‡ Chap. VI, "Of the Stationary State."

purpose of enabling it to support a larger, but not a better or a happier population, I sincerely hope, for the sake of posterity, that they will be content to be stationary, long before necessity compels them to it.

It is scarcely necessary to remark that a stationary condition of capital and population implies no stationary state of human improvement. There would be as much scope as ever for all kinds of mental culture, and moral and social progress; as much room for improving the Art of Living, and much more likelihood of its being improved, when minds ceased to be engrossed by the art of getting.

APPENDIX The Saving Ratio and Economic Growth

An interesting corollary to the growth model in which capital stock grows at the same rate as real aggregate demand is that the net saving ratio of the economy becomes determined.

Let the constant ratio between capital stock and real output be denoted by σ, the capital coefficient. That is

$$K = \sigma Y_R^*$$

From the assumed equality of g^* and g_K, we have

$$g^* = g_K = \frac{1}{K}\frac{dK}{dt} = \frac{I_{n,R}}{K}$$

since real net investment $I_{n,R}$ is the rate of change of capital stock K.

Replacing K by σY_R^* in the last equation, we find

$$g^* = \frac{1}{\sigma}\frac{I_{n,R}}{Y_R^*}$$

This says that if g^* and σ are otherwise determined, they imply a particular ratio of net investment to aggregate demand. Since net investment equals net saving, the expression may also be written

$$g^* = \frac{1}{\sigma}\frac{S_{n,R}}{Y_R^*} = \frac{s_n}{\sigma} \tag{20-9}$$

where s_n is the ratio of net saving to aggregate demand in real units.

Thus, if aggregate demand is to grow at the same rate as full-employment output, with capital stock growing at the same rate, we have, from Eqs. (20-8) and (20-9),

$$g^* = g_L + \frac{g_A}{f_L} = \frac{s_n}{\sigma}$$

Thus, in order to keep such an economy growing along such a dynamic equilibrium path, the net saving ratio must be

$$s_n = \sigma g^* = \sigma\left(g_L + \frac{g_A}{f_L}\right)$$

The required saving ratio will be higher if the capital coefficient is larger or if the growth rate of real aggregate demand and output is higher. If net saving flows based on private decision making do not equal this required level, then fiscal, monetary, or other policies must be used to ensure this proper net saving ratio if the growth rate for aggregate demand is to be maintained at the potential growth rates of labor supply (g_L) and of total factor productivity (g_A).

Approximate values for the United States in recent years are: σ, 1.6 years; g^*, 4.3 percent per year; g_L, 1 percent per year; g_A, 2.5 percent per year; and f_L, 0.75. Thus $s_n = 1.6(4.3) = 1.6[1 + (2.5/0.75)] = 6.9$ percent. This matches closely the ratio of net saving to GNP in recent years, though the value of capital consumption allowances used in computing net saving is of doubtful accuracy. Also, the assumption of a constant ratio of capital stock to real aggregate demand may be in error or subject to time lags in the adjustment of capital stock to desired levels.

Part Five

Forecasting the Economic Environment

Chapter 21

Forecasting National Economic Activity

THE ROLE OF FORECASTING IN DECISION MAKING AND POLICY FORMATION

Business managers attempt to achieve company objectives as fully as possible within an uncertain environment. The outcomes of their decisions depend on the effectiveness of their employment of company resources but also upon uncontrollable external events, such as future states of the economy, tastes and preferences of customers, and future actions of competitors. When managers can control activities, they make plans and issue orders; when they cannot control, they must make forecasts and "cooperate with the inevitable."

Since national economic activity influences incomes and behavior of customers and competitors, economic forecasts are important for successful adaptation to the uncontrollable economic environment, especially in areas of marketing, finance, purchasing, employment and union negotiations, and long-term planning of production facilities, distribution organization, mergers, and acquisitions.

In the Government Sector, economic forecasts are needed for short-term revenue estimates and for longer-term planning of the tax structure to meet future demands for government services. In addition, at the federal level, macroeconomic models are needed to permit intelligent choice of policies to control the economy in order to achieve national economic and social objectives. The outcomes of government policy actions are influenced by the responses of private decision makers in households, businesses, and financial institutions. So federal government and central bank economists need to develop quantitative models representing private sector economic behavior in order to forecast the outcomes of alternative fiscal and monetary policy actions. Only then can they choose policies which will probably guide economic activity toward their desired targets for the national economy.

TYPES OF ECONOMIC FORECASTS

The characteristics of national economic forecasts vary with the needs of users. For long-term business or government planning, annual forecasts extending five or ten years into the future are necessary. For short-term budget preparation, forecasts by quarters extending one or two years ahead are required.

If planning of manpower or facilities is involved, forecasts should be in physical volume units, or constant-dollar measures.   If the planning deals with income flows, expenditures, or balance-sheet items, then forecasts should be expressed as money flows or stocks.

Finally, the level of aggregation in the economic forecast will be determined by the detail required by users.   Perhaps forecasts of aggregate GNP will be an adequate indicator for one firm of the strength of its sales and of the economic environment in which it will be operating.   Another firm may wish to forecast disposable income, or plant and equipment expenditures, or interest rates, to assist its planning.   The federal government may want a forecast of personal and corporate incomes in order to project budget receipts, or it may want a forecast of employment and unemployment by industry and geographical region in order to select a desirable manpower policy.

In the following chapters we shall examine methods for preparing economic forecasts at two levels of aggregation:

1   Simple forecasts of major aggregates, such as real and money GNP, the GNP price deflator, aggregate labor force, employment, and unemployment

2   Forecasts of major components of aggregate demand, sector income flows, saving and borrowing, price indexes, and employment

NONCAUSAL AND CAUSAL MODELS FOR FORECASTING

There are two main approaches in forecasting future economic activity: (1) noncausal models, and (2) causal analysis.   Both approaches utilize historical data as a basis for forecasting the future—aside from clairvoyance, those are the only data we have—but they differ in the amount and nature of historical data needed and in the underlying assumptions regarding the degree of continuity between past and future economic developments.

NONCAUSAL FORECASTING

In this approach, the forecaster says: " Let's analyze the historical pattern of movement for the time series representing the economic activity we want to forecast. Then let's project that past pattern into the future as the forecast for that variable." This is the way astronomers predicted the future positions of the planets before Newton's laws of motion were formulated.   The growth of a population or an individual might be extrapolated in this way.

This approach is agnostic; no attempt is made to determine the causal forces at work.   It is assumed that the forces influencing the variable to be forecast, whatever they may be, will continue to change in the future much as they have in the past.   Because they involve a strong faith in the persistence of past patterns of change, predictions derived by noncausal approaches are sometimes referred to as " naïve forecasts."

The simplest forecast of all would be to say: " Next period's observation will be the same as the value of the variable last period." Such a forecast might be acceptable in a static economy with no uncertainty—but then forecasting would not be needed anyway! In most cases, last period's value will not be an adequate guide to the future for four reasons: (1) Economic activity may be changing because of *long-term trends* in population, productivity, or incomes. (2) *Short-term fluctuations* in demand components and the reinforcing feedback in the system may lead to cyclical movements around the long-term trend. (3) *Seasonal forces* may change from period to period for time intervals shorter than a year. (4) *Nonrepetitive, irregular forces*, such as changes in tax rates, transfer payments, money supply, major strikes, inventory adjustments, or shocks from international trade and payments flows, may alter the normal course of events.

It may be desirable to decompose the historical time series into the four component parts just listed. Each of the components may then be forecast separately. The combination of their projections provides the desired forecast of the composite variable.

Noncausal approaches are relatively simple, hence quick and inexpensive. The input data required are simply the historical record of the single series to be forecast. Little knowledge of the economic system and the behavioral responses of sectors is required. Forecast values of the given series depend on predictions of only one other variable—time. Its values can be forecast exactly at any future period!

However, noncausal approaches have shortcomings. Their assumption of a very strong continuity between past and future forces affecting sales may break down. Then future patterns will not repeat past movements. And naïve approaches do not give insight into the causal determinants of changes in the variable being forecast. Hence they are of no help to policy makers in deciding how changes in policy variables will affect economic activity, and noncausal models cannot be improved by tracking down the causal sources of error in past forecasts.

CAUSAL FORECASTING

The basic idea underlying causal forecasting is this: Changes in economic variables reflect changes in the economic decisions and behavior of demanders and suppliers in markets, and these decisions and behavior are changed by causal factors impinging on the sectors. If we can discover stable quantitative relations between changes in causal factors and the consequent changes in economic variables, we can derive the desired forecast from forecasts of changes in the underlying causal factors. If Y is the variable to be forecast and X_1, X_2, and X_3 are the underlying causal factors, then we try to discover some functional relation:

$$Y = f\{X_1, X_2, X_3\}$$

The form of the functional relation is chosen on the basis of economic theory and experience, and the coefficients in the relation are evaluated to make the equation fit the historical data closely.

Of course, this just pushes the forecasting problem back one step to the forecasting of X_1, X_2, and X_3. However, if they can be forecast more readily than Y, or if one or more of them are policy variables, then we have made a net gain.

As the causal approach is pushed back to derive equations explaining X_1, X_2, X_3, and other causal factors, we wind up with a set of equations describing the behavioral relations and institutional constraints which constitute the macroeconomic system. The culmination of this approach lies in the complete econometric models which contain as many independent equations as there are endogenous variables in the system. Assuming well-behaved functional relations, the forecaster can then calculate the values of all endogenous variables from known values of the exogenous variables, those whose values are determined "outside" the causal network described by the set of equations. If the equations involve some time lags in the effects of endogenous variables or if the exogenous variables fluctuate through time, it becomes possible to calculate the time paths of the endogenous variables—the equation system becomes a dynamic model. Many macroeconomic models such as this have been developed, ranging in size from the 8- to 11-equation models presented in this course up to computerized models of 200 to 300 equations. The larger models, of course, contain much more detail on income and expenditure components and on prices and flows in various markets, but it seems still to be an open question as to what size or form of model gives consistently better accuracy in forecasting major aggregates of economic activity.

As compared to "naïve" approaches, the causal forecasting models require more knowledge of economic theory, historical data on more variables, and more statistical expertise and calculations. Their logic entices the analyst to push ever further back along the causal linkages underlying macroeconomic variables until the model becomes very complex, hard to comprehend intellectually, and enormously time-consuming to analyze and forecast—just like the real world! The forecaster who adopts a causal approach faces difficult decisions in determining when to say: "This far and no farther. Beyond this set of equations and endogenous variables I shall either treat additional causal variables as exogenous or omit them and accept any loss of accuracy and consistency which that may entail."

Within the constraints imposed by resources available for forecasting, causal models can yield distinct benefits for their added costs. They offer prospects for greater accuracy of forecasts at cyclical turning points and at other times when the historical pattern of an economic variable is altered. If the break in continuity is caused by *forecastable* changes in causal variables, exogenous or otherwise, then causal models will outperform naïve models at crucial periods when uncertainty has been increased by abnormal changes in the economic environment. A causal model permits the analyst to build into the forecast his or her knowledge of prospective changes in fiscal or monetary policies or of lagged responses to known changes in economic conditions.

Policy makers must use causal models in order to assess the impact of their policy instruments on aggregate demand and employment. For example, they may need to know the multiplier effects of changes in government spending and tax rates, or the impact of changes in interest rates on residential construction.

Finally, the use of causal models permits the forecaster to make ex post runs on his or her model to identify sources of error in the forecast, whether of the causal variables or in the structure of the equations. Such analysis should permit progressive improvements in the forecaster's understanding of the economy and in his or her ability to forecast economic conditions accurately.

SIMPLE FORECASTS OF SOME MAJOR AGGREGATES

As an example of simple forecasting techniques, let us attack the problem of forecasting real GNP, total employment, the GNP price index, and current-dollar GNP for the year 1974.   We imagine that we are at the end of 1973 and have available correct data for the preceding years, including a good estimate for 1973 (fourth-quarter data not being reported till a month or two after year-end).   And we know about business conditions and prospects to the extent of information available at the end of 1973.

The proposed method of attack is:

1   A noncausal (time-series) determination of the long-term trend and pattern of change for each of the four variables

2   Qualitative consideration of prospective business conditions to estimate the percentage change in 1974

3   Reconciliation of the forecasts among themselves

As a guide to interpreting the postwar record of economic indicators, it is useful to keep at hand the following chronology of business-cycle expansions and contractions during the postwar years, as determined by the National Bureau of Economic Research.   (See Table 21-1.)

TABLE 21-1   United States Business-Cycle Expansions and Contractions, 1945–1973

Business-Cycle Reference Dates	Duration in Months	
	Expansion	Contraction
Trough—October 1945 Peak—November 1948	37	
Trough—October 1949 Peak—July 1953	45, W†	11
Trough—August 1954 Peak—July 1957	35	13, W†
Trough—April 1958 Peak—May 1960	25	9
Trough—February 1961 Peak—November 1969‡	105, W†	9
Trough—November 1970‡		12, W†
Average, all cycles	49	11
Average, peacetime cycles	32	10

† W indicates wartime expansions and subsequent contractions.
‡ Tentative and subject to revision.
Source: National Bureau of Economic Research, Inc. Available in *Business Conditions Digest*, U.S. Department of Commerce, May 1974, appendix E.

FORECAST OF REAL GNP

The beginning of wisdom in projecting a time series lies in examining its historical movement. Tabulating and graphing the data are the first steps. Postwar data for real GNP are recorded in column 1 of Table 21-2.

We would like to determine a *straight-line* trend in the historical data which can be extrapolated into the future as a guide to the forecast value. (Curves are notoriously hard to extrapolate accurately.) Column 2 indicates that real GNP has risen more rapidly in later postwar years, that is, that annual increments have recently tended to increase.

An accelerating rise in real GNP is to be expected because population, labor force, and labor productivity tend to grow geometrically, that is, by a constant percent per year rather than a constant increment per year. An exponential trend line rises at a constant percent per year:

$$Y_R = Y_{R,0}(10^{\beta t}) = Y_{R,0}(1 + g)^t$$

TABLE 21-2   Real GNP and Related Data, 1947–1973

Year	(1) Real GNP	(2) Change from Previous Year	(3) Log of Real GNP	(4) % Change from Previous Year
	(Billdar at 1958 Prices)			(%)
	Y_R	ΔY_R	Log Y_R	%ΔY_R
1947	309.9	–	2.4912	–
1948	323.7	13.8	2.5101	4.46
1949	324.1	0.4	2.5107	0.12
1950	355.3	31.2	2.5506	9.63
1951	383.4	28.1	2.5836	7.91
1952	395.1	11.7	2.5967	3.05
1953	412.8	17.7	2.6157	4.49
1954	407.0	−5.8	2.6096	−1.40
1955	438.0	31.0	2.6415	7.62
1956	446.1	8.1	2.6494	1.85
1957	452.5	6.4	2.6556	1.44
1958	447.3	−5.2	2.6506	−1.15
1959	475.9	28.6	2.6775	6.40
1960	487.7	11.8	2.6882	2.48
1961	497.2	9.5	2.6965	1.95
1962	529.8	32.6	2.7241	6.55
1963	551.0	21.2	2.7412	4.00
1964	581.1	30.1	2.7642	5.47
1965	617.8	36.7	2.7908	6.31
1966	658.1	40.3	2.8183	6.53
1967	675.2	17.1	2.8294	2.60
1968	706.6	31.4	2.8492	4.65
1969	725.6	19.0	2.8607	2.69
1970	722.5	−3.1	2.8588	−0.43
1971	745.4	22.9	2.8724	3.17
1972	790.7	45.3	2.8980	6.08
1973	837.4	46.7	2.9229	5.90

Source: For column 1, *Survey of Current Business*, July 1973, table A, p. 52, and recent issues; or *Economic Report of the President*, January 1974, appendix table C-2, p. 250.

where $1 + g = 10^\beta$. This curve may be transformed into a straight-line relation between log Y_R and t.

$$\log\ Y_R = \log\ Y_{R,0} + \beta t = \log\ Y_{R,0} + [\log(1 + g)]t$$

Column 3 in Table 21-2 contains values of log Y_R, and they are plotted together with time in Fig. 21-1(a). A least-squares straight line calculated for the points (1947–1971) is shown by the dashed line. Its equation is:

$$\log\ Y_R = 2.69 + 0.0159t$$

with t equal to 0 in 1959 at the midpoint of the historical time span for ease of calculation. This equation is equivalent to

$$Y_R = 489(1.0372)^t$$

It is a trend line which rises at a constant 3.72 percent per year, and it does seem to fit the historical data over the whole postwar period. In Fig. 21-1(b), the actual percentage change from the previous year is plotted and compared with the average 3.72 percent per year rise (horizontal dashed line). For the 12 years from 1948 through 1959, there are six points below and six points above the average line; the same is true for the 12 years from 1960 through 1971.

We seem to have found a good trend line, linear in log Y_R. Given that, how shall we forecast Y_R for 1974? Three possible rules are these:

1   No-change rule: Forecast next period's Y_R the same as the value last period. This rule does not use our knowledge that the series has an uptrend.

2   Trend-line rule: Forecast next period's Y_R as the trend-line value for next year. This should yield better forecasts, but does not make use of our knowledge that there are cyclical movements which persist above or below the trend line for several years at a time.

3   Average percent change rule: Forecast next period's Y_R as 3.72 percent higher than last period's value, that is, project ahead from last period's Y_R along a slope parallel to the trend line. Forecast points determined by this rule would be as far above or below the logarithmic trend line as last year's point was.

Application of these rules to a forecast of Y_R for 1974 yielded the results in Table 21-3.

If we were to choose from among these three rules solely on the basis of their "backcast" performance during 1948 through 1971, preference should be given to the average percent change rule. Over those 24 years, the mean absolute error† in predicting the year-to-year percent change of Y_R was 2.4 percentage points, whereas the comparable measure of error for the no-change rule was 4.0 percentage points, and for the trend-line rule, 3.0 percentage points.

---

† This is the average of the absolute values of differences between predicted and actual values of $\%\Delta Y_R$.

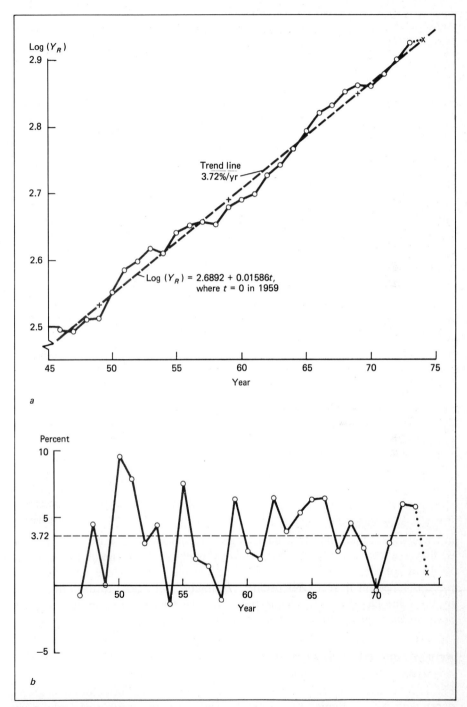

Figure 21-1 Postwar trend and cycles of real GNP.   (*a*) Log of real GNP.   (*b*) Percent change from previous year.

TABLE 21-3 Alternative Forecasts of Real GNP for 1974

	Y_R		ΔY_R	$\%\Delta Y_R$
	1973	1974		
	(billdar at 1958 prices)			%
1 No-change rule	837.4	837.4	0	0
2 Trend-line rule:				
$\log Y_R = 2.6892 + 0.01586(15)$				
$= 2.9271$	837.4	845.5	8.1	1.0
3 Average percent change				
rule: $Y_R = 837.4(1.0372)$	837.4	868.6	31.2	3.7

Is there any other basis for choosing among them—or for using some other value? If we have information regarding causal forces which will be influencing aggregate demand or supply in the forecast period, that information should be used in making the initial estimate. For example, the housing boom of 1971 and 1972 slowed in the latter part of 1973 from 2.36 million starts in 1972 to an annual rate under 1.6 million in the fourth quarter of 1973, and high interest rates on mortgages pointed to continued weakness. Inflation and the oil embargo had led to consumer uncertainty and loss of confidence, so that real consumer purchases of durable goods, notably automobiles, declined in the third and fourth quarters of 1973, and real nondurables buying dropped in the fourth quarter. Industrial production declined in December 1973, and the Arab oil embargo threatened to curtail utility and manufacturing output further in early 1974. Net exports rose smartly in 1973, but higher prices for oil imports threatened a decline in 1974. On the other hand, there were some favorable forces which indicated that any decline in business activity might be short and mild, assuming lifting of the oil embargo early in 1974. Federal budget estimates pointed to a renewed rise in federal government purchases, and state and local government purchases were expected to continue their strong rise. Surveys of business executives' investment plans and data on capital appropriations pointed to a strong rise in plant and equipment spending in 1974, perhaps 14 percent higher than in 1973.

Evidence such as this, along with published forecasts prepared by economists in universities, banks, industry, and management consultant organizations, suggested that weakness early in the year would probably limit the rise of real GNP in 1974 but not cause a decline for the year as a whole. On balance, an increase of perhaps 1 percent in real GNP seemed reasonable, very close to the value given by the trend-line rule. *The initial forecast value for real GNP in 1974 became $846 billdar at 1958 prices.* (See points marked X in Fig. 21-1.)

FORECAST OF TOTAL EMPLOYMENT

A noncausal forecast of employment can be developed in the same manner. In Fig. 21-2, total employment (including the armed forces) has been plotted on graph paper with a logarithmic vertical scale. This saves the trouble of looking up

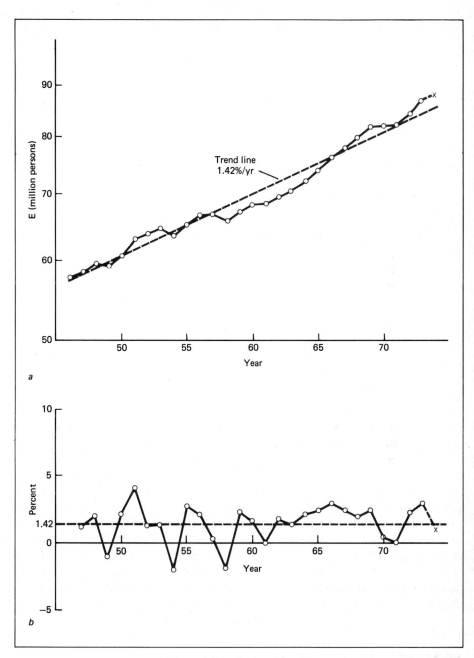

Figure 21-2   Postwar trend and cycles of total employment.   (*a*) Total employment (million persons, log scale).   (*b*) Percent change from previous year.

TABLE 21-4 Alternative Forecasts of Employment for 1974

		E		ΔE	%ΔE
		1973	1974		
		(million persons)			
1	No-change rule	86.74	86.74	0	0
2	Trend-line rule: log E = 1.8389 + 0.00613(15) = 1.9308	86.74	85.28	−1.46	−1.7
3	Average percent change rule: E = 86.74(1.0142)	86.74	87.97	1.23	+1.4

logs, and the ordinate values for total employment can be read directly. A straight-line trend has been passed through the historical points; its slope corresponds to a constant growth rate for employment at 1.42 percent per year. Its equation is:

$$\log E = 1.8389 + 0.00613t$$

or

$$E = 69.01(1.0142)^t$$

where t equals 0 in 1959, as before.

The bottom panel shows percentage changes from the previous year, fluctuating about the mean value 1.42 percent per year. In the later 1960s, employment rose faster than average during the Vietnam war build-up and because the generation born after World War II was expanding the labor force at a faster rate. Employment increased slowly during the recession of 1970 and 1971 and more rapidly in the ensuing recovery.

Given the 1973 value of 86.74 million persons employed, forecasts for 1974 might be as shown in Table 21-4.

In view of the subnormal growth in real GNP expected for 1974, a slow growth in employment would be expected, but not a decline. On balance, a rise of 1 percent seemed a good initial figure, leading to a *forecast of 87.6 million persons employed in 1974*. (See points marked X in Fig. 21-2.) This rise in employment seemed to stand in a reasonable relation to the estimated rise of 1 percent in real GNP, when compared with similar periods of slow growth in 1960 and 1970. However, it was high compared with the earlier recession years of 1949, 1954, and 1958, when employment declined.

FORECASTS OF GNP PRICE INDEX AND CURRENT-DOLLAR GNP

To round out the forecasts of major aggregates, we need to derive 1974 values for current-dollar GNP, which involves forecasting the GNP price index. Given the relation $Y = P_Y \times Y_R$, we could forecast any two of the variables and derive the

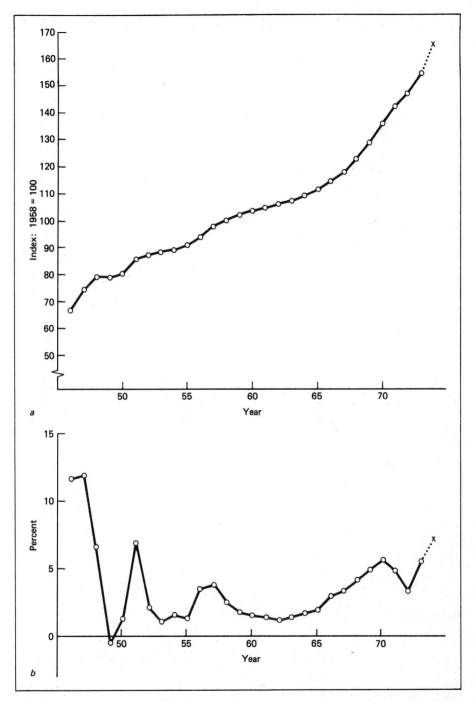

Figure 21-3 Postwar movements of GNP price index. (a) GNP implicit price index (1958 = 100).
(b) Percent change from previous year.

third.   Because inflation in recent years seems to have displayed considerable momentum of its own, a forecast value of P_Y was used to derive the forecast of Y.

Figure 21-3 displays the postwar record for the GNP price index and its pattern of change. P_Y has been plotted against an arithmetic scale in panel (*a*). The bottom panel shows percentage changes from the previous year, since is the most commonly used measure of inflation.

The four postwar periods of inflation stand out clearly: (1) the release of suppressed inflation from World War II (1946–1948); (2) the Korean war inflation (1951–1952); (3) the inflation associated with a durable goods boom (1956–1958); and (4) the Vietnam war inflation (1966 to date).   For such a relatively smooth series (high serial correlation for P_Y and its rate of change), none of the three forecasting rules described previously appears to be optimal.   Using the same inflation rate as in the previous year would be better.   But no noncausal approach will forecast the sudden changes in inflation rate which occurred in 1949, 1951, 1956, 1971, and 1973.

Probably the best system, short of a quantitative analysis of causal factors, is just to plot the data up through the most recent quarter available and then use qualitative knowledge of the prospective demand and cost pressures (plus controls, if any) to pick a percentage change which seems reasonable.   "Enlightened judgment" is a good name for the procedure.

The GNP price index in the fourth quarter of 1973 was 6.8 percent higher than a year earlier.   For 1974, demand was expected to rise moderately, but output to remain below full-employment levels.   The removal of wage and price controls, along with rises in petroleum prices, was expected to yield an inflationary "bulge" in the first half of 1974 and a slower rate of price rise thereafter.   On balance, a 7 percent price rise for 1974 seemed reasonable; this would lift the GNP price index to 164.7 for 1974, up from 153.9 for 1973.   See the forecast points marked X in Fig. 21-3.

The forecasts for P_Y and Y_R may be combined to derive a forecast of current-dollar GNP.

$$Y = P_Y \times Y_R = 1.647(846) = 1,393 \text{ billdar for } 1974$$

This involves a rise of 8.1 percent from the 1973 value of 1,289 billdar.

SUMMARY ON SIMPLE FORECASTS

We have now derived 1974 forecasts for four major aggregates, as follows:

Variable	1974 Forecast	% Change from 1973
Real GNP	846 billdar at 1958 prices	1.0
Employment	87.6 million persons	1.0
GNP price index	164.7 (1958 = 100)	7.0
Current-dollar GNP	1,393 billdar	8.1

The forecasts have been derived primarily from an examination and analysis of data for each series by itself, but the process was more than statistical extrapolation of time series.   Judgment entered at several points in the analysis—in deciding what historical span of data is relevant to the projections, in choosing among alternative forecasting rules, and in using knowledge of the current state of the economy to choose high or low values within the range of statistically acceptable forecasts.

These techniques are of general applicability to forecasting economic time series.   They are relatively simple—they involve a minimum of historical data, no complex statistical analysis, and little knowledge of macroeconomic theory or of causal relations within the economic system.   They place the forecast in the perspective of historical trends and cyclical fluctuations.   For some purposes this type of forecast may be adequate, in the sense of the old army maxim: "Never bring out more strength than the enemy requires."

From other viewpoints, the preceding approach is woefully inadequate. After all, *time* is not a causal variable, and pure statistical extrapolation of a time series is naïve, an admission of ignorance regarding the causal forces acting on a variable.   Policy makers need a causal representation of the economic system if they are to determine how changes in policy variables are related to changes in target variables.   Forecasters need to know what economic variables are exogenous and how they affect the rest of the system if they are to use their knowledge of propsective changes in government spending or taxing or in money supply to improve their economic projections.   Only a causal model of the economic system offers promise of improvement through time in one's understanding of the economy and in one's ability to forecast its variables accurately.

The next chapter presents a systematic approach for developing economic forecasts, which grows out of the framework of analysis developed earlier in this book.   It is a flexible approach which can accommodate noncausal methods for some sector incomes and expenditures, behavioral equations for other components, and choices of various stocks or flows as exogenous.   This approach leads quite naturally into complete econometric models of the economy.

Chapter 22

Short-Term Forecasting of Sector Incomes

NATURE OF SHORT-TERM GNP FORECASTING

Short-term economic forecasts usually present major components of gross national product and income flows for one to two years ahead. There is a strong seasonal pattern in their publication, with the peak flow in the fall. Business executives develop budgets and plans for the next calendar year at that time; annual revisions of GNP figures become available during the summer (usually in the July issue of the *Survey of Current Business*); and the overall picture of the current year comes into focus fairly well by the end of the third quarter.

Short-term forecasts are usually presented for quarterly periods, but at seasonally adjusted annual rates for comparability with yearly figures. Most forecasts are presented in current price values, though some models derive GNP and its components in constant-dollar units. The forecasting method described below will be based on current-dollar flows because incomes are reported only on that basis and because government budgets and much private decision making are couched in current-dollar measures. When constant-dollar magnitudes are needed for relating output to employment, estimates of price indexes will be used to deflate current-dollar values.

THE PLAN OF ATTACK

We shall build up forecasts of GNP by predicting expenditures of individual sectors and then adding them. A problem of circularity immediately emerges.

In the Personal Sector certainly, and to a lesser extent in other areas, expenditures depend on sector incomes. But sector incomes depend on aggregate output, which in turn depends on sector expenditures. The interaction is inherent in the fact that the national economic system does involve a circular flow of incomes and expenditures; it is a feedback system.

There are three possible ways to handle the problem of circularity in economic forecasting.

1 *Time lags.* If expenditures in a given period were determined by incomes of prior periods, or if aggregate output in a period were determined by demands of the previous period, then the circularity would be broken, since prior period

values are known input variables for the period being forecast. However, a quarter-year is quite surely a long enough period for significant interaction between incomes, expenditures, and production to occur within a given quarterly period.

2 *Simultaneous equations.* If the behavior functions for demand and production sectors are represented by a set of equations in which several of the endogenous variables have coincident timing, it may be possible to obtain solutions which allow for all feedback effects by solving the set of equations simultaneously to obtain joint solutions for the interdependent variables. This is the approach used in econometric models of the economy.

3 *Successive approximations.* It may be possible to obtain initial estimates of some of the interacting variables, say GNP and sector incomes. Then initial expenditure estimates by sectors can be calculated. The sum of sector expenditures yields revised estimates of GNP and sector incomes. Repeated cycling of income and expenditure estimates can lead to convergence on a consistent set of forecast values.

The third approach will be used in this chapter. The plan of attack consists of five steps:

1 Obtaining an initial estimate of current-dollar GNP

2 Breaking this down into estimates of income flows by sectors

3 Forecasting demand for GNP purchases by sectors, and totaling these components to obtain a revised forecast of GNP

4 Calculating saving or borrowing by sectors, as implied by the estimates in steps 2 and 3

5 Making any revisions and reconciliations in forecasts of sector incomes and expenditures to obtain a consistent set of values for incomes, expenditures, and saving or borrowing flows

At this stage the forecaster will have developed estimates for which aggregate demand and supply are equal and for which flows are consistent within each sector and among sectors. Such balance and consistency checks are a necessary condition for a correct forecast, but unfortunately not sufficient. The forecast figures may be wrong!

The following sections discuss each of the five steps in more detail. Beyond that, Chap. 24 describes methods for converting the forecast to real GNP, estimating the levels of employment and unemployment implied by the forecast, and forecasting prospective price changes.

INITIAL ESTIMATE OF CURRENT-DOLLAR GNP (step 1)

The preceding chapter derived a forecast of 1,193 billdar for GNP in 1974, based on time-series analysis and qualitative judgments as to overall business prospects for 1974. This is the kind of initial estimate of GNP which is needed to break into the circle of economic projection. As was evident, there is a range of forecast values which would lie within the bounds indicated by past experience. Supplementary sources of information may provide alternative estimates or help the analyst to decide on the most reasonable figure within the acceptable range.

It is vital, of course, to have on tap the latest available data on the national income accounts and related series. The principal "National Income and Product Tables" appear monthly in the *Survey of Current Business* and *Business Conditions Digest*, both published by the Bureau of Economic Analysis in the U.S. Department of Commerce, and in *Economic Indicators*, prepared by the President's Council of Economic Advisers. Those publications contain hundreds of other economic time series also. Important data may be secured before their publication if one is on the mailing list for copies of news releases for important series. Initial estimates of GNP and its components are normally available by the end of the month following each calendar quarter.

In addition, there are some series that serve as leading indicators of future business activity. The National Bureau of Economic Research analyzed thousands of series to determine the timing of their peaks and troughs with respect to business-cycle turning points. The series whose turning points have most consistently preceded turns in general business activity have been classified as "leading indicators." These series are reported and plotted monthly in *Business Conditions Digest*, along with coincident and lagging indicators which confirm the evidence of the leading series.

Forecasts for a given year can be located in many publications in the latter months of the preceding year and in the early months of the forecast year. Many will be found in the business press—*The Wall Street Journal*, *Business Week*, *Fortune*, The Conference Board's *Business Record*, *Forbes*, *Dun's Review*, *Commercial and Financial Chronicle*, and *Business Economics*, as well as the news magazines. The Federal Reserve Banks' publications, commercial bank letters, and some reports from insurance companies and investment advisory services contain explicit forecasts of GNP and its major components. Summaries of forecasts assembled from many sources are published in the *Predicast* publication, *Conference Board Statistical Bulletin*, the Federal Reserve Bank of Richmond's "Business Forecasts" (February), and occasionally in *Business Week*. University publications present the figures from econometric forecasting models at the University of Michigan, the University of California at Los Angeles, and the Wharton School at the University of Pennsylvania. The *Economic Report of the President* (February) contains the projections of the Council of Economic Advisers; other government publications, such as the *Survey of Current Business* and the *Federal Reserve Bulletin*, discuss business conditions but do not publish specific forecasts.

A consensus of such forecasts available around the end of 1973 stood somewhat as follows:

	1973	1974	Percent Change
Constant-dollar GNP (billdar at 1958 prices)	837.3	847.3	1.2
GNP price index (1958 = 100)	153.9	163.7	6.4
Current-dollar GNP (billdar)	1,288.2	1,386	7.6

Source: "Business Forecasts 1974," Federal Reserve Bank of Richmond, February 1974.

These estimates are close to the figures derived by extrapolation and judgment in the preceding chapter. The latter estimates will be used as initial estimates for this chapter, namely:

1974 Initial Estimates

Real GNP	846 billdar at 1958 prices
Price index	164.7 with 1958 = 100
GNP	1,393 billdar

ESTIMATES OF SECTOR INCOMES (step 2)

Once we have an initial estimate of GNP, we can break it down into income components on the basis of past relationships, allowing for expected changes in taxes and transfers and for income shifts during stages of the business cycle. Table 22-1 displays the tableau of major items comprising the net income of each of the four demand sectors. The four sector incomes plus the statistical discrepancy must equal GNP; that is, the entry in line 1 equals the sum of the entries in lines 2, 13, 16, 27, and 28. Data for three preceding years are tabulated; also, quarterly data for the most recent year is included to show current trends. Finally, there are the hungry cells in the column headed 1974^F—waiting to be filled with forecast figures.

PERSONAL SECTOR INCOME FLOWS

Let us start with the Personal Sector. Net disposable personal income (line 2) is to be obtained by estimating total personal income (line 8) and subtracting from it personal taxes, transfer, and interest payments (lines 9, 11, and 12). The personal income total (line 8) equals the factor payments to the Personal Sector (line 3) plus nonfactor interest and transfer payments received by persons (lines 4 through 7). Many of these items are small, and they change in a steady progression from year to year, so they can be extrapolated with adequate accuracy. But the items for national income to persons, government transfers, and personal tax payments require a little analysis.

First let us attack the item for national income to persons (line 3). We pro-

TABLE 22-1 Gross National Income Components (in billdar)

Line	Item	Symbol	1971	1972	1973	1974^F
1	Gross national product	GNP	1,055.4	1,155.2	1,289.1	1,393.0
2	Net disposable personal income	Y_d'	727.3	776.2	858.8	
3	National income to persons	NITP	739.3	803.6	880.8	
4	Interest paid by government	INTGP	13.4	13.0	14.6	
5	Interest paid by consumers	INTPP	17.6	19.7	22.5	
6	Business transfers to persons	TRBP	4.3	4.6	4.9	
7	Government transfers to persons	TRGP	88.9	98.3	112.6	
8	Personal income	Y_p	863.5	939.2	1,035.4	
9	Less personal tax payment	TPI	117.5	142.2	152.9	
10	Disposable personal income	Y_d	746.0	797.0	882.5	
11	Less personal transfers to foreigners	TRPF	1.0	1.0	1.2	
12	Less interest paid by consumers	INTPP	17.6	19.7	22.5	
13	Business gross retained earnings	GRE	111.8	124.4	135.2	
14	Capital consumption allowance	CCA	93.8	102.4	110.0	
15	Undistributed profits and inventory valuation adjustment	S_b	17.5	22.4	25.3	
15a	Wage accruals retained	WR	0.6	−0.5	−0.1	
16	Government net receipts	T_n	216.2	252.2	288.7	
17	Indirect business taxes	TIND	102.4	109.5	117.8	
18	Corporate profits taxes	TCP	37.4	42.7	55.8	
19	Personal tax payments	TPI	117.5	142.2	152.9	
20	Contributions for social insurance	TSS	64.6	73.7	92.1	
21	Gross tax receipts	T	322.0	368.2	418.6	
22	Subsidies less surplus government enterprises	SUB	1.2	1./	0.4	
23	Government transfers to persons	TRGP	88.9	98.3	112.6	
24	Government transfers to foreigners	TRGF	2.6	2.7	2.4	
25	Interest paid by government	INTGP	13.4	13.0	14.6	
26	Government transfers and interest	TRING	105.9	115.9	129.9	
27	Transfers to foreigners	TR_f	3.6	3.7	3.6	
28	Statistical discrepancy	SD	−3.4	−1.5	2.9	

Source: "National Income and Product Accounts," tables 1.9, 1.10, 2.1, 3.1–3.4, 4.1, *Survey of Current Business,* Dec. 1973; 4th quarter and year 1973 were added from Feb. 1974 issue.

ceed by deriving total national income from GNP and then break down national income into the components which flow to persons, to business capital accounts, and to governments. The derivation of national income from GNP is shown in Table 22-2, where lines 1 through 7 are drawn from Table 1.9 of the National Income and Product Accounts. Forecasts for 1974 have been entered and were derived by the following logic.

Line 1 GNP of 1,393 billdar, the initial forecast discussed in the preceding section.

Line 2 Capital consumption allowances (CCA) of 118.5 billdar. The 8.5 billdar rise from 1973 is an extrapolation of recent increments.

Line 3 Indirect taxes (TIND) of 125.3 billdar. The annual increment to indirect taxes has fluctuated in a range between 7 and 9 billdar for several years

1973			
1st Quarter	2nd Quarter	3rd Quarter	4th Quarter
1,242.5	1,272.0	1,304.5	1,337.5
829.4	846.6	867.0	892.2
848.3	867.1	890.5	917.2
13.5	14.0	14.9	15.9
21.2	22.0	23.0	23.8
4.8	4.9	5.0	5.1
108.8	110.8	113.7	116.9
996.6	1,019.0	1,047.1	1,078.9
145.1	149.3	156.0	161.1
851.5	869.7	891.1	917.8
0.9	1.0	1.1	1.8
21.2	22.0	23.0	23.8
131.5	132.0	136.9	140.6
106.9	109.0	110.5	113.5
24.6	23.1	26.4	27.1
0.0	−0.3	0.0	0.0
277.5	286.9	293.3	296.3
115.6	117.2	118.5	119.9
52.7	57.4	57.6	55.7
145.1	149.3	156.0	161.1
89.3	90.9	93.0	95.0
402.7	414.7	425.0	431.6
0.9	0.4	0.6	−0.2
108.8	110.8	113.7	116.9
2.1	2.3	2.5	2.7
13.5	14.0	14.9	15.9
125.2	127.8	131.7	135.3
3.0	3.3	3.5	4.5
1.1	3.2	3.7	3.7

with no uptrend, presumably because increased revenue sharing by the federal government has lessened the need for state and local governments to raise property and sales tax rates. A medium-sized increase of 8 billdar was projected for 1974.

Line 4 Business transfer payments (TRBP) of 5.2 billdar. This was a simple extrapolation of a small, steadily rising series.

Line 5 Statistical discrepancy (SD) of 4 billdar. This residual item had undergone a reversal from negative to positive in 1973. The figure for the fourth quarter of 1973 suggests that a positive value will continue in 1974.

Line 6 Subsidies less surplus government enterprises of −1.5 billdar. This item declined and finally reversed direction in the fourth quarter of 1973, presumably because of the termination of many farm subsidies. A moderate further decline was written in for 1974.

TABLE 22-2 Relation of GNP to National Income (in bildar)

Line	Item	Symbol	1971	1972	1973	1974^F	1973 1st Quarter	2nd Quarter	3rd Quarter	4th Quarter
1	Gross national product	GNP	1,055.4	1,155.2	1,289.1	1,393.0	1,242.5	1,272.0	1,304.5	1,337.5
	Less:									
2	Capital consumption allowances	CCA	93.8	102.4	110.0	118.5	106.9	109.0	110.5	113.5
3	Indirect business taxes	TIND	102.4	109.5	117.8	125.8	115.6	117.2	118.5	119.9
4	Business transfer payments	TRBP	4.3	4.6	4.9	5.2	4.8	4.9	5.0	5.1
5	Statistical discrepancy	SD	−3.4	−1.5	2.9	4.0	1.1	3.2	3.7	3.7
	Plus:									
6	Subsidies less surplus government enterprises	SUB	1.2	1.7	0.4	−1.5	0.9	0.4	0.6	−0.2
	Equals:									
7	National income	NI	859.4	941.8	1,053.9	1,138.0	1,015.0	1,038.2	1,067.4	1,095.1
8	Corporate profits and inventory valuation adjustment less dividends†	CPIVA −DIV	55.6	64.6	81.1	84.0	77.4	80.3	83.9	82.9
9	Contributions for social insurance	TSS	64.6	73.7	92.1	103.0	89.3	90.9	93.0	95.0
10	National income to persons	NITP	739.3	803.6	880.8	951.0	848.3	867.1	890.5	917.2
	Forecasting Aids									
11	Corporate profits and inventory valuation adjustment	CPIVA	80.1	91.1	109.0	114.5	104.3	107.9	112.0	111.9
12	Dividends	DIV	25.1	26.0	27.8	30.5	26.9	27.3	28.1	29.0
13	Ratio: line 9/(lines 9 and 10)	TSS/(TSS+ NITP)	8.04%	8.41%	9.46%	9.76%	9.53%	9.49%	9.46%	9.38%

† Plus small item for wage accruals retained.
Source: "National Income and Product Accounts," table 1.9, *Survey of Current Business.*

Now national income can be derived from GNP by making the indicated subtractions and additions of items on lines 2 through 6.   The next step is more interesting.   How will national income (or payments for factor services) be allocated among persons, business, and governments?   The three principal components to be forecast are:

Line 8   Corporate profits before tax with inventory valuation adjustment added and dividends subtracted.

Line 9   Contributions for social insurance by both employers and employees.

Line 10   National income to persons, which consists of compensation of employees (less contributions for social insurance), proprietors' income, rental incomes, interest from production units, and dividend payments.

The charts in Fig. 22-1 show how these components have fluctuated as a percentage of national income during the postwar years.   One striking feature is the rise in the share taken by contributions for social insurance starting in 1955, as coverage was broadened and rates were increased.   Because of this rise, there is a downtrend in the other shares, most notably in the profits component.   Also noteworthy are the inverse movements in shares of income going to the Personal Sector and to corporate profits, with the latter share at lows in the recession years 1949, 1954, 1958, 1960 and 1961, and 1970.   Given the slow growth of the economy expected for 1974, cyclical forces were forecast to depress the profits share in

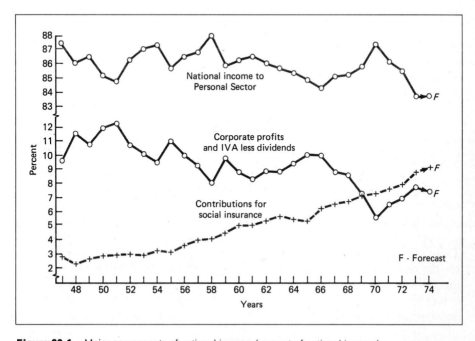

Figure 22-1   Major components of national income (percent of national income).

1974. The percentage taken for contributions to social insurance was expected to rise moderately in 1974 because the maximum amount of income subject to the tax had been raised. On balance, the share going to persons was expected to remain constant.

Calculations underlying the projected figures were made by first estimating corporate profits before tax plus inventory valuation adjustment (CPIVA) and the dividend payments (DIV) to arrive at line 8. The amount on line 8 was subtracted from national income, and the remainder was split between contributions for social insurance (TSS) and national income to persons (NITP).

To forecast changes in corporate profits, we may recall a basic identity from the national income accounts, relating GNP to the component costs, profits, and other charges paid by the Production Units Sector.

GNP *equals* labor costs *plus* profits *plus* nonlabor costs and nonfactor charges

or

$$Y = LC + CPIVA + NLC$$

This equation may be solved for CPIVA as follows:

$$CPIVA = Y - LC - NLC$$

Income shares are obtained by dividing both sides by Y.

$$\frac{CPIVA}{Y} = 1 - \frac{LC}{Y} - \frac{NLC}{Y}$$

On the right side of this equation, the labor share is perhaps three times as large as the proportion paid to nonlabor charges, and changes in the labor share will dominate changes in the ratio of CPIVA to Y. Note also that $LC/Y = LC/P_Y Y_R = ULC/P_Y$, where ULC is unit labor costs. As a rough approximation, then, we may say that percent changes in the corporate profits ratio will be largely determined by percent changes in the ratio of ULC to P_Y, the relation being a negative one. In terms of percent changes,

$$\%\Delta CPIVA - \%\Delta Y = \beta(\%\Delta P_Y - \%\Delta ULC)$$

Data for these variables were calculated for the private nonfarm sector of GNP, since corporate profits stem largely from this sector. The results are shown in the regression chart in panel (*b*) of Fig. 22-2, where the left side of the preceding equation becomes y and, on the right, the quantity in parentheses becomes x. There is a respectable correlation, with β about 6. It is noteworthy that the points for the recession years 1949, 1954, 1958, and 1970 fall well below the regression line; so a forecast point should be placed low on a judgment basis if a recession is forecast. The top panel in Fig. 22-2 shows a graph for the postwar years for actual percent changes in CPIVA and the values forecast from the predictor relation

$$\%\Delta CPIVA^F = \%\Delta Y + 6(\%\Delta P_Y - \%\Delta ULC)$$

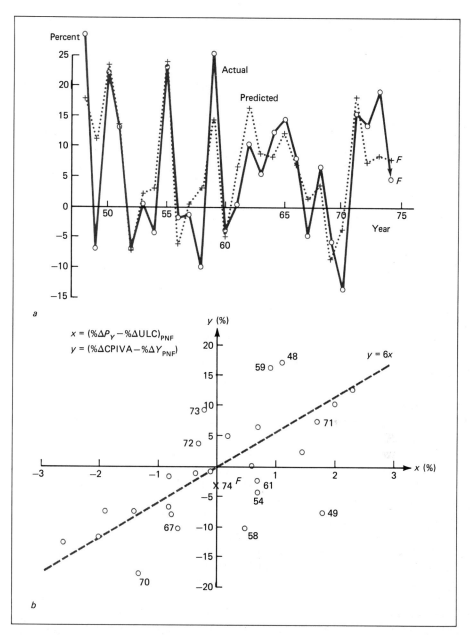

Figure 22-2 Predictor for percent change in corporate profits plus IVA (*a*) Percent change in CPIVA. (*b*) Regression.

where all values on the right refer to the private nonfarm sector. The predictor values tracked the actual change fairly well but were too high in recession years, as noted previously, and were low in 1972 and 1973, perhaps because of controls or inflation effects.

For 1974 an initial forecast of percent change in CPIVA (line 11) was made by assuming that private nonfarm GNP would rise in step with total GNP, that is, by 8 percent, and that prices would rise in step with unit labor costs, so that

$$x^F = \%\Delta P_Y{}^F - \%\,\Delta\text{ULC}^F = 0$$

On this basis CPIVA would be predicted to rise 8 percent for 1974. Because 1974 was expected to be a semirecession type of year, the forecast rise was lowered to 5 percent, yielding a value

CPIVA = 1.05(109.0) = 114.5 billdar for 1974

Next, to obtain line 8 in Table 22-2, we need a forecast of dividends (line 12). This is relatively easy, for dividends rise regularly and can be extrapolated with acceptable accuracy. Dividends were reported at 29 billdar in the fourth quarter of 1973. Removal of controls would permit some catch-up rises in 1974, but corporations would also be under pressure to retain earnings in an inflationary period. On balance, a relatively large increase to 30.5 billdar was projected for 1974. This permits us to calculate the item on line 8 of Table 22-2.

CPIVA − DIV = 114.5 − 30.5 = 84.0 billdar

Line 9 Contributions for social insurance (TSS) of 103 billdar. They were estimated by splitting the remainder of national income after subtracting the item on line 8. This can be done by looking at the historical pattern of the ratio TSS/ (NITP + TSS) and making allowance for any prospective increases in coverage, tax rates, or increase in maximum income subject to tax. In 1974 only the third factor was operating, and it was expected to shift contributions up by about 3 billdar. The ratios TSS/(NITP + TSS) for recent years are shown in line 13 of Table 22-2, and the 3 billdar increment amounts to about 0.3 percentage points of the ratio. Hence, the ratio was set at 9.76 percent for 1974, leading to

TSS = 0.976(1138.0 − 84.0) = 103.0 billdar

Line 10 National income payments to persons (NITP) of 951.0 billdar. The figure equals total national income less diversions to Business Capital Accounts and to the Government Sector.

NITP = 1138.0 − 84.0 − 103.0 = 951.0 billdar

This figure gave us the starting point for forecasting income flows to the Personal Sector. To complete the picture, we developed forecasts of items related to the Personal Sector in Table 22-3, which records values for 1973 and forecasts for 1974, together with the number of the source table in the "National Income and Product Accounts" where historical data may be found.

TABLE 22-3 Income Components for the Personal Sector (in billdar)

Line	Item	Symbol	1973	1974^F	Source Tables†
1	National income to persons	NITP	880.8	951.0	
2	Interest paid by government	INTGP	14.6	18.3	3.1, 3.3
3	Interest paid by consumers	INTPP	22.5	25.5	2.1
4	Business transfers to persons	TRBP	4.9	5.2	1.9
5	Government transfers to persons	TRGP	112.6	129.0	1.9
6	Personal income	Y_p	1,035.4	1,129.0	1.9, 2.1
	Less:				
7	Personal tax and nontax payments	TPI	152.9	169.5	2.1
	Equals:				
8	Disposable personal income	Y_d	882.5	959.5	2.1
	Less:				
9	Personal transfers to foreigners	TRPF	1.2	1.0	2.1
10	Interest paid by consumers	INTPP	22.5	25.5	2.1
	Equals:				
11	Net disposable personal income	Y_d'	858.8	933.0	7.1

† "National Income and Product Accounts," tables as indicated, *Survey of Current Business.*

Line 1 National income to persons (NITP) of 951.0 billdar is carried forward from Table 22-2, line 10.

Lines 2, 3, and 4 These are simply extrapolated, using historical annual and quarterly data in Table 22-1 plus judgment regarding trends in interest rates and outstanding debt in 1974.

Line 5 Government transfers to persons (TRGP) of 129.0 billdar include social security benefits, veterans benefits, unemployment insurance payments, government pensions, and state and local government welfare and relief outlays. Federal benefits under the old age, survivors, and disability insurance program (OASDI) account for about 40 percent of the total. For longer-range forecasts, correlations with unemployment rate and the consumer price index, combined with a steady uptrend, give a rather good explanation of changes in TRGP, but for short-term projections the trend rise, plus estimates of the effects of legislated changes in the OASDI benefits, gives adequate forecasts.

During the latter part of 1973, TRGP was rising at a rate of 3 billdar per quarter, with no rise in OASDI benefit rates or in unemployment percentage. If that rise should continue through 1974 and if we add on the estimated upshifts in OASDI benefits of 3.9 billdar in April and 2.4 billdar in July, the 1974 level for TRGP would average out close to 129 billdar. This would be a 14.5 percent rise from 1973, the same as the percent rise in 1973 and a little below the average annual percent rise from 1965 to 1973. In view of the expectations of a strong rise in the consumer price index and a moderate rise in the unemployment rate, this forecast may prove to have been conservative.

Line 6   Personal income (Y_P) of 1,129.0 billdar is just the sum of the entries in lines 1 through 5.

Line 7   Personal tax and nontax payments (TPI) of 169.5 billdar were estimated by projections of the ratios of these payments to the tax base, separately for federal and for state and local taxes.   The tax base chosen was national income to persons (NITP) plus personal contributions for social insurance (TSSP).   It was believed that transfer payments and government and consumer interest payments are largely nontaxable and should not be included in the tax base.   The value of TSSP has recently been running about 47 percent of total contributions by employers and employees (line 9, Table 22.2).   So the tax base in 1974 became

Tax base = NITP + TSSP = 951.0 + 0.47(103.0) = 999.5 billdar

Table 22-4 summarizes the recent movements in ratios of personal taxes to the tax base at the federal and at the state and local levels.   Federal tax rates were raised in 1968 and lowered in 1970 and 1971 near to the 1967 level.   After that, federal rates rose again because of overwithholding in 1972 and the effects of prosperity and inflation that shifted taxpayers into higher brackets.   No legislated changes were expected for 1974, so the slow trend rise in the tax ratio evident during the quarters of 1973 was extrapolated.   The 1974 ratio was set at 12.75 percent.   For state and local governments the tax ratio on personal incomes rose about one-fourth percentage point per year from 1967 to 1972 and then slowed

TABLE 22-4   Forecast of Personal Tax and Nontax Payments (in billdar)

	(1)	(2)	(3)	(4) (5) Personal Tax and Nontax Payments		(6) (7) Average Tax Ratios, in Percent	
Year	National Income to Persons (NITP)	Personal Contributions for Social Insurance (TSSP)†	Tax Base (NITP + TSSP)	Federal†	State & Local†	Federal, (4)/(3)	State & local (5)/(3)
1967	553.9	20.5	574.4	67.5	15.5	11.75	2.70
1968	603.3	22.8	626.1	79.7	18.3	12.72	2.92
1969	656.4	26.3	682.7	94.8	21.7	13.90	3.18
1970	698.3	28.0	726.3	92.2	24.4	12.70	3.36
1971	739.3	30.9	770.2	89.9	27.7	11.69	3.60
1972	803.6	34.7	838.3	107.9	34.3	12.88	4.09
1973	880.8	43.1	923.9	114.5	38.4	12.40	4.16
1973							
1st qtr.	848.3	41.9	890.2	108.5	36.6	12.19	4.11
2nd qtr.	867.0	42.6	909.6	111.4	37.9	12.25	4.17
3rd qtr.	890.5	43.6	934.1	116.9	39.1	12.51	4.19
4th qtr.	917.4	44.2	961.6	121.0	40.1	12.59	4.17
1974ᶠ	950.5	48.0	999.0	127.5	42.0	12.75	4.20

† "National Income and Product Accounts," *Survey of Current Business*: col. (2), table 2.1; col. (4), tables 3.1 and 3.2; col. (5), tables 3.3 and 3.4.

in 1973, perhaps because increased revenue sharing by the federal government was easing the financial pressures on state and local governments.   The slow rise continued through all quarters of 1973 and led to a forecast of 4.2 percent for the average tax ratio during 1974.

When the estimated tax ratios were applied to the forecast tax base of 999.5 billdar, the 1974 forecast for personal tax and nontax payments became:

Federal   0.1275(999.5) = 127.5 billdar
State and local   0.042(999.5) = 42.0 billdar
Total   = 169.5 billdar

The total was entered on line 7 of Table 22-3,   The rest of the forecast items in Table 22-3 come easily.

Line 8   Disposable personal income (Y_d) of 959.5 billdar equals personal income (line 6) less personal tax and nontax payments (line 7).

Line 9   Personal transfer payments to foreigners (TRPF) of 1.0 billdar, a small item with no trend, was estimated as equal to recent levels.

Line 10   Interest paid by consumers (INTPP) of 25.5 billdar, the same as line 3.

Line 11   Net disposable personal income (Y'_d) of 933.0 billdar equals the item on line 8 minus the items on lines 9 and 10.   This is the net income of the Personal Sector out of GNY, which we set out to obtain.

All 1974 forecasts for lines 2 through 12 in Table 22-1 have now been obtained. Next comes the Business Capital Accounts Sector.

BUSINESS CAPITAL ACCOUNTS INCOME FLOWS

Only three items comprising gross retained earnings are shown in Table 22-1— capital consumption allowances, undistributed profits and inventory valuation adjustment (IVA), and wage accruals retained.   However, forecasting the second of these items involves developing a detailed analysis of corporate profits and their distribution.   Table 22-5 presents historical data and sources for the component series.   The forecast figures were developed as follows.

Line 1   Capital consumption allowances (CCA) of 118.5 billdar were forecast previously and appear on line 2 of Table 22-2.

Line 2   Corporate profits and IVA combined (CPIVA) of 115.0 billdar were forecast previously and appear on line 11 of Table 22-2.

Line 3   Inventory valuation adjustment (IVA) of −19.0 billdar.   In recent years there has been a reasonably close negative correlation between IVA (in billdar) and percent changes in the wholesale price index (WPI), as may be seen in Table 22-6.

The 1974 forecast of %ΔWPI at 15 percent was based on the 1973 rise of 13.8 percent and on the fourth-quarter rise of 17.3 percent from the previous year.

TABLE 22-5 Business Capital Accounts Income Flows (in billdar)

Line	Item	Symbol	1971	1972	1973	1974^F	1973				Source Tables†
							1st Quarter	2nd Quarter	3rd Quarter	4th Quarter	
1	Capital consumption allowances	CCA	93.8	102.4	110.0	118.5	106.9	109.0	110.5	113.5	1.9
2	Corporate profits and inventory valuation adjustment	CPIVA	80.1	91.1	109.0	115.0	104.3	107.9	112.0	111.9	1.9, 1.10
3	Inventory valuation adjustment	IVA	−4.9	−6.9	−17.3	−19.0	−15.4	−21.1	−17.0	−15.5	1.10
4	Corporate profits before tax	CPBT	85.1	98.0	126.3	134.0	119.6	128.9	129.0	127.4	1.10
5	Corporate profits tax liability	TCP	37.4	42.7	55.8	59.0	52.7	57.4	57.6	55.7	1.10
6	Dividends	DIV	25.1	26.0	27.8	30.5	26.9	27.3	28.1	29.0	1.10
7	Undistributed profits	CPUN	22.5	29.3	42.6	44.5	40.0	44.2	43.4	42.6	1.10
8	Undistributed profits and inventory valuation adjustment	S_b	17.5	22.4	25.3	25.5	24.6	23.1	26.4	27.1	—
9	Wage accruals retained	WR	0.6	−0.5	−0.1	0.0	0.0	−0.3	0.0	0.0	1.9
10	Gross retained earnings	GRE	111.8	124.4	135.2	144.0	131.5	132.0	136.9	140.6	7.1
	Forecasting aids										
11	Ratio of inventory valuation adjustment to %ΔWPI	IVA/ %ΔWPI	−1.53	−1.50	−1.25	−1.30					
12	Corporate profits tax ratio	TCP/ CPBT	0.440	0.436	0.441	0.440	0.440				

† "National Income and Product Accounts," tables as indicated, *Survey of Current Business*.

TABLE 22-6 Relation between IVA and Percent Change in the Wholesale Price Index

Year	IVA	%ΔWPI	Ratio
1968	−3.3	2.5	−1.32
1969	−5.1	3.9	−1.31
1970	−4.8	3.7	−1.30
1971	−4.9	3.2	−1.53
1972	−6.9	4.6	−1.50
1973	−17.3	13.8	−1.25
1974^F	−19	15	−1.3

Also, it was believed that the wholesale price index would rise strongly in the first half of 1974 as controls were removed and as petroleum price rises worked their way through the economy but that increased output of farm products and fuels would slow the rise later in the year. That estimate, together with a medium ratio of −1.3, led to the forecast for IVA.

$$IVA = -1.3(15) = -19 \text{ billdar}$$

Line 4 Corporate profits before tax (CPBT) of 134.0 billdar are equal to line 2 minus line 3.

Line 5 Corporate profits tax liability (TCP) of 59.0 billdar has stayed close to 44 percent of CBPT in recent years, as shown on line 12. Use of that ratio for 1974 yields

$$TCP = 0.44(134.0) = 59.0 \text{ billdar}$$

Line 6 Dividend payments (DIV) of 30.5 billdar were estimated previously and appear in line 12 of Table 22-2.

Line 7 Undistributed corporate profits (CPUN) of 44.5 billdar equal line 4 minus lines 5 and 6.

Line 8 Undistributed profits corrected for inventory valuation adjustment (S_b) of 25.5 billdar, line 7 plus line 3, turned out to be nearly identical with the 1973 value.

Line 9 Wage accruals less disbursements (WR) were assumed to be zero.

Line 10 Gross retained earnings (GRE) of 144.0 billdar equals the sum of lines 1, 8, and 9. The forecast rise from 1973 is attributable to increased capital consumption allowances.

This provided the forecasts of income flows to the Business Capital Accounts which were required in lines 13 through 15*a* of Table 22-1.

GOVERNMENT SECTOR INCOMES, TRANSFERS, AND INTEREST PAYMENTS

With the exception of a small flow of transfer payments to foreigners, all the component flows to and from the Government Sector were forecast in preparation of the preceding tables. Historical data for 1973 and forecasts for 1974 are brought

TABLE 22-7 Government Sector Incomes, Transfers, and Interest (in billdar)

Line	Item	Symbol	1973	1974^F	Source of Forecast† Table	Line
1	Indirect business tax	TIND	117.8	125.8	22-2	3
2	Corporate profits tax	TCP	55.8	59.0	22-5	5
3	Personal tax and nontax payments	TPI	152.9	169.0	22-3	7
4	Contributions for social insurance	TSS	92.1	103.0	22-2	9
5	Gross tax receipts	T	418.6	456.8	(Sum of lines 1 through 4)	
6	Subsidies less surplus of government enterprises	SUB	0.4	−1.5	22-2	6
7	Government transfers to persons	TRGP	112.6	129.0	22-3	5
8	Government transfers to foreigners	TRGF	2.4	3.0	(Extrapolation)	
9	Interest paid by government	INTGP	14.6	18.3	22-3	2
10	Government transfers and interest	TRING	129.9	148.8	(Sum of lines 6 through 9)	
11	Government net receipts	TN	288.7	308.0	(Line 5 minus line 10)	

† Tables in this chapter.

together in Table 22-7, along with references to the sources of the forecasts. Historical data are found in NIPA Tables 1.9, 3.1, and 3.3.

SUMMARY ON FORECASTS OF SECTOR INCOMES

The data for sector incomes were assembled in Table 22-8, to serve as input to forecasting GNP purchases by the demand sectors. The process described earlier included a forecast of the statistical discrepancy as well as sector incomes; so we are assured in advance that our forecast components will total to GNP.

TABLE 22-8 Summary of Forecast of Sector Incomes in 1974

Component	Symbol	1973	1974^F	Percent Change	Source of Forecast† Table	Line
		(in billdar)				
Net disposable personal income	Y_d'	858.8	933.0	8.6	22–3	11
Gross retained earnings	GRE	135.2	144.0	6.5	22–5	10
Government net receipts	TN	288.7	308.0	6.7	22–7	11
Transfers to foreigners	TRF	3.6	4.0	11.1	22–3 / 22–7	9 / 8
Statistical discrepancy	SD	2.9	4.0	—	22–2	5
Gross national product	GNP	1,289.1	1,393.0	8.1%		

† Tables in this chapter.

It is noteworthy that net disposable personal income was expected to rise faster than GNP, largely because of the forecast rise of 14.5 percent in transfer payments from government.   The large rise in transfers accounted for the below-average rise in net government receipts also, for gross tax receipts rose by 9.1 percent.   The slow rise in gross retained earnings reflects weakness of profits in a sluggish economy.   In short, the forecasts of sector incomes seemed consistent with the business outlook.

Chapter 23

Forecasting Sector Purchases of GNP

FORECASTING GNP PURCHASES BY SECTORS

Step 3 in the forecasting procedure outlined in the preceding chapter is the heart of the forecast, the prediction of aggregate demands. We shall use the national product accounts framework and attempt to forecast the major components of spending for each of the four final demand sectors.

A basic decision must be made on whether to develop the demand forecasts in current or in constant dollars, with a price forecast added in the latter approach. To be sure, most microeconomic demand analyses focus on physical quantities, and it may be that some macroeconomic expenditure series show a more stable long-term growth pattern when measured in constant dollars. However, it seems best to develop demand components in current dollars initially—for several reasons. Sector income flows are reported in current-dollar terms, and, as this text has argued earlier, spending and saving decisions may well be made in current rather than constant dollars. Some of our advance indicators of spending are plans expressed in current-dollar terms, for example, government budgets and expenditure plans for plant and equipment. Also, we shall want to make our sector reconciliations of incomes, expenditures, and saving or borrowing in terms of money flows. However, one need not be dogmatic about this. In some cases the constant-dollar series may be easier to project. Or we may want to work some forecasts available in physical units into our forecast, for example, automobiles purchased or housing starts. Such forecasts may be worked back into the current-dollar framework by extrapolating price indexes in the areas of demand involved. Note also that we shall need to develop some overall price indexes eventually, anyway, in order to calculate a forecast of real GNP to reconcile with a prediction of employment.

The general method of attack on each component of aggregate demand will be to:

1 Assemble and analyze historical data for the series over the past 10 years, or two business cycles, including quarterly data for the most recent year. The analysis may involve plotting the data, computing percent changes from the previous year, and estimating the accuracy of forecasts based on some noncausal rule.

TABLE 23-1 Gross National Product Components (in billdar)

Line	Item	Symbol	1971	1972	1973	1974ᶠ	1973 1st Quarter	2nd Quarter	3rd Quarter	4th Quarter
1	Gross national product	GNP	1,055.5	1,155.2	1,289.1		1,242.5	1,272.0	1,304.5	1,337.5
2	Personal consumption expenditures	C	667.2	726.5	804.0		779.4	795.6	816.0	825.2
3	Durables	CD	103.6	117.4	130.8		132.2	132.8	132.8	125.6
4	Nondurables	CND	278.7	299.9	335.9		322.2	330.3	341.6	349.6
5	Services	CS	284.9	309.2	337.3		325.0	332.6	341.6	350.0
6	Gross private domestic investment	I_d	153.2	178.3	202.1		194.5	198.2	202.0	213.9
7	Nonresidential fixed investment	INRF	104.4	118.2	136.2		130.9	134.1	138.0	141.8
8	Structures	INRS	37.9	41.7	48.4		45.3	47.2	49.5	51.7
9	Producers' durable equipment	IPDE	66.5	76.5	87.8		85.5	86.9	88.6	90.1
10	Residential structures	IRS	42.7	54.0	58.0		59.0	59.6	59.2	54.0
11	Change in business inventories	ICBI	6.1	6.0	8.0		4.6	4.5	4.7	18.0
12	Government purchases of goods and services	G	234.3	255.0	277.1		268.6	275.3	279.0	285.6
13	Federal	GF	98.1	104.4	106.6		105.5	107.3	106.8	106.8
14	National defense	GFD	71.6	74.4	73.9		74.3	74.2	74.2	73.0
15	Other federal	GFO	26.5	30.1	32.7		31.2	33.1	32.7	33.8
16	State and local	GSL	136.2	150.5	170.5		163.0	168.0	172.2	178.8
17	Net exports	X_n	0.8	−4.6	5.8		0.0	2.8	7.6	12.8
18	Exports	X	66.3	73.5	102.0		89.7	97.2	104.5	116.4
19	Imports	Z	65.5	78.1	96.2		89.7	94.4	97.0	103.6
20	GNP price index†	P_Y	141.60	146.10	153.94		149.81	152.46	155.06	158.36
21	GNP in constant dollars‡	Y_R	745.4	790.7	837.4		829.3	834.3	841.3	844.6
22	Potential GNP	$Y_{R,FE}$	787.3	818.8	851.6	885.7	839.1	847.3	855.7	864.1

† Billdar at 1958 prices.
‡ 1958 = 100.
Sources: Lines 1 through 21, "National Income and Product Accounts," tables 1.1, 1.2, 8.1, *Survey of Current Business*, Dec. 1973; 4th quarter and year 1973 were added from Feb. 1974 issue. Line 22, President's Council of Economic Advisers, published in *Business Conditions Digest*, March 1974, p. 95.

2   Search for some causal, behavioral basis for forecasting the given expenditure series.   Perhaps there exist some foreshadowing indicators of future spending, such as government budgets, capital appropriations for plant and equipment, housing starts, construction contract awards, order backlogs, or inventory-to-sales ratios.   Perhaps some variables causally linked to the expenditures can be forecast, such as sector incomes.   In such cases, . the noncausal forecast should be modified or replaced on the basis of the best causal information and relations available.

This approach has been illustrated already in the development of forecasts of real GNP, employment, current-dollar GNP, and sector incomes.   An attempt will be made to bring in more causal determination in the expenditure forecasts. But there will still be a lot of room for judgment.   The procedure suggested here is a kind of half-way house between naïve extrapolation methods and statistically derived equations in a complete econometric model.

The question of circularity may well be raised again at this point.   If sector demands are forecast on the basis of time-series extrapolations modified by judgment which is based on knowledge of business conditions and on the initial forecasts of GNP and sector incomes, will not the demand components which we obtain reflect merely the original assumptions and add up to the initial GNP?   Is not the initial GNP forecast "driving" the demand components, so that we obtain a forced fit between it and sector purchases?

Not so.   As we shall see, some of the demand components are determined, at least in part, by autonomous forces.   To the extent that this is true, the causal chain is reversed.   Exogenous elements in demand "drive" production in their sectors and, via feedback loops, influence incomes and hence such "induced" demand components as consumer nondurables and services, inventory investment, and imports.   So our demand forecasts will have a considerable degree of independence.   They will not automatically total to the initial GNP forecast.   Revisions and reconciliations will be needed.

With this introduction we turn to the task of forecasting the major components of aggregate demand.   Table 23-1 lists the demand components which are to be forecast and presents historical data for 1971 through 1973, including quarterly data for 1973.

In the column headed 1974[F] are the hungry cells to be filled with forecast figures.   Potential GNP can be entered because it is currently rising 4 percent per year.   After that, it is good strategy to forecast the more autonomous components of aggregate demand first.   So we begin with government purchases of goods and services.

GOVERNMENT PURCHASES OF GOODS AND SERVICES

Government purchases of goods and services are reported in "National Income and Product Accounts" (NIPA), table 1.1, under three categories: (1) state and local government purchases, (2) federal national defense, and (3) other federal purchases.   Recent figures for these components are recorded here in Table 23-1.

1   Since there are no national summary budget figures for state and local governments, this component was forecast by a noncausal approach, with judgment added.   Figure 23-1 presents two bases for extrapolating the series: (*a*) the percent change from the preceding year, and (*b*) the percent ratio to GNP.   These graphs reveal a strong rate of rise in state and local purchases in recent years—faster than

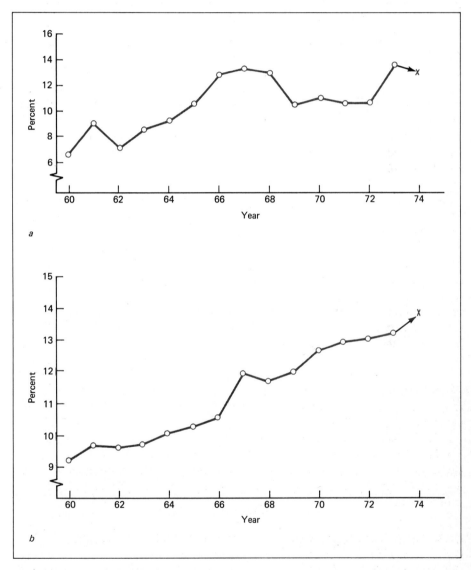

Figure 23-1   Analysis of state and local government purchases.   (*a*) Percent change from preceding year.   (*b*) Percent ratio to GNP.

TABLE 23-2 Federal Purchases of Goods and Services (in billdar)

Calendar Periods	NIPA Data			Federal Budget Estimates†	
	Half-Year	Annual Averages		February 1973	June 1973
		Calendar	Fiscal		
National defense					
1971, 2nd half-year	71.4	71.6			
1972, 1st half-year	76.6		74.0	74.3	
1972, 2nd half-year	72.2	74.4			
1973, 1st half-year	74.2		73.2	74.6	75.1
1973, 2nd half-year	73.6	73.9			
1974, 1st half-year^F	76.0		74.8	75.9	74.9
1974, 2nd half-year^F	78.0	77.0			
Other federal purchases					
1971, 2nd half-year	28.4	26.5			
1972, 1st half-year	29.8		29.1	28.8	
1972, 2nd half-year	30.4	30.1			
1973, 1st half-year	32.2		31.3	31.1	31.4
1973, 2nd half-year	33.2	32.7			
1974, 1st half-year^F	36.0		34.6	35.6	36.2
1974, 2nd half-year^F	38.0	37.0			

† Estimates on the basis of "National Income and Product Accounts," *Survey of Current Business*, February and June, 1973.

the rise in GNP. The ratio to GNP increases, and the rise is particularly marked in years when the GNP grew slowly—1961, 1967, and 1970. In the fourth quarter of 1973, the percent change from the preceding year stood at 13.2 percent and the ratio to GNP at 3.4 percent. Continued strength was anticipated in 1974. Federal revenue sharing had placed state and local governments in a strong financial position since 1971, as evidenced by a large surplus of receipts over expenditures in the NIPA. Although pressures for expansion of educational facilities were lessening and though capital-funds markets were expected to be tight in early 1974, it was felt that demands for growing expenditures for pollution control, waste disposal, urban renewal, and other needs, plus rises in pay rates for government employees, would maintain upward pressures on government spending. On balance, a 12.9 percent rise in state and local purchases was forecast for 1974, leading to a figure of 192.5 billdar, or 13.8 percent of the forecast GNP of 1,393 billdar. These values are plotted in Figure 23-1, and the large increase in the ratio to GNP was justified because a relatively slow rise in GNP was being forecast.

For federal government purchases, budget estimates are available to guide the forecasts. Articles in the *Survey of Current Business* in February and around midyear present budget estimates converted to an NIPA basis. (See the data in Table 23-2.)

2 National defense purchases are reported in the top half of the table.†

† Additional data on Defense Department obligations, contract awards, and new orders are available in publications of that department and in the monthly *Business Conditions Digest*.

Quarterly data from NIPA Table 1.1 are combined into half-yearly and then annual averages for both calendar and fiscal years in the three left-hand columns. The right-hand columns present budget estimates converted to an NIPA basis, by fiscal years. These data and the quarterly figures in Table 23-1 revealed that national defense purchases had fluctuated with no marked trend for the past three years. The budget as of June 1973 estimated continued stability through fiscal 1974. Purchases declined in the second half of 1973 largely because shipments of arms to Israel out of existing stocks counted as negative purchases in the NIPA. Replacement purchases will presumably raise expenditures in 1974. So the forecast figure in the first half of 1974 was set at 76 billdar to bring the average for fiscal 1974 close to the June budget estimate of 74.9 billdar. A continued rise to 78 billdar was predicted for the second half of 1974 in view of prospective effects of inflation and new procurement programs on defense spending. On a calendar-year basis, 1974 purchases for national defense were forecast at 77 billdar, up 4.2 percent from the 1973 level of 73.9 billdar and declining as a percent of GNP.

3. Other federal purchases are reported in the bottom half of Table 23-2 and also in Table 23-1. They display a rising trend, about in step with GNP. The budget estimate of 36.2 billdar for fiscal 1974, taken with the reported figure of 33.2 billdar for the second half of 1973, would imply a huge jump in spending to 39.2 billdar in the first half of 1974. This seemed too large and therefore was cut back to 36 billdar, but it did suggest a continued rise in the last six months of 1974. The 1974 forecast level was 37 billdar, up 13.1 percent from the 32.7 billdar in 1973. As a percent of GNP, the rise was from 2.54 percent in 1973 to 2.65 percent in 1974.

In summary, the forecasts for government purchases were as follows.

	1973	1974F	Percent Change
	(billdar)		
Government purchases (total)	227.1	306.5	10.6
Federal national defense	73.9	77	4.2
Other federal purchases	32.7	37	13.1
State and local purchases	170.5	192.5	12.9

NONRESIDENTIAL FIXED INVESTMENT

Business expenditures for structures and producers' durable equipment (NRFI) are motivated by expectations of future profitability of investment projects. Consequently, they involve forecasts of many variables, such as customer preferences and incomes, competitors' actions, production functions, costs and availability of inputs, and interest rates. The decisions are complex and subject to rapid alteration as expectations of economic conditions change.

However, major capital expenditures involve considerable advance planning and long time lags between the start and completion of projects. So it may be possible to obtain lead indications of capital spending by tapping information early in the decision-spending sequence. Since World War II, several surveys have aimed to do just that. The Conference Board now surveys the thousand largest U.S. manufacturers quarterly and obtains data on their capital appropriations, capital expenditures, and the level and net changes of backlogs of appropriated funds. These series are reported in the Board's publications and in *Business Conditions Digest*. The McGraw-Hill Information Systems Company (F. W. Dodge Division) publishes data on construction contracts and on contracts and orders for plant and equipment. These series appear in that company's publications, and some are contained in *Business Conditions Digest*.

Finally, we now have data from surveys of business firms' plans to purchase new plant and equipment, ranging up to two years ahead. There are two sources of this expectational data. First, the McGraw-Hill Economics Department surveys business executives in the fall and publishes in early November an estimate of total national expenditures for plant and equipment in the next calandar year, classified by some 25 component industries. The survey findings are updated twice the next year, and are reported in *Business Week* in mid-February and late April. The Bureau of Economic Analysis in the U.S. Department of Commerce and the Securities and Exchange Commission jointly conduct comparable surveys several times a year and publish the findings in the *Survey of Current Business* issues for January, March, June, September, and December. The December report runs through the first two quarters of the next year; the January figures cover the coming year's total; the March findings are broken down by two quarters and the second half of the year; the June and September figures cover the current year by quarters. Data are also provided relating to manufacturers' expected inventories and sales one quarter ahead, to their evaluation of their inventory and capacity conditions, and to starts and carryover of plant and equipment projects to date.

Two causal approaches are thus available for forecasting nonresidential fixed investment: (1) a model which attempts to quantify the influences of important causal forces and to project ahead on the basis of forecasts of those variables; and (2) use of the series on business executives' capital-spending plans, perhaps modified by short-term influences which may cause spending to deviate from plans. The second approach has been adopted here, using the McGraw-Hill survey findings published in November as the basis for a forecast of investment spending for the following year.

It should be noted that plant and equipment expenditures are not identical with the nonresidential fixed investment component in the GNP accounts. The plant and equipment series is less comprehensive in that it does not include the capital expenditures of farm enterprises, professional persons, nonprofit institutions, and real estate operators. Also, GNP private investment includes oil-well drilling costs charged to current expense, purchases of passenger cars by salaried persons who receive reimbursement for use of their cars, net purchases of used capital goods from the government, and dealers' margins on the purchase of used capital. It would probably be more accurate to use National Income and Product tables 5.2 through 5.5 to break nonresidential fixed investment into a plant and equipment series and other categories, then use the survey findings to forecast just the plant and equipment portion. The method to be described here simply assumes that the year-to-year percent change in plant and equipment spending is a good forecaster for the percent change in total nonresidential fixed investment.

Figure 23-2 shows a graph of percent change from the previous year for the nonresidential fixed-investment series, to be designated %ΔNRFI. The recession years 1949, 1954, 1958, and 1961 stand out as years of declines in investment; 1967 and 1970 were years of slow rise in dollar spending but of declines in constant-dollar investment.

As a first attempt at causal explanation, let us use as a forecast of %ΔNRFI the expected percent change in plant and equipment expenditures (%ΔPEX) as reported in the McGraw-Hill survey in November of the preceding year. To get away from the capital shortages of the early postwar years and the Korean war period, the analysis is restricted to the historical period from 1956 to date. A regression diagram covering these years shows that positive values of %ΔPEX are associated with positive values of %ΔNRFI, and negative with negative. However, the mean absolute value of the difference $y = $ %ΔNRFI $-$ %ΔPEX is 4.5 percentage points.

Presumably, we could improve the predictions if we could discover some variable which accounts for the deviation of actual investment expenditures from the planned values as of November in the preceding year. Several variables were investigated—percent changes in GNP, in money supply, and in corporate profits. Best results were obtained by using the percent change in corporate profits and inventory valuation adjustment (%ΔCPIVA).

In the lower part of Fig. 23-2, the value of $y = $ %ΔNRFI $-$ %ΔPEX has been plotted together with %ΔCPIVA. There is strong evidence of positive correlation, and the dashed line through the origin with slope 0.4 would seem to represent the regression well. The remaining error of forecast, after allowing for %ΔCPIVA, will be the vertical distances from the circles to the dashed line. The mean absolute error for these values is only 2.5 percentage points, a substantial reduction of error of prediction.

Our predictor equation is now

$$(\%\Delta\text{NRFI})^F = \%\Delta\text{PEX} + 0.4(\%\Delta\text{CPIVA})$$

The predicted values were calculated year by year from reported values of the variables on the right side of the equation. In the top of Fig. 23-2, the predictor values have been plotted as pluses (+'s) joined by a dotted line along with the actual values. Of course, some of this accuracy would be lost in practice because of errors in forecasting corporate profits.

The McGraw-Hill survey of November 1973 showed a planned rise in plant and equipment expenditures by 13.6 percent for the following year. The forecast

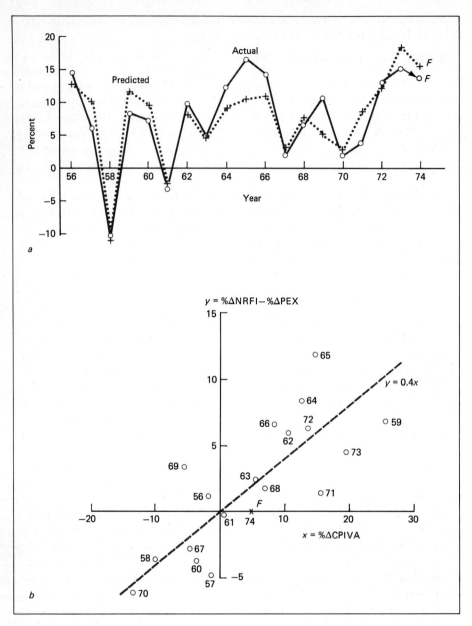

Figure 23-2 Forecasting nonresidential fixed investment. (*a*) Percent change from preceding year. (*b*) Predictor relation for percent change of NRFI.

of 1974 corporate profits made in the preceding chapter was %ΔCPIVA as 5 percent. Thus our forecast for NRFI comes out as follows:

$$(\%\Delta NRFI)^F = \%\Delta PEX + 0.4(\%\Delta CPIVA) = 13.6\% + 0.4(5\%) = 15.6\%$$

Later in the forecasting process, this initial forecast was revised down to 13.6 percent—to allow for the upward bias of the predicted values in recent years and for expected unfavorable effects of supply bottlenecks and tight money. Given the 1973 level of 136.2 billdar, the forecast of nonresidential fixed investment for 1974 became

NRFI = 1.136(136.2) = 154.7 billdar

By January of the forecast year and in succeeding months, the findings of additional surveys of expected plant and equipment expenditures are published by the Securities and Exchange Commission and the Department of Commerce, and by the McGraw-Hill Economics Department. Later plans usually approach the the true value for the year more and more closely, and hence permit progressive improvement of the forecast as the year progresses.

INVESTMENT IN RESIDENTIAL STRUCTURES

The residential-structures component of GNP measures the flow of value added in private housing-construction activity in a given time period. It covers farm and nonfarm residences, single-family and multiple-dwelling units, seasonal and year-round houses; it excludes group quarters such as dormitories, transient accommodations like hotels and motels, and mobile homes. The series is estimated from monthly surveys of building permits issued by local authorities. Permits are converted to housing starts, and starts are multiplied by an average value figure (based on permit data) to yield construction value. This value of residential housing starts is then spread over a seven-month span according to a normal pattern of on-site construction activity, and the resulting series for value added is seasonally adjusted.

To forecast construction activity on residential structures (RS), it would seem logical to obtain a forecast of housing starts HS and to relate residential construction expenditures to that series. There is no expectational series for housing starts comparable to that for nonresidential fixed investment. So a multivariate statistical approach is required, as exemplified by equations in some of the econometric models, or else the housing starts series must be extrapolated, with judgmental modifications based on the expected economic conditions affecting housing markets in the year ahead. The second of these approaches was used in preparing our forecast here.

In Table 23-3, column 1 presents the data on private housing starts, and column 4, the data on GNP expenditure for residential structures.

First, the housing starts series needs to be adjusted in timing to yield a measure of physical construction activity occurring in each period as a result of the starts.

TABLE 23-3   Residential Construction Expenditures

Year	(1) Private Housing Starts Annual (HS)	(2) Private Housing Starts 4th Quarter (HS$_{4Q}$) †	(3) Housing Starts Construction (HSC)	(4) Investment in Residential Structures (RS)	(5) Average Value per Start (RS/HSC)
	(million starts per year)			(in billdar)	(in $000 per start)
1959	1.517	0.313	1.539	25.5	16.6
1960	1.252	0.263	1.282	22.8	17.8
1961	1.313	0.304	1.289	22.6	17.5
1962	1.463	0.344	1.439	25.3	17.6
1963	1.610	0.377	1.590	27.0	17.0
1964	1.529	0.346	1.547	27.1	17.5
1965	1.473	0.339	1.474	27.2	18.5
1966	1.165	0.210	1.246	25.0	20.1
1967	1.292	0.334	1.218	25.1	20.6
1968	1.508	0.364	1.490	30.1	20.2
1969	1.467	0.302	1.504	32.6	21.7
1970	1.434	0.389	1.382	31.2	22.5
1971	2.052	0.505	1.982	42.7	21.6
1972	2.357	0.553	2.328	54.0	23.2
1973	2.045	0.371	2.155	58.0	27.0
1973 1st qtr.	2.393		2.395	59.0	24.7
2nd qtr.	2.212		2.320	59.6	25.7
3rd qtr.	2.009		2.148	59.2	27.6
4th qtr.	1.584		1.859	54.0	29.0
1974^F	1.700	0.454	1.650	51.3	31.0

† Unadjusted for seasonal factors.
Sources: Cols. 1 and 2, *Economic Report of the President*, February 1974, appendix table C-39, and *Survey of Current Business*, p. S-10. Col. 3, for annual data:
$$HSC_t = HS_t + 0.6(HS_{4Q})_{t-1} - 0.6(HS_{4Q})_t$$
For quarterly data:
$$HSC_t = 0.1HS_{t-2} + 0.5HS_{t-1} + 0.4HS_t$$
Col. 4, "National Income and Product Accounts," table 1.1, *Survey of Current Business*.

On the average, about 40 percent of the work on starts during a given quarter-year is performed in that quarter, 50 percent in the next quarter, and 10 percent in the second quarter following. Applying this time pattern of activity to quarterly data on housing starts, we can obtain a series for construction activity resulting from housing starts. This series will be designated "housing starts construction" (HSC), and its derivation is shown in columns 2 and 3 in Table 23-3. The quarterly data for 1973 in column 1 indicate that a sharp decline in housing starts was under way in late 1973, with the fourth-quarter level almost 34 percent below the first quarter. Such a decline is normal during a period of strong economic expansion when nonresidential investment rises rapidly and competes capital funds and construction resources away from housing activity. Interest rates on mortgages rose rapidly in 1973, but the competition for funds drove short-term rates up faster and slowed the flow of funds into savings banks and savings and loan associations, which are the principal source of mortgage funds. Also, of course, the record boom in residential construction during 1972 to 1973 presumably brought the nation's stock of residences nearly up to desired levels, so that some increases in vacancies and weakening in demand were to be expected. At year-end 1973, it was expected that the shortage of funds and adequacy of stocks would continue during the first half of 1974, but that an easing of interest rates and renewed flow of mortgage funds later in the year would bring activity back toward the level of 2 million units per year, a level believed to be near the long-term trend. On balance, 1.7 million starts was a reasonable forecast for 1974 as a whole, with a rising level during the year. The housing starts construction figure (HSC) was placed at 1.65 million units because fewer units in process were carried into 1974 than were expected to be carried into 1975.

The next step was to estimate the average value per start in 1974. Column 5 in Table 23-3 shows the historical data for 1959 through 1973, including the quarterly figures in 1973. In general, of course, there has been an upward trend in average value per start. There has also been a tendency for unit value to decline when housing activity is strong (1963, 1968, 1971) and to rise faster than average when housing activity is weak (1960, 1966, and 1969 and 70). Presumably the mix shifts toward lower-priced homes in boom times and toward high-priced units when construction resources and mortgage funds are withdrawn from residential markets.†
This pattern was broken somewhat during 1972 when a persistent inflation drove average value up as housing rose. The rising tide of average value per start ran even more strongly during 1973, with increases of nearly $1,500 per start in each quarter of 1973. Continued weakness in the housing industry will tend to slow the rise in 1974, but cost pressures will probably continue to be strong. On balance, a 15 percent rise in average value per unit, to a level of $31,000 per start, was estimated for 1974.

The product of the forecasts for housing starts construction HSC and average value per start led to the desired forecast of expenditures for residential structures RS in the GNP.

$$RS = (1.65 \times 10^6) \times (31 \times 10^3) = 51.3 \times 10^9 = 51.3 \text{ billdar}$$

EXPORTS AND IMPORTS

About two-thirds of U.S. exports are merchandise shipments. The other third are sales of services to foreigners—transportation services, financial services, foreign tourist expenditures in the United States, income on foreign investments

---

† In addition, there are some expenditures included in HS, such as repairs and alterations, nonhousekeeping units, and brokers' commissions, which do not fluctuate as widely as do housing starts.

(services of U.S. capital funds employed abroad), and so on.   Some of these components depend strongly on economic prosperity in countries with which we trade. So forecasting U.S. exports would logically involve prediction of business conditions in foreign countries.   Such an analysis is beyond the scope of this approach.   So the noncausal percent change technique was used, modified on the basis of qualitative knowledge of prospects for causal factors.

Figure 23-3 shows that U.S. exports were very volatile in the early postwar years, when the nation was helping to rebuild war-devastated economies in Europe and Japan, and during the Korean war (mid-1950 to mid-1953).   For the years 1954 through 1971, the year-to-year percent changes fluctuated around a median of 8 percent and had mean absolute deviation around the median of 4.9 percentage points.   Since then, exports have been influenced by a number of exogenous

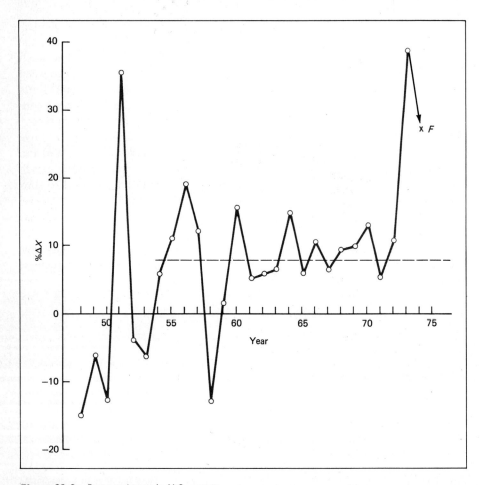

Figure 23-3   Percent change in U.S. exports.

influences that stimulated them beyond the range of fluctuations during the previous decade. In mid-1971, the United States suspended payment of gold for dollar balances held by foreign governments, and the dollar was devalued by about 8 percent against the currencies of major trading partners late in the year. Price controls in the United States were imposed in August 1971 and continued on some commodities through 1973—encouraging exports to higher-priced foreign markets. A worldwide grain shortage in 1972 led to a huge increase in exports of farm products. In February 1973 came a second devaluation of the dollar, this time by 10 percent, followed by a floating exchange rate against principal foreign currencies. Inflation did continue in the United States through 1972 and 1973, but not so rapidly as abroad. All these changes combined in unknown proportions to raise exports by 39 percent from 1972 to 1973. The rate of gain from the preceding year rose each quarter of 1973 to 46 percent in the fourth quarter. While continued rises were expected in 1974, some slowing of the rate of rise was predicted. A figure between 25 and 30 percent was selected, though it must be admitted that a large error of forecast should be attached to the estimate. The forecast of exports X in 1974 became

$$X = 1.275(102.0) = 130.0 \text{ billdar}$$

The figure was later checked against the forecast of imports to be sure that a reasonable level of net exports emerges from the forecasts.

Imports (Z) is the first of our demand components which can be regarded as induced by levels of GNP. United States imports of consumer and capital goods, and expenditures for tourist and other services, should increase as U.S. aggregate demand rises, and vice versa. Also, U.S. imports of raw materials and components should fluctuate in phase with changes in U.S. production. This expectation is borne out in Fig. 23-4 by the regression chart for the percent change in imports ($\%\Delta Z$) plotted against the percent change in GNP ($\%\Delta GNP$). A line averaged through the points indicates that imports would fall if GNP rose more slowly than 4 percent per year. But as the growth rate of GNP increases, the rate of rise of imports is about 3 times as great, in percentage points. That is, the equation for the regression line is approximately

$$\%\Delta Z = -12.3 + 3.2(\%\Delta GNP)$$

For the forecast rise of GNP by 8.1 percent in 1974, the calculated rise in imports would have been

$$\%\Delta Z = -12.3 + 3.2(8.1) = 13.6\%$$

Given the 1973 value of imports as 96.2 billdar, the 1974 forecast would have been $Z = 1.136(96.2) = 109.3$ billdar

Unfortunately, this forecasting system was confidently expected to fail in 1974. Oil shortages during the embargo which started in November 1973 and

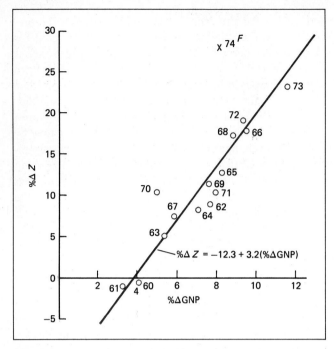

Figure 23-4   Imports related to GNP (percent changes from preceding year).

subsequent price rises for imports of petroleum and its products were expected to raise import payments by perhaps 12 to 16 billdar in 1974. If we arbitrarily add a figure of 14 billdar to the imports calculated above, the 1974 level of imports becomes

$$Z = 109.3 + 14.0 = 123.3 \text{ billdar}$$

The combined forecasts for exports and imports yield a 1974 figure for net exports.

	Billdar		Percent Change
	1973	1974	
Exports	102.0	130.0	+27
Imports	96.2	123.3	+28
Net exports	5.8	6.7	

The figure for net exports does seem reasonable. The export surplus reported for the fourth quarter of 1973 was 12.8 billdar, but this was expected to shrink in 1974 as higher prices of petroleum raised imports.

PERSONAL CONSUMPTION EXPENDITURES

Consumption expenditures grow at nearly the same rate as net disposable income in the long run, but the average propensity to consume may fluctuate in the short run because of the influence of previous income, inflation, credit conditions, adequacy of stocks of consumer durables and financial assets, or expectations regarding future incomes and prices. The forecasting approach used was to examine the historical ratios to Y_d' for personal saving and components of consumption expenditures, and then to predict those ratios in the forecast year, using judgment to estimate how nonincome factors would change the allocation of income. Finally, multiplication of the projected ratios by the forecast of Y_d' in the preceding chapter yielded forecasts of personal consumption and saving flows in 1974.

Figure 23-5 shows the consumption and saving ratios as percentages of net disposable income for 1948 through 1973, with horizontal dashed lines to show average levels. In each year the three percentages must, of course, sum to 100 percent. Several patterns of movement are significant. There is a tendency for the share of income spent on durable goods to decline in recession years—1949, 1954, 1958, 1961, 1967, 1970. In the early postwar recessions, this decline was offset by a rise in the proportion of income spent for nondurables and services, but in 1967 and 1970 it was a rise in saving ratio which offset declines in the shares of the other two components. There is some tendency for durables and saving ratios to move inversely, presumably because some purchases of durables are financed by consumer borrowing, which is a deduction from personal saving. Finally, the pattern for spending on nondurables and services shows waves lasting 10 to 15 years, thus implying inverse waves of accumulation of durable goods and financial assets (saving).

For 1974 the ratio for nondurables and services was expected to continue its recovery toward the average level near 80 percent. A figure of 79.0 percent was chosen, up from 78.4 percent in 1973. Durables spending was expected to decline as a percent of Y_d' for several reasons. Automobile buying had been strong for 2 years and prospective shortages and conservation of gasoline in 1974 were expected to reduce auto purchases and to lead to a market shift toward small cars. In addition, low housing construction is usually accompanied by weakness in purchases of household appliances, furniture, and carpets. So the consumer durables ratio was cut back from 15.2 percent in 1973 to 14.0 percent in 1974. These forecasts for consumption expenditures implied a saving ratio of 7.0 percent. As may be seen in Fig. 23-5, this level is well within the range of recent experience and involves a moderate rise from 6.4 percent in 1973.

More detailed analyses may be made to evaluate the influence of nonincome factors on individual components of consumer spending and saving, as is done in econometric models. Automobiles, other durables, and saving have probably received most attention. However the approach described here provides an overall perspective and a reasonable judgment forecast. The dollar figures are shown in Table 23-4.

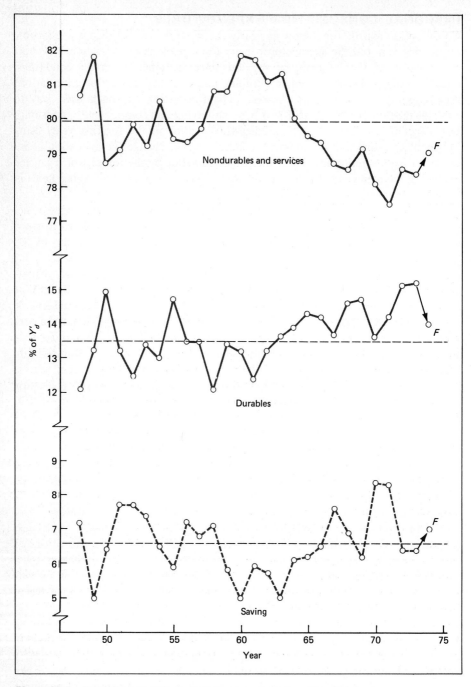

Figure 23-5 Personal consumption and saving as percent of net disposable income.

TABLE 23-4 1974 Forecast of Consumer Spending and Saving

	Percent of Y_d'		Billdar		Percent Change
	1973	1974	1973	1974	1973–1974
Net disposable personal income (Y_d')	100	100	858.8	933.0	8.6
Consumer Expenditures					
Nondurables and services (CND + CS)	78.4	79.0	673.2	737.0	9.5
Durables (CD)	15.2	14.0	130.8	130.6	−0.2
Personal saving (S_P)	6.4	7.0	54.8	65.4	19.4

CHANGE IN BUSINESS INVENTORIES

The final component of GNP to be forecast is the change in business inventories. Investment in business inventories equals the annual rate of change in physical quantities of farm and nonfarm business inventories valued at their average prices during the given year or quarter. This differs from the rate of change in book value of inventories by the inventory valuation adjustment IVA. The latter subtracts from book-value changes the capital gains (or loss) component arising from changes in prices of inventories of goods during the time they are held by businesses.

To forecast the change in business inventories (CBI), we cannot base the analysis on year-to-year percent changes, because CBI values can go to zero or very small values and make such changes huge. Instead, we shall examine the ratio CBI/GNP, with both numerator and denominator in current dollars. This ratio variable is plotted for the postwar years in Fig. 23-6. It was rather erratic during the early postwar years and the Korean war (1950 to 1953) and during the Vietnam war build-up (1965 and 1966). Also, the ratio dropped markedly in most recession years (1949, 1954, 1958, 1961, 1967, 1970).

Until 1970 an accelerator type of relation expressing inventory investment as a function of the *change* in GNP proved to be a helpful predictor. (This relation was explained and pictured in Chap. 9.) However, in recent years the inventory investment ratio has been lower than the 0.9 percent average for 1948 through 1971, and fluctuations of the ratio have been much reduced. Many analysts believe that this reduction and stabilization of inventory ratios have resulted from increased use of operations research models and computers to monitor inventories and control purchasing policy. So a ratio of 0.5 percent was chosen for 1974, in line with levels of the ratio in the past 4 years and appropriate for a semirecession period. The dollar-value forecast becomes

$$CBI = 0.005(1,393) = 7.0 \text{ billdar}$$

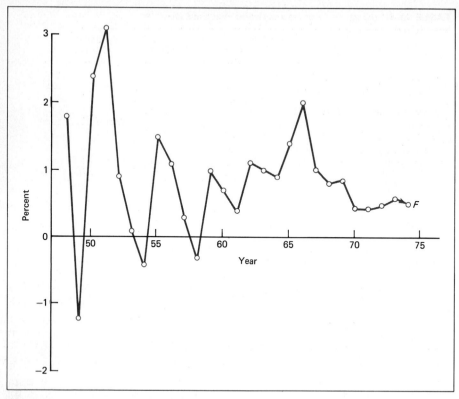

Figure 23-6   Change in business inventories (CBI) as percent of GNP.

SUMMARY OF FORECASTS OF GNP COMPONENTS (step 3)

The forecasts of GNP components for 1974, developed in the preceding sections, are summarized in Table 23-5, which provides the figures to fill the hungry cells back in Table 23-1.   Total GNP turns out to be 1,393.5 billdar, practically identical with the initial estimate of 1,393 billdar arrived at in Chap. 21.   In practice, this close correspondence is not achieved in the first go-around, and it becomes necessary to use this GNP forecast to derive new sector income estimates (Chap. 22, step 2) and then revise the sector demand forecasts in this chapter as necessary. The forecasts of GNP components presented here were the product of such an iterative procedure, so that the sums of the sector incomes and of the sector demands were equal to each other and to the initial GNP projection.

Presumably, the forecasts of GNP components could be improved by further causal statistical analysis.   More causal variables could be introduced, including monetary influences.   Quarterly data could be used in order to determine time lags more accurately.   Income-expenditure feedback loops could be written into a system of equations to permit simultaneous solution for several endogenous

TABLE 23-5 Summary of Forecasts of GNP Components for 1974

	1973	1974F	Percent Change
	(in billdar)		
Gross national product	1,289.1	1,393.5	8.1
Personal consumption expenditures	804.0	867.6	7.9
Durables	130.8	130.6	−0.2
Nondurables and services	673.2	737.0	9.5
Gross private domestic investment	202.1	212.7	5.2
Nonresidential fixed investment	136.2	154.7	13.6
Residential structures	58.0	51.0	−12.0
Change in business inventories	8.0	7.0	—
Government purchases of goods and services	277.1	306.5	10.6
Federal national defense	73.9	77.0	4.2
Other federal	32.7	37.0	13.1
State and local	170.5	192.5	12.9
Net exports	5.6	6.7	—
Exports	102.0	130.0	27
Imports	96.2	123.3	28

variables. And both income and expenditure components could be subdivided and forecasts prepared for smaller categories. Such are the lines of development in econometric models.

The overall picture indicated that consumer spending was expected to rise about in step with GNP, but with demands for nondurables and services rising faster and with durables purchases unchanged in current dollars. In the investment area, plant and equipment spending was expected to rise about 14 percent and residential construction to decline by 12 percent, with inventory investment down a little. Government spending was forecast to rise moderately faster than GNP—national defense spending slowly (4 percent), but state and local and other federal purchases more rapidly (13 percent). Both exports and imports were forecast to rise by more than 25 percent, with about a 20 percent rise in net exports. The rise in GNP is 8 percent, and in the next chapter, that increase will be analyzed into price rise and growth in real GNP.

SAVING AND BORROWING BY SECTORS (step 4)

As a check for possible inconsistencies in the forecasts of sector incomes and expenditures, and as a guide to making revisions when the forecasts of demand components do not add up to the initial estimate of GNP, it is desirable to calculate the saving or borrowing by sectors. The necessary figures are brought together in Table 23-6 on the following page. Historical data were obtained from the NIPA source tables cited in connection with Tables 22-1 and 23-1.

TABLE 23-6 Saving, Borrowing, and Investment Components (in billdar)

Line	Item	Symbol	1971	1972	1973	1974ᶠ	1973 1st Quarter	2nd Quarter	3rd Quarter	4th Quarter
	Personal Sector									
1	Net disposable personal income	Y'_d	727.3	776.2	858.8	933.0	829.4	846.6	867.0	892.2
2	Personal consumption expenditure	C	667.2	726.5	804.0	867.6	779.4	795.5	816.0	825.2
3	Personal saving	S_p	60.2	49.7	54.8	65.4	50.0	51.0	51.1	67.1
	Business Capital Accounts									
4	Gross retained earnings	GRE	111.8	124.4	135.2	144.0	131.5	132.0	136.9	140.6
5	Gross private domestic investment	I_d	153.2	178.3	202.1	212.7	194.5	198.2	202.0	213.9
6	Business borrowing	BOR_b	41.4	53.9	66.9	68.7	63.0	66.2	65.1	73.3
	Government Sector									
7	Government net receipts	T_n	216.2	252.2	288.7	308.0	277.5	286.9	293.3	296.3
8	Government purchases of goods and services	G	234.3	255.0	277.1	306.5	268.6	275.3	279.0	285.6
9	Government surplus (lines 14 and 15)	S_g	−18.1	−2.8	11.4	1.5	8.9	11.6	14.3	10.8
	Foreign Sector									
10	Transfers to foreigners (net)	TR_f	3.6	3.7	3.6	4.0	3.0	3.3	3.5	4.5
11	Net exports	X_n	0.8	−4.6	5.8	6.7	0.0	2.8	7.6	12.8
12	Foreign borrowing	BOR_f	−2.8	−8.4	2.2	2.7	−3.0	−0.5	4.0	8.3
13	Statistical discrepancy	SD	−3.4	−1.5	2.9	4.5	1.1	3.2	3.7	3.7
14	Federal surplus	S_f	−22.2	−15.9	0.9		−5.0	0.0	4.0	4.7
15	State and local surplus	S_{sl}	4.0	13.1	10.5		13.9	11.5	10.4	6.0

Personal saving (line 5) seemed reasonable. Not surpisingly, because its ratio to Y_d' was checked against historical levels when the forecast of Y_d' was split into personal consumption expenditures and saving in a previous section.

The forecast for business borrowing (line 6) may be suspect. The rise from the 1973 level is very small, because gross retained earnings rise almost as fast as investment expenditures. The situation is not unprecedented. Borrowing actually declined in all the postwar recession years. The figure for the fourth quarter of 1973 had shown a large rise, because of the high rate of accumulation of inventories. A decline below that level for 1974 was considered reasonable but worth watching carefully as the year unfolded.

The government surplus for 1974 (line 9) worked out at only 1.5 billdar, following a large surplus of 11.4 billdar in 1973. The quarterly pattern during 1973 showed a peak in the third quarter. Separate data for surpluses of the federal government and the state and local governments are shown in lines 14 and 15. The quarterly patterns are opposite. If the surplus for state and local governments continues to decline or becomes negative, and if the federal surplus stabilizes or declines, then the forecast decline in the Government Sector surplus could occur. But it was another forecast which had to be watched as further data became available.

Foreign Sector borrowing (line 12) reflects the moderate rise in export surplus expected in 1974. Note that it is below the levels in the second half of 1973.

REVISIONS AND RECONCILIATIONS (step 5)

The forecasts developed in Chaps. 22 and 23 showed sector incomes and expenditures which yielded GNY and GNP totals as very nearly equal. And the check on sector saving and borrowing flows (Table 23-6) revealed two areas of possible adjustment. That is, some reallocation of incomes or expenditures could have been made to increase the Government Sector surplus and to raise business borrowing by an equal amount. On the whole, however, these forecasts seemed to be reasonable and consistent at the time they were made. Time will tell whether they were accurate. (If readers ever check these forecasts against later reported figures, they should compare forecast with reported values for year-to-year *changes* of variables, rather than forecast with actual *levels*. This is because 1973 data will be revised by the time 1974 figures are reported, and the 1974 forecasts would have been different if revised 1973 data had been used as a base. However, *changes* in levels, in billdar or percentages, should be less affected by revisions of historical levels. Also, predicting changes in levels is really what forecasting is all about.)

When an adequate reconciliation of total GNP and of sector incomes, expenditures, and saving or borrowing has been achieved, then it is time to "quarterize" the forecast, if the quarterly pattern needs to be forecast. The sequence of estimation is much the same as for the initial forecast. Exogenous GNP components should be estimated first, in such a way as to embody any information concerning the prospective pattern of movement during the year but keeping the average of the four quarterly annual rates equal to the annual total. The quarterly movement of the endogenous demand components—consumer nondurables and services, imports, inventory change—can then be set in a reasonable relation to the quarterly totals of the exogenous components. Thus the quarterly pattern of GNP demanded can be established.

Each quarter's GNP can then be allocated to sectors on the income side, with allowance being made for any jumps in tax rates or transfer payments at specific points in time. Then a table should be drawn up for each quarter with sector saving or borrowing calculated, and necessary adjustments should be made to achieve a reasonable balance of the accounts. (If several persons work independently on components of the forecast, discrepanicies will be evidenced also in abnormal shifts in the statistical discrepancy, calculated as a residual.)

Additional details may be added to the model, but at a rising cost in terms of computational complexity. As more variables and relations (equations) are added to the model, it becomes more desirable to develop an explicit econometric model so that solutions may be carried through by computers, either simultaneous solutions for jointly dependent endogenous variables or step-by-step solutions for successive time periods. An open question remains of how much the accuracy of forecasts is improved by use of more disaggregated models, but computerization of the calculations becomes well-nigh mandatory if finer details or many simulations with alternative exogenous inputs are to be developed.

Examination of the variables and relations contained in econometric models may provide suggestions for improvements in the forecasting approach described in this book.

Chapter 24

Forecasting Real GNP, Employment, and Prices

Now that we have a forecast for current-dollar flows of incomes and GNP purchases, we are ready to complete the last four steps of this forecasting approach:

Step 6: Forecast of real GNP

Step 7: Estimate of levels of employment and unemployment

Step 8: Forecast of price changes

Step 9: Final review and adjustments

Real GNP was forecast on a noncausal basis back in Chap. 21. Now we can modify, and hope to improve, on that initial estimate by using our revised forecast of current-dollar GNP along with an initial projection of P_Y. Real GNP is calculated in this way and then divided among three major producing sectors: general government, farm, and private nonfarm components. In step 7 we shall project total labor force, use productivity estimates to derive employment from the real GNP forecasts for the three major producing sectors, and then calculate unemployment. In step 8 we estimate rises in wage rates and unit labor costs to arrive at a revised estimate of price changes in 1974. And finally, the whole forecast is reviewed and modified as needed in light of the new information and forecasts developed in these last three steps.

FORECAST OF REAL GNP (step 6)

In previous chapters, on the basis of historical patterns, knowledge of the business outlook in general, and forecasts published in the business press, we arrived at the following projections for 1974:

	1973	1974	Percent Change
Y (billdar)	1,289.1	1,393.0	8.1
P_Y (1958 = 100)	153.94	164.7	7.0
Y_R (billdar at 1958 prices)	837.4	846.0	1.0

In Chap. 23 we developed a revised forecast of GNP in current dollars. If that had differed from the original estimate, then, using the previous price projection, we would calculate a revised estimate for Y_R.

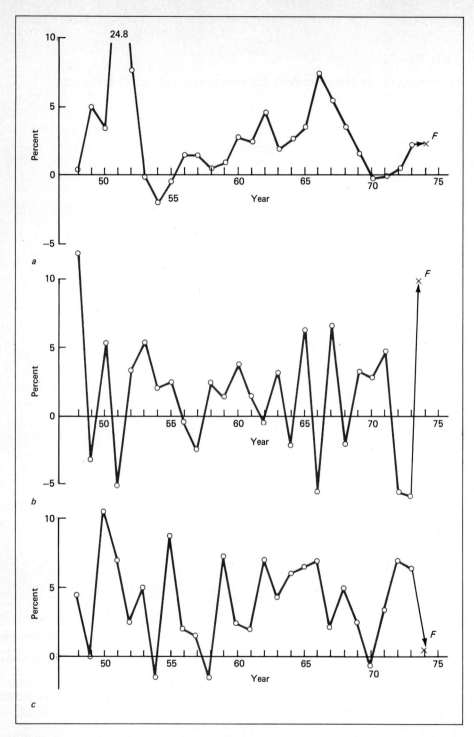

Figure 24-1 Percent changes for real GNP by sectors. (*a*) General government GNP. (*b*) Farm GNP. (*c*) Private nonfarm GNP.

Table 24-1 1974 Forecast for Real GNP by Producing Sectors

	Billdar at 1958 Prices		Percent Change
	1973	1974F	
Total real GNP	837.4	846.0	1.0
Government	62.5	64.0	2.4
Farm	23.2	25.5	10.0
Private nonfarm	751.7	756.5	0.6

In order to develop employment forecasts from the forecast of real GNP, it is desirable to split Y_R into three major components because productivity movements are different for government, farm, and private nonfarm GNP.   Figure 24-1 presents percent changes from the previous year for the three components during the postwar period, based on data in table 1.8 in the NIPA. We proceed to forecast 1974 percent changes for the two smaller components (the government and farm sectors).

General government GNP consists of compensation of civilian and military employees except those working in government enterprises.   It rose rapidly during the Korean war and the Vietnam war expansions, but the rate of increase has been below 5 percent in other years.

A forecast of 2.4 percent rise was selected and is indicated by an X in panel (a) of the chart.

Farm GNP has been quite erratic, exhibiting alternate-year rises and declines since 1960.   High demand and large price rises for farm output in 1973 were expected to lead to a large increase in 1974 output.   As indicated by the X mark, a forecast of 10 percent rise in farm production was chosen.

From these two percent-change forecasts we can compute the 1974 levels of real GNP for the government and farm sectors, then obtain private nonfarm GNP by subtraction from the 1974 forecast of Y_R, as shown in Table 24-1.

FORECAST OF EMPLOYMENT AND UNEMPLOYMENT (step 7)

With output forecast by producing sectors, we proceed to estimate employment in those sectors.   Subtraction of employment from a projection of total labor force will yield an unemployment figure, which can be checked for political feasibility and which can give some indication of prospective actions by monetary and fiscal policy makers.

First, we need a forecast of total labor force including the armed forces.   The postwar historical record of year-to-year changes is plotted in Fig. 24-2.   The increased annual rate of births which began in 1946 is now causing a larger growth rate of the labor force than the rate in the postwar years to 1963.

In addition to its growth trend, the labor force displays marked fluctuations in annual increments, as is evident in the chart.   Increases tend to be large in

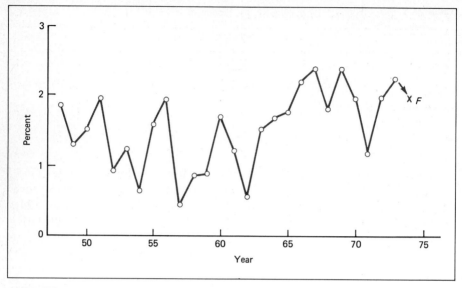

Figure 24-2   Percent change in total labor force (including armed forces).

years of strong business activity and/or when the armed forces are increasing rapidly (1948, 1950 and 1951, 1955 and 1956, 1966 through 1969). Slower growth appears after a year or two of above-normal increases or during recession years (1949, 1952, 1954, 1957 and 1958, 1962, 1971). For 1974 a medium rate of growth would seem likely, since recent gains would not seem excessive but slow growth in output is anticipated. Tentatively, a rise of about 2 percent was forecast, from 91.04 million in 1973 to 92.86 million in 1974.

To organize the estimation of employment and unemployment, a tabular layout, such as Table 24-2, is recommended. Lines 1 through 10 present data on labor force, employment by producing sectors, and unemployment. Lines 11 through 15 refer to the real GNP components, which we have forecast above. Lines 16 through 19 present productivity by sector in terms of output per employed person, that is, the ratios of real GNP to corresponding sector employment from preceding lines in the table. In forecasting employment, one estimates these productivity ratios as a means for bridging from the forecasts of real GNP to employment forecasts for producing sectors. Lines 20 through 25 indicate a more detailed analysis of output per man-hour and man-hours worked in the private nonfarm sector, used as a basis for developing the estimate on line 19—and also for calculating unit labor costs later. Notes below the table give sources or calculating procedures for the historical entries.

The sequence of steps in forecasting 1974 items is as follows.

Item 1   Labor force equals 92.86 million persons, as projected above from Fig. 24-2.

TABLE 24-2   Output, Productivity, and Employment

Line	Item		1969	1970	1971	1972	1973	1974F
1.	Labor force	(millions)	84.24	85.90	86.93	88.99†	91.04‡	92.86
2.	Armed forces	"	3.51	3.19	2.82	2.45	2.33	2.23
3.	Civilian labor force	"	80.73	82.72	84.11	86.54†	88.71‡	90.63
4.	Civilian employed	(millions)	77.90	78.63	79.12	81.70†	84.41‡	85.57
5.	Government	"	12.20	12.54	12.86	13.29	13.66	14.10
6.	Private	"	65.70	66.09	66.26	68.41†	70.75‡	71.47
7.	Farm	"	3.61	3.46	3.39	3.47	3.45	3.70
8.	Nonfarm	"	62.09	62.63	62.87	64.94†	67.30‡	67.77
9.	Unemployed	(millions)	2.83	4.09	4.99	4.84†	4.30	5.06
10.	Percent of civilian labor force	"	3.5%	4.9%	5.9%	5.6%	4.9%	5.6%
11.	GNP (billdar at 1958 prices)		725.6	722.5	745.4	790.7	837.4	846.0
12.	Government		60.7	60.7	60.7	61.1	62.5	64.0
13.	Private		664.9	661.7	684.7	729.5	774.9	782.0
14.	Farm		24.1	24.8	26.0	24.6	23.2	25.5
15.	Nonfarm		640.8	639.6	658.7	704.9	751.7	756.5
	Real GNP per employed person							
16.	Government		3,865	3,860	3,870	3,880	3,910	3,920
17.	Private							
18.	Farm		6,675	7,170	7,670	7,090	6,725	6,890
19.	Nonfarm		10,320	10,170	10,480	10,855	11,170	11,160
	Percent changes for private nonfarm items							
20.	Output		2.8	−0.6	3.4	7.0	6.6	0.6
21.	Output per man-hour		−0.2	0.7	4.3	4.2	3.0	1.1
22.	Man-hours		3.0	−1.3	−0.9	2.7	3.6	−0.5
23.	Persons employed		3.1	0.9	0.4	2.8	3.6	0.7
24.	Hours per employed person		−0.1	−2.2	−1.0	−0.1	0.0	−1.2
25.	Output per employed person		−0.3	−1.5	2.6	4.0	2.9	−0.1

† Values were raised in 1972 because of adjustment to the 1970 census: line 1, up 0.33; lines 4, 6, and 8, up 0.30; line 9, up 0.03.

‡ Values were raised an additional 0.05 million in 1973.

Sources:
Lines 1–10: *Economic Indicators*, pp. 10 and 13 (for line 5), or *Survey of Current Business*, p. S-13.
Lines 11–15: "National Income and Product Accounts," table 1.8, *Survey of Current Business.*
Line 16:   Line 12 ÷ (line 2 + line 5).
Line 17:   Line 13 ÷ line 6.
Line 18:   Line 14 ÷ line 7.
Line 19:   Line 15 ÷ line 8.
Line 20:   Percent change for line 15.
Line 21, 22: From indexes in Dept. of Labor releases on "Productivity, Wages, and Prices."
Line 23:   Percent change for line 8.
Line 24:   Line 22 minus line 23 (approximation).
Line 25:   Percent change for line 19.

Item 2   Armed forces of 2.23 million persons for the annual average, according to federal budget estimates.

Item 3   Civilian labor force of 90.63 million, the difference between item 1 and item 2.

Items 11 through 15   Real GNP by producing sector was forecast in Table 24-1.

Item 16   The national income accountants, lacking measurements of productivity changes for government employees, assume no change in output per employed person here.   Small fluctuations occur because of changes in the proportions of employees in different grades and occupations, but we cannot go far wrong in taking a figure of $3,920 at 1958 prices per employee.

Item 5   Government employment equals government output (item 12) divided by employee productivity (item 16).

$$\frac{64.0 \times 10^9}{3.92 \times 10^3} = 16.33 \text{ million employees}$$

Subtraction of 2.23 million in the armed forces (item 2) leaves 14.10 million civilian employees of the government (item 5).

Item 7   Farm employment had shown a relatively steady decline, but the strong rise in farm output forecast for 1974 was expected to require increased labor inputs.   Purely on a judgment basis, the 10 percent rise in output was split between a 7.2 percent increase in employment and a 2.5 percent gain in productivity.

Item 18   Farm productivity becomes the ratio of output to employment. Farm real GNP per employed person equals $(25.5 \times 10^9)/(3.70 \times 10^6)$, or $6,890 at 1958 prices per employed person.

Item 19   Private nonfarm output per employed person might be projected by relating its percent change to the percent change in real private nonfarm GNP. But it is more useful to push the analysis further in this sector to forecast output per man-hour and average annual hours per employed person, then to combine them to obtain item 19.   This is done in items 20 through 25.

Item 20   The percent change from the previous year for real private nonfarm GNP is calculated from the forecast in item 15.   For 1974, $\%\Delta Y_R$ equals 0.6 percent.

Item 21   Output per man-hour (private nonfarm) has risen about 2.7 percent per year compounded during the postwar years, but the annual changes have ranged from 6.2 percent to -0.6 percent, with a mean absolute deviation from the median of a little over 1 percentage point for the years 1960 through 1971.   See circles joined by solid lines in Fig. 24-3a.

Probably cyclical fluctuations in output cause part of the changes in output per man-hour (OMH).   When output rises rapidly, overhead labor—management, marketing, accounting, research, and other staff personnel—does not rise as rapidly as output; hence OMH rises.   Conversely, OMH would be expected to decline in the early stages of a recession because overhead labor is not reduced rapidly. However, following such an unexpected change in output, business firms will probably adjust overhead labor and cause a lagged reaction on OMH in the opposite direction.

This logic suggests an explanatory relation, such as

$$\%\Delta OMH_t = a + b(\%\Delta Y_{R,t}) - c(\%\Delta Y_{R,t-1})$$

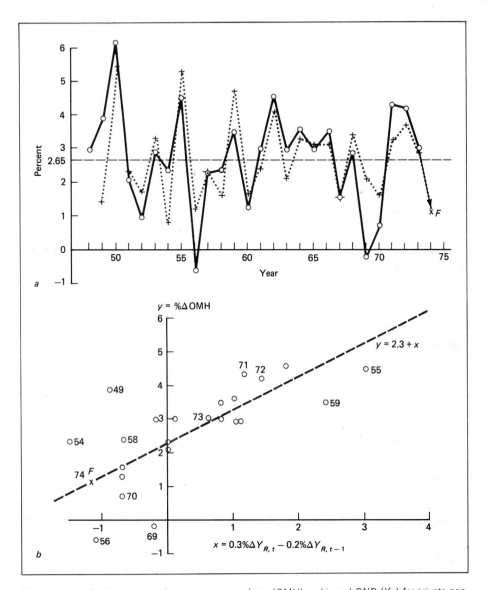

Figure 24-3 Percent changes in output per man-hour (OMH) and in real GNP (Y_R) for private non-farm sector.   (*a*) Percent change in output per man-hour.   (*b*) Percent change in OMH versus percent change in Y_R at t and $t-1$.

Exploratory graphical analysis indicated that good weights on current and lagged output change would be 0.3 and 0.2.   In Fig. 24-3*b*, %ΔOMH is plotted in a regression chart against the composite causal variable

$$x_t = 0.3(\%\Delta Y_{R,t}) - 0.2(\%\Delta Y_{R,t-1})$$

A line of relation

$$\%\Delta\text{OMH}_t = 2.3 + x_t$$

fits the points reasonably well and reduces the mean absolute error from 1.0 to 0.6 percentage point for the years 1960 through 1971.   Values calculated from this predictor equation during the postwar years are shown in Fig. 24-3*a* by plus signs joined by dotted lines.   Unexplained variations are still fairly large.   They may arise from shifts between years in the proportions of output produced by industries with different levels of OMH, from use of too simple a time-lag structure, from effects of different expectations of business managers regarding future output, from nonlinearities of the relation at very high or low levels of output in relation to capacity, or other factors.   However, this relation is simple, utilizes explanatory variables which have been forecast, and reduces forecast error considerably.

For 1974, we calculate

$$\%\Delta\text{OMH} = 2.3 + 0.3(0.6) - 0.2(6.6) = 1.1\%$$

Item 22   Aggregate man-hours worked per year (MH) is calculated by dividing output by OMH.   For percent changes, we have

$$100 + \%\Delta\text{MH} = 100 \frac{100 + \%\Delta Y_R}{100 + \%\Delta\text{OMH}} = 100 \frac{100.6}{101.1} = 99.5$$

$$\%\Delta\text{MH} = -0.5\%$$

(As an approximation, $\%\Delta\text{MH} = \%\Delta Y_R - \%\Delta\text{OMH} = 0.6 - 1.1 = -0.5\%$.)

Item 23   Private nonfarm employment (E) may be derived from MH if we know how the average of annual hours per employed person (H) changes.

$$E = \frac{\text{MH}}{H}$$

To split the change in MH between E and H, we may plot the regression chart relating %ΔE to %ΔMH, as in Fig. 24-4*b*.   The relation is reasonably "tight" and would have yielded the points marked as pluses joined by dotted lines in Fig. 24-4*a*.   The reduction of H in recessions leads to a smaller decline in E than in MH; the increase of H when output rises strongly leads to smaller rises in E than in MH in boom times.

For 1974, we have forecast that %ΔMH equals −0.5 percent.   So %ΔE = 0.75 + 0.67 (−0.5) = 0.4 percent,   In view of the fact that the values of %ΔE have exceeded the calculated values in 1970 through 1973, it was decided to raise %ΔE to 0.7 percent in 1974.

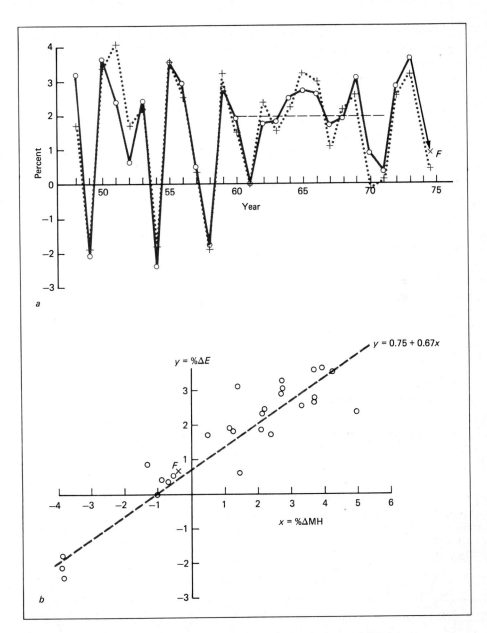

Figure 24-4 Percent changes in employed persons (*E*) and man-hours (MH) for private nonfarm sector. (*a*) Percent change in employed persons. (*b*) Percent change in *E* versus percent change in MH.

Item 24   Hours per employed person (*H*) may be calculated from the relation $H = \text{MH}/E$.   In terms of percent changes,

$$\%\Delta H \simeq \%\Delta\text{MH} - \%\Delta E = -0.5 - 0.7 = -1.2\%$$

Item 25   Output per employed person may be calculated either from Y_R/E or by multiplying OMH by *H*.   Determination of percent change from the second relation for 1974 results in:

$$\%\Delta(Y_R/E) \simeq \%\Delta\text{OMH} + \%\Delta H = 1.1 - 1.2 = -0.1\%$$

Item 19   Private nonfarm real GNP per employed person in 1974 can now be forecast at 0.1 percent below the 1973 figure.   The result is $11,160 at 1958 prices per employed person.

Item 8   Now we can obtain the missing figure for civilians employed in private nonfarm jobs.   For 1974, it is

$$E = \frac{Y_R}{Y_R/E} = \frac{756.5 \times 10^9}{11.160 \times 10^3} = 67.77 \text{ million persons}$$

(The same result could have been obtained from the item 23 forecast of a 0.7 percent rise in *E*, since 1.007 times 67.30 equals 67.77 million persons also.)

Item 6   Private employment becomes 71.47 million persons for 1974, the sum of items 7 and 8.

Item 4   Total civilian employment in 1974 becomes 85.57 million persons, the sum of items 5 and 6.

Item 9   Unemployment is the difference of item 3 minus item 4.   For 1974, the calculated value is 5.06 millions, higher than in 1973.

Item 10   The unemployment rate turns out to be 5.6 percent in 1974, that is, item 9 divided by item 3.

The unemployment rate seemed reasonable—in the perspective of recent levels of unemployment and in view of the fact that the forecast growth of real GNP in 1974 was slower than the expected rise in potential output.   "Potential GNP" is the output which would be produced in a year with 96 percent of the labor force employed and with productivity and average hours at levels corresponding to this "full-employment" situation.   According to the Council of Economic Adviser's calculations, potential GNP in 1973 was 851.6 billdar at 1958 prices and increasing at about 4.0 percent per year.   The above 1974 forecast of Y_R (846 billdar at 1958 prices) would be 4.4 percent below the prospective potential real GNP of 1.04 (851.6), or 885.7 billdar at 1958 prices in that year.   According to Okun's law,† the percentage of unemployment will exceed 4 percent

---

†Arthur M. Okun discovered this empirical relation during his years as a member and later chairman of the Council of Economic Advisers, between 1965 and 1968.

by about one-third of the percent gap of actual output below potential. This relation yields

$$\%U = 4.0 + \frac{1}{3.2}(4.4) = 5.4\%$$

which agrees well with our forecast of 5.6 percent for the unemployment rate.

If the initial estimates for labor force and employment had given an unrealistic figure for unemployment, as is often the case, it would have been necessary to repeat the estimation process for items in Table 24-2 and to use judgment to shift estimates a little in such directions as to yield an acceptable level of unemployment.

The figures in the 1974^F column of Table 24-2 did seem to present a consistent and plausible picture of the employment situation for 1974, given the forecast of output levels developed previously. As a final step, we need to check on the initial projection of GNP price level to see that the real GNP figures are reasonably consistent with the forecast of aggregate demand in money measure.

FORECAST OF PRICE CHANGES (step 8)

The overall GNP price index may well be derived from forecasts of percent changes in the price indexes for the three producing sectors—government, farm, and private nonfarm. The postwar data for changes in these price indexes are shown in Fig. 24-5. (Note that the scales are different in the three panels.) The price index for private nonfarm GNP has the greatest weight in the overall GNP price index, and its rate of change fluctuates least, from 0.9 percent to 6.9 percent. The government GNP price index has a weight about one-seventh of the preceding component, but its annual changes range from 2.6 percent to 10.3 percent. The price index for farm GNP is much more volatile, showing annual changes ranging from -17 percent to $+21$ percent. But it is the least important component, with weight about one-thirtieth of that for private nonfarm GNP.†

As was brought out in the chapter on inflation, the price index for private nonfarm GNP is closely correlated with unit labor costs (ULC) in private nonfarm production. An attempt to forecast price change on this basis is illustrated in Fig. 24-6. Panel (b) displays a positive correlation when $\%\Delta ULC_{NF}$ is positive, but indicates that prices continue to rise about 1 percent per year when ULC_{NF} declines. So we use as our forecaster:

$$\%\Delta P_{NF} = \begin{cases} 1 + 0.61(\%\Delta ULC_{NF}) & \text{for } \%\Delta ULC_{NF} \geq 0 \\ 1 & \text{for } \%\Delta ULC_{NF} < 0 \end{cases}$$

---

† If we assume that the sector shares of real GNP stay constant from one year to the next, then

$$\%\Delta P_Y = \frac{Y_{NF}}{Y}(\%\Delta P_{NF}) + \frac{Y_G}{Y}(\%\Delta P_G) + \frac{Y_F}{Y}(\%\Delta P_F)$$

In 1970 the sector shares of current-dollar GNP were 0.85, 0.12, and 0.03 respectively. So we calculate for 1971:

$$\%\Delta P_Y = 0.85(4.6) + 0.12(7.5) + 0.03(-0.7) = 4.6\%$$

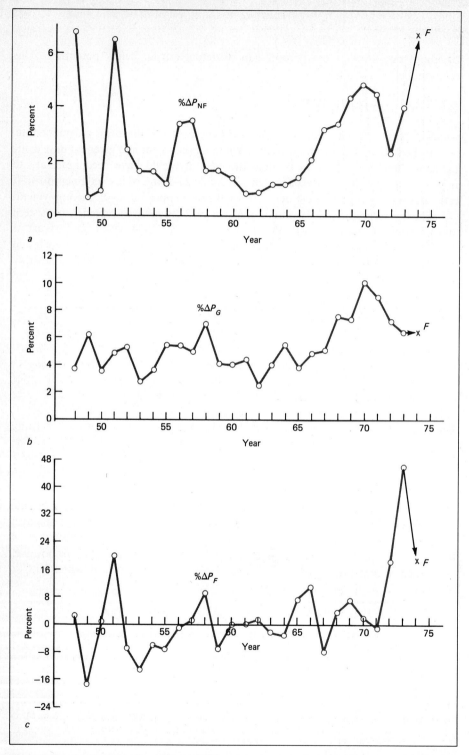

Figure 24-5 Percent changes in price indexes for GNP by producing sectors. (*a*) Private nonfarm GNP price. (*b*) Government GNP price. (*c*) Farm GNP price.

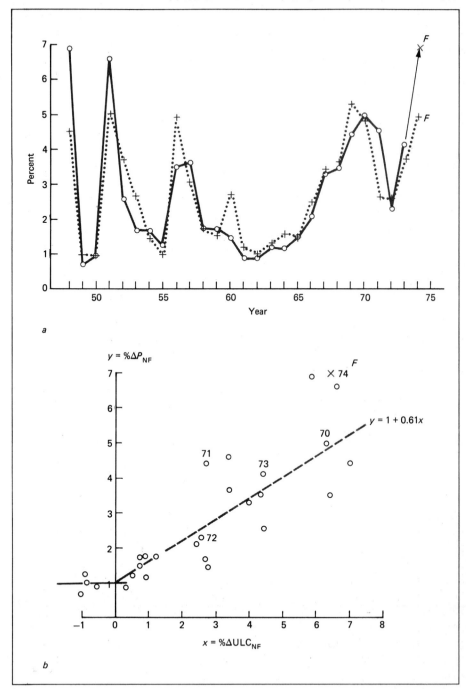

Figure 24-6 Percent change in price (*P*) and in unit labor cost (ULC) for private nonfarm GNP. (*a*) Percent change in price. (*b*) Percent change in P_{NF} versus percent change in ULC_{NF}.

Application of this forecasting relation throughout the postwar years yields the predicted points marked by plus signs in panel (a).

Now we are left with the problem of forecasting ULC_{NF}. Note that ULC equals W/OMH, where W is labor compensation per man-hour (wage rates plus fringe benefits) and OMH is output per man-hour. In terms of percent changes, we have approximately

$$\%\Delta ULC = \%W - \%\Delta OMH$$

We have already forecast $\%\Delta OMH$ for private nonfarm GNP as 1.1 percent for 1974 (item 21 in Table 24-2). So we need only add a value for $\%\Delta W$.

In econometric models, changes in wage rates are sometimes related to unemployment rate (inversely), to the consumer price index, and to profits ratios, usually with different time lags for the various causal factors. This approach will be left as an exercise for the reader. We shall project the rise in W by examining its historical record and applying judgment based on our expectations of the economic environment in 1974. Figure 24-7 (top panel) shows $\%\Delta W$ from 1948 through 1973. Its rise has held close to 7 percent per year for six years. During 1974 higher unemployment would tend to reduce the rate of rise, but continuing rapid rises in consumer prices and removal of controls would tend to maintain or increase it. On balance, compensation rates W were forecast to rise 7.5 percent in 1974, on the high side of the range experienced from 1968 through 1973.

The postwar record for $\%\Delta OMH$, along with its 1974 forecast of 1.1 percent, is also shown in the top panel of Fig. 24-7. The vertical distance between the two curves is an approximate measure of $\%\Delta ULC$. In many recent years, $\%\Delta ULC$ has been large both because $\%\Delta W$ has been high and because $\%\Delta OMH$ has been low. In the bottom panel we have the picture of fluctuations in $\%\Delta ULC$ during the postwar years. The above forecasts lead to a value for 1974:

$$\%\Delta ULC \simeq \%\Delta W - \%\Delta OMH = 7.5 - 1.1 = 6.4\%$$

When this forecast for unit labor cost is substituted into the price forecasting equation, we obtain for 1974:

Calculated $\%P_{NF} = 1 + 0.61(6.4) = 4.9\%$

There were persuasive reasons for believing that the actual price rise would be greater than indicated by this "normal" relation to unit labor costs. At the end of 1973 business managers regarded profit margins as low. Thus it was expected that they would attempt to raise prices faster than unit labor costs after price controls were removed in 1974, in order to increase margins. Also, some nonlabor cost components were rising strongly, for example, interest rates and prices of raw materials and imported goods, notably petroleum. Consequently, the forecast rise in nonfarm prices was set moderately higher than the 6.4 percent rise in unit labor costs.

$\%\Delta P_{NF}$ equals 6.8 percent, leading to a level of $P_{NF} = 1.068(145.5) = 155.4$ for the 1974 average.

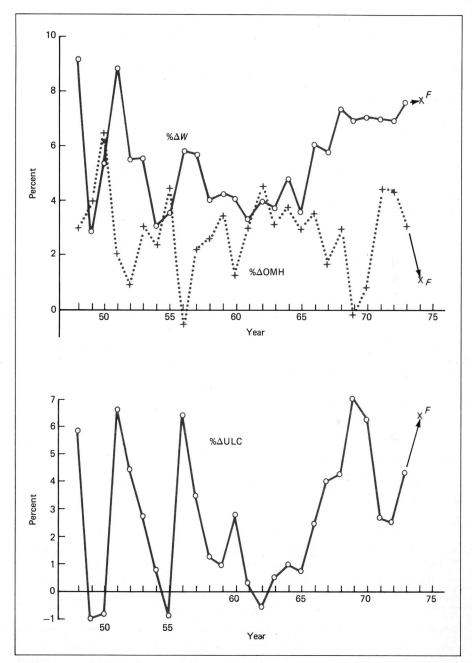

Figure 24-7 Percent changes in compensation per man-hour (*W*), in OMH, and in ULC for private nonfarm GNP.

The government GNP price index (middle panel of Fig. 24-5) reflects levels of compensation per man-hour for general government employees, since no rise in OMH is recognized for production of government services. The growing strength and militancy of unions of public service employees have kept wage rises in state and local employment abreast of those in private firms in recent years. A 6.6 percent increase in compensation of government employees was forecast for 1974, the same rate as in 1973. This translates directly into $\%\Delta P_G$ equal to 6.6 percent.

Rises in farm prices were moderate until 1972 and then skyrocketed with a 19 percent increase in 1972 and a 47.5 percent surge in 1973. The trend within 1973 was strongly upward, with the fourth-quarter level nearly 55 percent above the year earlier. Further increases in agricultural prices were expected, but slowing in the latter part of 1974. A value of $\%\Delta P_F$ equal to 20 percent for 1974 seemed reasonable, but the margin for error was large. Thus, $P_F = 1.20(205.7) = 247.0$ for 1974.

The forecasts of percent changes in GNP price indexes by sectors are shown by the X's for 1974 in Fig. 24-5. They may be combined into a forecast of the percent rise in total GNP price index by use of the weights that reflect the recent proportions of current-dollar GNP produced in each sector. (See Table 24-3.)

TABLE 24-3   1974 Forecasts of Price Indexes for GNP and Components

Component	Indexes: 1958 = 100		$\%\Delta P$	Weight	Product
	1973	1974			
Private nonfarm GNP	145.5	155.4	6.8	0.85	5.78
Government GNP	236.1	251.7	6.6	0.12	0.79
Farm GNP	205.7	246.8	20.0	0.03	0.60
Total GNP	153.9	165.0	7.17	1.00	7.17

The percent rise in the GNP price index worked out at 7.17 percent, implying an index of 165.0 in 1974 (with 1958 = 100). This was well above the rise of 5.4 percent in 1973, but it seemed fully justified in view of trends at the time; for example, the index in the fourth quarter of 1973 was 7.3 percent above the fourth-quarter index in 1972. The increase checked well with the initial estimate of a 7.0 percent rise, derived by extrapolation and judgment back in Chap. 21.

Were it not so, adjustments would be needed to real GNP—or perhaps back to the forecast of components of GNP and GNY. If real GNP were altered, then compensating adjustments would be needed in employment or in output per employed person in one or more sectors. Unless the price discrepancy is large, such adjustments can be accommodated by changes for a few variables within their errors of forecast and without a major overhaul of the interlocking network of forecasts.

Once the GNP price index has been determined, price indexes for the demand components of GNP may be derived in a consistent manner by noting their historical rates of rise compared to $\%\Delta P_Y$ and by bringing in any other relevant information. Then constant-dollar demand components of GNP can be calculated and reconciled with aggregate real GNP.

FINAL REVIEW AND ADJUSTMENTS (step 9)

The forecasts for the complete, interacting current-account flows are now in view. Demand-sector incomes and expenditures have been forecast and reconciled with each other. Money GNP has been split into price and real output forecasts. The implications of real GNP for employment in producing sectors have been worked out, and employment has been reconciled with labor force and unemployment. In addition, wage and productivity rises have been used to derive forecasts of GNP price changes.

At this point, one should stand back and see whether the projected picture of the economic system is intellectually satisfying. Are the parts harmonious, or is there any incongruity that makes one gag mentally? For example, are the wage and price rises reasonable for a year of slow growth, petroleum embargo, and removal of price and wage controls? Is the unemployment rate reasonable for an election year during which the labor force is growing rapidly and the armed forces are being reduced? Are the personal saving rate and the corporate profits share of national income consistent with the economic environment of inflation below full employment? How does the forecast year fit into the longer-term picture of growth, full employment, inflation, income shares, and international balance of payments?

Most of these questions have been taken care of by exercise of judgment in choosing forecast values of variables, often deviating from regression lines or average rates of change But one additional area may be cited for further checking— money and capital markets.

MONEY AND CAPITAL MARKETS

Could monetary influences upset the forecast? Policy statements from the Federal Reserve System indicated its intention to increase the money supply at a moderate rate and maintain credit conditions to slow inflation. Presumably, continued federal government borrowing and rising demand for capital funds as businesses increased their investments in fixed capital and inventories would lead to some rises in interest rates. Moderate rises may be welcomed to stimulate international flows of funds favorable to the U.S. balance of payments, but it is assumed that money supply will be increased fast enough to prevent higher interest rates from causing a recession. And the expected moderation of inflation late in 1974 should weaken one factor which had led to high long-term interest rates.

The absence of forecasting relations for money demand and supply or for flows of funds through capital markets in this model prevents derivation of forecasts for money supply and interest rates. This is a weakness of the present model, but until we know better how money supply and interest rates affect aggregate output, employment, and prices, possible benefits do not seem to be enough to justify the costs of modeling those areas of the economy.

CLOSING COMMENTS

The preceding chapters have presented detailed procedures for developing a complete and consistent forecast for major components of the national income and product accounts, plus employment, productivity, and price level.

The model has a well-articulated skeleton provided by the national income and product accounts. Exogenous information and judgment can be applied to this framework at many places during development of the forecast, but the variables are constrained to consistency by the quantitative linkages provided by the causal relations, by the sequence for estimation of components, and by an iterative revision of initial estimates for GNP and prices. Reported data will become readily available at later dates to check on the accuracy of all parts of the forecast, and, it is hoped, to permit improvement of the model over the years.

If you do check up on these forecasts, however, please *check forecast changes against later reported values for year-to-year changes instead of comparing absolute levels for the series.* Historical levels of GNP and its components are revised for the three most recent years in July each year in the National Income Issue of the *Survey of Current Business.* These revisions alter historical data on which the forecast was based but do not affect the annual increments or percent changes so much. Thus, it is more meaningful to compare forecast changes from the previous year with the actual changes as reported later. Predicting changes is what forecasting is all about, in any event.

Econometric models embodied in systems of simultaneous equations are a logical outgrowth of a model such as this. Those models contain blocks of equations for determining (1) GNP demand components; (2) income, tax, and transfer flows; (3) production and employment; (4) prices; and sometimes (5) money and capital flows. The equations in each block would embody causal relations between variables, those entering the present model plus others which can be added in a more elaborate computerized model. The process of solving the equations of the system would be shortened because simultaneous solution of the equations for endogenous variables eliminates the steps of successive approximation used here for GNP and prices.

As to accuracy, the choice between the simpler models and the econometric systems of 100 or more equations is not yet clear. Correctness of forecasts depends heavily on the exogenous inputs fed into the model and on ensuring that the causal relations well represent up-to-date behavior of the decision makers in the economy. There is no reason why the masters of econometric models should command better information or judgment regarding exogenous inputs. Their dependence on historical patterns to determine constants in the behavioral and institutional equations may prevent their catching changed relationships quickly, though the operators of such models do shift constants on a judgment basis to reflect recent or expected deviations from past relations.   But the computerized equation models do facilitate the calculation of the consequences of alternative exogenous inputs or shifts of relations, so that an array of forecasts, based on different assumptions, can be quickly generated.

On the whole, the simple model described here does generate consistent forecasts of acceptable accuracy and exposes in a straightforward manner the structure and interrelations of the macroeconomic system.   And it is well adapted to a group forecasting project with individuals or teams working on separate parts of the forecast and coordinating their outputs into a coherent whole.

Appendix

Symbols are useful. After once being defined in word language, they call back to mind quickly and unambiguously concepts which are frequently used, and they can be combined in the symbolic logic of equations. However, in a subject so broad and complex as macroeconomics, the number of frequently used concepts becomes so large that the gain of brevity from using symbols may be outweighed by the danger of forgetting their definitions.

This appendix brings together in one place the important symbols used in this text, along with their names or interpretations, their units of measure, and reference pages in the text where they are introduced or explained. The first section contains symbols for quantities (nouns)—flows, stocks, coefficients, and other ratio variables. The second section contains symbols for modifiers (adjectives) designating certain qualitative characteristics for subclasses of the main quantities listed in the first section; these symbols appear as prefixes, subscripts, or superscripts appended to the symbols for the main quantities.

In each section the symbols are listed alphabetically, and for each letter the ordering is: single capital letters, single small letters, compounds beginning with the letter, and corresponding Greek letters. Usually the symbols resemble abbreviations.

Some symbols with modifiers are included in the first section of the appendix. They are important, frequently used variants of the quantity designated by the main symbol. Less important variants of a quantity may not be listed individually, but their interpretation may be recalled to mind by combining the meaning of the main variable in the first section of this appendix and of the modifiers in the second section.

Special symbols appearing only in the forecasting part of the text (Chaps. 21 through 24) have not been included here, because of their restricted use.

VARIABLES, CONSTANTS, AND COEFFICIENTS

Symbol	Interpretation	Units	Reference pages
A	Total factor productivity measure	Billdar at 1958 prices	33
A	Value of financial asset, value of nonmoney financial assets	$ billions	160, 264
a	Constant in equation relating C^* to Y_d'	Billdar	31, 116

Symbol	Interpretation	Units	Reference pages
a'	Constant in equation relating C^* to Y_s	Billdar	143
a_0	Constant in equation relating C^* to P_Y and Y_s	Billdar	305
a_1	Coefficient of P_Y in equation relating C^* to P_Y and Y_s	Billdar at 1958 prices	305
a_2	Coefficient of Y_s in equation relating C^* to P_Y and Y_s	Dimensionless	305
APC	Average propensity to consume	Dimensionless	45, 116
APL	Average productivity of labor	$ at 1958 prices per man-hour	25
APS	Average propensity to save	Dimensionless	45, 117
α	Constant in equation relating K^* to $\hat{Y}^*$	$ billion	192
α	Constant in equation relating c to y_d	$ per year at 1958 prices per person	128
α	Constant in equation relating NFC to R	Billdar	101
B	Value of bonds	$ billion	361
b	Coefficient of Y_d' in equation relating C^* to Y_d'	Dimensionless	31, 116
b'	Coefficient of Y_s in equation relating C^* to Y_s	Dimensionless	143
Billdar	Acronym for billion dollars at annual rate	$ billion per year	48
BOR	Borrowing, inflow of capital funds	Billdar	32, 59
BR	Borrowed reserves	$ billion	264
β	Coefficient of R in equation relating NFC to R	Dimensionless	101, 177
β	Coefficient of y_d in equation relating c to y_d	Dimensionless	128
β	Coefficient of $\hat{Y}^*$ in equation relating K^* to $\hat{Y}^*$	Years	192
C	Personal consumption expenditures	Billdar	31, 59
c	Real per capita consumption expenditures	$ per year at 1958 prices per person	119
CC	Current outlay or cost of an investment	Dollars	160
CPI	Consumer price index	$ per base-year unit	376

Symbol	Interpretation	Units	*Reference pages*
CU	Currency holdings	$ billion	264
D	Capital consumption allowances, depreciation	Billdar	59
D	Deposits in banks	$ billion	264
DD	Demand deposits, checking account balances	$ billion	264
DIV	Dividend payments	Billdar	361
E	Employment	Million persons	25
E	Value of equity claims	$ million	361
E	Elasticity, ratio of percent changes of two variables	Dimensionless	45
e	Constant in equation relating Y_d^l to Y_s	Billdar	30
EM_s	Excess supply of money	$ billion	356
ER	Excess reserves of commercial banks	$ billion	268
F, f	Flow of goods, services, or money	Units per year	39
F_{sw}	Flow of services from an item of wealth	Billdar at 1958 prices	359
f	Coefficient of Y_s in equation relating Y_d^l to Y_s	Dimensionless	30
f_L, f_K	Income shares of labor and capital	Dimensionless	436
Φ_1	Constant for transfers to foreigners	Billdar	224
G	Government purchases of goods and services	Billdar	32
G_0	Constant in equation relating G^* to P_Y and Y_s	Billdar	305
g	Growth rate	Percent per year	27, 166
g	Constant in equation relating I_d^* to i	Billdar	183
g'	Constant in equation relating I_d^* to Y_s	Billdar	43, 189
g''	Constant in equation relating I_d^* to Y_s and i	Billdar	192
g_1	Coefficient of P_Y in equation relating G^* to P_Y and Y_s	Billdar at 1958 prices	305
g_2	Coefficient of P_Y in equation relating G^* to P_Y and Y_s	Dimensionless	305
GNP	Gross national product	Billdar	24, 48
GNY	Gross national income	Billdar	49
GO	Gold and foreign reserve assets	$ billion	271

Symbol	Interpretation	Units	Reference pages
GRE	Gross retained earnings, cash flow to business capital accounts	Billdar	31, 59
γ_1	Constant in equation relating GRE to Y_s	Billdar	224
γ_2	Coefficient of Y_s in equation relating GRE to Y_s	Dimensionless	224
H	Average hours worked per year by employed persons	Hours per year	25
h	Coefficient of i in equation relating I_d^* to i	Billdar per % per year	183
h'	Coefficient of Y_s in equation relating I_d^* to Y_s	Dimensionless	43, 189
h_0	Coefficient of P_Y in equation relating I_d^* to i, P_Y and Y_s	Billdar at 1958 prices	305
h_1	Coefficient of iP_Y in equation relating I_d^* to i, P_Y, and Y_s	Billdar at 1958 prices per % per year	305
h_2	Coefficient of Y_s in equation relating I_d^* to i, P_Y, and Y_s	Dimensionless	305
I	Gross investment flow	Billdar	32, 63
I_d	Gross private domestic investment	Billdar	32, 59
I_n	Net investment, that is, net of capital consumption	Billdar	64
I_u	Unplanned investment in inventories	Billdar	88, 148
i	Interest rate, discount factor	Percent per year	39, 160
INJ	Injections	Billdar	63, 139
INJ_n	Net injections, that is, externally financed	Billdar	289
INT	Interest payments	Billdar	361
INV	Inventory stock	$ billion	39, 91
K	Aggregate stock of capital goods	$ billion at 1958 prices	33
k, k'	Autonomous expenditure multiplier	Dimensionless	139, 191
k	Coefficient of y_p in equation relating c_P to y_p	Dimensionless	127
k	Firm's stock of capital goods	$ billion at 1958 prices	101
L	Aggregate flow of labor services	Billion man-hours per year	25, 99
L	Loans outstanding for commercial banks	$ billion	284

Symbol	Interpretation	Units	*Reference pages*
L^*	Demand for labor, aggregate flow of labor services desired by employers	Billion man-hours per year	107, 316
L_s	Labor supply, aggregate flow the labor force desires to provide	Billion man-hours per year	346
ℓ	Flow of labor services to a firm	Man-hours per year	101
LF	Labor force	Million persons	24
M	Stock of money	$ billion	38, 263
M^*	Demand for money stocks	$ billion	257
M_1^*	Transactions and precautionary demand for money	$ billion	258
M_2^*	Asset demand for money	$ billion	258
$M_{2,0}$	Constant term in equation relating M^* to Y and i	$ billion	257
M_s	Money supply, stock desired by monetary authorities	$ billion	282
M'	Money supply plus U.S. deposits at commercial banks	$ billion	289
MC	Marginal costs	$ per unit of output	96
MEI	Marginal efficiency of investment	Percent per year	161
MNRL	Marginal net revenue of labor	$ per man-hour	102
MPC	Marginal propensity to consume	Dimensionless	39, 116
MPD	Marginal productivity of durable goods	Fraction per year	361
MPK	Marginal productivity of capital	Fraction per year	176
MPL	Marginal productivity of labor	$ at 1958 prices per man-hour	102
MPM	Marginal productivity of money stock	Fraction per year	362
MPS	Marginal propensity to save	Dimensionless	45, 117
MPW	Marginal productivity of an item of wealth	Fraction per year	359
MR	Marginal revenue	$ per unit of output	96
μ_1	Coefficient of Y in equation relating M^* to Y and i	Years	257
μ_2	Coefficient of i in equation relating M^* to Y and i	$ billion per % per year	257
μ, μ'	Stock adjustment coefficient	Fraction per year	192, 202

Symbol	Interpretation	Units	*Reference pages*
μ_0	Coefficient of P_Y in equation relating M^* to i, P_Y, and Y_S	$ billion at 1958 prices	309
μ_2'	Coefficient of iP_Y in equation relating M^* to i, P_Y, and Y_S	$ billion at 1958 prices per % per year	309
N	Population	Million persons	129
NCG	Net capital gain	Billdar	67
NFC	Nonfactor charges	Billdar	101
NI	National income	Billdar	54
NNP	Net national product	Billdar	74
NR	Net receipts of investment project in specified time period	Dollars	160
NW	Net worth	$ billion	264
ONFC	Other nonfactor charges	Billdar	99
P, p	Price	$ per physical unit	41, 402
P_c	Price of consumer goods	$ per unit of consumer goods	129
P_K	Price of capital goods	$ per unit of capital goods	161
P_Y	Price of GNP	$ per base-year unit	98
POP	Population	Million persons	25
PROP	Value of real property owned	$ million	264
PV	Present value of a future asset or income stream	$ billion	160
π	Profit	$ per year	97
Q, q	Physical quantity of good or service, real inventory holdings	$ billion at 1958 prices	39, 174
R	Total revenue	$ per year	97
R	Reserves of commercial banks	$ billion	265
R	Exchange rate	Francs per dollar	402
r	Rate of return on capital goods or wealth	Percent per year	99, 161, 359
r_w'	Marginal rate of return on an item of wealth	Percent per year	360
r_D	Reserve ratio required for demand deposits	Dimensionless	265
r_T	Reserve ratio required for time deposits	Dimensionless	265
RR	Required reserves	$ billion	265
ρ	Depreciation rate	Fraction per year	193

Symbol	Interpretation	Units	Reference pages
ρ	Real rate of return on capital stock	Fraction per year	435
S	Saving flow	Billdar	32, 64
S_b	Business saving, that is, undistributed profits	Billdar	59
S_g	Government saving	Billdar	59
S_p	Personal saving	Billdar	59, 116
S_n	Net saving, that is, net of capital consumption	Billdar	64
s_u	Ratio of net saving to aggregate demand	Dimensionless	445
s_p	Real per capita personal saving	$ per year at 1958 prices per person	124
SD	Statistical discrepancy	Billdar	56
SV	Scrap value	Dollars	160
σ	Coefficient of Y_R^* in equation relating K to Y_R^*	Years	445
T, t	Time or periods	Years	39
T	Gross tax receipts	Billdar	59
T_n	Net tax receipts, that is, net of government transfer and interest payments	Billdar	31
TC	Total costs	$ per year	97
TD	Time deposits	$ billion	264
TECH	Technology, techniques of production	Dimensionless	303
TR	Transfer payments	Billdar	31
τ_1	Constant in equation relating T_n to Y_s	Billdar	224
τ_2	Coefficient of Y_s in equation relating T_n to Y_s	Dimensionless	224
U	Unemployment	Billion man-hours per year	316
ULC	Unit labor cost	$ per unit of output	335, 383
UO	Unfilled orders	$ billion	91
V	Income velocity for total money stock	Number per year	39, 258
V_1	Income velocity of transactions balances	Number per year	258
V^*	Desired, long-term velocity of money	Number per year	366
VAL	Money value of a stock	$ billion	41

Symbol	Interpretation	Units	Reference pages
W	Money wage rate	$ per man-hour	39, 99
W	Wealth, net worth	$ billion	67
w	Real wage rate	$ per man-hour at 1958 prices	435
WDL	Withdrawals	Billdar	63, 141
WPI	Wholesale price index	$ per base-year unit	377
X	Exports	Billdar	59
X_0	Constant in equation relating X^* to P_Y	Billdar	305
X_n	Net exports, exports minus imports	Billdar	32, 59
x_1	Coefficient of P_Y in equation relating X^* to P_Y	Billdar at 1958 prices	305
Y	Gross national product, GNP	Billdar	34
Y^*	Aggregate demand, purchases desired by demand sectors	Billdar	32
Y_d	Disposable personal income	Billdar	30, 57
Y_d'	Net disposable personal income, that is, net of personal transfer and interest payments	Billdar	59
Y_{FE}	Full-employment GNP, or potential GNP	Billdar	228, 233
Y_p	Personal income	Billdar	57
Y_R	Real GNP, or GNP at constant prices	Billdar at 1958 prices	25
$Y_{R,s}$	Aggregate real supply, physical flow of output desired by production units	Billdar at 1958 prices	98
Y_s	Aggregate supply, value of flow of output expected by production units	Billdar	30, 98
y, y_d	Real per capita disposable income	$ per year at 1958 prices per person	119, 124
$y_p, \overline{Y}$	Permanent income	Billdar	127, 366
y_R	Real output of a firm	Base-year units	96
Z	Imports	Billdar	59
Z_0	Constant in equation relating Z^* to P_Y and Y_s	Billdar	305
z_1	Coefficient of P_Y in equation relating Z to P_Y and Y_s	Billdar at 1958 prices	305
z_2	Coefficient of Y_s in equation relating Z to P_Y and Y_s	Dimensionless	305

PREFIXES, SUBSCRIPTS, AND SUPERSCRIPTS

Symbol	Interpretation	Reference pages
A	Relating to total factor productivity	435
AUT	Autonomous, that is, not varying with changes in Y_s	185
α	Exponent for labor input in production function	33
α	Subscript designating particular equilibrium state	35
α	Coefficient of Y_R in equation relating Q_R to Y_R	174
B	In the base year, in base-year units	41
B	Relating to borrowing sectors of economy	358
b	Of, from, or to the business capital accounts	59
β	Exponent for capital stock in production function	33
β	Subscript designating particular equilibrium state	35
C	Relating to commercial banks	264
c	Relating to consumer goods	129
D	Demanded	335
d	Disposable or domestic	30, 32, 57
δ	Change from initial to final equilibrium value	35, 140
Δ	Change between values at two observation points	44
EX	Exogenous, not dependent on Y or i	293
EXP	Expected value of	257
f	Of, from, or to the Foreign Sector	31, 60
FE	At full employment	99
FR	Relating to Federal Reserve System	264
g	Of, from, or to the Government Sector	59
IND	Induced, that is, varying with changes in Y_s	185
ind	Indirect	99
IS	Investment-saving equilibrium line	187
j	General subscript for sectors or classes of goods	42
K	Relating to capital stock	99, 193
L	Relating to labor	40, 435
ℓ	Long-term	124
LM	Money demand and supply equilibrium line	310
M	Measured in money units	39
max	Maximum	310
n	Net	31
NB	Nonbank sectors of economy	264
o, O	Exogenous or initial value	33
p	Of, from, or to the Personal Sector	57
p	Permanent	127
priv	Of, from, or to Private Sectors	65
Q	Measured in physical units	39, 176
R	Real, in physical units	25
ρ	Relating to replacement investment	193
S	Supplied, produced	30

Symbol	Interpretation	Reference pages
s	Short-term	124
t	Transitory	127
t	Relating to time period t	41
u	Unplanned, undesired	88
US	Relating to U.S. government	264
$\hat{\ }$	Forecast or predicted value of variable	192
$'$	Adjusted or changed value	59, 104
$*$	Demanded, desired	31
$\cdot$	Derivative with respect to time, that is, $d(\ )/dt$	67

Index